Quantum chemistry

An introduction

Quantum chemistry

An introduction

R. L. FLURRY, Jr

Professor of Chemistry
University of New Orleans

PRENTICE-HALL, INC.

Englewood Cliffs, New Jersey 07632

Library of Congress Cataloging in Publication Data

Flurry, Robert L., Jr.
Quantum chemistry.

Includes bibliographies and index.
1. Quantum chemistry. I. Title.
QD462.F583 1983 541.2'8 82-7715
ISBN 0-13-747832-1 AACR2

Printed in the United States of America

Production: Nicholas Romanelli
Cover design: Edsal Enterprises
Manufacturing buyer: John Hall

ISBN 0-13-747832-1

Prentice-Hall International, Inc., *London*
Prentice-Hall of Australia Pty. Limited, *Sydney*
Editora Prentice-Hall do Brazil, Ltda, Rio de Janeiro
Prentice-Hall of Canada, Inc., *Toronto*
Prentice-Hall of India Private Limited, *New Delhi*
Prentice-Hall of Japan, Inc., *Tokyo*
Prentice-Hall of Southeast Asia Pte. Ltd., *Singapore*
Whitehall Books Limited, *Wellington, New Zealand*

Contents

chapter three:

ROTATION AND ANGULAR MOMENTUM 29

chapter four:

VIBRATIONS 63

chapter five:

THE HYDROGEN ATOM 75

chapter six:

APPROXIMATION METHODS IN QUANTUM CHEMISTRY 86

chapter thirteen:

POINT GROUPS 222

chapter fourteen:

FURTHER EXAMPLES FROM HÜCKEL THEORY 243

chapter fifteen:

TRANSITION-METAL COMPOUNDS 265

chapter sixteen:

MOLECULAR VIBRATIONS 276

APPENDICES

Preface

Humans seem to have a basic need to explain all phenomena in terms of past experience. In ancient times this need led people to develop elaborate mythologies, in which a myriad of gods, each with very human qualities, controlled human destinies. In modern science this same need leads us to devise models, based upon our previous experiences, for explaining, by analogy, phenomena that we cannot explain from first principles. These models range from ill-defined, vague concepts through physical models to refined mathematical theories. Whatever form they take, however, they are efforts to bridge over gaps in our fundamental knowledge. A good model must satisfy two criteria: it must be able to adequately explain the phenomenon under consideration to the one proposing the model, and it must be usable to convey the explanation to others.

Quantum mechanics provides us with a mathematical model for describing chemistry at the atomic and molecular level. Most of the concepts involved in explaining the major features of atomic and molecular structure are within the mathematical experience that the average chemist has or can easily obtain. On the other hand, attempts at "conceptualizing" quantum mechanics lead quickly to problems that cannot be modeled on the basis of our past experience, simply because our past experience is based upon the Newtonian mechanics of macroscopic objects, rather than upon the quantum mechanics of microscopic objects.

This book offers a first introduction to the basic concepts of quantum mechanics for chemists (i.e., quantum chemistry), primarily in terms of the mathematical experience of upper-level undergraduates or beginning graduate students in a typical chemistry curriculum. It presents some less familiar mathematical concepts, but very little effort has been made to present physical or conceptual models. The primary purpose is to teach the principles of quantum chemistry, especially as applied to chemical bonding and to spectroscopy. A secondary aim is to prepare the student for courses in modern inorganic chemistry and modern organic chemistry and for more advanced study in quantum theory.

Overall, the material covered is relatively traditional, but there are some unique aspects. Since qualitative developments can be misleading, a number of topics that often are treated in a qualitative manner in introductory texts are given more detailed treatments here (the Pauli principle, for example). For other topics, however (such as the radial equation of the hydrogen atom) the formal treatment is only outlined. In such cases the reader is referred to a more complete treatment. Simple applications of formal concepts are given both in the text and in the end-of-chapter problems. Very few of the problems draw directly on formal derivations; the reason is my personal conviction that the applications are what make most students appreciate the formalism. I do not expect students to be able to duplicate the formal derivations on their first exposure to the material. That will come in more advanced courses for those who go further.

Both the Heinsenberg and Schrödinger approaches to quantum mechanics are introduced in the early chapters. This should help the student appreciate the meaning of the various matrices that are routinely used in quantum-chemical calculations.

Group theory is introduced and applied throughout the text wherever it is useful, rather than in a separate chapter (where most texts put it, if they include it at all). The presentation of group theory is largely expository, instead of developmental. Hopefully, the examples of the utility of group theory will motivate the more interested students to further study, while presenting the practical aspects to those who are less interested in the formal development. The first introduction to group theory is to the three-dimensional rotation group (Chapter 3). The symmetric permutation group is introduced next (Chapter 7). These groups, which are less familiar to many chemists, are introduced before point groups (Chapter 13) in order to alert the reader to the importance of symmetries other than point or spatial symmetries. (Special unitary groups are briefly mentioned in Chapter 17.)

Chapters 1 through 6 of the text present the basic tools and concepts of quantum chemistry. Chapters 7 and 8 present atomic structure. Chapters 9 through 15 develop molecular electronic structure. Much emphasis is placed on Hückel theory, since it is so "transparent." (It provides a semiquantitative description of bonding without being clouded by the mathematical complexities of difficult integral evaluations or by iteration procedures.) Chapter 16 treats molecular vibrations and Chapter 17 magnetic properties (primarily magnetic resonance). These two chapters illustrate problems not involving electronic wave functions. Furthermore, if magnetism is treated in a spin-only approximation, a basis set that is complete, within the approximation, can be used for constructing the wave functions. Chapter 18, on chemical reactivity, illustrates that much important information can be obtained from a totally qualitative application of quantum-chemical concepts.

If used as a text, the book contains enough material for a two-semester course. In a one-semester course it would be logical to emphasize either the fundamentals or molecular bonding. For the first option, Chapters 1 through 6 should be emphasized, with suitable applications being chosen from the other chapters. For the second option, most of Chapter 3, all of Chapter 4, much of Chapters 7 and 8, and all of Chapters 16 and 17 can be omitted without destroying the continuity in the development of molecular electronic theory.

One learns quantum chemistry by doing it; for this reason problems are provided at the end of each chapter. Almost all of these can be done with the aid of no more than a pocket calculator. Appendix 4 provides the answers to those problems preceded by an asterisk.

I am indebted to many people for their help in getting this work into print. First are the students whose responses helped shape the form of presentation—my students in Chemistry 4311 during 1976-80 at the University of New Orleans, and the Chemistry 81 class at Dartmouth College during the fall of 1980. Professors Gordon A. Gallup, Hans H. Jaffe, Peter Lykos, Stanley C, Neely, Robert G. Parr, and L. G. Pedersen, and Mr. Mark Rosenberg read and offered valuable comments on the entire manuscript. Numerous other people offered suggestions on various parts also. Most of the final editing was done while I was a visting professor at Dartmouth. I express my gratitude to Dr. Roger Soderberg and the entire department for their hospitality. Finally. I thank the editorial and production staff of Prentice-Hall, especially Betsy Perry and Nicholas Romanelli, for their efforts.

R. L. FLURRY, JR.

University of New Orleans

chapter one

Introduction to quantum theory

1.1 ELECTROMAGNETIC RADIATION

The study of *electromagnetic radiation*, its properties, and its interaction with matter is of prime importance to quantum chemistry. A study of the emission of electromagnetic radiation by a hot body led Max Planck, in 1900, to postulate the quantization of energy. Atomic spectroscopy—the study of the absorption and emission of electromagnetic radiation by a system—led Niels Bohr to propose the first acceptable theory of an atom (the hydrogen atom) in 1913. Virtually all our detailed knowledge about the structure of atoms and molecules comes from our studies of their interaction with electromagnetic radiation.

Much of our knowledge of the macroscopic world also depends upon electromagnetic radiation. Light, the form of electromagnetic radiation most familiar to us, is the basis of our vision. Radio waves, another form of electromagnetic radiation, bring us much of our information. Furthermore, all the energy that reaches us from our sun—the energy that sustains life on our planet—arrives in the form of electromagnetic radiation.

Electromagnetic radiation is commonly discussed in terms of *waves*. We are familiar with many types of waves and wave motion. At the seashore we see ocean waves. On a bowed violin string we see standing waves, and we hear the tone transported to our ears by sound waves. All these waves imply some sort of oscillatory motion. This motion is characterized by an *amplitude*, a *frequency* or a *wavelength*, and, for waves moving through space, a *velocity of propagation*. The last three of these characteristics are related by the equation

$$\lambda\nu = c \qquad \textbf{1.1}$$

Table 1.1

RADIATION FREQUENCY AND WAVELENGTH RANGES OF THE ELECTROMAGNETIC SPECTRUM

Range	Typical Frequency[a]	Typical Wavelength
Radio	$\nu \sim 1 \times 10^5$ Hz	$\lambda \sim 3 \times 10^3$ m
Microwave	$\nu \sim 3 \times 10^9$ Hz	$\lambda \sim .1$ m
Infrared	$\nu \sim 3 \times 10^{13}$ Hz	$\lambda \sim 1 \times 10^{-5}$ m
Visible light	$\nu \sim 3.75 \times 10^{14}$ Hz	$\lambda \sim 8 \times 10^{-7}$ m
	$\nu \sim 7.5 \times 10^{14}$ Hz	$\lambda \sim 4 \times 10^{-7}$ m
Ultraviolet	$\nu \sim 1.5 \times 10^{15}$ Hz	$\lambda \sim 2 \times 10^{-7}$ m
Vacuum UV	$\nu \sim 3 \times 10^{15}$ Hz	$\lambda \sim 1 \times 10^{-7}$ m
X-ray	$\nu \sim 3 \times 10^{18}$ Hz	$\lambda \sim 1 \times 10^{-10}$ m
Gamma ray	$\nu \sim 3 \times 10^{20}$ Hz	$\lambda \sim 1 \times 10^{-12}$ m

[a] The dimensions of frequency are $(\text{time})^{-1}$. When time is in seconds, the unit is designated as a *hertz* (after Heinrich Hertz) and abbreviated Hz.

where λ is the wavelength, the spatial distance between like positions on successive waves (crests on the ocean waves, for example); ν is the frequency, the number of wavelengths that pass an observer in a given period of time; and c is the velocity of propagation.

We know of ways to detect and to produce electromagnetic radiation over a very wide range of wavelengths or frequencies. Waves in different frequency bands are commonly given names such as radio waves, visible light, and X-rays; however, they are all manifestations of the same phenomena, differing only in wavelength or frequency. The other quantity in Eq. 1.1, the velocity, when the medium is a vacuum, is one of the fundamental constants of nature. The best value for this constant, the speed of light in a vacuum, is

$$c = 2.99792250 \times 10^8 \text{ m sec}^{-1} \tag{1.2}$$

(This is more easily remembered as 3×10^8 m sec^{-1} or 3×10^{10} cm sec^{-1}. The speed of light is medium-dependent.) Table 1.1 lists the common divisions of the electromagnetic spectrum, along with a typical frequency and wavelength for each.

1.2 BLACK-BODY RADIATION

An idealized *black body* is an object that emits or absorbs all frequencies of electro-magnetic radiation with equal probability. For emission of thermally produced radiation this condition can be achieved, to a good approximation, by a closed furnace. Studies can be made of the radiation emitted from a small slit in the furnace. The experimental information that ultimately led to the postulation of the quantization of energy was derived by studying the density of energy emitted by a black body, at a given temperature, as a function of the frequency or the wavelength of the radiation. The total energy density, ρ, is defined as

$$\rho = \frac{4}{c} E \tag{1.3}$$

where c is the speed of light and E is the total emissive power (energy per unit area, per unit time). The energy density per unit frequency, ρ_ν, is

$$\rho_\nu = \frac{4\pi}{c} B_\nu \tag{1.4}$$

where B_ν is the emissive power per hertz at frequency ν, per meter per degree (solid angle), and per second in a direction normal (perpendicular) to the surface of the black body. The quantities E, B_ν, ρ, and ρ_ν are all functions of the temperature. (Note that ρ equals $\int \rho_\nu \, d\nu$.) The experimental result is schematically illustrated in Figure 1.1.

During the latter part of the nineteenth century a number of people attempted to explain theoretically the energy density curve. One of the earliest results was the Stephan-Boltzmann law, proposed by Stephan in 1879 and derived by Boltzmann in 1884. This says that the total energy emitted by a black body is proportional to the fourth power of the absolute temperature,

$$E = \sigma T^4 \tag{1.5}$$

where σ is the proportionality constant. This gives the area under the curves but not the shape, since it makes no reference to frequency. A partially successful treatment of the frequency was offered by Wien in 1896. Recognizing the similarity of the shape of the ρ_ν curve to the distribution of molecular velocities of a gas, he proposed a law of the form

$$\rho_\nu = \alpha\nu^3 \exp\left(-\frac{\beta\nu}{T}\right) \tag{1.6}$$

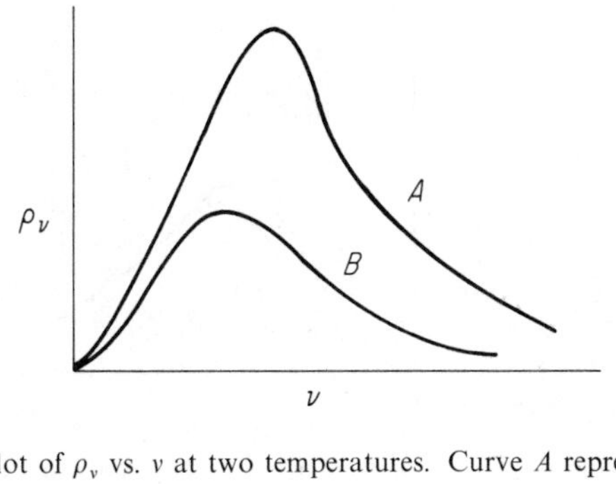

Figure 1.1 A plot of ρ_ν vs. ν at two temperatures. Curve A represents the higher temperature.

where α and β are empirical constants. (It is interesting to note that Max Planck published one of the derivations of this law.) The Wien displacement law, as it is called, gives excellent agreement with experiment at high frequencies or short wavelengths; however, the agreement is poor at the other end of the scale.

Lord Rayleigh recognized some errors in the derivations of the Wien displacement law. Correcting these errors (and applying J. H. Jeans' further corrections to Rayleigh's result) yielded the relation

$$\rho_\nu = \frac{8\pi\nu^2 kT}{c^3} \qquad \textbf{1.7}$$

where k is the Boltzmann constant. The Rayleigh-Jeans law gives excellent agreement at low frequencies (long wavelengths) but completely fails at higher frequencies, since it continually increases with increasing frequency. Figure 1.2 shows the Wien law and the Rayleigh-Jeans law in comparison with experiment.

In 1900 Max Planck arrived at an equation that completely fitted the entire curve:

$$\rho_\nu = \frac{8\pi\nu^3}{c^3} \frac{h}{\exp(h\nu/kT) - 1} \qquad \textbf{1.8}$$

Basically, the equation was derived by "curve fitting." Planck knew how the curve behaved at low frequencies and at high frequencies. The problem was to join them. When the one new constant, h, is evaluated, the equation fits the experimental curve exactly. (Planck's value of 6.55×10^{-27} erg sec for h, derived from the black-body data, is remarkably close to the best modern value for *Planck's constant*, $h = 6.626196 \times 10^{-27}$ erg sec or 6.626196×10^{-34} J sec.) The most significant result, however, was not the equation itself, but the fact that its derivation required that the frequency, ν, could not be continuous, but rather had to assume discrete, or *quantized*, values. Since $h\nu$ has the dimensions of energy, this implies that *energy must be quantized.*

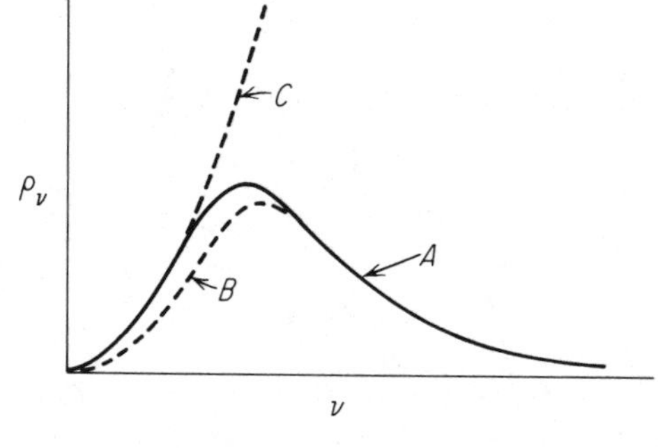

Figure 1.2 The experimental black-body radiation curve (curve A) compared with the Wien displacement law (curve B) and the Rayleigh-Jeans law (curve C).

Many people, apparently including Planck himself, thought for several years that the quantization requirement was a mathematical artifact with no physical significance. However, Einstein's 1905 work on the photoelectric effect and his and Debye's work on the heat capacity of solids overcame most of the doubts. These two phenomena appeared to be unrelated to the black-body radiation problem and to each other, yet both were explained by incorporating energy quantization into the theories. One curve fitting that required energy quantization might have been an artifact; however, if the theoretical explanation of three unrelated phenomena required quantization, it must have a physical significance. Today, the concept of the quantization of energy and the equation

$$E = h\nu \qquad \textbf{1.9}$$

are deeply ingrained in much of modern science.

In the light of Eq. 1.9, let us consider the frequencies corresponding to some more familiar energy units. We have

$$\begin{aligned} 1\ \text{eV/molecule} &= 2.41805 \times 10^{14}\ \text{Hz} = 8.06573 \times 10^{3}\ \text{cm}^{-1} \\ 1\ \text{erg/molecule} &= 1.50931 \times 10^{26}\ \text{Hz} = 5.03451 \times 10^{15}\ \text{cm}^{-1} \\ 1\ \text{J/molecule} &= 1.50931 \times 10^{33}\ \text{Hz} = 5.03451 \times 10^{22}\ \text{cm}^{-1} \\ 1\ \text{kcal/mole} &= 1.04855 \times 10^{13}\ \text{Hz} = 3.49758 \times 10^{2}\ \text{cm}^{-1} \end{aligned} \qquad \textbf{1.10}$$

In Eqs. 1.10 we have used the unit cm^{-1} as well as Hz. These numbers correspond to the energies expressed in *wave numbers*. Frequency equals c/λ and is a very large number for energies of importance to chemists. The wave number, $\bar{\nu}$, is defined as $1/\lambda$, and has the units cm^{-1}. These numbers are of a more reasonable size for chemically important energies. They are usually used by spectroscopists.

1.3 BOHR THEORY

The simplest of all atoms is that of hydrogen. Its spectrum, like that of every other atom, consists of very sharp lines, representing the absorption or emission of electromagnetic radiation at discrete frequencies. There are only four lines in the visible region of the hydrogen spectrum. These occur at $1.52 \times 10^4\ \text{cm}^{-1}$ (4.57×10^{14} Hz), $2.06 \times 10^4\ \text{cm}^{-1}$ (6.17×10^{14} Hz), $2.30 \times 10^4\ \text{cm}^{-1}$ (6.91×10^{14} Hz), and $2.44 \times 10^4\ \text{cm}^{-1}$ (7.32×10^{14} Hz). This series, known as the Balmer series, can be fitted to a very simple equation,

$$\bar{\nu} = R\left(\frac{1}{4} - \frac{1}{n^2}\right) \qquad \textbf{1.11}$$

where n is a whole number greater than 2 and R is a numerical constant known as the *Rydberg constant*, having the numerical value of $1.0973732 \times 10^5\ \text{cm}^{-1}$. By the first decade of this century a second series, known as the *Paschen series* and occurring in the infrared region, was also known. The Paschen series can be fitted to the equation

$$\bar{\nu} = R\left(\frac{1}{9} - \frac{1}{n^2}\right) \qquad \textbf{1.12}$$

where R is the same numerical constant and n is now greater than 3. Such a simple spectrum should be explainable by a simple theory. Niels Bohr supplied the theory in 1913.

After receiving his Ph.D. in 1911, Bohr went to Cambridge to work with J. J. Thomson. Sharp disagreements with Thomson over Thomson's classical model for the atom led Bohr to leave Cambridge to work with Rutherford at Manchester. Rutherford had shown that an atom consisted of a very small, relatively massive, positively charged nucleus surrounded by very light, negatively charged electrons.

Rutherford had proposed a planetary model of the atom, in which the electrons revolved about the nucleus in definite orbits, much as the planets revolve about the sun. The motion was assumed to be such that the centrifugal force on the moving electrons was balanced by the electrostatic attraction between the positively charged nucleus and the negatively charged electrons:

$$\text{negative of electrostatic force} = \frac{+Ze^2}{r^2}$$

$$= \frac{mv^2}{r} = \text{centrifugal force} \qquad \textbf{1.13}$$

where $+Ze$ is the charge on the nucleus, $-e$ is the charge on the electron, r is the radius of the orbit, m is the mass of the electron, and v is its velocity. We know that energy can be expressed as the sum of kinetic energy, T, and potential energy, V. The kinetic energy is $\frac{1}{2}mv^2$, while for an electron in a one-electron atom the potential energy is $-Ze^2/r$. Thus, we have

$$E = T + V = \frac{1}{2}mv^2 - \frac{Ze^2}{r} \qquad \textbf{1.14}$$

But from Eq. 1.13 we see that

$$\frac{Ze^2}{r} = mv^2 \qquad \textbf{1.15}$$

so we have

$$E = -\frac{1}{2}mv^2 = \frac{-Ze^2}{2r} \qquad \textbf{1.16}$$

Note, incidentally, that

$$E = -T = \frac{1}{2}V \qquad \textbf{1.17}$$

This is the *virial theorem* of classical mechanics.

This argument is perfectly good when applied to planets and suns, with the force being gravitational, rather than electrostatic; however, there is a major flaw in it when applied to electrons and nuclei. Electrons are charged particles. Electrodynamics tells us that charges moving in a circular path emit energy. Thus, in the Rutherford model, the electron should be constantly losing energy and consequently slowing down and spiraling into the nucleus. (For the hydrogen atom, the duration of this process should be of the order of 10^{-8} sec.) This is contrary to the fact that an atom is perfectly stable.

Shortly after returning to his native Denmark, Bohr supplied the missing piece to the theory for the hydrogen atom. He added quantization. His hypothesis was that angular momentum, p_ϕ, was quantized:

$$p_\phi = mvr = \frac{nh}{2\pi} \tag{1.18}$$

where n is an integer. Equation 1.18 can be solved for r to give

$$r = \frac{nh}{2\pi mv} \tag{1.19}$$

Substituting this into Eq. 1.15, we have

$$mv^2 = \frac{2\pi Ze^2 mv}{nh} \tag{1.20}$$

Solving for the velocity,

$$v = \frac{2\pi Ze^2}{nh} \tag{1.21}$$

If Eq. 1.21 is substituted into Eq. 1.19, we obtain a value for r in terms of fundamental constants and the integer, n:

$$r = \frac{n^2h^2}{4\pi^2 Ze^2 m} \tag{1.19a}$$

This states that the orbits can have only certain values for their radius. Consequently, there can be no spiraling between orbits. On the other hand, if we substitute Eq. 1.21 into Eq. 1.16, we obtain

$$E = -\frac{1}{2}mv^2 = \frac{-2\pi^2 mZ^2e^4}{n^2h^2} \tag{1.22}$$

—an equation for E as a function of the integer, n. If we add a subscript to E to indicate the value of n, we can express energy changes accompanying spectral transitions, which Bohr identified with $h\nu$, as

$$\Delta E_{ji} = E_j - E_i \tag{1.23}$$

or, in terms of frequencies,

$$\nu_{ji} = \frac{1}{h}(E_j - E_i) \tag{1.24}$$

If E_j is the final state and E_i the initial, a positive ΔE represents an absorption and a negative an emission of energy.

As we see from Eq. 1.19, Bohr's theory tells us that the electron in the hydrogen atom can be in orbits having only certain radii. All other values of r are forbidden. Thus, the electron can jump from one orbit to another, with the appropriate changes of energy, but it cannot spiral in. Furthermore, from Eq. 1.22 we see that if n were to equal zero, the atom would have an infinitely large negative energy. Thus, there is a lowest finite energy (corresponding to what is called the *ground state*) and a minimum radius that the electron can have. Note, however, that the theory does not say how the charged electron can move in a circular path and not emit energy.

Let us look again at the spectral frequencies of Eq. 1.24. If we substitute in Eq. 1.22 for the energy, we have

$$\begin{aligned} \nu_{ji} &= \frac{2\pi^2 m Z^2 e^4}{h^3}\left[-\frac{1}{n_j^2} - \left(-\frac{1}{n_i^2}\right)\right] \\ &= \frac{2\pi^2 m Z^2 e^4}{h^3}\left(\frac{1}{n_i^2} - \frac{1}{n_j^2}\right) \end{aligned} \tag{1.25}$$

or

$$\bar{\nu}_{ji} = \frac{2\pi^2 m Z^2 e^4}{h^3 c}\left(\frac{1}{n_i^2} - \frac{1}{n_j^2}\right) \tag{1.25a}$$

The quantity before the parentheses equals the Rydberg constant when it is evaluated with $Z = 1$, for hydrogen. Thus, the Balmer series is just Eq. 1.25 when n_i equals 2. The Paschen series results when n_i equals 3. We know, however, that n_i can also equal 1. Evaluating the frequency for the first of these, $n_i = 1$ and $n_j = 2$, we see that it occurs in the ultraviolet region of the spectrum at $8.23 \times 10^4\ \text{cm}^{-1}$. The other lines of this series would occur at higher frequencies. Very shortly after the publication of Bohr's paper, Lyman observed this ultraviolet series, which now bears his name.

Bohr's theory correctly predicts the spectrum of the hydrogen atom and also of any one-electron ion (if Z is changed to the proper value) in the absence of any external electric or magnetic fields. Corrections can be made to handle a one-electron atomic system in the presence of external fields. However, attempts to extend the theory to handle many-electron atoms, or to use it to describe chemical bonding, met with complete failure.

Two simultaneous developments of "modern quantum theory," which overcame these difficulties, occurred about a dozen years later. The *matrix mechanics* of W. Heisenberg (1925) and the *wave mechanics* of E. Schrödinger (1926) appeared at first to be different, owing to the differences in the mathematics involved. Heisenberg's theory is based upon matrix algebra (which he developed himself, not realizing that the mathematician Cayley had done so 70 years earlier), while Schrödinger's theory uses differential equations. Schrödinger showed the equivalence of the two approaches. Often the two are combined in modern applications of quantum theory, although for qualitative discussions many people feel more comfortable with the Schrödinger representation.

Before leaving our discussion of Bohr theory, we should point out that energy units and distance units from Bohr theory are useful in atomic and molecular quantum mechanics. The most common energy unit is twice the magnitude of the energy of the hydrogen-atom ground state (i.e., the magnitude of the hydrogen-atom potential energy) and is called the *Hartree atomic energy* unit:

$$1 \text{ Hartree} = mv^2 = \frac{Ze^2}{r} = \frac{4\pi^2 mZ^2e^4}{h^2} = 27.211652 \text{ eV} \qquad \textbf{1.26}$$

The distance unit is the *Bohr radius* of the ground state of the hydrogen atom:

$$1 \text{ Bohr} = \frac{h^2}{4\pi^2 mZe^2} = .52917715 \text{ Å} \equiv a_0 \qquad \textbf{1.27}$$

Both of these are frequently referred to simply as 1 atomic unit (a.u.), with energy or distance being implied by the context. These are equivalent to setting $h/2\pi$, e, and m all equal to unity.

1.4 HEISENBERG'S MATRIX MECHANICS

A verbal description of Heisenberg's development of quantum mechanics sounds relatively simple, if his basic assumptions are accepted without justification. Heisenberg started by assuming that there is a matrix (Appendix 2) that corresponds to each observable of a system. The quantum rules were derived from matrix algebra. Particular importance was placed upon the commutation properties of the matrices.

The *commutator*, $[\mathbf{A}, \mathbf{B}]$, of two matrices $\mathbf{A}$ and $\mathbf{B}$ is defined as

$$[\mathbf{A}, \mathbf{B}] = \mathbf{AB} - \mathbf{BA} \qquad \textbf{1.28}$$

Equation 1.28 does not necessarily equal zero when **A** and **B** are matrices. Consider two general 2×2 matrices

$$\mathbf{A} = \begin{bmatrix} a & b \\ c & d \end{bmatrix} \qquad \textbf{1.29}$$

$$\mathbf{B} = \begin{bmatrix} e & f \\ g & h \end{bmatrix} \qquad \textbf{1.30}$$

The commutator is

$$[\mathbf{A}, \mathbf{B}] = \begin{bmatrix} a & b \\ c & d \end{bmatrix}\begin{bmatrix} e & f \\ g & h \end{bmatrix} - \begin{bmatrix} e & f \\ g & h \end{bmatrix}\begin{bmatrix} a & b \\ c & d \end{bmatrix}$$

$$= \begin{bmatrix} ae + bg & af + bh \\ ce + dg & cf + dh \end{bmatrix} - \begin{bmatrix} ea + fc & eb + fd \\ ga + hc & gb + dh \end{bmatrix} \qquad \textbf{1.31}$$

Obviously, this will equal zero only in restricted cases. Matrices are said to *commute* when their commutator does equal zero.

The quantum rules arise when we postulate two particular commutator relations. First, the commutator of any observable with the energy (the matrix representing the energy is called the *Hamiltonian matrix*, **H**) gives the time variation of the observable, multiplied by $i\,(=\sqrt{-1})$ and by Planck's constant divided by 2π ($\hbar$ is defined as $h/2\pi$):

$$[\mathbf{A}, \mathbf{H}] = i\hbar \frac{dA}{dt} \equiv i\hbar\dot{\mathbf{A}} \qquad 1.32$$

If **A** commutes with **H**, the commutator is zero. Properties that commute with the Hamiltonian do not vary with time and are called *constants of motion.* The other postulated relation is the fundamental form of the famous *Heisenberg uncertainty principle*:

$$\Delta\mathbf{A}\,\Delta\mathbf{B} = -\frac{i}{2}[\mathbf{A}, \mathbf{B}] \qquad 1.33$$

where $\Delta\mathbf{A}$ and $\Delta\mathbf{B}$ are the minimum uncertainties in the observables A and B. The common statement that $\Delta E\,\Delta t$ equals $\hbar/2$ is but one specific example of this. We will not attempt to justify Heisenberg's theory here. For such justification the reader is referred to the works by Jammer, Gamow, and Troup listed in the bibliography. We will, however, illustrate the use of the Heisenberg method in Chapter 4.

1.5 THE DE BROGLIE HYPOTHESIS AND SCHRÖDINGER'S WAVE MECHANICS

Einstein's treatment of the photoelectric effect required that a quantum of electromagnetic radiation (which he called a photon) have a momentum associated with it. In other words, it behaved as a moving particle. In his 1925 doctoral thesis from the University of Paris, Louis de Broglie reasoned that if, under certain circumstances, electromagnetic radiation behaved as a particle instead of as a wave, there might be circumstances under which particles of matter would behave as waves. Put very simply, we have, from Einstein's theory of relativity:

$$E = mc^2 \qquad 1.34$$

On the other hand, the Planck relationship is

$$E = h\nu = \frac{hc}{\lambda} \qquad 1.35$$

Equating the two and dividing through by c, we obtain

$$mc = \frac{h}{\lambda} \qquad 1.36$$

This implies that for a particle of matter we might write

$$mv = \frac{h}{\lambda} \qquad \textbf{1.37}$$

or

$$\lambda = \frac{h}{mv} = \frac{h}{p} \qquad \textbf{1.38}$$

where m is the mass of the particle at velocity v (or mv is its linear momentum, p). The wave nature of matter was soon confirmed by Davisson and Germer, who showed that a beam of electrons could be diffracted by the regular spacings of the atoms in a crystal, just as light is diffracted by the regular spacings of the lines in a diffraction grating.

Bohr's quantization condition for angular momentum can immediately be derived from de Broglie's relationship. If the electron in a Bohr orbit is behaving as a wave, the orbit must be such that a *standing wave* is formed; i.e., the circumference of the orbit must be an integer multiple of the wavelength, or else destructive interference will occur. We have

$$2\pi r = n\lambda \qquad \textbf{1.39}$$

or

$$\lambda = \frac{2\pi r}{n} \qquad \textbf{1.39a}$$

but from the de Broglie relationship

$$mv = \frac{h}{\lambda} = \frac{nh}{2\pi r} \qquad \textbf{1.40}$$

Rearranging, we get Bohr's quantization condition

$$mvr = p_\phi = \frac{nh}{2\pi} \qquad \textbf{1.41}$$

Schrödinger developed his wave mechanics by taking the wave equations for classical electromagnetic radiation and substituting the de Broglie relationship into them. Maxwell's equation for the propagation of a wave in one dimension is

$$\frac{\partial^2 \Psi}{\partial x^2} = \frac{1}{v^2}\frac{\partial^2 \Psi}{\partial t^2} \qquad \textbf{1.42}$$

where Ψ is the function describing the wave (the *wave function*), x is the direction of propagation, v is the velocity of propagation, and t is time. The most general solution of the second-order differential equation that is Eq. 1.42 is

$$\Psi(x, t) = a \exp\left[2\pi i\left(\frac{x}{\lambda} - vt\right)\right] \qquad \textbf{1.43}$$

where a is the amplitude. Two other acceptable solutions are

$$\Psi(x, t) = a \sin 2\pi \left(\frac{x}{\lambda} - \nu t\right) \tag{1.44}$$

$$\Psi(x, t) = a \cos 2\pi \left(\frac{x}{\lambda} - \nu t\right) \tag{1.45}$$

Let us consider the exponential form of the solution to the wave equation. This can be written

$$\Psi = a e^{2\pi i x/\lambda} e^{-2\pi i \nu t} \tag{1.46}$$

Let us designate $a \exp(2\pi i x/\lambda)$ as $\psi(x)$. We have

$$\Psi = \psi(x) e^{-2\pi i \nu t} \tag{1.47}$$

Differentiating this twice with respect to x yields

$$\frac{\partial^2 \Psi}{\partial x^2} = \frac{\partial^2 \psi(x)}{\partial x^2} e^{-2\pi i \nu t} \tag{1.48}$$

Differentiating with respect to time, we have

$$\frac{\partial \Psi}{\partial t} = -\psi(x) 2\pi i \nu e^{-2\pi i \nu t} \tag{1.49}$$

and, differentiating again,

$$\frac{\partial^2 \Psi}{\partial t^2} = -\psi(x) 4\pi^2 \nu^2 e^{-2\pi i \nu t} \tag{1.50}$$

Substituting Eqs. 1.48 and 1.50 into 1.42 yields

$$\frac{\partial^2 \psi(x)}{\partial x^2} e^{-2\pi i \nu t} = -\frac{1}{v^2} \psi(x) 4\pi^2 \nu^2 e^{-2\pi i \nu t} \tag{1.51}$$

Canceling the exponential terms and equating ν^2/v^2 to λ^{-2}, we have

$$\frac{\partial^2 \psi(x)}{\partial x^2} = -\frac{4\pi^2}{\lambda^2} \psi(x) \tag{1.52}$$

Let us now substitute in the de Broglie relation for λ (Eq. 1.37). This gives

$$\frac{d^2 \psi}{dx^2} = -\frac{4\pi^2 m^2 v^2}{h^2} \psi \tag{1.53}$$

(Note that we have gone to the total differential in our notation, since ψ is a function of only one variable.)

Now we will relate Eq. 1.53 to the energy of the system. Energy is the sum of kinetic energy, T, and potential energy, V:

$$E = T + V$$
$$= \frac{1}{2}mv^2 + V \tag{1.54}$$

Solving for v^2, we have

$$v^2 = \frac{2}{m}(E - V) \tag{1.55}$$

Substituting into Eq. 1.53 gives us

$$\frac{d^2\psi}{dx^2} = -\frac{8\pi^2 m}{h^2}(E - V)\psi \tag{1.56}$$

and rearranging,

$$\left(-\frac{\hbar^2}{2m}\frac{d^2}{dx^2} + V\right)\psi = E\psi \tag{1.57}$$

where we have used $\hbar$ for $h/2\pi$.

Equation 1.57 is the *time-independent Schrödinger equation* in one dimension. We can extend it to two or three dimensions by simply adding in the second derivative with respect to y and/or z. The quantity in parentheses is commonly referred to as the *Hamiltonian operator*, $\hat{H}$. The equation is then written

$$\hat{H}\psi = E\psi \tag{1.58}$$

This has the form of an *eigenvalue equation.* An operator (in this case $\hat{H}$) operating on a function (called the *eigenfunction*) yields a constant (in this case the energy *eigenvalue*) times the eigenfunction. The Hamiltonian operator is the operator representing energy. Note that the potential-energy term is unchanged from the form in the classical expression (Eq. 1.54); however, the kinetic energy has a new form. Classically, kinetic energy can be expressed in terms of momentum, p, as $p^2/2m$. Confining our attention to the x direction (and p_x) and relating the classical and quantum-mechanical expressions for kinetic energy, we see

CLASSICAL		QUANTUM-MECHANICAL	
$\frac{p_x^2}{2m}$	$\Rightarrow$	$-\frac{\hbar^2}{2m}\frac{d^2}{dx^2}$	1.59
p_x	$\Rightarrow$	$\pm\left(i\hbar\frac{d}{dx}\right)$	1.60

(The negative sign of Eq. 1.60 is conventionally chosen.) One way of setting up the proper quantum-mechanical operator for a system is to first set up the classical operator and then replace each linear momentum by $-i\hbar d/dx$.

For time-dependent problems, the *time-dependent Schrödinger equation* is required. Consider again the general solution to the wave equation shown in Eq. 1.43:

$$\Psi(x, t) = a \exp\left[2\pi i \left(\frac{x}{\lambda} - \nu t\right)\right] \tag{1.61}$$

Differentiate this with respect to time:

$$\frac{\partial \Psi(x, t)}{\partial t} = -2\pi i \nu a \exp\left[2\pi i\left(\frac{x}{\lambda} - \nu t\right)\right] \tag{1.62}$$

Substituting in the Planck relation for energy and frequency, we get

$$\frac{\partial \Psi(x, t)}{\partial t} = -\frac{iE\Psi(x, t)}{\hbar} \tag{1.63}$$

or, replacing E with the operator $\hat{H}$ and rearranging, we get the time-dependent Schrödinger equation:

$$\hat{H}\Psi = i\hbar \frac{\partial \Psi}{\partial t} \tag{1.64}$$

If we compare Eq. 1.64 with Eq. 1.32 from Heisenberg's theory, we see that Heisenberg assigns time variation to the observable itself, while Schrödinger assigns it to the wave function.

1.6 THE POSTULATES OF QUANTUM MECHANICS

As with every deductive science, the formal development of quantum mechanics is based upon fundamental postulates. These postulates are not as intuitively self-evident as are many scientific postulates; however, they are accepted as correct because the theories based upon them agree with experiment. They are the cornerstones of our mathematical model for describing chemistry at the atomic and molecular level.

The first postulate is that any system can exist only in specific states (*eigenstates*), and that each state is characterized by a wave function (*eigenfunction*) in the Schrödinger development or by a state vector (*eigenvector*) in the Heisenberg development. Although chemists commonly think primarily in terms of energy states, a state may be defined with respect to any observable quantity. In fact, the complete specification of a state requires the specification of all mutually compatible (i.e., commuting) observables. In practice, it is quite common to refer to incompletely specified states of a system as if they were the true states. This is reasonable when the observables pertinent to the experiment under consideration are specified, as, for example, for a spectroscopy experiment when states are specified in terms of energy only. However, one should always be aware of any incomplete specification of a state.

The second postulate has to do with the observables. To every observable there corresponds an operator, in the Schrödinger representation, or a matrix, in the Heisenberg representation (these matrices themselves can be considered to be constructed from an operator and a set of basis functions or basis vectors). If the operators or matrices commute, the wave function or state vector can be constructed so that it is simultaneously an eigenfunction or eigenvector of the commuting observables.

The third postulate concerns the probabilistic interpretation of quantum mechanics. If ψ_j is a wave function or $\boldsymbol{\psi}_j$ is a state vector for a system in state j, and $\hat{p}$ is a Schrödinger operator or $\mathbf{p}$ a Heisenberg matrix representing some observable, then the *expectation value* (the quantum-mechanical average or expected mean), $\langle p \rangle_j$, of the observable in that state is

$$\langle p \rangle_j = \frac{\int \psi_j^* \hat{p} \psi_j \, d\tau}{\int \psi_j^* \psi_j \, d\tau} \qquad \textbf{1.65a}$$

(where $d\tau$ is a volume element over all variables and the integration is over all values of all variables), or

$$\langle p \rangle_j = \frac{\boldsymbol{\psi}_j^\dagger \mathbf{p} \boldsymbol{\psi}_j}{\boldsymbol{\psi}_j^\dagger \boldsymbol{\psi}_j} \qquad \textbf{1.65b}$$

Usually, wave functions or state vectors are normalized so that

$$\int \psi_j^* \psi_j \, d\tau = 1 \qquad \textbf{1.66a}$$

or

$$\boldsymbol{\psi}_j^\dagger \boldsymbol{\psi}_j = 1 \qquad \textbf{1.66b}$$

In the wave description the quantity $\psi_j^* \psi_j$, evaluated at a particular set of values for the variables, is the probability of finding the system at that set of variables. Thus, Eq. 1.66a is just the statement that there is unit probability of finding the system somewhere in the space defined by the variables.

BIBLIOGRAPHY

ATKINS, P. W., *Molecular Quantum Mechanics*. Clarendon Press, Oxford, 1970.

GAMOW, G., *Thirty Years that Shook Physics*. Anchor Books, Garden City, N. Y., 1966.

JAMMER, M., *The Conceptual Development of Quantum Mechanics*. McGraw-Hill Book Company, New York, 1966.

KARPLUS, M., and PORTER, R. N., *Atoms and Molecules*. W. A. Benjamin, Inc., New York, 1970.

LEVINE, I. N., *Quantum Chemistry*. Allyn & Bacon, Boston, 2d ed., 1974.

SLATER, J. C., *Quantum Theory of Atomic Structure*, Vol. 1. McGraw-Hill Book Company, New York, 1960.

TROUP, G., *Understanding Quantum Mechanics*. Methuen & Co., London, 1968.

PROBLEMS

1.1 Plot the Wien displacement law, the Rayleigh-Jeans law, and Planck's law for a temperature of 1200°K and up to a frequency of 6×10^{14} Hz. In the Wien law, let $\alpha = 8\pi h/c^3$ and $\beta = h/k$.

***1.2** Show that the low-frequency limit of Planck's law yields the Rayleigh-Jeans law and that the high-frequency limit yields Wien's law. [*Hint:* Use a series expansion for the exponential term at low frequency.]

***1.3** Calculate the kinetic energy and the de Broglie wavelength for each of the following:

(a) A 10-gm bullet moving at a velocity of 500 m sec^{-1}.

(b) A 1-kg turtle moving at 1 cm sec^{-1}.

(c) A 90-kg man moving at 2 m sec^{-1}.

(d) A 5000-kg airplane moving at 100 m sec^{-1}.

(e) The earth as it moves around the sun.

(f) An electron moving at 7×10^6 cm sec^{-1} (this corresponds to a thermal velocity of about 325°K).

(g) An electron moving at 2×10^{10} cm sec^{-1} (the approximate velocity of an electron in the $n = 1$ Bohr orbit of zinc).

(h) A neutron moving at 1×10^7 cm sec^{-1}.

***1.4** Find the velocities of an electron in the $n = 1$, $n = 2$, and $n = 3$ Bohr orbits of He^+.

***1.5** The spectra and ionization potentials of the alkali metals (family 1A of the periodic table) can be fairly well approximated by Bohr theory if an "effective quantum number," $n' = (n - d)$, where d is a "quantum defect," is assumed to replace n. From the first ionization potential, calculate the quantum defect for the ns electron and the energy of the $(n + 1)s \leftarrow ns$ transition for Li ($n = 2$; I.P. = 5.363 eV) and for Na ($n = 3$; I.P. = 5.137 eV). Use the hydrogen atom value for the Rydberg constant (i.e., assume that the inner-shell electrons completely screen the nucleus). (The experimental transition for Na is 25,730 cm^{-1}.)

1.6 Another theoretical model for the spectra and ionization potentials of the alkali metals is to assume that the single ns valence electron is in a hydrogenlike orbit with a "screened" nucleus (i.e., a nuclear charge modified by the inner-shell electrons) with an effective charge of $(Z - s)$, where s is the "screening constant." From the ionization potential for the ns electron, calculate the screening constant and the energy of the $(n + 1)s \leftarrow ns$ transitions for Li and Na.

***1.7** (a) By using Eq. 1.46 of the text, the de Broglie relation (Eq. 1.38), and the fact that the energy is the sum of the kinetic (T) and potential (V) energies, rewrite the general one-dimensional time-dependent wave function in terms of energy.

(b) What is the behavior of this wave function if a particle is moving in a field of a constant, finite potential that is greater than E?

(c) Qualitatively, what would happen if a particle moving in the positive x direction encountered a barrier of finite $V(> E)$ and finite thickness? How does this differ from what you expect from a classical viewpoint?

***1.8** (a) From Eq. 4.9 of the text, deduce the uncertainty relation for position and momentum.

(b) Find the uncertainty in momentum for each of the following, assuming the given precision in measuring the position:
(1) A 90-kg man moving at 2 m sec^{-1} (Δx = 1 mm).
(2) An electron moving at 2×10^8 m sec^{-1} (Δx = 1 Å).

chapter two

Constant potentials and potential wells

2.1 INTRODUCTION

Equation 1.57 presents the time-independent Schrödinger equation for a single particle moving in one dimension. It is a second-order differential equation in one variable. Its solution, either for the energy or for the wave function, obviously will depend upon the functional form of the potential, V. For certain forms of the potential, exact closed-form, analytical solutions can be found. For other forms of the potential, the equation must be solved numerically.

The general solution to the motion of one particle requires three dimensions, x, y, and z (or some other three-dimensional coordinate system). The corresponding Schrödinger equation is a second-order differential equation in three variables. A system of N particles would require three coordinates for each particle, or a total of $3N$ to describe the system. Thus, the general Schrödinger equation for N particles is a second-order differential equation in $3N$ variables. Further, if there are interactions between the particles, the variables are coupled; i.e., the motion of each particle affects that of the others. We see that the problem quickly becomes very complicated.

It turns out that for only a very few problems is an exact, analytical, closed-form quantum-mechanical solution known. In principle, any other system should be solvable by direct numerical methods; in practice, however, most quantum-mechanical problems are solved by the use of approximate methods. Most of this text will draw heavily on such methods. In this chapter, though, we will present closed-form solutions for a particle moving in a constant potential and for a particle moving in a potential well. In the next three chapters we present the closed-form solutions of three other problems.

2.2 PARTICLE MOVING IN FREE SPACE

A particle moving in a field-free space has no potential field acting on it; consequently, the Hamiltonian contains only the kinetic-energy operator. If we use the general solution for the one-dimensional wave equation (Eq. 1.43),

$$\Psi(x, t) = a \exp\left[2\pi i\left(\frac{x}{\lambda} - \nu t\right)\right] \tag{2.1}$$

we have, for the time-dependent Schrödinger equation,

$$-\frac{\hbar^2}{2m}\frac{\partial^2 \Psi}{\partial x^2} = i\hbar\frac{\partial \Psi}{\partial t} \tag{2.2}$$

Carrying out the differentiation of Ψ in Eq. 2.1 with respect to x, and substituting into Eq. 2.2, we have

$$-\frac{\hbar^2}{2m}\left(-\frac{4\pi^2}{\lambda^2}\Psi\right) = i\hbar\frac{\partial \Psi}{\partial t} \tag{2.3}$$

or

$$i\hbar\frac{\partial \Psi}{\partial t} = \frac{h^2}{2m\lambda^2}\Psi \tag{2.4}$$

$$= \frac{p^2}{2m}\Psi \tag{2.4a}$$

$$= E\Psi \tag{2.4b}$$

where we have substituted in the de Broglie relation to get 2.4a and recognized that we have only kinetic energy to get 2.4b. This is nothing more than a restatement of the time-dependent equation we obtained in Chapter 1. If the particle is moving in a field of constant potential, V, the solution is the same, except that the E in Eq. 2.4b is replaced by $(E - V)$.

Let us consider the consequences of Heisenberg's uncertainty principle as applied to a particle in free space. We have for the minimum uncertainty

$$\Delta E\ \Delta t = \frac{\hbar}{2} \tag{2.5}$$

Substituting in the Planck relationship, we get

$$h\ \Delta\nu\ \Delta t = \frac{\hbar}{2} \tag{2.6}$$

or

$$\Delta\nu\,\Delta t = \frac{1}{4\pi} \qquad \textbf{2.6a}$$

Thus, since the product, νt, occurs with a multiplier of 2π in Eq. 2.1, there is a phase uncertainty of $\frac{1}{2}$ in the wave function for a particle moving in free space. This means that, even though the moving particle can be treated mathematically by a wave equation, we cannot find maxima and minima in the amplitude of the wave, as we can with ordinary wave patterns.

Although the free-particle problem may seem trivial, it is actually quite important. For example, the quantum-mechanical scattering problem is based upon the free-particle wave function.

2.3 PARTICLE IN A ONE-DIMENSIONAL POTENTIAL WELL (SCHRÖDINGER TREATMENT)

Most of the problems to which chemists apply quantum mechanics involve bound states; i.e., the motion of the particle is restricted to certain regions of space. The form of the potential portion of the Hamiltonian determines whether or not bound states exist. If the potential is zero or constant over all space, bound states do not occur unless there are other restrictions, such as continuity, on the wave function. If, on the other hand, there are finite regions of space where the potential is lower than in other regions, bound states may exist. Perhaps the simplest such problem is that of the particle in a one-dimensional potential well (or "particle in a box"). In this problem the potential is assumed to be zero (or a finite constant) in a finite region of a one-dimensional space and infinitely large outside this region (Figure 2.1).

The solution to the particle-in-a-box problem is very simple in the Schrödinger representation. We accomplish it, as in all exactly solvable problems in the Schrödinger treatment, by requiring that the wave function behave properly. Specifically, a wave function must be *continuous*, *single-valued*, and *finite* throughout the configuration space of the system. For the one-dimensional box, the allowed wave functions

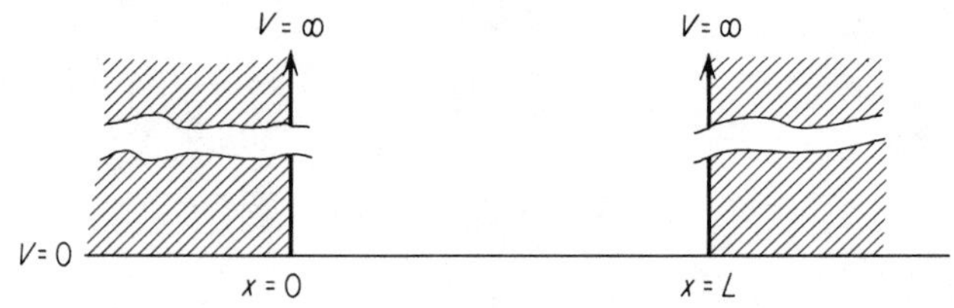

Figure 2.1 The potential energy of a one-dimensional potential well. The potential, V, is zero between $x = 0$ and $x = L$ and becomes abruptly infinite for all $x < 0$ and all $x > L$.

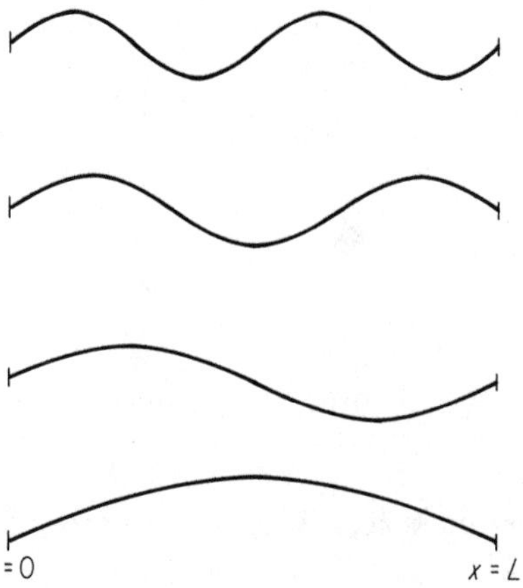

Figure 2.2 Allowed vibrations of a violin string.

are analogous to the allowed modes of vibration of a violin string (see Figure 2.2). The allowed wavelengths of the indicated vibrations are

$$\lambda = 2L \tag{2.7a}$$

$$\lambda = L \tag{2.7b}$$

$$3\lambda = 2L \tag{2.7c}$$

$$2\lambda = L \tag{2.7d}$$

or in general

$$\lambda = \frac{2L}{n} \tag{2.8}$$

where n is a whole number. This result can be used with the de Broglie relation to obtain the energy immediately. We have

$$\frac{h}{mv} = \frac{2L}{n} \tag{2.9}$$

$$v = \frac{nh}{2mL} \tag{2.10}$$

$$v^2 = \frac{n^2h^2}{4m^2L^2} \tag{2.11}$$

$$E = \frac{1}{2}mv^2 = \frac{n^2h^2}{8mL^2} \tag{2.12}$$

We did not actually use the Schrödinger equation in obtaining Eq. 2.12. In order to relate it to other solutions of the Schrödinger equation, however, let us go through a solution based upon the Schrödinger equation. Consider first the Hamiltonian. Inside the box the potential is zero, and we have

$$\hat{H}(0 \leq x \leq L) = -\frac{\hbar^2}{2m}\frac{d^2}{dx^2} + 0 \qquad \textbf{2.13}$$

Outside the potential well the potential is infinite, and we have

$$\hat{H}(x < 0; x > L) = -\frac{\hbar^2}{2m}\frac{d^2}{dx^2} + \infty \qquad \textbf{2.14}$$

However, we know that the Schrödinger equation

$$\hat{H}\psi = E\psi \qquad \textbf{2.15}$$

must hold over all space. The only way that Eq. 2.15 can hold over all space for a finite energy is for the wave function, ψ, to equal zero outside the potential well. For the wave function to be continuous, it must have the same value at x equals zero and L as it does outside the well; i.e.

$$\psi(0) = \psi(L) = 0 \qquad \textbf{2.16}$$

A sine function has a value of zero for an argument of zero. Thus, we might expect ψ to be a sine function. A sine function also equals zero if its argument is an integer multiple of π. We have

$$\psi(0) = A \sin (0) \qquad \textbf{2.17}$$

and

$$\psi(L) = A \sin (n\pi) \qquad \textbf{2.18}$$

where A is the amplitude of the function. The argument of the sine function must be a continuous function of x. The function

$$\psi_n(x) = A \sin \frac{n\pi x}{L} \qquad \textbf{2.19}$$

where n is an integer greater than zero, satisfies all our requirements. Note that n cannot have a value of zero, or else the wave function would equal zero over the entire space. By convention, n is chosen to be positive. A negative n would only change the sign of ψ.

We can now apply the Hamiltonian of Eq. 2.13 to the wave function of Eq. 2.19:

$$-\frac{\hbar^2}{2m}\frac{d^2}{dx^2} A \sin \frac{n\pi x}{L} = EA \sin \frac{n\pi x}{L} \qquad \textbf{2.20}$$

Carrying out in steps the operations on the left, we have

$$\frac{d}{dx} A \sin \frac{n\pi x}{L} = \frac{n\pi}{L} A \cos \frac{n\pi x}{L} \qquad \textbf{2.21}$$

$$\frac{d^2}{dx^2} A \sin \frac{n\pi x}{L} = -\frac{n^2\pi^2}{L^2} A \sin \frac{n\pi x}{L} \qquad \textbf{2.22}$$

$$-\frac{\hbar^2}{2m}\frac{d^2}{dx^2} A \sin \frac{n\pi x}{L} = \frac{n^2h^2}{8mL^2} A \sin \frac{n\pi x}{L} \qquad \textbf{2.23}$$

Equation 2.19 is, thus, an eigenfunction of the Hamiltonian. Comparing Eqs. 2.20 and 2.23, we again see that

$$E_n = \frac{n^2h^2}{8mL^2} \tag{2.12}$$

In order to evaluate the constant, A, of Eq. 2.19, we need to give a physical interpretation to the wave function. The wave function ψ is interpreted as a probability amplitude. Its square, $|\psi|^2$ (or $\psi^*\psi$ if the function is complex), is interpreted as a probability function. In other words, $|\psi(x)|^2$ is the probability that the particle will be at position x—i.e., between x and $x + dx$. (This interpretation can be verified from the Davisson-Germer electron-diffraction experiment.) If the function is time-dependent, then $|\Psi(x, t)|^2$ is the probability that the particle will be at position x at time t. In the present case we know that our particle must be somewhere in space. The total probability is unity. This implies that

$$\int_{-\infty}^{\infty} |\psi|^2\, dx = 1 \tag{2.24a}$$

In the present case the wave function equals zero, except when $0 \leq x \leq L$, giving

$$\int_{-\infty}^{0} 0\, dx + \int_{0}^{L} |\psi|^2\, dx + \int_{L}^{\infty} 0\, dx = 1 \tag{2.24b}$$

or

$$\int_{0}^{L} |\psi|^2\, dx = 1 \tag{2.24c}$$

A similar requirement holds for any single-particle wave function. The integral of the square of the function over the available configuration space must equal unity. This is known as the *normalization* condition.

If we substitute Eq. 2.19 into Eq. 2.24c, we have

$$A^2 \int_{0}^{L} \sin^2 \frac{n\pi x}{L}\, dx = 1 \tag{2.25}$$

The integral has the value $L/2$. Thus

$$A^2 \frac{L}{2} = 1 \tag{2.26}$$

$$A^2 = \frac{2}{L} \tag{2.26a}$$

$$A = \sqrt{\frac{2}{L}} \tag{2.26b}$$

and

$$\psi_n(x) = \sqrt{\frac{2}{L}} \sin \frac{n\pi x}{L} \tag{2.27}$$

It is instructive to multiply together two different eigenfunctions, corresponding to different values of n, and integrate the result over our allowed space. We have

$$\int_{-\infty}^{\infty} \psi_n(x)\psi_{n'}(x)\,dx = \frac{2}{L}\int_0^L \sin\frac{n\pi x}{L}\sin\frac{n'\pi x}{L}\,dx$$
$$= \frac{2}{L}\left[\frac{\sin\left[\frac{\pi}{L}(n-n')x\right]}{\frac{2\pi}{L}(n-n')} - \frac{\sin\left[\frac{\pi}{L}(n+n')x\right]}{\frac{2\pi}{L}(n+n')}\right]\Bigg|_0^L$$
$$= 0 \tag{2.28}$$

The functions are said to be *orthogonal*. Any two different eigenfunctions (corresponding to different quantum numbers) to a given problem will always be orthogonal. (If the functions are complex, the integrand must be $\psi_n^*\psi_{n'}$.) In effect, the complete set of eigenfunctions to a problem forms a complete set of linearly independent functions. These can be used to define a function space that forms a basis for a vector algebra. In this way the Heisenberg and Schrödinger representations of quantum mechanics are connected.

The extension of the particle-in-a-potential-well problem to two or three orthogonal (i.e., perpendicular) dimensions is simply a generalization of the one-dimensional treatment. Quantization occurs in each direction. For three dimensions, we have

$$E_{n_x n_y n_z} = \frac{h^2}{8m}\left(\frac{n_x^2}{a^2} + \frac{n_y^2}{b^2} + \frac{n_z^2}{c^2}\right) \tag{2.29}$$

where a, b, and c are the dimensions in the x, y, and z directions, respectively, and

$$\psi_{n_x n_y n_z} = \sqrt{\frac{8}{V}}\sin\frac{n_x\pi x}{a}\sin\frac{n_y\pi y}{b}\sin\frac{n_z\pi z}{c} \tag{2.30}$$

where V is the volume of the box. The problem can also be solved in a straightforward manner for wells having other regular shapes—a circle (two dimensions), a sphere (three dimensions), and so on.

In the first paragraph of this section we stated that the potential can be zero or a finite constant within the potential well. If the potential is a nonzero constant, all the energy levels will be shifted by the magnitude of the constant. There is no change in the wave functions.

2.4 FREE-ELECTRON THEORY OF THE SPECTRA OF CONJUGATED SYSTEMS

The particle in a one-dimensional well is the basis for the "free-electron" description of the π-electron systems of conjugated linear polyenes. The framework of the conjugated system is assumed to comprise a one-dimensional well with a constant potential inside and an infinite potential outside. The length of the box is usually assumed to be the length of the conjugated chain, with one bond length added at each end. (The extra length is to place the *nodes*, the positions where the wave function equals zero, at the proper places.) Each solution to the problem is considered to be an "orbital" that can accommodate two electrons. The ground state is constructed by assigning two electrons to each successively higher-energy orbital until all the π electrons are accounted for. Electronic spectral transitions can be considered to involve the promotion of an electron from an occupied orbital to a vacant orbital.

The foregoing is illustrated schematically for ethylene, butadiene, and hexatriene in Figure 2.3. The total π-electron energies of the ground states are 2, 10, and 28, respectively, in units of $h^2/(8mL^2)$. (Obviously, however, the value of L is different for each.) If there are N electrons in the π system, $N/2$ levels are occupied in the ground state. The first transition corresponds to an orbital transition from n equals $N/2$ to n equals $(N/2 + 1)$. This can be expressed in closed form. Since each atom contributes one π electron, N electrons correspond to N atoms, and the length of the

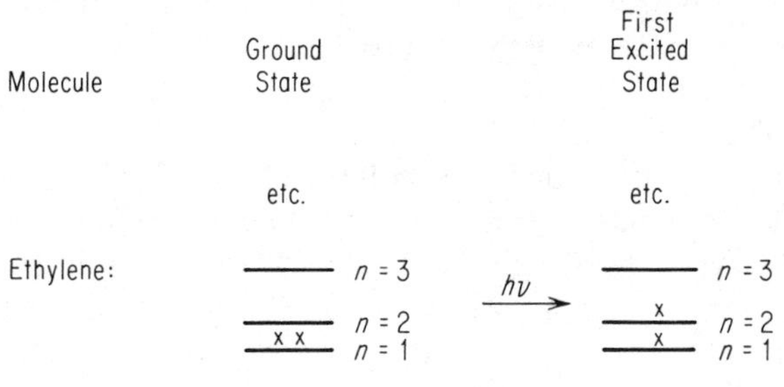

Figure 2.3 Schematic diagram of the free-electron energy levels and orbital occupancy for the ground and first excited states of the π-electron systems of ethylene, butadiene, and hexatriene.

box is $(N + 1)b$, where b is the average bond length. Thus, the energy of the first transition is

$$\Delta E = \frac{[(N/2 + 1)^2 - (N/2)^2]h^2}{8m[(N + 1)b]^2}$$

$$= \frac{h^2}{8mb^2(N + 1)} \qquad 2.31$$

If the average bond length is assumed to be 1.4 Å (1.4×10^{-10} m), the transition, expressed in wave numbers, is

$$\bar{\nu} = \frac{154{,}739\ cm^{-1}}{N + 1} \qquad 2.32$$

Table 2.1 presents the calculated transitions for a few members of the polyene series, along with the experimental values. Note that the numerical agreement is not good (one should not expect it to be for such a crude model), but there is a linear relationship between $\bar{\nu}_{calc}$ and $\bar{\nu}_{exp}$. This is emphasized if a least-squares fit is applied to the data. The least-squares equation is

$$\bar{\nu}_{fit} = .893\bar{\nu}_{calc} + 17{,}153\ \text{cm}^{-1} \qquad 2.33$$

The least-squares values (last column of Table 2.1) are very close to the experimental values. The least-squares equation would be very useful to approximate the spectrum of an unknown polyene. Similar correlations can be carried out for other types of conjugated systems. Polyenes with heteroatoms use the same type of equation. In fact, the results for the cyanine dyes are even better than for the polyenes. Aromatic systems may use either rings or two-dimensional boxes, and so on. Before the advent of computers, free-electron theory was used by scientists in the dye industry to design dyes with specific colors.

Table 2.1

CALCULATED AND OBSERVED FIRST SPECTRAL TRANSITIONS FOR SOME LINEAR POLYENES

N^a	$\bar{\nu}_{calc}{}^b$	$\bar{\nu}_{exp}{}^c$	$\bar{\nu}_{fit}{}^d$
2	51,580 cm^{-1}	61,500 cm^{-1}	63,207 cm^{-1}
4	30,948	46,080	44,785
✓6	22,106	39,750	36,891
8	17,193	32,900	32,504
10	14,067	29,940	29,713
20	7,369	22,371	23,732

[a] Number of atoms in the conjugated chain.
[b] Equation 2.32.
[c] Experimental.
[d] Equation 2.33.

Particle-in-a-box models are not limited to spectral properties. For example, we derive the translational partition function from statistical mechanics by considering the quantized translational energy levels of a molecule in a three-dimensional box. Radioactive decay can be modeled by a particle in a box with a wall of finite thickness. The decay process is then considered to be a manifestation of quantum-mechanical *tunneling*. There are many other applications as well.

BIBLIOGRAPHY

ANDERSON, J. M., *Introduction to Quantum Chemistry*. W. A. Benjamin, Inc., New York, 1969.

ATKINS, P. W., *Molecular Quantum Mechanics*. Clarendon Press, Oxford, 1970.

FLURRY, R. L., JR., *Molecular Orbital Theories of Bonding in Organic Molecules*. Marcel Dekker, New York, 1968.

HAMEKA, H. F., *Introduction to Quantum Theory*. Harper & Row, Publishers, New York, 1967.

LEVINE, I. N., *Quantum Chemistry*. Allyn & Bacon, Boston, 2d ed., 1974.

PAULING, L., and WILSON, E. B., JR., *Introduction to Quantum Mechanics*. McGraw-Hill Book Company, New York, 1935.

PLATT, J. R., RUEDENBERG, K., SCHERR, C. W., HAM, N. S., LABHART, H., and LICHTEN, W., *Free-Electron Theory of Conjugated Molecules*. John Wiley & Sons, New York, 1964.

PROBLEMS

2.1 Consider a 1.5-gm ping-pong ball constrained to move in one dimension the length of a 2.5-m table.

(a) Solve the particle in a one-dimensional-box problem for the energy and wave function.

(b) Find the approximate quantum number if the ball is moving at a velocity of 50 km hr^{-1}.

(c) Find the energy difference between this quantum level and the next higher one.

2.2 The diphenylpolyenes have the general formula $C_6H_5(CH{=}CH)_nC_6H_5$. The first electronic transitions of the first few members of the series are:

n	λ (Å)
1	3060
2	3340
3	3580
4	3840
5	4030
6	4200
7	4350

Use the one-dimensional free-electron theory to predict the wavelengths of the first electronic transitions for $n = 11$ and for $n = 15$. Assume that each phenyl group contributes three carbon atoms (and three electrons) to the conjugated system. (The experimental values are 5300 Å and 5700 Å, respectively.)

***2.3** Another "particle-in-a-box" type problem that can be solved is that of the particle moving on a ring of constant radius, subject to a constant potential on the ring and an infinite potential off the ring. The Schrödinger equation, after the proper variables have been substituted, is

$$-\frac{\hbar^2}{2mr^2}\frac{d^2\psi(\phi)}{d\phi^2} = (E - V)\psi(\phi)$$

where r is the fixed radius and ϕ is the angular variable.

(a) Solve this problem for the energy levels and wave functions.

(b) This problem has been used as a model for the spectra of polynuclear aromatic molecules. The perimeter of the molecule defines the circumference of the ring. Using this model, calculate the first $\pi^* \leftarrow \pi$ transitions for benzene, naphthalene, anthracene, tetracene, and pentacene. [*Hint:* The circumference of the ring will be $N \times 1.4$ Å, where N is the number of atoms. In this case, zero is an allowed quantum number (why?) and all the other levels are twofold degenerate.]

(c) The band classified as ^1L_b by Platt is observed at the indicated wavelengths:

Molecule	$\bar{\nu}$ (cm^{-1})
Benzene	3.8×10^4
Naphthalene	3.2×10^4
Pentacene	2.4×10^4

The ^1L_b bands for anthracene and tetracene are obscured by the much stronger ^1L_a band. Predict the ^1L_b transition for these two molecules.

***2.4** Platt [*J. Chem. Phys.*, **22**, 1448 (1954)] has used a two-dimensional "free-electron" theory to describe the spectra of aromatic hydrocarbons. Assume that all bond lengths are 1.4 Å and that you have a rectangular, two-dimensional box whose dimensions extend one bond length (in the valence directions) beyond the carbon atoms; calculate the first two electronic transitions for naphthalene and anthracene.

2.5 Use the wave functions from the preceding problem and make a contour plot of the electron density in the ground state of naphthalene. Use a computer or programmable calculator. Evaluate the wave functions for a grid of about .28 Å spacings. Note that, owing to symmetry, only one-fourth of the points must be calculated.

chapter three

Rotation and angular momentum

3.1 INTRODUCTION

Gas-phase molecules are in continuous motion. The kinetic theory of gases tells us that this motional energy is thermal energy. If the molecules are considered to be rigid, structured particles, they can have two types of motion, translational and rotational. In Chapter 2 we saw that for a particle with translational energy, but confined to a finite region of space (to a container), the available energy levels are quantized. For molecules moving at thermal velocities in laboratory-sized containers, these energy levels are so closely spaced as to be, for all practical purposes, a continuum.

The rotational motion of a molecule is also quantized, because there are periodic restrictions on the motion. That is, $\psi_R(2\pi)$ must equal $\psi_R(0)$, where ψ_R is the rotational wave function for motion about some axis. The energy spacings between the rotational energy levels of molecules are of a magnitude to be observable in the microwave spectral region. The rotations of a linear molecule can be described, to a good approximation, as those of a rigid rotor consisting of two point masses connected by a massless rigid rod. The rotational behavior of nonlinear molecules is more complicated but is related to the rigid-rotor problem.

In this chapter we will present the quantum-mechanical solution to the rigid-rotor problem in full detail, using the Schrödinger approach. This presentation will illustrate the Schrödinger treatment of a fairly complicated problem; it also will provide the tools for discussing the microwave spectra of linear molecules and the starting point for discussing the rotational behavior of nonlinear molecules. The angular-momentum behavior obtained from the rigid-rotor problem is the same as that for any system (such as an atom) having spherical symmetry. Also, angular-momentum theory is intimately related to group theory, so group theory is introduced in this chapter.

3.2 THE RIGID ROTOR (SCHRÖDINGER TREATMENT)

The rotations of a linear molecule can be described, to a good approximation, as those of a rigid rotor. The motion of a rotating object can best be described in spherical polar coordinates. Such a system for a rotor consisting of two masses is shown in Figure 3.1. In terms of spherical polar coordinates, the Cartesian coordinates are

$$x = r \sin\theta \cos\phi \qquad \textbf{3.1a}$$

$$y = r \sin\theta \sin\phi \qquad \textbf{3.1b}$$

$$z = r \cos\theta \qquad \textbf{3.1c}$$

where r is the distance from the origin, θ is the angle from the positive z axis, and ϕ is the angle of a projection of r in the xy plane from the positive x axis (see Figure 3.1). The allowed values of the variables are

$$0 \le r \le \infty \qquad \textbf{3.2a}$$

$$0 \le \theta \le \pi \qquad \textbf{3.2b}$$

$$0 \le \phi \le 2\pi \qquad \textbf{3.2c}$$

The kinetic energy for a particle, in Cartesian coordinates, is

$$T = \tfrac{1}{2}mv^2 = \tfrac{1}{2}m(\dot{x}^2 + \dot{y}^2 + \dot{z}^2) \qquad \textbf{3.3}$$

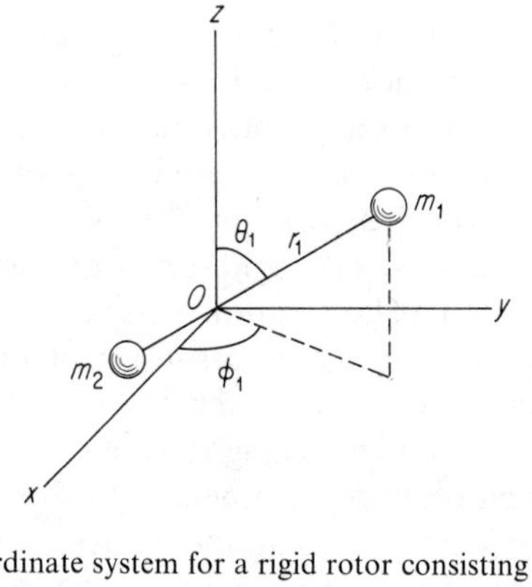

Figure 3.1 The coordinate system for a rigid rotor consisting of two masses rigidly joined. The origin is on the center of mass. Spherical polar coordinates are indicated for particle 1.

where $\dot{x}$, $\dot{y}$, and $\dot{z}$ are the time derivatives of x, y, and z. Substituting Eqs. 3.1 into 3.3 yields

$$T = \tfrac{1}{2}m(\dot{r}^2 + r^2\dot{\theta}^2 + r^2\dot{\phi}^2 \sin^2 \theta) \qquad 3.4$$

By the definition of a rigid rotor, r is a constant for each particle; thus for each particle the kinetic energy is

$$T = \tfrac{1}{2}mr^2(\dot{\theta}^2 + \dot{\phi}^2 \sin^2 \theta) \qquad 3.5$$

or for two particles, since the changes in θ and in ϕ are the same for each particle,

$$T = \tfrac{1}{2}(m_1 r_1^2 + m_2 r_2^2)(\dot{\theta}^2 + \dot{\phi}^2 \sin^2 \theta) \qquad 3.6$$

The *moment of inertia* for a collection of particles is defined as

$$I = \sum_i m_i r_i^2 \qquad 3.7$$

In terms of the moment of inertia, the classical kinetic energy is

$$T = \tfrac{1}{2}I(\dot{\theta}^2 + \dot{\phi}^2 \sin^2 \theta) \qquad 3.8$$

The term in the parentheses is the square of the angular velocity, ω.

For the quantum-mechanical solution to the rigid-rotor problem in the Schrödinger representation, we need the appropriate Schrödinger equation. The three-dimensional time-independent Schrödinger equation can be written in the general form

$$\nabla^2\psi + \frac{2m}{\hbar^2}(E - V)\psi = 0 \qquad 3.9$$

The operator, ∇^2, the *Laplacian* operator (or simply the Laplacian), represents the second derivative with respect to the coordinates. In three-dimensional Cartesian coordinates, it is

$$\nabla^2 = \frac{\partial^2}{\partial x^2} + \frac{\partial^2}{\partial y^2} + \frac{\partial^2}{\partial z^2} \qquad 3.10$$

Its form will be different in other coordinate systems; however, there is a systematic method of finding the form in any system. If the new coordinates are labeled u, v, and w, then in the new system (Pauling and Wilson, Sec. IV-16)

$$\nabla^2 = \frac{1}{q_u q_v q_w}\left[\frac{\partial}{\partial u}\left(\frac{q_v q_w}{q_u}\frac{\partial}{\partial u}\right) + \frac{\partial}{\partial v}\left(\frac{q_u q_w}{q_v}\frac{\partial}{\partial v}\right) + \frac{\partial}{\partial w}\left(\frac{q_u q_v}{q_w}\frac{\partial}{\partial w}\right)\right] \qquad 3.11$$

where

$$q_u^2 = \left(\frac{\partial x}{\partial u}\right)^2 + \left(\frac{\partial y}{\partial u}\right)^2 + \left(\frac{\partial z}{\partial u}\right)^2 \qquad 3.12a$$

$$q_v^2 = \left(\frac{\partial x}{\partial v}\right)^2 + \left(\frac{\partial y}{\partial v}\right)^2 + \left(\frac{\partial z}{\partial v}\right)^2 \qquad 3.12b$$

$$q_w^2 = \left(\frac{\partial x}{\partial w}\right)^2 + \left(\frac{\partial y}{\partial w}\right)^2 + \left(\frac{\partial z}{\partial w}\right)^2 \qquad 3.12c$$

Choosing spherical polar coordinates as the new coordinates, we obtain

$$\nabla^2 = \frac{\partial^2}{\partial r^2} + \frac{2}{r}\frac{\partial}{\partial r} + \frac{1}{r^2 \sin\theta}\frac{\partial}{\partial\theta}\left(\sin\theta\frac{\partial}{\partial\theta}\right) + \frac{1}{r^2\sin^2\theta}\frac{\partial^2}{\partial\phi^2} \tag{3.13}$$

For the rigid rotor, r is constant. Thus, for the rigid rotor, the Laplacian has the form

$$\nabla^2 = \frac{1}{r^2}\left[\frac{1}{\sin\theta}\frac{\partial}{\partial\theta}\left(\sin\theta\frac{\partial}{\partial\theta}\right) + \frac{1}{\sin^2\theta}\frac{\partial^2}{\partial\phi^2}\right] \tag{3.14}$$

Substituting this into Eq. 3.9, and using the definition of the moment of inertia, we find that we must solve the equation

$$\frac{1}{\sin\theta}\frac{\partial}{\partial\theta}\left(\sin\theta\frac{\partial\psi}{\partial\theta}\right) + \frac{1}{\sin^2\theta}\frac{\partial^2\psi}{\partial\phi^2} + \frac{2I}{\hbar^2}E\psi = 0 \tag{3.15}$$

In order to solve Eq. 3.15, we will let our wave function $\psi(\theta, \phi)$, be a product of two functions, $T(\theta)$ and $F(\phi)$, each of which is a function of only one coordinate. (This is a trial solution in the independent coordinates, θ and ϕ.)

$$\psi(\theta, \phi) = T(\theta)F(\phi) \tag{3.16}$$

Substituting this into Eq. 3.15, we have

$$\frac{F(\phi)}{\sin\theta}\frac{\partial}{\partial\theta}\left[\sin\theta\frac{\partial T(\theta)}{\partial\theta}\right] + \frac{T(\theta)}{\sin^2\theta}\frac{\partial^2 F(\phi)}{\partial\phi^2} + \frac{2IE}{\hbar^2}T(\theta)F(\phi) = 0 \tag{3.17}$$

Let us now multiply through by $(\sin^2\theta)/FT$. This gives

$$\frac{\sin\theta}{T}\frac{\partial}{\partial\theta}\left(\sin\theta\frac{\partial T}{\partial\theta}\right) + \frac{1}{F}\frac{\partial^2 F}{\partial\phi^2} + \left(\frac{2I}{\hbar^2}\sin^2\theta\right)E = 0 \tag{3.18}$$

or, rearranging,

$$\frac{\sin\theta}{T}\frac{\partial}{\partial\theta}\left(\sin\theta\frac{\partial T}{\partial\theta}\right) + \left(\frac{2I}{\hbar^2}\sin^2\theta\right)E = -\frac{1}{F}\frac{\partial^2 F}{\partial\phi^2} \tag{3.19}$$

The left-hand side of Eq. 3.19 depends only upon θ and is independent of ϕ. The converse is true for the right-hand side. The equality can hold for all values of θ and ϕ only if each side equals a constant (the same constant for each side). Let us call this constant M^2. We have succeeded in separating Eq. 3.15 into two independent equations:

$$-\frac{1}{F}\frac{d^2F}{d\phi^2} = M^2 \tag{3.20}$$

and

$$\frac{\sin\theta}{T}\frac{d}{d\theta}\left(\sin\theta\frac{dT}{d\theta}\right) + \frac{2I}{\hbar^2}E\sin^2\theta = M^2 \tag{3.21}$$

The $F(\phi)$ equation (3.20) is easily solved. We have

$$\frac{d^2F(\phi)}{d\phi^2} = -M^2F(\phi) \tag{3.22}$$

The general solution is

$$F(\phi) = Ne^{\pm iM\phi} \tag{3.23}$$

where N is a normalizing constant. Taking a particular sign choice (say the negative), we can evaluate N by requiring that $F(\phi)$ be normalized. We have (note that we must use F^*F, since F is complex)

$$1 = \int_0^{2\pi} F^*F \, d\phi \tag{3.24a}$$

$$= N^2 \int_0^{2\pi} e^{iM\phi}e^{-iM\phi} \, d\phi \tag{3.24b}$$

$$= N^2 \int_0^{2\pi} d\phi \tag{3.24c}$$

$$= 2\pi N^2 \tag{3.24d}$$

or

$$N = \frac{1}{\sqrt{2\pi}} \tag{3.25}$$

and

$$F = \frac{1}{\sqrt{2\pi}} e^{\pm iM\phi} \tag{3.26}$$

The quantum number, M, must be an integer. We can show this by writing $F(\phi)$ in real form. We have

$$F_r(\phi) = \frac{1}{2}[F_+(\phi) + F_-(\phi)]$$

$$= \frac{N}{2}[\exp(iM\phi) + \exp(-iM\phi)]$$

$$= N \cos M\phi \tag{3.26a}$$

Now, for the wave function to be continuous and single-valued, the value of $F_r(\phi)$ must be the same as $F_r(2\pi)$. This can occur only when the argument of the cosine function is an integral multiple of 2π—i.e., when M is an integer.

The $T(\theta)$ equation (3.21) is much more difficult to solve. A standard mathematical technique often employed is to try to express the unknown function as a series expansion. This technique is used in solving several problems in the Schrödinger representation—notably the harmonic-oscillator problem, the rigid-rotor problem, and the hydrogen-atom problem. We will illustrate it in detail for the rigid rotor. (This is the same as the angular portion of the hydrogen-atom problem or any other three-dimensional angular-momentum problem.)

Equation 3.21 can be rewritten as

$$\sin\theta \frac{d}{d\theta}\left(\sin\theta \frac{dT}{d\theta}\right) + \left(\frac{2IE}{\hbar^2}\sin^2\theta\right)T - M^2T = 0 \qquad 3.27$$

In order to avoid carrying the constants throughout the derivation, let us define

$$\beta = \frac{2IE}{\hbar^2} \qquad 3.28$$

If we substitute this into Eq. 3.27 and divide through by $\sin^2\theta$, we have

$$\frac{1}{\sin\theta}\frac{d}{d\theta}\left(\sin\theta\frac{dT}{d\theta}\right) - \frac{M^2}{\sin^2\theta}T + \beta T = 0 \qquad 3.29$$

Let us now make a change of variables. Let

$$z = \cos\theta \qquad 3.30$$

and

$$P(z) = T(\theta) \qquad 3.31$$

[Note that $P(z)$ is the same function, expressed in terms of the new variable.] Then

$$\sin^2\theta = 1 - z^2 \qquad 3.32$$

and

$$\frac{d}{d\theta} = \frac{dz}{d\theta}\frac{d}{dz}$$

$$= -\sin\theta\frac{d}{dz} \qquad 3.33$$

Substituting Eqs. 3.33 and 3.31 into Eq. 3.29, we have

$$\frac{1}{\sin\theta}\left(-\sin\theta\frac{d}{dz}\right)\left(-\sin^2\theta\frac{dP}{dz}\right) - \frac{M^2}{\sin^2\theta}P + \beta P = 0 \qquad 3.34$$

Simplifying and using Eq. 3.32 gives

$$\frac{d}{dz}\left[(1 - z^2)\frac{dP(z)}{dz}\right] + \left[\beta - \frac{M^2}{1 - z^2}\right]P(z) = 0 \qquad 3.35$$

Owing to the $(1 - z^2)$ term in the denominator, Eq. 3.35 becomes infinite (has *singularities*) at z values of ± 1 (i.e., $\cos\theta$ of ± 1 or θ of $n\pi$). The wave function must be well behaved over all space, including at the singularities. Singularities are more easily handled at zero or infinity; consequently, we will again do a change of variables to study the behavior at a z value of -1, and still again at $+1$. First, let

$$x = 1 + z \qquad 3.36$$

$$(1 - z^2) = x(2 - x) \qquad 3.36a$$

Note that x goes to zero as z goes to -1. Further, let

$$R(x) = P(z) \qquad 3.37$$

Substituting these into Eq. 3.35, we have

$$\frac{d}{dx}\left[x(2-x)\frac{dR(x)}{dx}\right]+\left[\beta-\frac{M^2}{x(2-x)}\right]R(x)=0 \qquad 3.38$$

Now we will attempt to solve for our unknown function, $R(x)$, by a series expansion. Let us use a power series of the form

$$p(x)=\sum_{v=0}^{\infty}a_v x^v \qquad 3.39$$

where the a_v are numerical coefficients of the powers of x. In the present case, however, if the $R(x)$ of Eq. 3.38 is to remain well behaved at the singularity, $x = 0$, $R(0)$ must equal zero. To insure this, our series expansion of Eq. 3.39 should be multiplied by some power of x, say x^s, to give

$$R(x)=x^s\sum_{v=0}^{\infty}a_v x^v=x^s p(x) \qquad 3.40$$

Substituting this into Eq. 3.38 yields

$$\frac{d}{dx}\left\{x(2-x)\frac{d}{dx}\left[x^s\sum_{v=0}^{\infty}a_v x^v\right]\right\}+\left[\beta-\frac{M^2}{x(2-x)}\right]\left[x^s\sum_{v=0}^{\infty}a_v x^v\right]=0 \qquad 3.41$$

Let us let

$$p=\sum_{v=0}^{\infty}a_v x^v=a_0+a_1x+a_2x^2+a_3x^3+\ldots \qquad 3.42$$

$$p'=\frac{dp}{dx}=a_1+2a_2x+3a_3x^2+\ldots \qquad 3.42a$$

$$p''=\frac{d^2p}{dx^2}=2a_2+6a_3x+\ldots \qquad 3.42b$$

The equation to be solved is

$$\frac{d}{dx}\left[x(2-x)\frac{dx^sp}{dx}\right]+\left[\beta-\frac{M^2}{x(2-x)}\right]x^sp=0 \qquad 3.43$$

Consider the first term. We have

$$\frac{dx^sp}{dx}=sx^{s-1}p+x^sp' \qquad 3.44a$$

$$x(2-x)\frac{dx^sp}{dx}=2sx^sp-sx^{s+1}p+2x^{s+1}p'-x^{s+2}p' \qquad 3.44b$$

$$\frac{d}{dx}\left[x(2-x)\frac{dx^sp}{dx}\right]=2s^2x^{s-1}p+2sx^sp'-s(s+1)x^sp-sx^{s+1}p'+2(s+1)x^sp'$$
$$+2x^{s+1}p''-(s+2)x^{s+1}p'-x^{s+2}p'' \qquad 3.44c$$

Substituting into Eq. 3.43 and collecting terms yields

$$[2s^2x^{s-1} - s(s+1)x^s]p + [(4s+2)x^s - (2s+2)x^{s+1}]p' + (2x^{s+1} - x^{s+2})p'' + \left[\beta - \frac{M^2}{x(2-x)}\right]x^s p = 0 \quad 3.45$$

If we now expand p, p', and p'' as in Eqs. 3.42, we get, on collecting like a_v,

$$a_0(2s^2x^{s-1} - s^2x^s) + a_1[2(s+1)^2x^s - (s+1)(s+2)x^{s+1}] + \ldots + \left[\beta - \frac{M^2}{x(2-x)}\right]x^s(a_0 + a_1x + a_2x^2 + \ldots) = 0 \quad 3.46$$

Now let us multiply by $x(2-x) = 2x - x^2$ to remove it from the denominator, and again do some collecting of terms to get

$$2a_0(2s^2x^s - s^2x^{s+1}) - a_0(2s^2x^{s+1} - s^2x^{s+2}) + \ldots + 2a_0\beta x^{s+1} - a_0\beta x^{s+2} + \ldots - a_0M^2x^s - a_1M^2x^{s+1} + \ldots = 0 \quad 3.47$$

In order for Eq. 3.47 to equal zero for any arbitrary value of x, either all a_v must equal zero, or the coefficients of each power of x must equal zero. The first option, which would yield a function equal to zero over all space, is unacceptable; consequently the second must be chosen. Requiring that the term containing each power of x equal zero, we have

$$a_0(4s^2 - M^2)x^s = 0 \quad 3.48a$$

$$(-4a_0x^2 + 2a_0\beta - a_1M^2)x^{s+1} = 0 \quad 3.48b$$

and so on. The first of these is sufficient to determine s. Since neither a_0 nor x in general is zero, we have

$$s = \frac{|M|}{2} \quad 3.49$$

We repeat the procedure for the other singularity by defining a y that equals $(1 - z)$ and an $R(y)$ as a polynomial:

$$R(y) = y^r \sum_{v=0}^{\infty} a_v y^v \quad 3.50$$

This leads to identically the same treatment, with the conclusion that

$$r = \frac{|M|}{2} \quad 3.51$$

Thus, our $P(z)$ function can be expressed

$$P(z) = x^{|M|/2}y^{|M|/2}G(z) \quad 3.52$$

where the x and y terms take care of the singularities and $G(z)$ is some new function of z for which we must solve. Substituting for x and y, we find

$$P(z) = (1 - z^2)^{|M|/2} G(z) \tag{3.53}$$

If we substitute Eq. 3.53 into our $P(z)$ equation (3.35), we obtain

$$\frac{d}{dz}\left\{(1 - z^2)\frac{d}{dz}\left[(1 - z^2)^{|M|/2} G(z)\right]\right\} + \left(\beta - \frac{M^2}{1 - z^2}\right)(1 - z^2)^{|M|/2} G(z) = 0 \tag{3.54}$$

Carrying out the indicated differentiation, dividing through by $(1 - z^2)^{|M|/2}$, and collecting terms, we obtain

$$(1 - z^2)G'' - 2(|M| + 1)zG' + [\beta - |M|(|M| + 1)]G = 0 \tag{3.55}$$

where the primes have the same meaning as in Eqs. 3.42.

We will now use a power series for G. Let

$$G = \sum_{v=0}^{\infty} a_v z^v = a_0 + a_1 z + a_2 z^2 + a_3 z^3 + a_4 z^4 + \ldots \tag{3.56}$$

$$G' = \sum_{v=0}^{\infty} v a_v z^{v-1} = a_1 + 2a_2 z + 3a_3 z^2 + 4a_4 z^3 + \ldots \tag{3.56a}$$

$$G'' = \sum_{v=0}^{\infty} v(v - 1)a_v z^{v-2} = 1\cdot 2a_2 + 2\cdot 3a_3 z + 3\cdot 4a_4 z^2 + \ldots \tag{3.56b}$$

Substituting these into Eq. 3.55 yields

$$\sum_{v=0}^{\infty} \{(1 - z^2)v(v - 1)a_v z^{v-2} - 2(|M| + 1)zva_v z^{v-1} + [\beta - |M|(|M| + 1)]a_v z^v\} = 0 \tag{3.57}$$

or

$$\begin{aligned} &1\cdot 2a_2 + 2\cdot 3a_3 z + 3\cdot 4a_4 z^2 + \ldots \\ &-1\cdot 2a_2 z^2 - 2\cdot 3a_3 z^3 - 3\cdot 4a_4 z^4 - \ldots \\ &- 2(|M| + 1)a_1 z - 2\cdot 2(|M| + 1)a_2 z^2 - 2\cdot 3(|M| + 1)a_3 z^3 - \ldots \\ &+ [\beta - |M|(|M| + 1)]a_0 + [\beta - |M|(|M| + 1)]a_1 z + \ldots = 0 \end{aligned} \tag{3.57a}$$

Again, the coefficients of the individual powers of the variable, z, must vanish in order for this to hold for any arbitrary value of z. We have

$$1\cdot 2a_2 + [\beta - |M|(|M| + 1)]a_0 = 0 \tag{3.58a}$$

$$2\cdot 3a_3 + \{[\beta - |M|(|M| + 1)] - 2(|M| + 1)\}a_1 = 0 \tag{3.58b}$$

$$3\cdot 4a_4 + \{(\beta - |M|(|M| + 1)] - 2\cdot 2(|M| + 1) - 1\cdot 2\}a_2 = 0 \tag{3.58c}$$

and so on. Equations 3.58 define a *recursion relation* between the coefficients

$$(v + 1)(v + 2)a_{v+2} + \{[\beta - |M|(|M| + 1)] - 2v(|M| + 1) - v(v - 1)\}a_v = 0 \qquad \textbf{3.59}$$

or

$$a_{v+2} = \frac{(v + |M|)(v + |M| + 1) - \beta}{(v + 1)(v + 2)} a_v \qquad \textbf{3.60}$$

An infinite series with the relation of Eq. 3.60 between alternate coefficients converges if $-1 < z < 1$, but diverges (becomes infinite) if z equals ± 1. Cos θ (i.e., z) can have the values of ± 1; consequently, if the infinite series is maintained, the wave function becomes infinite at $\cos\theta$ values of ± 1. [For example, if $|M| = 0$, $\beta = 0$, $z = 1$, and $a_0 = 1$, the series becomes $\sum_v v/(v + 2)$. This series goes to infinity if an infinite summation is retained.] The wave function must remain finite over all space, so the series expansion must be terminated after a finite number of terms. (To reiterate, this termination—which, as we shall see, forces quantization—is necessary to keep the wave function well behaved.) Termination can be accomplished if there is some value of v, say v', such that a_{v+2} equals zero. If this coefficient vanishes, all the higher ones in either the odd or even series (depending upon whether v is odd or even) also will vanish. Looking at Eq. 3.60, we see that this will happen if the numerator of the right-hand side of the equation equals zero for v equal to v'—i.e., if

$$(v' + |M|)(v' + |M| + 1) = \beta \qquad \textbf{3.61}$$

This will terminate either the even or the odd series. We make the other series vanish by choosing either a_0 or a_1 to be zero. The expansion for a particular $G(z)$ contains only even or only odd powers of z.

Let us define a quantum number J

$$J = v' + |M| \qquad \textbf{3.62}$$

We have

$$\beta = J(J + 1) = \frac{2IE}{\hbar^2} \qquad \textbf{3.63}$$

or

$$E_J = \frac{\hbar^2}{2I} J(J + 1) \qquad \textbf{3.64}$$

—a rather simple expression for such a tedious derivation. J is the energy-related quantum number. From Eq. 3.62 we see that the allowed values of J are $|M|$, $|M| + 1$, $|M| + 2$, and so on. On removing the absolute-magnitude sign from M and solving for M in terms of J, we see that

$$-J \leq M \leq J \qquad \textbf{3.65}$$

There are $(2J + 1)$ allowed values of M for every value of J. The energy does not depend upon M. Thus each E_J level can be represented by $(2J + 1)$ different wave functions, each corresponding to a different value of M. The energy states are said to be $(2J + 1)$-fold *degenerate*. The function $T(\theta)$, defined as

$$T(\theta) = (1 - z^2)^{|M|/2} G(z) \qquad \textbf{3.66}$$

is a function known as the *associated Legendre polynomial.* Table 3.1 presents some of the associated Legendre polynomials.

Table 3.1

SOME ASSOCIATED LEGENDRE POLYNOMIALS (NORMALIZED TO UNITY)[a]

J	M	$T_{JM}(\theta)$
0	0	$\frac{\sqrt{2}}{2}$
1	0	$\frac{\sqrt{6}}{2} \cos\theta$
1	± 1	$\frac{\sqrt{3}}{2} \sin\theta$
2	0	$\frac{\sqrt{10}}{4}(3\cos^2\theta - 1)$
2	± 1	$\frac{\sqrt{15}}{2} \sin\theta\cos\theta$
2	± 2	$\frac{\sqrt{15}}{4} \sin^2\theta$
3	0	$\frac{3\sqrt{14}}{4}\left(\frac{5}{3}\cos^3\theta - \cos\theta\right)$
3	± 1	$\frac{\sqrt{42}}{8} \sin\theta(5\cos^2\theta - 1)$
3	± 2	$\frac{\sqrt{105}}{4} \sin^2\theta\cos\theta$
3	± 3	$\frac{\sqrt{70}}{8} \sin^3\theta$

[a] For normalization, remember that the integration variable is $\cos\theta$, not θ. The normalizing factor is

$$\left[\frac{2J + 1}{2}\frac{(J - |M|)!}{(J + |M|)!}\right]^{1/2}$$

Table 3.2

SOME SPHERICAL HARMONICS[a]

$$
\begin{aligned}
Y_{0,0} &= N_{0,0} \\
Y_{1,-1} &= N_{1,-1} \sin\theta e^{-i\phi} \\
Y_{1,0} &= N_{1,0} \cos\theta \\
Y_{1,1} &= N_{1,1} \sin\theta e^{i\phi} \\
Y_{2,0} &= N_{2,0}(3\cos^2\theta - 1) \\
Y_{2,\pm 1} &= N_{2,1} \sin\theta \cos\theta e^{\pm i\phi} \\
Y_{2,\pm 2} &= N_{2,2} \sin^2\theta e^{\pm 2i\phi} \\
Y_{3,0} &= N_{3,0}(\tfrac{5}{3}\cos^3\theta - \cos\theta) \\
Y_{3,\pm 1} &= N_{3,1} \sin\theta(5\cos^2\theta - 1)e^{\pm i\phi} \\
Y_{3,\pm 2} &= N_{3,2} \sin^2\theta \cos\theta e^{\pm 2i\phi} \\
Y_{3,\pm 3} &= N_{3,3} \sin^3\theta e^{\pm 3i\phi}
\end{aligned}
$$

[a] $N_{L,M}$ represents the normalizing constant. Various normalizing conventions appear in the literature.

Table 3.3

REAL FORMS OF SOME SPHERICAL HARMONICS

	Transforms as	Degeneracy	Letter Designation
$Y_{0,0} = N$[a]	Sphere	1	S
$Y_{1,0} = N\cos\theta$	z	3	P
$\frac{1}{2}(Y_{1,1} + Y_{1,-1}) = N\sin\theta\cos\phi$	x		
$\frac{i}{2}(Y_{1,-1} - Y_{1,1}) = N\sin\theta\sin\phi$	y		
$Y_{2,0} = N(3\cos^2\theta - 1)$	$3z^2 - r$	5	D
$\frac{1}{2}(Y_{2,1} + Y_{2,-1}) = N\sin\theta\cos\theta\cos\phi$	xz		
$\frac{i}{2}(Y_{2,-1} - Y_{2,1}) = N\sin\theta\cos\theta\sin\phi$	yz		
$\frac{1}{2}(Y_{2,2} + Y_{2,-2}) = N\sin^2\theta\cos 2\phi$	$x^2 - y^2$		
$\frac{i}{2}(Y_{2,-2} - Y_{2,2}) = N\sin^2\theta\sin 2\phi$	xy		

[a] N is the normalizing constant (it depends on L and M).

The overall wave function for the rigid rotor is the product $T(\theta)F(\phi)$. These products are frequently referred to as the *spherical harmonics*, $Y_{LM}(\theta, \phi)$. Some of them are listed in Table 3.2. Because $F(\phi)$ contains i, all the spherical harmonics with M not equal to zero are complex. By combining the $e^{iM\phi}$ and $e^{-iM\phi}$ terms, we can obtain real forms of the spherical harmonics. Some of these are listed in Table 3.3. The functions are plotted on a Cartesian axis system in Figure 3.2. Plots such as these frequently appear in the literature. Note that M is not a valid quantum number for the real functions, since each real function contains both $+M$ and $-M$. This seems a trivial point, since the energy of the rigid rotor does not depend upon M; however, if an external field is present, the perturbation causes an M dependence in the energy.

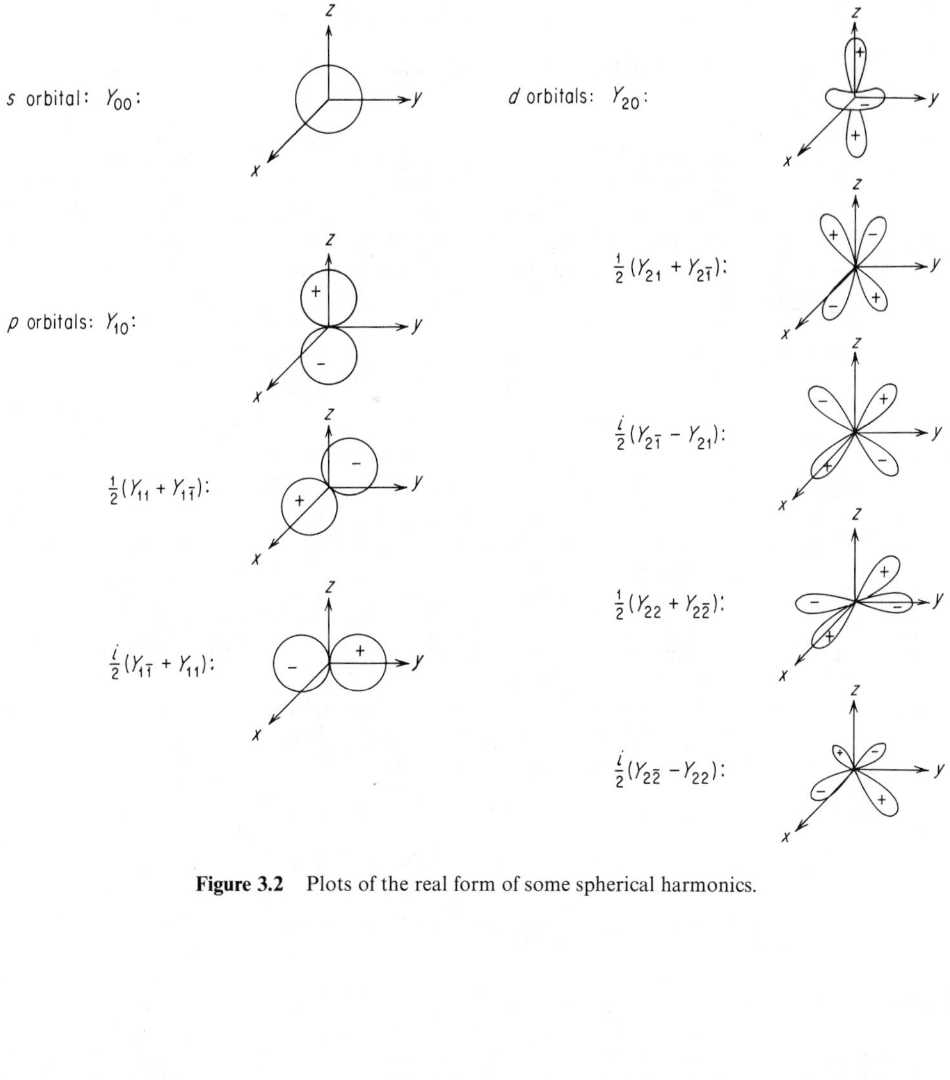

Figure 3.2 Plots of the real form of some spherical harmonics.

3.3 ANGULAR MOMENTUM

In Eq. 3.8 we saw that the kinetic energy of the rigid rotor could be expressed in terms of the moment of inertia, I, and the angular velocity, ω:

$$T = \tfrac{1}{2}I\omega^2 \tag{3.67}$$

Now, the angular momentum of a rotating body, l, is defined as

$$l = I\omega \tag{3.68}$$

Combining Eqs. 3.67 and 3.68, we see that the kinetic energy can be written

$$T = \frac{l^2}{2I} \tag{3.69}$$

The time-independent Schrödinger equation is

$$\hat{H}\psi = E\psi \tag{3.70}$$

The Hamiltonian operator is the sum of the kinetic-energy operator and the potential-energy operator; however, for the rigid-rotor problem the potential is zero. Thus, if the Hamiltonian of Eq. 3.70 is the rigid-rotor Hamiltonian, if ψ is the rigid-rotor wave function, and if the energy is the rigid-rotor energy (Eq. 3.64), we have, writing the angular momentum as an operator,

$$\frac{\hat{l}^2\psi}{2I} = \frac{\hbar^2}{2I}J(J+1)\psi \tag{3.71}$$

or

$$\hat{l}^2\psi = J(J+1)\hbar^2\psi \tag{3.72}$$

In other words, the wave function is an eigenfunction of the square of the total angular momentum, having $J(J+1)\hbar^2$ as the eigenvalue. In fact, the solution to the rigid-rotor problem is the solution to the generalized quantum-mechanical angular-momentum problem. Equation 3.72 is valid whenever we encounter a quantized angular momentum.

Consider now the z component of the angular momentum, which we will call l_z. This corresponds classically to rotation about the z axis (see Figure 3.1). Thus, according to Eq. 1.60, the quantum-mechanical operator for l_z (or p_ϕ) is

$$\hat{l}_z \Rightarrow -i\hbar\frac{\partial}{\partial\phi} \tag{3.73}$$

This depends only on the coordinate ϕ; consequently, it operates only on the function

$F_\pm(\phi)$ of Eq. 3.26. Operating on $F_+(\phi)$, we have

$$\hat{l}_z F_+(\phi) = -i\hbar \frac{\partial}{\partial \phi} \frac{1}{\sqrt{2\pi}} e^{iM\phi}$$

$$= \hbar M \frac{1}{\sqrt{2\pi}} e^{iM\phi}$$

$$= M\hbar F_+(\phi) \qquad \textbf{3.74}$$

In other words, $F(\phi)$ [and consequently $\psi(\theta, \phi)$] is an eigenfunction of $\hat{l}_z$, having the eigenvalue $M\hbar$. This is again general for quantum-mechanical angular-momentum problems. Any acceptable wave function for a system in a stationary state must be an eigenfunction of the total $\hat{l}^2$ and $\hat{l}_z$ operators for the system. If the relevant symmetry of the system is spherical, the eigenvalue equations are Eqs. 3.72 and 3.74.

The function, $F_\pm(\phi)$, of Eq. 3.26 is the function describing rotation in two dimensions (in a plane). It shows up in many places. For example, it is the solution to the "particle-on-a-ring" free-electron model for aromatic systems (see Problem 2.3). It is also required for describing the electronic and vibrational wave functions of linear molecules.

3.4 MICROWAVE SPECTRA OF LINEAR MOLECULES

The rotational energy levels of a rigid linear molecule are those of the rigid rotor. The rotational spectra of molecules occur in the microwave range. *Selection rules* tell which transition may, theoretically, be observed. The rotational selection rule for absorption or emission of radiation is

$$\Delta J = \pm 1 \qquad \textbf{3.75}$$

This selection rule does not appear directly in the solution of the problem in the Schrödinger representation. We will derive it in another context in Section 3.6. Let us consider the energy difference between a state characterized by J and one by $J' = J + 1$. We have

$$\Delta E = \frac{\hbar^2}{2I}[J'(J' + 1) - J(J + 1)]$$

$$= \frac{\hbar^2}{2I}[(J + 1)(J + 2) - J(J + 1)]$$

$$= 2\frac{\hbar^2}{2I}(J + 1) \qquad \textbf{3.76}$$

The transition is commonly reported in wave numbers as

$$\bar{\nu} = 2B(J + 1) \qquad \textbf{3.77}$$

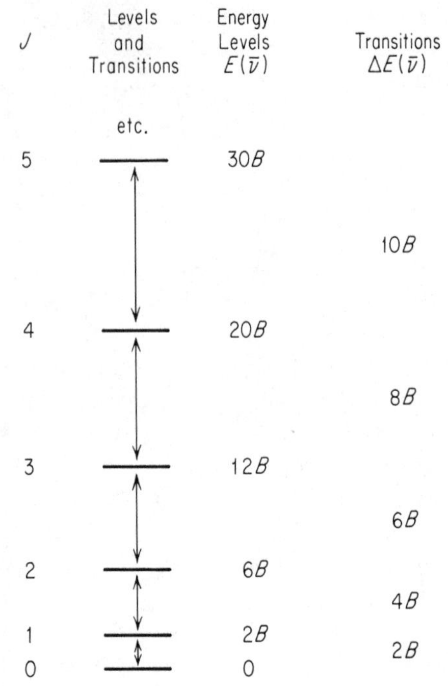

Figure 3.3 Schematic representation of the energy levels and transitions for a rigid rotor.

where B is the *rotational constant*

$$B = \frac{\hbar}{4\pi c I} \tag{3.78}$$

Figure 3.3 shows schematically the energy levels of a rotor. The rotational levels of a molecule are sufficiently close that many are thermally populated at normal temperatures; consequently, spectral transitions are observed starting from many different levels. The actual energy spacings that would be seen in a spectrum are shown in Figure 3.4. The spectrum should consist of a series of equally spaced lines,

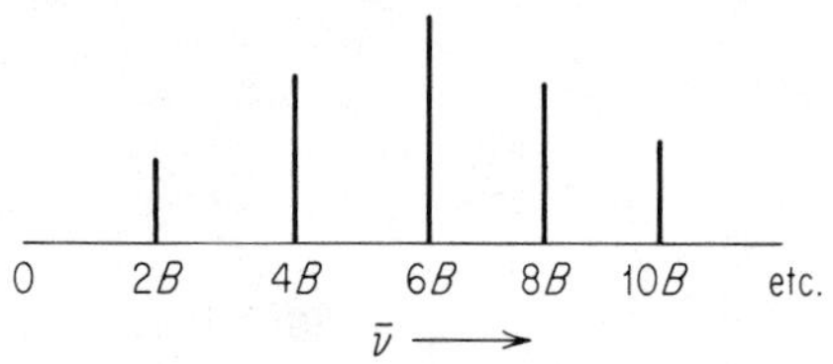

Figure 3.4 Schematic representation of the microwave spectrum of a rigid rotor. The intensities are arbitrary.

Table 3.4

MICROWAVE SPECTRAL DATA FOR HYDROGEN FLUORIDE

Origin of Transition (J)	$\bar{\nu}$ (cm^{-1})	$\Delta\bar{\nu}$	B	r_{HF} (Å)[a]
0	47.08			
		41.11	20.56	.929
1	82.19			
		40.96	20.48	.931
2	123.15			
		40.85	20.43	.932
3	164.00			
		40.62	20.31	.935
4	204.62			
		40.31	20.16	.938
5	244.93			

[a] Internuclear separation.

the separation between them being $2B$. In fact, the spacings are not quite equal, owing primarily to centrifugal distortion caused by the rotation. Table 3.4 lists microwave spectral data for the hydrogen fluoride molecule.

The internuclear separation is found from the moment of inertia. For a diatomic molecule, we have

$$I = m_1r_1 + m_2r_2 \tag{3.79}$$

or, for a linear polyatomic molecule,

$$I = \sum_i m_ir_i \tag{3.79a}$$

The r_i are the distances of the nuclei from the center of mass. The center of mass is defined by requiring, for a diatomic molecule,

$$m_1|r_1| = m_2|r_2| \tag{3.80}$$

of for a polyatomic,

$$\sum_i m_ir_i = 0 \tag{3.80a}$$

where r_i carries a sign. Substituting Eq. 3.80 into 3.79, we have, for a diatomic,

$$\begin{aligned} I &= m_2r_2r_1 + m_1r_1r_2 \\ &= r_1r_2(m_1 + m_2) \end{aligned} \tag{3.81}$$

But the total internuclear separation, r, equals $(r_1 + r_2)$; thus

$$m_1r_1 = m_2r_2 = m_2(r - r_1) \tag{3.82}$$

or

$$r_1 = \frac{m_2 r}{m_1 + m_2} \tag{3.83}$$

Similarly

$$r_2 = \frac{m_1 r}{m_1 + m_2} \tag{3.84}$$

and

$$I = \frac{m_1 m_2}{m_1 + m_2} r^2$$
$$= \mu r^2 \tag{3.85}$$

where μ is the *reduced mass.* Thus

$$r = \left(\frac{I}{\mu}\right)^{1/2}$$
$$= \left(\frac{\hbar}{4\pi c B \mu}\right)^{1/2} \tag{3.86}$$

Evaluating the constants, we have, with μ in atomic mass units,

$$r = 4.1059 \times 10^{-10}(B\mu)^{-1/2} \text{ m} \tag{3.87}$$

Only one structural parameter is obtained; consequently, this method is directly useful only for diatomic molecules. More complicated linear molecules can be treated with isotopic substitution. We will discuss nonlinear molecules, which have three moments of inertia, in Section 3.7.

3.5 THE THREE-DIMENSIONAL ROTATION GROUP

It is frequently convenient to discuss angular momentum in terms of *group theory.* Although this text will not develop group theory, it will introduce various concepts and demonstrate their utility when it is natural to do so. Readers who would like a more complete development are referred to the works in the bibliography.

Group theory, in the abstract sense, is a branch of mathematics first conceived of in 1832 by Evarist Galois in his studies of the solutions of equations. In the general case, a *group* is a set of arbitrary mathematical *elements* (which can be almost anything), with a defined combining law, which exhibits *associativity* [i.e., $A(BC) = (AB)C$, and so on] in the combination and *closure* in the set (all members of the set can be generated from combinations of other members of the set). The law of combination of the elements is referred to as *multiplication.* A *multiplication table* can be defined for the elements. A *matrix representation* (usually a matrix representation is implied whenever the word *representation* is used) is a set of matrices that obey the same multiplication table as the elements of the group. The *irreducible representations* of a group are the simplest possible sets of representations. The *character* of an element in a representation is the *trace* (or *spur*—the sum of the diagonal elements) of the matrix representing that element in that representation. Every group contains the *identity* element, the element that, when multiplied by any other element, leaves that element unchanged. The representation of the identity element is the identity, or unit, matrix (a matrix having 1's along the diagonal and 0's everywhere else). Thus the character of the identity for any representation is the dimension of the matrices for that representation. For any irreducible representation, the character of the identity is also the *degeneracy* of any function that transforms as (behaves under the operations of the group as) that irreducible representation.

There are a number of mathematical constructions for the group describing angular momentum. The most convenient for our purposes is the group of Euclidean rotations, plus the inversion through the origin. Mathematically, this is the group of three-dimensional orthogonal matrices, **O**(3). [Sometimes the group is called $\mathbf{R}_h(3)$, the **R**(3) signifying the three-dimensional rotations and the h subscript the inclusion of the inversion.] The representations of **O**(3) are labeled as D_g^j or D_u^j, where j corresponds to the generalized angular momentum and the g or u tells whether the representation is symmetric (g for the German *gerade*) or antisymmetric (u for *ungerade*) with respect to the inversion. The dimension of a given D^j is $(2j + 1)$. Thus, a system with a j value of zero is nondegenerate, one with a j of one is threefold degenerate, and so on. In the most general case, j can be half-integer as well as integer.

When applied to the quantum-mechanical angular-momentum problem, the index j corresponds to the appropriate angular-momentum quantum number. For example, the integer j values correspond to the integer J values of the rigid rotor. Thus, each energy level of a rigid rotor is associated with a different irreducible representation of the rotation group. The half-integer j values, as we shall see later, allow us to describe the "spin" of the electron. Most of the properties of the **O**(3) group that we will need can be derived from the rotations only—i.e., from **R**(3). [**R**(3) is said to be the rotational *subgroup* of **O**(3).] The character table for a group presents the characters of each group element (in this case, the identity and the rotation operations) for each irreducible representation. Many of the applications of group theory require only the character table.

The character table for **R**(3) needs only the behavior for the identity operation, E, and for an arbitrary rotation, $C(\phi)$. Any other arbitrary rotation (there are an infinite number of them) has the same behavior. The character table for **R**(3) is given in Table 3.5. Its form is that of any character table. The rows are labeled by the names of the irreducible representations (in this case there are an infinite number of them). The columns are labeled by the group operations. Each entry is the character of the indicated operation in the indicated representation.

Products of irreducible representations are important in the applications of group theory. The characters of a product are just the product of the characters

Table 3.5

CHARACTER TABLE FOR **R**(3)

(Group name) **R**(3)		(Group operations) E	$C(\phi)$
(Representation Labels)	D^0	1	1
	D^1	3	$1 + 2\cos\phi$
	D^2	5	$1 + 2\cos\phi + 2\cos 2\phi$
	D^j	$2j + 1$	$1 + \sum_{k=1}^{j} 2\cos k\phi$
	$D^{1/2}$	2	$2\cos\frac{1}{2}\phi$
	$D^{3/2}$	4	$2\cos\frac{1}{2}\phi + 2\cos\frac{3}{2}\phi$
	D^j	$2j + 1$	$\sum_{k=1/2}^{j} 2\cos k\phi$

(Characters)

for the individual representations. Product representations that are not irreducible representations can be resolved into a sum of irreducible representations. For most groups, rules exist that let us avoid going through taking the product. For **R**(3), the rule is

$$D^j \otimes D^{j'} = \sum_{k=|j-j'|}^{j+j'} D^k \qquad \textbf{3.88}$$

This is known as the *Clebsch-Gordan rule.*

Notice that the first representation listed in Table 3.5 has $+1$ for both characters. Every group has a representation with $+1$ for all characters. This representation is the *totally symmetric irreducible representation* of the group. Any function or operator transforming as the totally symmetric irreducible representation of a group is unchanged by any of the operations of the group. Any *scalar* property of a system is unchanged by any operation of the group. Consequently, scalar properties, and the operators that describe them, must transform as the totally symmetric irreducible representation of the group describing the system.

The character table for the **R**(3) group requires only the characters for the identity and the rotation. All arbitrary rotations about any axis have the same form for their characters; consequently, the group contains an infinite number of $C(\phi)$ rotations. In the character table we list only one of them. The **O**(3) character table also requires the characters of other operations. In finite spatial symmetry groups (or *point groups*, as they are usually called), five types of symmetry operations are encountered (see Chapter 13). Two of these are the identity, E, and the rotation operation (or proper rotation), $C(\phi)$. The others are the *inversion*, symbolized by i;

the *reflection* through a plane, σ; and the *improper rotation*, $S(\phi)$. The improper rotation involves an ordinary rotation accompanied by a reflection in a plane perpendicular to the axis of rotation. (An alternative construction involves a rotation accompanied by an inversion.) There are again an infinite number of σ's and $S(\phi)$'s. The inversion is equivalent to an improper rotation in the special case where the angle of rotation is 180°. The reflection is equivalent to the improper rotation when the angle is zero. Thus, two types of operation are sufficient to generate the other types.

Table 3.6 presents the character table for the **O**(3) group. (Note that D_g^0 is the totally symmetric irreducible representation.) The **R**(3) group is a subgroup of **O**(3). The identity and $C(\phi)$ operations are all that is needed. The character table is just the first three columns of Table 3.6. The g and u subscripts are meaningless for **R**(3), since the group does not contain the inversion.

From Table 3.5 or 3.6 we see that the integer-j valued representations (odd dimensions) are listed separately from the half-integer valued representations (even

Table 3.6

CHARACTER TABLE FOR THE **O**(3) POINT GROUP[a]

O(3)	E	$C(\phi)$	i	$S(\phi)$	σ
D_g^0	1	1	1	1	1
D_g^1	3	$1 + 2\cos\phi$	3	$1 - 2\cos\phi$	-1
D_g^2	5	$1 + 2\cos\phi + 2\cos 2\phi$	5	$1 - 2\cos\phi + 2\cos 2\phi$	1
D_g^j	$2j + 1$	$1 + \sum_{k=1}^{j} 2\cos k\phi$	$2j + 1$	$1 + \sum_{k=1}^{j} (-1)^k 2\cos k\phi$	$(-1)^j$
D_u^0	1	1	-1	-1	-1
D_u^1	3	$1 + 2\cos\phi$	-3	$-1 + 2\cos\phi$	1
D_u^2	5	$1 + 2\cos\phi + 2\cos 2\phi$	-5	$-1 + 2\cos\phi - 2\cos 2\phi$	-1
D_u^j	$2j + 1$	$1 + \sum_{k=1}^{j} 2\cos k\phi$	$-(2j + 1)$	$-1 - \sum_{k=1}^{j} (-1)^k 2\cos k\phi$	$(-1)^{j+1}$
$D_g^{1/2}$	2	$2\cos\frac{1}{2}\phi$	2	$2\sin\frac{1}{2}\phi$	0
$D_g^{3/2}$	4	$2\cos\frac{1}{2}\phi + 2\cos\frac{3}{2}\phi$	4	$2\sin\frac{1}{2}\phi - 2\sin\frac{3}{2}\phi$	0
D_g^j	$2j + 1$	$\sum_{k=1/2}^{j} 2\cos k\phi$	$2j + 1$	$\sum_{k=1/2}^{j} (-1)^{k-(1/2)} 2\sin k\phi$	0
$D_u^{1/2}$	2	$2\cos\frac{1}{2}\phi$	-2	$-2\sin\frac{1}{2}\phi$	0
$D_u^{3/2}$	4	$2\cos\frac{1}{2}\phi + 2\cos\frac{3}{2}\phi$	-4	$-2\sin\frac{1}{2}\phi + 2\sin\frac{3}{2}\phi$	0
D_u^j	$2j + 1$	$\sum_{k=1/2}^{j} 2\cos k\phi$	$-(2j + 1)$	$\sum_{k=1/2}^{j} (-1)^{k+(1/2)} 2\sin k\phi$	0

[a] The **R**(3) group is the first three columns [containing the representation labels and the characters for E and $C(\phi)$], with the g, u distinction removed from the representations.

dimensions). The latter are referred to as the *double-valued* representations. The reason can be seen from the characters, χ, of the $C(\phi)$. Consider the D^2 representation of **R**(3) for the special value of 2π for the angle

$$\begin{aligned}\chi(C(2\pi)) &= 1 + 2\cos(2\pi) + 2\cos(4\pi)\\ &= 5\\ &= \chi(E)\end{aligned} \quad \textbf{3.89}$$

We see, as would be expected, that a rotation by 2π is equivalent to the identity. Now consider the $D^{3/2}$ representation for a ϕ value of 2π. We have

$$\begin{aligned}\chi(C(2\pi)) &= 2\cos\tfrac{1}{2}(2\pi) + 2\cos\tfrac{3}{2}(2\pi)\\ &= 2\cos\pi + 2\cos 3\pi\\ &= -4\\ &= -\chi(E)\end{aligned} \quad \textbf{3.90}$$

Even though a physical rotation by 2π returns a system to the starting orientation, the character in this case is not the character of the identity, but rather its negative. It takes a ϕ value of 4π to return the character of $C(\phi)$ to that of the identity. There are two values for the character of the physical identity operation for the double-valued representations. This is a situation that we do not encounter in our "life-sized" world. At the quantum level, however, it is encountered for any particle having a half-integer "spin."

A consideration of the characters of **O**(3) allows us to derive multiplication rules for the irreducible representations of this group. The j rule is the same as for **R**(3) (Eq. 3.88). The rule for the subscripts is

$$g \times g = u \times u = g \quad \textbf{3.91a}$$

$$g \times u = u \times g = u \quad \textbf{3.91b}$$

Thus, for example, the product $D_g^1 \times D_u^2$ is

$$D_g^1 \times D_u^2 = D_u^1 + D_u^2 + D_u^3 \quad \textbf{3.92}$$

Chained products can be obtained by chaining the binary products (for convenience, the circle on the product symbol is dropped)—e.g.,

$$\begin{aligned}D_u^{3/2} \times D_u^3 \times D_u^2 &= D_u^{3/2} \times (D_g^1 + D_g^2 + D_g^3 + D_g^4 + D_g^5)\\ &= (D_u^{1/2} + D_u^{3/2} + D_u^{5/2}) + (D_u^{1/2} + D_u^{3/2} + D_u^{5/2} + D_u^{7/2})\\ &\quad + (D_u^{3/2} + D_u^{5/2} + D_u^{7/2} + D_u^{9/2}) + (D_u^{5/2} + D_u^{7/2} + D_u^{9/2} + D_u^{11/2})\\ &\quad + (D_u^{7/2} + D_u^{9/2} + D_u^{11/2} + D_u^{13/2})\\ &= 2D_u^{1/2} + 3D_u^{3/2} + 4D_u^{5/2} + 4D_u^{7/2} + 3D_u^{9/2} + 2D_u^{11/2} + D_u^{13/2}\end{aligned} \quad \textbf{3.93}$$

Note particularly that, from the form of Eqs. 3.88 and 3.91, the only time that a product can contain the totally symmetric D_g^0 [or D^0 in **R**(3)] is when the product involves a representation times itself.

Often representations are obtained, by methods other than by simple products of irreducible representations, that are not irreducible. To be acceptable, they must be expressible as sums of irreducible representations. (Such representations are said to be *reducible*.) Reducible representations of the **R**(3) and **O**(3) groups can be reduced in a simple manner if we consider the character under the rotation. If a representation, Γ, is reducible in **R**(3), it can be expressed as

$$\Gamma = \sum_q a_q D^q \qquad \textbf{3.94}$$

where the a_q are numerical coefficients. The character under the rotation will have the form

$$\chi(C(\phi)) = \sum_q a_q \left(1 + \sum_{k=1}^{q} 2 \cos k\phi\right) \qquad \textbf{3.95a}$$

for integer indices, or

$$\chi(C(\phi)) = \sum_q a_q \left(\sum_{k=1/2}^{q} 2 \cos k\phi\right) \qquad \textbf{3.95b}$$

for half-integer indices. The highest coefficient of ϕ in the result will determine the largest value of q in the irreducible representation. The coefficient of $\cos q\phi$ will be $2a_q$. Thus we reduce the representations by subtracting out the succesively highest-indexed representations until the representation is exhausted (Appendix 5). The procedure works, even if the reducible representation contains a mixture of even- and odd-indexed terms.

As an example, consider the following representation in **R**(3):

R(3)	E	$C(\phi)$
Γ	12	$4 + 4\cos\frac{1}{2}\phi + 2\cos\phi + 2\cos 2\phi$

3.96

The last term under $C(\phi)$ tells us that the representation contains D^2. Since the coefficient of $\cos 2\phi$ is 2, D^2 is present only one time. Subtracting the characters of D^2 from 3.96, we have

	E	$C(\phi)$
$\Gamma - D^2$	7	$3 + 4\cos\frac{1}{2}\phi$

3.97a

The only cosine term is $\cos\frac{1}{2}\phi$. This arises from $D^{1/2}$. The coefficient of 4 tells us that this occurs twice. Subtracting this out, we have

	E	$C(\phi)$
$\Gamma - D^2 - 2D^{1/2}$	3	3

3.97b

This is simply $3D^0$. Thus

$$\Gamma = 3D^0 + 2D^{1/2} + D^2 \tag{3.97c}$$

Any other reducible representation in **R**(3) can be reduced in a similar manner. In **O**(3), the g, u behavior can be determined by checking the character of $S(\phi)$ for each D^j.

3.6 SELECTION RULES

We are now in a position to use group theory to derive the $\Delta J = \pm 1$ selection rule for the microwave spectroscopy of linear molecules. In order for any direct absorption or emission of electromagnetic radiation to be observed, the *transition dipole* connecting the two energy states must be nonvanishing. The transition dipole, μ_{ij}, (see Section 6.7) is defined as

$$\mu_{ij} = \int \psi_i^* \hat{\mu} \psi_j \, dv \tag{3.98}$$

where ψ_i and ψ_j are the wave functions for the initial and final states, $\hat{\mu}$ is the ordinary dipole operator (a vector operator, charge times the distance vector), and the integration is over all space. We see immediately that, in the rigid-rotor approximation for a rotating molecule, if there is no charge separation (i.e., no permanent dipole moment), the transition dipole will vanish.

If an integral yields a scalar quantity, as it must if it represents an observable, the group-theoretical representation of the integral must be that of a scalar. The only scalar irreducible representation of a group is the totally symmetric irreducible representation, the representation with $+1$ for all its characters. In **R**(3) this is D^0, and it is D_g^0 in **O**(3). We can obtain the representation spanned by an integral (the representation according to which the integral transforms) by taking the product of the representations of the various factors of the integrand. Thus, the representation spanned by μ_{ij} is the product

$$\Gamma_{\mu_{ij}} = \Gamma_i \times \Gamma_\mu \times \Gamma_j \tag{3.99}$$

where we have used the general symbol Γ for the representations, and the subscripts have the obvious meaning.

We have seen that the only time D^0 occurs in a product of representations in **R**(3) is when a representation is multiplied times itself. Thus, if $\Gamma_{\mu_{ij}}$ is to contain the D^0 representation, the product of any two of the representations on the right-hand side of Eq. 3.99 must contain the third representation. In particular, let

$$\Gamma_\mu \subset \Gamma_i \times \Gamma_j \tag{3.100}$$

The representations Γ_i and Γ_j are just D^J and $D^{J'}$, where J and J' are the rotational quantum numbers of the initial and final states. The distance vector is the nonscalar portion of the dipole operator, μ. This is a three-dimensional vector quantity; consequently, it transforms as the three-dimensional D^1 of $\mathbf{R}(3)$. The selection rule, then, requires that

$$D^J \times D^{J'} \supset D^1 \tag{3.101}$$

But, from the Clebsch-Gordan rule,

$$D^J \times D^{J'} = \sum_{k=|J-J'|}^{J+J'} D^k \tag{3.102}$$

For a nonzero J value, this contains D^1 only if J' equals J or $J \pm 1$. For a zero J value, J' must equal $J \pm 1$. If J' equals J, the two rotational states are the same, and there is no transition. Thus, the valid selection rule is

$$\Delta J = \pm 1 \tag{3.103}$$

as stated in Section 3.4.

The spectral selection rule of Eq. 3.100 is valid in any group. In fact the generalized selection rule for any integral, which states that

$$\int f_i f_j \ldots \hat{O}_\alpha \hat{O}_\beta \ldots dv \neq 0 \tag{3.104}$$

only if

$$\Gamma_i \times \Gamma_j \times \ldots \times \Gamma_\alpha \times \Gamma_\beta \times \ldots \supset \Gamma_{\text{sym}} \tag{3.105}$$

where $f_i, f_j, \ldots$ are any functions, $\hat{O}_\alpha, \hat{O}_\beta, \ldots$ are any operators, and Γ_{sym} is the totally symmetric irreducible representation in the appropriate group, is one of the most important of all rules for applying group theory to quantum chemistry.

3.7 ROTATIONAL PROPERTIES OF NONLINEAR MOLECULES

Nonlinear molecules have three moments of inertia. They are, by convention, labeled I_a, I_b, and I_c with $I_a < I_b < I_c$, if they are all different. Molecules with three different moments are called *asymmetric tops*. If the molecule has threefold, or higher, rotational symmetry about one and only one axis (see Chapter 13), two of these moments of inertia will be equal. Such molecules are called *symmetric tops*. Depending upon the shape of the molecule, the unique moment of inertia of a symmetric top can be either greater than or less than the two equal moments. Those having the largest unique moment of inertia are called *oblate* tops; those having the smallest are called *prolate* tops. In linear molecules, one of the moments of inertia is zero; consequently, linear molecules are the extreme case of the prolate tops. Planar symmetric tops are the extreme case of the oblate tops. In molecules that have threefold or higher rotational symmetry about two or more different axes, all three moments of inertia are equal. Such molecules are called *spherical tops*.

The symmetry group for describing the rotations of nonlinear molecules is not just the **R**(3) group. For describing such systems, two coordinate reference frames are required—one within the molecule, and the external laboratory coordinate system. The moments of inertia interact to yield a net overall moment of inertia within the internal coordinate system. If the molecule is a spherical top, **R**(3) is the appropriate internal group; if the molecule has lower symmetry, however, the internal group is of lower symmetry than **R**(3). The overall moment of inertia behaves in the external coordinate system just as does the angular momentum of a rigid rotor. The appropriate external group is, thus, always **R**(3). The overall group, **G**, is the product of the groups acting on the two coordinate systems—the external **R**(3) and the internal group (which we will call $\mathbf{G}_I$), whatever it may be:

$$\mathbf{G} = \mathbf{R}(3) \times \mathbf{G}_I \qquad \textbf{3.106}$$

For the simple rigid-rotor problem (linear molecules) the rotational energy levels are $(2J + 1)$-fold degenerate in the absence of external fields. For nonlinear molecules there can be higher degeneracies, depending upon the nature of $\mathbf{G}_I$. For example, in a spherical top, where $\mathbf{G}_I$ is **R**(3), the levels are $(2J + 1)^2$-fold degenerate.

The energy levels for the spherical top and for the symmetrical top can be determined from our previous discussion of angular momentum (Section 3.3). The general classical energy (which is only kinetic energy) for a rotating body is

$$E = \frac{l_a^2}{2I_a} + \frac{l_b^2}{2I_b} + \frac{l_c^2}{2I_c} \qquad \textbf{3.107}$$

For the spherical top the three moments of inertia are equal, so this becomes

$$E = \frac{l^2}{2I} \qquad \textbf{3.108}$$

where

$$l^2 = l_a^2 + l_b^2 + l_c^2 \qquad \textbf{3.109}$$

We know from Eq. 3.72 that our wave function must be an eigenfunction of the operator corresponding to the square of the total angular momentum, $\hat{l}^2$. Thus, we have

$$\begin{aligned} \hat{H}\psi &= E\psi \\ &= \frac{\hat{l}^2}{2I}\psi \\ &= \frac{\hbar^2}{2I}J(J + 1)\psi \end{aligned} \qquad \textbf{3.110}$$

or

$$E_J = \frac{\hbar^2}{2I}J(J + 1) \qquad \textbf{3.111}$$

This is the same as the energy of the rigid rotor; however, the degeneracy is $(2J + 1)^2$, rather than $(2J + 1)$ as in the rigid rotor. Direct microwave absorption is not seen for spherical tops, however, since such molecules do not have permanent dipole moments.

The classical energy of the symmetric top is again that of Eq. 3.107. Now, however, two of the moments of inertia are equal and one is different. If we assume that I_a is the unique moment, we can rewrite Eq. 3.107 as

$$\begin{aligned} E &= \frac{l_a^2}{2I_a} + \frac{l_b^2 + l_c^2}{2I_b} \\ &= \frac{l_a^2}{2I_a} + \frac{l^2 - l_a^2}{2I_b} \\ &= \frac{l^2}{2I_b} + \frac{l_a^2}{2}\left(\frac{1}{I_a} - \frac{1}{I_b}\right) \end{aligned} \tag{3.112}$$

Using the operators for angular momentum, we obtain the Hamiltonian as

$$\hat{H} = \frac{\hat{l}^2}{2I_b} + \frac{\hat{l}_a^2}{2}\left(\frac{1}{I_a} - \frac{1}{I_b}\right) \tag{3.113}$$

Now the $\hat{l}^2$ operator satisfies Eq. 3.72, and the internal a axis can be chosen as the internal z axis so that Eq. 3.74 can be used for $\hat{l}_a$. (We will call the quantum number K, rather than M, since it refers to an internal axis system. It has the same range of values as M.) We have

$$\begin{aligned} \hat{H}\psi &= E\psi \\ &= \left[\frac{\hat{l}^2}{2I_b} + \frac{\hat{l}_a^2}{2}\left(\frac{1}{I_a} - \frac{1}{I_b}\right)\right]\psi \\ &= \left[\frac{\hbar^2}{2I_b}J(J + 1) + \frac{K^2\hbar^2}{2}\left(\frac{1}{I_a} - \frac{1}{I_b}\right)\right]\psi \end{aligned} \tag{3.114}$$

If we designate the unique moment of inertia as $I_{a'}$, whether it is the largest or the smallest, the general energy expression for a symmetric top is

$$E_{JK} = \frac{\hbar^2}{2I_b}J(J + 1) + \frac{K^2\hbar^2}{2}\left(\frac{1}{I_{a'}} - \frac{1}{I_b}\right) \tag{3.115}$$

Note that if $I_{a'}$ is the smallest moment of inertia, the energy increases with increasing $|K|$, while if it is the largest, the energy decreases with increasing $|K|$. Note, also, that for $|K|$ not equal to zero, the states characterized by K and by $-K$ are degenerate, since the energy depends on the square of K and, hence, its magnitude, and not its sign.

Many symmetric tops have permanent dipole moments. For those that do, microwave spectra can be observed. The selection rule for ΔJ is the same as for linear molecules: ΔJ must be ± 1. The K selection rule is

$$\Delta K = 0 \tag{3.116}$$

(This selection rule arises because the permanent dipole moment of a symmetric top lies along the axis of the unique moment of inertia.) The energy, in wave numbers, is frequently written

$$\bar{v}_{JK} = BJ(J+1) + (A' - B)K^2 \tag{3.117}$$

where A' and B are the rotational constants for the two moments of inertia (see Eq. 3.78). For linear molecules only one structural parameter, the single moment of inertia, could be obtained from the microwave spectrum. For symmetric tops it would appear that two rotational constants and, consequently, two moments of inertia can be obtained. Unfortunately, however, if centrifugal distortion is ignored, only the value of B can be determined, owing to the $\Delta K = 0$ selection rule. Consequently the microwave spectrum of a symmetric top is very similar to that for a linear molecule, but with the absorptions broadened a bit owing to centrifugal distortion (since it affects A' and B differently).

The energy levels for the asymmetric top cannot be obtained in closed form, because no moment of inertia has a unique relationship within the rotating molecule.

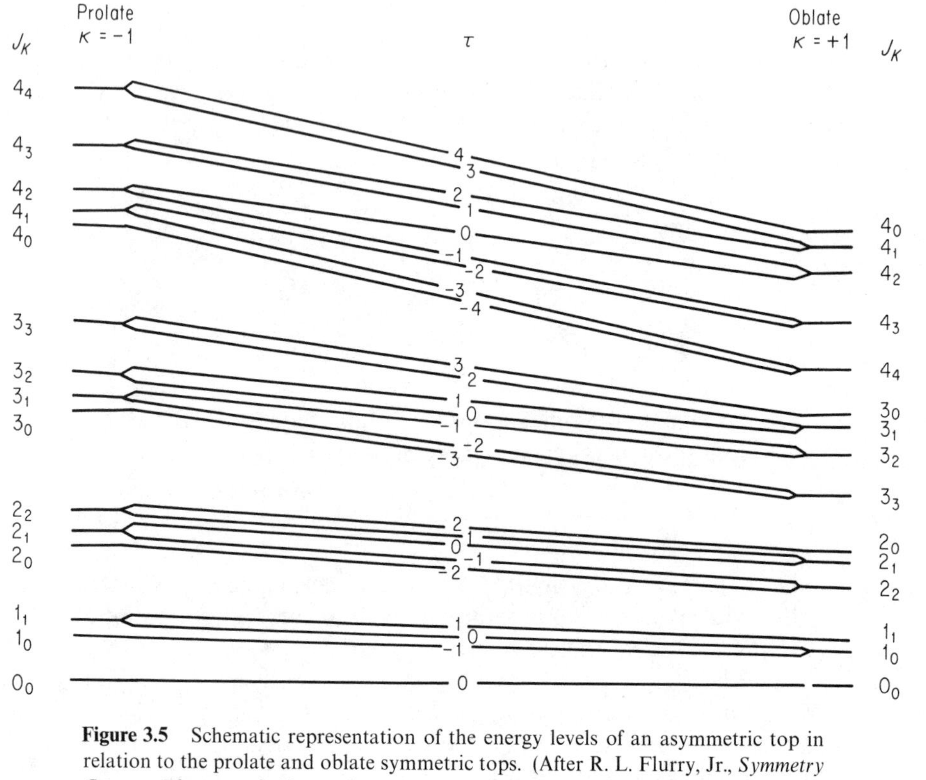

Figure 3.5 Schematic representation of the energy levels of an asymmetric top in relation to the prolate and oblate symmetric tops. (After R. L. Flurry, Jr., *Symmetry Groups: Theory and Chemical Applications*, Prentice-Hall, Inc., Englewood Cliffs, N.J., 1980. By permission.)

The classical energy is still as in Eq. 3.107. This can be rewritten, in terms of the rotational constants, A, B, and C, as

$$E = \tfrac{1}{2}(A + C)l^2 + \tfrac{1}{2}(A - C)(l_a^2 - l_c^2 + \kappa l_b^2) \qquad \textbf{3.118}$$

where E is in wave numbers, and κ is the *asymmetry parameter*, defined as

$$\kappa = \frac{2B - A - C}{A - C} \qquad \textbf{3.119}$$

The quantum-mechanical solution is usually expressed as

$$E = \tfrac{1}{2}(A + C)J(J + 1) + \tfrac{1}{2}(A - C)E(\kappa) \qquad \textbf{3.120}$$

The $E(\kappa)$ are obtained numerically from the solutions of a symmetric top by using perturbation theory (see Chapter 6). For a prolate top κ has the value of -1, while for an oblate top it is $+1$. For the asymmetric top κ lies between these values. Furthermore, the twofold degeneracy of the symmetric top is lifted. The $(2J + 1)$ different levels having the same J value are labeled by an index τ, which is not a quantum number. The relation of the asymmetric-top energy levels to those of prolate and oblate tops is shown schematically in Figure 3.5.

3.A *APPENDIX: MATRIX REPRESENTATIONS AND CHARACTERS FOR* **R**(3)

In order to illustrate the concept of the character of a matrix representation, let us consider the three-dimensional identity matrix, **E**:

$$\mathbf{E} = \begin{bmatrix} 1 & 0 & 0 \\ 0 & 1 & 0 \\ 0 & 0 & 1 \end{bmatrix} \qquad \textbf{3.A1}$$

and the matrix corresponding to a rotation of ϕ about the z axis, $\mathbf{C}(\phi)$:

$$\mathbf{C}(\phi) = \begin{bmatrix} \cos\phi & \sin\phi & 0 \\ -\sin\phi & \cos\phi & 0 \\ 0 & 0 & 1 \end{bmatrix} \qquad \textbf{3.A2}$$

An arbitrary vector

$$\mathbf{r} = a\mathbf{i} + b\mathbf{j} + c\mathbf{k} \qquad \textbf{3.A3}$$

where a, b, and c are scalars and **i**, **j**, and **k** are the unit Cartesian vectors, can be expressed as a row vector of the coefficients

$$\mathbf{r} = [a \quad b \quad c] \qquad \textbf{3.A4}$$

The products **rE** and **rC**(ϕ) are

$$\mathbf{rE} = [a \quad b \quad c]\begin{bmatrix} 1 & 0 & 0 \\ 0 & 1 & 0 \\ 0 & 0 & 1 \end{bmatrix}$$

$$= [a \quad b \quad c] \qquad \mathbf{3.A5}$$

and

$$\mathbf{rC}(\phi) = [a \quad b \quad c]\begin{bmatrix} \cos\phi & \sin\phi & 0 \\ -\sin\phi & \cos\phi & 0 \\ 0 & 0 & 1 \end{bmatrix}$$

$$= [(a\cos\phi - b\sin\phi) \quad (a\sin\phi + b\cos\phi) \quad c] \qquad \mathbf{3.A6}$$

Equation 3.A5 leaves the vector **r** unchanged, while Eq. 3.A6 rotates **r** by the angle ϕ in the counterclockwise direction about the z axis.

Because of the zeros in **C**(ϕ), it can be factored into the *direct sum* (see Appendix 2) of two matrices

$$\mathbf{C}(\phi) = \begin{bmatrix} \cos\phi & \sin\phi \\ -\sin\phi & \cos\phi \end{bmatrix} \oplus [1] \qquad \mathbf{3.A7}$$

The matrix **E** can be similarly factored:

$$\mathbf{E} = \begin{bmatrix} 1 & 0 \\ 0 & 1 \end{bmatrix} \oplus [1] \qquad \mathbf{3.A8}$$

(Note that **E** could actually be factored into three one-dimensional unit matrices.) If we let **C**(ϕ) be the matrix representation for the arbitrary rotation operation of the **R**(3) group, we can express it as a three-dimensional representation as in Eq. 3.A2, or as a direct sum of a two-dimensional and a one-dimensional representation as in Eq. 3.A7. (Note, however, that this is for rotation about the z axis only.) If we let **E** be the matrix representation for the identity operation of **R**(3) (the operation that does nothing), it can be expressed as a three-dimensional representation (Eq. 3.A1), the direct sum of a two-dimensional and a one-dimensional representation (Eq. 3.A8), or the direct sum of three one-dimensional representations. If, however, we wanted the representations to be characteristic of the entire group, the three-dimensional representation would have to be associated with the three-dimensional representation of **C**(ϕ), the two-dimensional with the two-dimensional representation of **C**(ϕ), and the one-dimensional with the one-dimensional. The **C**(ϕ) matrix could be factored into one-dimensional components only in the special cases when ϕ equals $n\pi$.

We have said that the character of a representation is the trace (the sum of the diagonal elements) of a representation matrix. The characters of the one-dimensional, two-dimensional, and three-dimensional representations for the identity are 1, 2, and 3, respectively, while those for the rotation are 1, 2 cos ϕ, and (1 + 2 cos ϕ). These are actually representations of the same rotation in a one-dimensional, a two-dimensional, or a three-dimensional space. In the three-dimensional space we also have rotations about x and y. The matrix representations of each individual rotation can be factored into a one-dimensional and a two-dimensional matrix. However, the three rotation matrices cannot simultaneously undergo the same one-dimensional and two-dimensional factorization.

The matrix representations we have discussed until now have been for a three-dimensional vector. If the system being described is to have spherical symmetry, the coefficients of x, y, and z must be the same; consequently the representation describing the x, y, z behavior must be a single three-dimensional representation. We need to be able to describe a general function with spherical symmetry in our three-dimensional space. Any general function can be written as a polynomial in x, y, and z:

$$f(x, y, z) = \sum_k \sum_l \sum_m a_{klm} x^k y^l z^m \qquad \textbf{3.A9}$$

The exponents, k, l, and m can have any integer value, including zero. The symmetry behavior of terms with negative exponents will be the same as that for the terms with the corresponding positive exponents. Thus, we can restrict our attention to non-negative values of k, l, and m. If spherical symmetry is to be retained, the coefficients of all terms with the same value of $(k + l + m)$ must be the same. We can rewrite Eq. 3.A9 as

$$f(x, y, z) = \sum_{j=0} a_j x^k y^l z^m \qquad \textbf{3.A10}$$

where j equals $(k + l + m)$.

For a j value of zero in Eq. 3.A10 we have a single term, a constant. The constant is unchanged by the rotation. Its behavior under both the identity and $\mathbf{C}(\phi)$ can be described by the 1×1 matrix whose only element is $+1$. The characters are also $+1$. The representation is one-dimensional. As we have already stated, the representation for a j value of one is three-dimensional. The character for the identity is 3, while that for $\mathbf{C}(\phi)$ is $(1 + 2 \cos \phi)$. If j equals two, we have the six terms, x^2, y^2, z^2, xy, xz, and yz. These are not all independent, however, since $(x^2 + y^2 + z^2)$ equals r^2. We have six terms, but one relation among them. Thus, the appropriate representation is five-dimensional. The representation for the j value of two (which we shall call D^2) should be derivable from that for a j value of one (D^1), since the terms leading to D^2 are binary products of those leading to D^1. We could do this by taking the direct products (see Appendix 2) of the matrices of Eqs. 3.A1 and 3.A2 with themselves and reducing the result. It is sufficient, however, to use the characters, since the trace of the direct product of two matrices is the product of the traces. We have (calling the characters χ)

$$\begin{aligned} \chi(E \otimes E) &= 3 \times 3 \\ &= 9 \end{aligned} \qquad \textbf{3.A11}$$

and

$$\begin{aligned} \chi(C(\phi) \otimes C(\phi)) &= (1 + 2 \cos \phi)(1 + 2 \cos \phi) \\ &= 1 + 4 \cos \phi + 4 \cos^2 \phi \\ &= 3 + 4 \cos \phi + 2 \cos 2\phi \end{aligned} \qquad \textbf{3.A12}$$

The result of Eq. 3.A11 is not 5; obviously, then, the result will be reducible. The reduction is apparent on inspection of Eq. 3.A12. We see that it contains the character of the rotation in D^1 (perhaps twice). If we subtract out the characters of D^1, we have

	E	$C(\phi)$
$D^1 \otimes D^1$	9	$3 + 4\cos\phi + 2\cos 2\phi$
D^1	3	$1 + 2\cos\phi$
$D^1 \otimes D^1 - D^1$	6	$2 + 2\cos\phi + 2\cos 2\phi$

3.A13a

The resulting representation is six-dimensional. If we subtract out D^0, we are left with a representation of the proper dimension:

	E	$C(\phi)$
$D^1 \otimes D^1 - D^1$	6	$2 + 2\cos\phi + 2\cos 2\phi$
D^0	1	1
$D^1 \otimes D^1 - D^1 - D^0 \equiv D^2$	5	$1 + 2\cos\phi + 2\cos 2\phi$

3.A13b

We have, in fact, done two things. We have found the characters for D^2 and we have shown that

$$D^1 \otimes D^1 = D^0 + D^1 + D^2 \qquad \textbf{3.A14}$$

In general, there are $(2j + 1)$ independent terms for a given j value, leading to a $(2j + 1)$-dimensional irreducible representation. For each, the character for the identity is simply

$$\chi(E) = 2j + 1 \qquad \textbf{3.A15}$$

By successive direct products of D^1 with the D^j, we can verify that the character of a general $C(\phi)$ is

$$\chi(C(\phi)) = 1 + \sum_{k=1}^{j} 2\cos k\phi \qquad \textbf{3.A16}$$

These generate a set of representations having only odd dimensions. Even-dimensional representations can also be generated by allowing j to have half-integer values. These representations are useful for describing the properties of electrons and other particles having half-integer "spins." For these, the character for the identity is still $(2j + 1)$, while that for the rotation is

$$\chi(C(\phi)) = \sum_{k=1/2}^{j} 2\cos k\phi \qquad \text{(half-integer } j\text{)} \qquad \textbf{3.A17}$$

(There are other constructions for the rotation characters, but for our purposes those given here will be easier to work with.) The character table for **R**(3) was given in Table 3.5.

BIBLIOGRAPHY

ANDERSON, J. M., *Introduction to Quantum Chemistry*. W. A. Benjamin, Inc., New York, 1969.

ATKINS, P. W., *Molecular Quantum Mechanics*. Clarendon Press, Oxford, 1970.

FLURRY, R. L., JR., *Symmetry Groups: Theory and Chemical Applications*. Prentice-Hall, Inc., Englewood Cliffs, N. J., 1980.

LEVINE, I. N., *Quantum Chemistry*. Allyn & Bacon, Boston, 2d. ed., 1974.

LINNETT, J. W., *Wave Mechanics and Valency*. Methuen & Co., London, 1960.

PAULING, L., and WILSON, E. B., JR., *Introduction to Quantum Mechanics*. McGraw-Hill Book Company, New York, 1935.

PROBLEMS

***3.1** The rotational constants, B, for $^1H^{35}Cl$ and $^1H^{80}Br$ are 10.5909 cm^{-1} and 8.473 cm^{-1}, respectively. For each, calculate

(a) The moment of inertia.

(b) The bond length.

(c) The relative population of the levels having $0 \leq J \leq 10$ at 300°K, using the Boltzmann distribution function.

***3.2** The first few rotational lines in the microwave spectrum of the linear OCS molecule are, for the indicated isotopic species:

$^{16}O^{12}C^{32}S$	$^{16}O^{12}C^{34}S$
.81142 cm^{-1}	.79162 cm^{-1}
1.21713	—
1.62284	1.58317
2.02854	—

(a) Calculate the moment of inertia of each species.

(b) Find the bond lengths in OCS. [*Hint:* Shifting the axis of rotation from the center of mass by a distance d changes the moment of inertia by an amount $\sum_i m_i d^2$.]

***3.3** Construct the following products of representations within either **O**(3) or **R**(3) (as indicated by the notation).

(a) $D_g^1 \times D_u^2$.

(b) $D_g^{3/2} \times D_g^{1/2}$.

(c) $D_u^{3/2} \times D_u^1 \times D_g^{1/2}$.

(d) $D_g^{1/2} \times D_u^{3/2} \times D_g^{5/2}$.

(e) $D_g^0 \times D_u^1 \times D_g^2 \times D_u^3$.

3.4 In Raman spectroscopy, the selection rule depends upon a nonvanishing quadrupole operator. The quadrupole operator transforms as D_g^2 within **O**(3). Show that the ΔJ selection rule for the rotational Raman spectroscopy of a linear molecule is $\Delta J = \pm 2$.

* **3.5** Consider the following molecules:

Hydrogen cyanide
Formaldehyde
Acetylene
Ammonia
Water
Benzene
Ozone
Ethane (staggered conformation)
1,1,1-Trichloroethane (staggered conformation)
Sulfur hexafluoride

(a) Classify each according to its "top" type.

(b) Which would exhibit observable microwave spectra?

***3.6** The operator corresponding to the square of angular momentum, $\hat{L}^2$, is

$$-\left(\frac{1}{\sin\theta}\frac{\partial}{\partial\theta}\sin\theta\frac{\partial}{\partial\theta} + \frac{1}{\sin^2\theta}\frac{\partial^2}{\partial\phi^2}\right)\hbar^2$$

That for the z component of angular momentum, $\hat{L}_z$, is

$$-i\hbar\frac{\partial}{\partial\phi}$$

Show that the functions $Y_{0,0}$, $Y_{1,-1}$, and $Y_{2,2}$ from Table 3.2 are eigenfunctions of $\hat{L}^2$ and $\hat{L}_z$. What is the general form for the eigenvalues of $\hat{L}^2$ and $\hat{L}_z$ (in terms of the L and M quantum numbers)?

***3.7** For ammonia, NH_3, r(N—H) is 1.014×10^{-10} m and the H—N—H angle is 106°47′.

(a) Calculate the moments of inertia and rotational constants for NH_3.

(b) Calculate the relative thermal populations of the first several levels from part (a) at 25°C, using the Boltzmann distribution function,

$$\frac{N_j}{N_i} = \frac{g_j}{g_i}\exp\left(-\frac{\Delta E_{ij}}{kT}\right).$$

chapter four

Vibrations

4.1 THE HARMONIC OSCILLATOR (HEISENBERG TREATMENT)

The vibrations of a diatomic molecule can, to a good approximation, be described by a harmonic oscillator. Those of polyatomic molecules can be described by coupled harmonic oscillators. Consequently, the quantum-mechanical harmonic-oscillator problem is of interest to chemists. Furthermore, it is a problem that is exactly solvable in closed form. To illustrate the Heisenberg approach, we will solve the harmonic-oscillator problem in detail by matrix mechanics. Although the mathematics involved is totally different from that involved in the Schrödinger treatment, the level of difficulty is about the same. (The particle-in-a-box problem, by contrast, is much simpler in the Schrödinger treatment.)

The typical harmonic oscillator is represented by a mass, m, supported by a spring from a rigid support (Figure 4.1). (The physical model here should be idealized; i.e., the spring should be massless and perfectly elastic.) Two masses in free space connected by a spring (the model of a vibrating diatomic molecule) give rise to the same equations, if the single mass is replaced by the reduced mass, μ, where

$$\mu = \frac{m_1 m_2}{m_1 + m_2} \qquad \textbf{4.1}$$

and m_1 and m_2 are the two connected masses. In our treatment of this problem, instead of using x, y, and/or z as our coordinates, we will use the generalized displacement coordinate, q, for displacement from the equilibrium position.

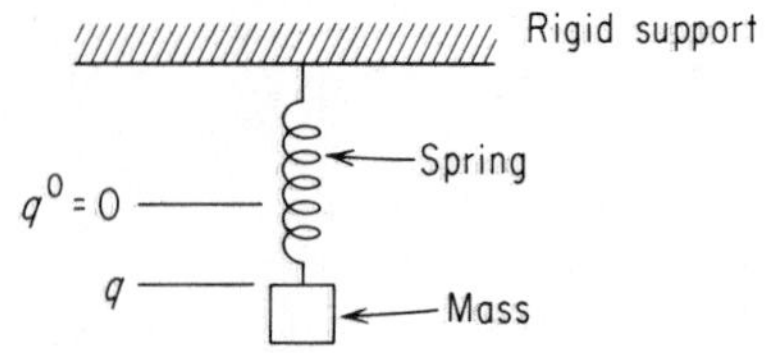

Figure 4.1 Simple harmonic oscillator. The q is the displacement from the equilibrium position, q^0.

The suspended mass of Figure 4.1, when displaced by q, is subject to a restoring force, F (Hooke's law):

$$F = -kq \tag{4.2}$$

where k is the Hooke's law constant (or force constant). The classical equation of motion for the oscillator is

$$m\ddot{q} = -kq \tag{4.3}$$

where $\ddot{q}$ is an abbreviated notation for d^2q/dt^2. The potential energy is $\frac{1}{2}kq^2$ and the total classical energy is

$$E = \frac{p^2}{2m} + \frac{kq^2}{2} \tag{4.4}$$

where p is the momentum. For convenience, let us substitute in for k in terms of the classical frequency of oscillation:

$$\mathrm{k} = m\omega^2 \tag{4.5}$$

where ω is $2\pi\nu$, and ν is the classical frequency. This gives

$$E = \frac{p^2}{2m} + \frac{m\omega^2 q^2}{2} \tag{4.6}$$

Now, according to Heisenberg, each of the observables E, p, and q, has a matrix corresponding to it. We will call these **H**, **P**, and **Q**. The Heisenberg Hamiltonian is

$$\mathbf{H} = \frac{1}{2m}\mathbf{P}^2 + \frac{m\omega^2}{2}\mathbf{Q}^2 \tag{4.7}$$

The rows and columns of the matrices are assumed to be labeled by the states of the system. In effect, the states provide the *basis vectors* for the *vector algebra* according to which the matrices are constructed. If the **H** matrix is in diagonal form (all elements off the diagonal, H_{ij}, $i \neq j$, are zero), the diagonal elements (H_{ii}) will be the energies of the states of the system. (Crudely, H_{ii} has contributions only from state i; however, if any H_{ij} are not equal to zero, states i and j are mixed.) Thus, to solve for our energy states, we must find the form of the matrix **H** and require that **H** be a diagonal matrix. To do this, we will make use of two commutation relations.

The first we have already had:

$$[\mathbf{A}, \mathbf{H}] = i\hbar\dot{\mathbf{A}} \tag{4.8}$$

This can be rearranged to give

$$\dot{\mathbf{A}} = \frac{i}{\hbar}[\mathbf{H}, \mathbf{A}] \tag{4.8a}$$

The second, which can be derived from the first (Green, Sec. 2.5), is

$$[\mathbf{Q}^n, \mathbf{P}] = ni\hbar\mathbf{Q}^{n-1} \tag{4.9}$$

The equation of motion to be solved is the quantum-mechanical analog of Eq. 4.3. Classically, we know that the momentum, p, is equal to mv. However, the velocity, v, is just the time derivative of the coordinate, q. Thus, we have

$$\dot{\mathbf{Q}} = \frac{1}{m}\mathbf{P} = \frac{i}{\hbar}[\mathbf{H}, \mathbf{Q}] \tag{4.10}$$

This comes in handy in solving for $\ddot{\mathbf{Q}}$. We have

$$\begin{aligned}\ddot{\mathbf{Q}} &= \frac{i}{\hbar}[\mathbf{H}, \dot{\mathbf{Q}}] \\ &= \frac{i}{\hbar}\left[\mathbf{H}, \left(\frac{1}{m}\mathbf{P}\right)\right] \\ &= \frac{1}{m}\dot{\mathbf{P}} \qquad\qquad 4.11\end{aligned}$$

But,

$$\dot{\mathbf{P}} = \frac{i}{\hbar}[\mathbf{H}, \mathbf{P}] \tag{4.12a}$$

$$= \frac{i}{\hbar}\left[\left(\frac{1}{2m}\mathbf{P}^2 + \frac{m\omega^2}{2}\mathbf{Q}^2\right), \mathbf{P}\right] \tag{4.12b}$$

$$= \frac{i}{2m\hbar}[\mathbf{P}^2, \mathbf{P}] + \frac{im\omega^2}{2\hbar}[\mathbf{Q}^2, \mathbf{P}] \tag{4.12c}$$

$$= \frac{im\omega^2}{2\hbar}(2i\hbar\mathbf{Q}) \tag{4.12d}$$

$$= -m\omega^2\mathbf{Q} \tag{4.12e}$$

or, from Eq. 4.11,

$$\ddot{\mathbf{Q}} = -\omega^2\mathbf{Q} \tag{4.13}$$

(In going from 4.12c to 4.12d, we use the fact that any quantity commutes with any power of itself to eliminate the first term and use Eq. 4.9 for the second term.) Equation

4.13 is the quantum-mechanical analog of Eq. 4.3. We can rearrange Eq. 4.13 to give

$$\ddot{\mathbf{Q}} + \omega^2\mathbf{Q} = \mathbf{0} \tag{4.14}$$

where $\mathbf{0}$ is the *null matrix*, a matrix having zeros for all elements.

Let us consider the form of the matrix elements for the left-hand side of Eq. 4.14. Here we make use of Heisenberg's *time development of states.* He assumed that the time-dependent behavior represents an oscillation between states. In other words, the elements of $\mathbf{Q}$ are

$$q_{ij} = q_{ij}^0 \exp(i\omega_{ij}t) \tag{4.15}$$

where q_{ij}^0 is an amplitude and the exponential term gives the time behavior. A general element of the left-hand side of Eq. 4.14 is

$$(\ddot{\mathbf{Q}} + \omega^2\mathbf{Q})_{ij} = \ddot{q}_{ij} + \omega^2 q_{ij} = 0 \tag{4.16}$$

But

$$\ddot{q}_{ij} = \frac{d^2}{dt^2}(q_{ij}^0 \exp(i\omega_{ij}t))$$

$$= -\omega_{ij}^2 q_{ij}^0 \exp(i\omega_{ij}t) \tag{4.17}$$

Substituting Eqs. 4.15 and 4.17 into 4.16, we have

$$(\omega^2 - \omega_{ij}^2)q_{ij}^0 \exp(i\omega_{ij}t) = 0 \tag{4.18}$$

Since the exponential does not, in general, equal zero, this equation can be satisfied only if q_{ij}^0 equals zero, or if ω_{ij} equals $\pm\omega$. We make the latter choice, since the first gives us no information. The indices i and j can have any value; however, for convenience, let us choose the indices such that $q_{j+1,j}^0$ is associated with $+\omega$ and $q_{j-1,j}^0$ is associated with $-\omega$ [i.e., such that i equals $(j + 1)$ and $(j - 1)$, respectively]. This yields a $\mathbf{Q}$ matrix with nonzero elements only at the first positions off the diagonal:

$$\mathbf{Q} = \begin{bmatrix} 0 & q_{01} & 0 & 0 & \cdots \\ q_{10} & 0 & q_{12} & 0 & \cdots \\ 0 & q_{21} & 0 & q_{23} & \cdots \\ \cdots & \cdots & \cdots & \cdots & \cdots \end{bmatrix} \tag{4.19}$$

In order to construct the Hamiltonian matrix of Eq. 4.7, we also need the momentum matrix, $\mathbf{P}$. According to Eq. 4.10,

$$\mathbf{P} = m\dot{\mathbf{Q}} \tag{4.20}$$

or

$$p_{ij} = m\frac{d}{dt}q_{ij} \tag{4.21a}$$

$$= im\omega_{ij}q_{ij}^0 \exp(i\omega_{ij}t) \tag{4.21b}$$

$$= im\omega_{ij}q_{ij} \tag{4.21c}$$

where we have used Eq. 4.15 for the definition of q_{ij}. Thus,

$$\mathbf{P} = im\begin{bmatrix} 0 & \omega_{01}q_{01} & 0 & 0 & \cdots \\ \omega_{10}q_{10} & 0 & \omega_{12}q_{12} & 0 & \cdots \\ 0 & \omega_{21}q_{21} & 0 & \omega_{23}q_{23} & \cdots \\ \cdots & \cdots & \cdots & \cdots & \cdots \end{bmatrix} \qquad \textbf{4.22}$$

But we have defined i such that ω_{ij} equals ω if i equals $(j + 1)$ and $-\omega$ if i equals $(j - 1)$. This gives

$$\mathbf{P} = im\omega\begin{bmatrix} 0 & -q_{01} & 0 & 0 & \cdots \\ q_{10} & 0 & -q_{12} & 0 & \cdots \\ 0 & q_{21} & 0 & -q_{23} & \cdots \\ \cdots & \cdots & \cdots & \cdots & \cdots \end{bmatrix} \qquad \textbf{4.23}$$

Squaring **P**, we have

$$\mathbf{P}^2 = m^2\omega^2\begin{bmatrix} q_{01}q_{10} & 0 & -q_{01}q_{12} & \cdots \\ 0 & q_{01}q_{10} + q_{12}q_{21} & 0 & \cdots \\ -q_{21}q_{10} & 0 & q_{12}q_{21} + q_{23}q_{32} & \cdots \\ \cdots & \cdots & \cdots & \cdots \end{bmatrix} \qquad \textbf{4.24}$$

and squaring **Q** gives

$$\mathbf{Q}^2 = \begin{bmatrix} q_{01}q_{10} & 0 & q_{01}q_{12} & \cdots \\ 0 & q_{01}q_{10} + q_{12}q_{21} & 0 & \cdots \\ q_{21}q_{10} & 0 & q_{12}q_{21} + q_{23}q_{32} & \cdots \\ \cdots & \cdots & \cdots & \cdots \end{bmatrix} \qquad \textbf{4.25}$$

Substituting Eqs. 4.24 and 4.25 into 4.7 gives, for **H**,

$$\mathbf{H} = m\omega^2\begin{bmatrix} q_{01}q_{10} & 0 & 0 & \cdots \\ 0 & q_{01}q_{10} + q_{12}q_{21} & 0 & \cdots \\ 0 & 0 & q_{12}q_{21} + q_{23}q_{32} & \cdots \\ \cdots & \cdots & \cdots & \cdots \end{bmatrix} \qquad \textbf{4.26}$$

This matrix is diagonal, as we desired. The diagonal elements of **H** are the energies. They have the general form

$$E_n = H_{nn} = m\omega^2(q_{n,\,n+1}q_{n+1,\,n} + q_{n,\,n-1}q_{n-1,\,n}) \qquad \textbf{4.27}$$

Evaluation of the energies requires the evaluation of the q_{ij}'s. This can be accomplished from the commutation relation (from Eq. 4.9)

$$[\mathbf{Q}, \mathbf{P}] = i\hbar\mathbf{I} \qquad \textbf{4.28}$$

where **I** is the *unit matrix*, having elements of unity along the diagonal and elements of zero everywhere else. Substituting in Eqs. 4.19 and 4.23 for **Q** and **P**, we have

$$\mathbf{QP} = im\omega \begin{bmatrix} 0 & q_{01} & 0 & 0 & \cdots \\ q_{10} & 0 & q_{12} & 0 & \cdots \\ 0 & q_{21} & 0 & q_{23} & \cdots \\ \cdots & \cdots & \cdots & \cdots & \cdots \end{bmatrix} \begin{bmatrix} 0 & -q_{01} & 0 & 0 & \cdots \\ q_{10} & 0 & -q_{12} & 0 & \cdots \\ 0 & q_{21} & 0 & -q_{23} & \cdots \\ \cdots & \cdots & \cdots & \cdots & \cdots \end{bmatrix}$$

$$= im\omega \begin{bmatrix} q_{01}q_{10} & 0 & -q_{01}q_{12} & \cdots \\ 0 & q_{12}q_{21} - q_{10}q_{01} & 0 & \cdots \\ q_{21}q_{10} & 0 & q_{23}q_{32} - q_{21}q_{12} & \cdots \\ \cdots & \cdots & \cdots & \cdots \end{bmatrix} \quad \textbf{4.29}$$

$$\mathbf{PQ} = im\omega \begin{bmatrix} 0 & -q_{01} & 0 & 0 & \cdots \\ q_{10} & 0 & -q_{12} & 0 & \cdots \\ 0 & q_{21} & 0 & -q_{23} & \cdots \\ \cdots & \cdots & \cdots & \cdots & \cdots \end{bmatrix} \begin{bmatrix} 0 & q_{01} & 0 & 0 & \cdots \\ q_{10} & 0 & q_{12} & 0 & \cdots \\ 0 & q_{21} & 0 & q_{23} & \cdots \\ \cdots & \cdots & \cdots & \cdots & \cdots \end{bmatrix}$$

$$= im\omega \begin{bmatrix} -q_{01}q_{10} & 0 & -q_{01}q_{12} & \cdots \\ 0 & q_{10}q_{01} - q_{12}q_{21} & 0 & \cdots \\ q_{21}q_{10} & 0 & q_{21}q_{12} - q_{23}q_{32} & \cdots \\ \cdots & \cdots & \cdots & \cdots \end{bmatrix} \quad \textbf{4.30}$$

Thus

$$[\mathbf{Q}, \mathbf{P}] = 2im\omega \begin{bmatrix} q_{01}q_{10} & 0 & 0 & \cdots \\ 0 & q_{12}q_{21} - q_{10}q_{01} & 0 & \cdots \\ 0 & 0 & q_{23}q_{32} - q_{21}q_{12} & \cdots \\ \cdots & \cdots & \cdots & \cdots \end{bmatrix} \quad \textbf{4.31}$$

$$= i\hbar\mathbf{I} \quad \textbf{4.31a}$$

In other words, the matrix on the right side of Eq. 4.31, which equals $(-i/2m\omega)[\mathbf{Q}, \mathbf{P}]$, equals $\hbar\mathbf{I}/2m\omega$. Each diagonal element equals $\hbar/2m\omega$. We have

$$q_{01}q_{10} = \frac{\hbar}{2m\omega} \quad \textbf{4.32}$$

$$q_{12}q_{21} - q_{10}q_{01} = \frac{\hbar}{2m\omega} \quad \textbf{4.32a}$$

Substituting Eq. 4.32 into 4.32a (the **Q** matrix elements q_{01} and q_{10} commute) gives

$$q_{12}q_{21} = \frac{2\hbar}{2m\omega} \quad \textbf{4.33}$$

From the third diagonal element, we have

$$q_{23}q_{32} - q_{21}q_{12} = \frac{\hbar}{2m\omega} \tag{4.34}$$

Substituting into Eq. 4.33 gives

$$q_{23}q_{32} = \frac{3\hbar}{2m\omega} \tag{4.35}$$

The pattern quickly becomes obvious:

$$q_{n,\,n+1}q_{n+1,\,n} = \frac{(n+1)\hbar}{2m\omega} \tag{4.36}$$

Therefore, from Eq. 4.27,

$$\begin{aligned} E_n &= m\omega^2\left[\frac{(n+1)\hbar}{2m\omega} + \frac{n\hbar}{2m\omega}\right] \\ &= \frac{(2n+1)\hbar\omega}{2} \\ &= \left(n + \frac{1}{2}\right)\hbar\omega \end{aligned} \tag{4.37}$$

This is more commonly expressed in terms of the frequency, ν, than in terms of the angular velocity, ω:

$$E_n = \left(n + \frac{1}{2}\right)h\nu_0 \tag{4.38}$$

(The reason for designating the frequency as ν_0 will be apparent shortly.) Note that if n equals zero, there is still a vibrational energy of $\frac{1}{2}h\nu_0$. This is known as the *zero-point energy*. Physically, this implies that the quantum-mechanical harmonic oscillator is never at rest, but always vibrating with at least the zero-point energy. (If the oscillator were at rest, the uncertainty principle would be violated. The position and the momentum would be known exactly.) When applied to a diatomic molecule, this model says that we cannot have a fixed, definite internuclear separation of the atoms. What we call the internuclear distance must be an average distance between the atoms.

Derivations of the concept of an absolute zero of temperature that are based upon the kinetic theory of gases tell us that at absolute zero all molecular motion ceases. Equation 4.38 tells us that this cannot be the case. If we have a crystalline substance at absolute zero, all modes of vibration for the crystal are still occurring at their zero-point frequency. One method of trying to obtain temperatures near

0°K involves finding crystalline substances that have a phase transition near 0°K. If the phase transition is endothermic (owing to differences in the zero-point energies for the two crystalline phases), the temperature is further lowered.

4.2 VIBRATIONAL SPECTRA OF DIATOMIC MOLECULES

The difference between two energy levels for the harmonic oscillator is

$$\Delta E = h\nu_0(n_2 - n_1) \tag{4.39}$$

That is, at the harmonic-oscillator level of approximation, vibrational spectral transitions (which occur in the infrared spectral region) can only be integer multiples of $h\nu_0$. Transitions with different values of $(n_2 - n_1)$ will obviously occur at different frequencies. This is why the fundamental frequency is designated as ν_0. Remembering our original definition of the force constant in terms of frequency (Eq. 4.5), we see that

$$\nu_0 = \frac{1}{2\pi}\sqrt{\frac{k}{\mu}} \tag{4.40}$$

where μ is the reduced mass (Eq. 4.1). For molecules, k is usually expressed in units of 10^2 N m^{-1} (10^5 dyne cm^{-1} or millidynes per angstrom), and μ is expressed in atomic mass units. Numerically evaluating the constants, we have

$$\nu_0 = 3.906 \times 10^{13}\left(\frac{k}{\mu}\right)^{1/2} \text{ sec}^{-1} \tag{4.41a}$$

$$\bar{\nu}_0 = 1302.8\left(\frac{k}{\mu}\right)^{1/2} \text{ cm}^{-1} \tag{4.41b}$$

$$h\nu_0 = 2.59 \times 10^{-13}\left(\frac{k}{\mu}\right)^{1/2} \text{ erg}$$

$$= 2.59 \times 10^{-20}\left(\frac{k}{\mu}\right)^{1/2} \text{ J} \tag{4.41c}$$

and

$$k = \left(\frac{\bar{\nu}_0}{1302.8}\right)^2 \mu \times 10^2 \text{ N m}^{-1} \tag{4.42}$$

Table 4.1 presents vibrational spectral data for some diatomic molecules.

In spectroscopy, the selection rules tell us that certain transitions are theoretically allowed while others are theoretically forbidden. The rules depend upon the particular type of experiment. For example, infrared spectroscopy is a direct absorption of electromagnetic radiation. The selection rule again depends upon the transition

Table 4.1

VIBRATIONAL SPECTRAL DATA FOR SOME DIATOMIC MOLECULES

Molecule and Isotopic Composition	$\bar{\nu}_0$ (cm^{-1})	k (10^2 N m^{-1})
$^1H_2$	4401.21	5.7510[a]
$^1H^2H$	3813.15	5.7543[a]
$^2H_2$	3115.50	5.7588[a]
$^1H^{35}Cl$	2990.95	5.1631
$^{12}C^{16}O$	2169.81	19.0185
$^{14}N_2$	2358.57	22.9478
$^{16}O_2$	1580.19	11.7658

[a] Note the small differences in these force constants. This is common for isotopic substitution.

dipole, as in Eq. 3.98:

$$\mu_{ij} = \int \psi_i^* \hat{\mu} \psi_j \, dv \qquad \textbf{3.98}$$

The selection rule states that there must be a nonvanishing component of the dipole (or a coordinate, since $\hat{\mu}$ equals er or, in the present notation, eq) connecting the two states of the transition. In contrast to microwave spectroscopy, the coordinates of the atoms (and, consequently, the dipole) change during the course of a vibration. Because we have already worked out the Heisenberg matrix for **Q**, we know which vibrational states have coordinate components connecting them, and we can immediately deduce the infrared selection rules for the harmonic oscillator. From Eq. 4.19 we see that **Q** has off-diagonal elements connecting only adjacent states. For a transition to be infrared-allowed we must have

$$n_2 = n_1 \pm 1 \qquad \textbf{4.43}$$

Transitions having values of Δn other than ± 1 are observed experimentally (but with low intensity), owing to anharmonicity in the vibrations of real molecules.

Vibrational spectroscopy is also studied by the *Raman* technique. Raman spectroscopy is an inelastic light-scattering experiment. The energy of the scattered light differs from that of the incident light by amounts corresponding to the energy of the vibrational excitation. The interaction depends upon the *polarizability*. The appropriate operator for determining the selection rule is the *quadrupole* operator, which contains the square of the coordinates. Equation 4.25 gives the Heisenberg matrix for $\mathbf{Q}^2$. This matrix has nonvanishing elements on the diagonal and at two positions from the diagonal. At first glance, it might appear that Δn would be ± 2; however, an examination of the matrix elements shows that they depend only on the nonvanishing elements of **Q**. Therefore, the Raman selection rule, in terms of Δn, in the harmonic-oscillator approximation is the same as the infrared selection rule. We will see later, however, that there are symmetry-induced selection rules that are different for infrared spectroscopy and Raman spectroscopy.

4.3 VIBRATIONS OF POLYATOMIC MOLECULES

Consider a collection of N unconnected atoms. Each atom will have three degrees of translational freedom lying along the local x, y, and z coordinates. There will be a total of $3N$ independent motions for the collection of atoms. Now assume that the atoms are rigidly attached together to form a three-dimensional structure. This structure, as a whole, will have three degrees of translational freedom. It will also have three degrees of rotational freedom, since it has structure (only two, if the structure is linear). A molecule falls in between these two extremes. The atoms are bonded together to give a three-dimensional molecule; however, the bonds are not completely rigid. The atoms can vibrate with respect to each other. There are again a total of $3N$ possible independent motions. Three of these motions will correspond to translations of the entire molecule. Three others (or two, if the molecule is linear) will correspond to rotations of the entire molecule. The remaining $3N - 6$ (or $3N - 5$ if linear) will correspond to the vibrations of the molecule. For example, a diatomic molecule has $(3 \times 2 - 5)$, or one, degree of vibrational freedom. A nonlinear triatomic molecule (such as H_2O) has $(3 \times 3 - 6)$, or three, degrees of vibrational freedom. A linear triatomic (such as CO_2) has four degrees of vibrational freedom, and so on. These cases are illustrated in Figure 4.2.

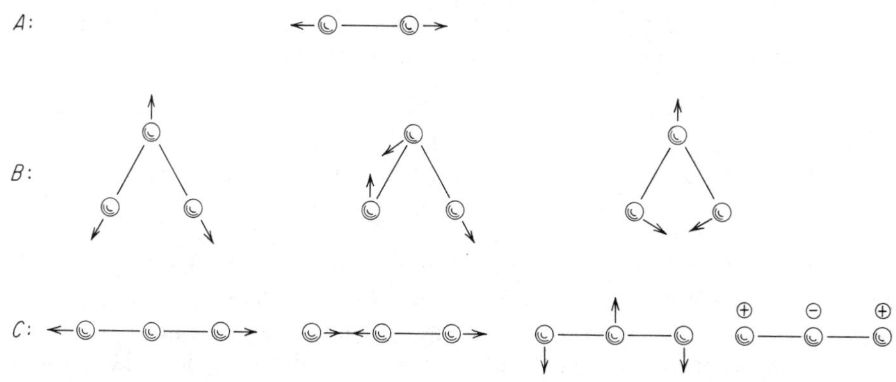

Figure 4.2 The independent vibrations of A, a diatomic molecule; B, a nonlinear triatomic molecule; and C, a linear triatomic molecule. The fourth vibration of the linear triatomic is perpendicular to the third one shown, with the atoms moving out of the plane of the drawing.

For simple molecules, such as those shown in Figure 4.2, the quantum-mechanical treatment is relatively simple. The diatomic molecule is treated as a one-dimensional harmonic oscillator. The bond-stretching vibrations of the triatomics can, to a good approximation, be treated as simple linear combinations of two-center harmonic oscillators, while the bending motions can be treated with a single harmonic potential corresponding to the bend. For example, in the symmetrical stretching vibration of a linear A—B—A molecule, the central atom is not displaced (see Figure 4.2c). The problem is that of two simple harmonic oscillators. The wave function can be written

$$\psi = \frac{1}{\sqrt{2}}(\psi_1 + \psi_2) \qquad \textbf{4.44}$$

where ψ_1 and ψ_2 are one-dimensional harmonic-oscillator functions. The Hamiltonian is the sum of two identical harmonic-oscillator functions. The energy is

$$E = \int \psi^* \hat{H} \psi \, dv \qquad \textbf{4.45a}$$

$$= \int \psi^*(\hat{H}_1 + \hat{H}_2)\psi \, dv \qquad \textbf{4.45b}$$

$$= \frac{1}{2}\left[\int \psi_1^* \hat{H}_1 \psi_1 \, dv + \int \psi_2^* \hat{H}_2 \psi \, dv\right] \qquad \textbf{4.45c}$$

$$= \left(n + \frac{1}{2}\right) h\nu_0 \qquad \textbf{4.45d}$$

$$= \left(n + \frac{1}{2}\right) \hbar \sqrt{\frac{k}{\mu}} \qquad \textbf{4.45e}$$

where μ is the AB reduced mass. (The crossed terms in Eq. 4.45c vanish because ψ_1 and ψ_2 are functions of different coordinates.)

The other functions are more complicated, because all atoms move. The potential functions depend on the coordinates of more than one atom. Consequently, the energy equation cannot be factored into sums of single-atom terms. We will develop general methods for handling polyatomic molecules in Chapter 16.

BIBLIOGRAPHY

GREEN, H. S., *Matrix Mechanics.* P. Noordhoff, Ltd., Groningen, Netherlands, 1965.

LAWDEN, D. F., *The Mathematical Principles of Quantum Mechanics.* Methuen & Co., London, 1967.

TROUP, G., *Understanding Quantum Mechanics.* Methuen & Co., London, 1968.

PROBLEMS

*4.1 Calculate the force constants and zero-point energies for the following molecules:

Molecule	$\bar{\nu}_0$ (cm^{-1})
(a) H_2	4401.21
(b) $^2H_2$ (deuterium)	3115.50
(c) $^3H_2$ (tritium)	2546.47
(d) Li_2	351.43
(e) Na_2	159.12
(f) K_2	92.02
(g) F_2	916.64
(h) O_2	1580.19
(i) N_2	2358.57
(j) $^1H^{19}F$	4138.32
(k) $^1H^{35}Cl$	2990.95
(l) $^1H^{80}Br$	2648.98
(m) $^1H^{177}I$	2309.01

*4.2 Examine the force constants for the above four series of molecules, (a)–(c), (d)–(f), (g)–(i), and (j)–(m). What is the apparent relation between bond strength and force constant? Is there a relation between vibrational frequency (or wave number) and bond strength?

*4.3 Calculate the relative Boltzmann populations of the $v = 0$ and $v = 1$ vibrational levels for H_2, N_2, F_2, Na_2, and K_2.

*4.4 Using commutation relations, find the expression for $\dot{\mathbf{P}}$ if the Hamiltonian has the following forms:

(a) $\hat{H} = p^2/2m + 1/q$.

(b) $\hat{H} = p^2/2m + q$.

*4.5 Calculate the energy levels for the cubic three-dimensional harmonic oscillator.

*4.6 Calculate the energy levels for the isotropic (spherically symmetrical) three-dimensional harmonic oscillator with $V(r) = (k/2)r^2$.

*4.7 In spectroscopy experiments, electronic transitions normally occur at higher energies than vibrational transitions, which occur at higher energies than rotational transitions. Spectral transitions involve transitions between states in which all three (electronic, vibrational, and rotational) degrees of excitation must be specified. In a transition any of the quantities may change. The overall selection rules involve the combination of the selection rules for each. Deduce the overall selection rule for a transition in the infrared spectral region for a diatomic molecule.

chapter five

The hydrogen atom

5.1 INTRODUCTION

We come now to the quantum-mechanical solution of the hydrogen-atom problem. This exactly solvable, closed-form problem can be handled in either the Heisenberg or the Schrödinger representation. We will outline the Schrödinger treatment here. We have a single electron and a nucleus. The electron has a charge of $-e$. For generality, we will let the nucleus have a charge of $+Ze$, where Z is the atomic number. The potential energy is a function only of the distance of separation of the electron and the nucleus:

$$V(r) = -\frac{Ze^2}{r} \qquad \textbf{5.1}$$

Since the potential has spherical symmetry (i.e., there is no angular dependence), we will find it useful to work again in spherical polar coordinates as we did for the rigid rotor. The Hamiltonian operator for the hydrogen atom is

$$\hat{H} = -\frac{\hbar^2}{2\mu}\nabla^2 - \frac{Ze^2}{r} \qquad \textbf{5.2}$$

Note that we have used the reduced mass, μ, of the electron and the nucleus, rather than the mass of the electron, since the nucleus moves slightly, relative to the center of mass, as the electron moves. Using the electron mass would introduce an error of only about .05 percent in the energy.

5.2 SEPARATION OF VARIABLES

The Schrödinger equation to be solved is

$$\left(-\frac{\hbar^2}{2\mu}\nabla^2 - \frac{Ze^2}{r}\right)\psi(r, \theta, \phi) = E\psi(r, \theta, \phi) \tag{5.3}$$

In Chapter 3 we constructed the Laplacian operator in spherical polar coordinates (Eq. 3.13). Substituting this into Eq. 5.3 and rearranging yields

$$\left[\frac{1}{r^2}\frac{\partial}{\partial r}\left(r^2\frac{\partial\psi}{\partial r}\right) + \frac{1}{r^2\sin\theta}\frac{\partial}{\partial\theta}\left(\sin\theta\frac{\partial\psi}{\partial\theta}\right) + \frac{1}{r^2\sin^2\theta}\frac{\partial^2\psi}{\partial\phi^2}\right] + \frac{2\mu}{\hbar^2}[E - V(r)]\psi = 0 \tag{5.4}$$

where $V(r)$ is $-Ze^2/r$. We were able to solve the rigid-rotor problem by separating variables. We will use the same approach here. Let

$$\psi(r, \theta, \phi) = R(r)T(\theta)F(\phi) \tag{5.5}$$

Substituting into Eq. 5.4, we obtain

$$\frac{1}{r^2}\frac{\partial}{\partial r}\left(r^2\frac{\partial R}{\partial r}\right)TF + \frac{1}{r^2\sin\theta}\frac{\partial}{\partial\theta}\left(\sin\theta\frac{\partial T}{\partial\theta}\right)RF + \frac{1}{r^2\sin^2\theta}\frac{\partial^2 F}{\partial\phi^2}RT + \frac{2\mu}{\hbar^2}[E - V(r)]RTF = 0 \tag{5.6}$$

Now divide through by RTF to get

$$\frac{1}{R}\left[\frac{1}{r^2}\frac{\partial}{\partial r}\left(r^2\frac{\partial R}{\partial r}\right)\right] + \frac{1}{T}\left[\frac{1}{r^2\sin\theta}\frac{\partial}{\partial\theta}\left(\sin\theta\frac{\partial T}{\partial\theta}\right)\right] + \frac{1}{F}\left(\frac{1}{r^2\sin^2\theta}\frac{\partial^2 F}{\partial\phi^2}\right) + \frac{2\mu}{\hbar^2}[E - V(r)] = 0 \tag{5.7}$$

and multiply by $r^2\sin^2\theta$ to give

$$\frac{\sin^2\theta}{R}\frac{\partial}{\partial r}\left(r^2\frac{\partial R}{\partial r}\right) + \frac{\sin\theta}{T}\frac{\partial}{\partial\theta}\left(\sin\theta\frac{\partial T}{\partial\theta}\right) + \frac{1}{F}\frac{\partial^2 F}{\partial\phi^2} + \frac{2\mu r^2\sin^2\theta}{\hbar^2}[E - V(r)] = 0 \tag{5.8}$$

We can separate the ϕ-dependent term from the others, which are not ϕ-dependent:

$$\frac{-1}{F}\frac{\partial^2 F}{\partial\phi^2} = \frac{\sin^2\theta}{R}\frac{\partial}{\partial r}\left(r^2\frac{\partial R}{\partial r}\right) + \frac{\sin\theta}{T}\frac{\partial}{\partial\theta}\left(\sin\theta\frac{\partial T}{\partial\theta}\right) + \frac{2\mu r^2\sin^2\theta}{\hbar^2}[E - V(r)] \tag{5.9}$$

As in the rigid-rotor problem, this equality can hold for all values of all variables if and only if both sides of the equation equal the same constant. This time we will use the lower-case letter and call the constant m^2. We have

$$\frac{1}{F}\frac{\partial^2 F}{\partial\phi^2} = -m^2 \tag{5.10}$$

Substituting this into Eq. 5.7, we get

$$\frac{1}{R}\left[\frac{1}{r^2}\frac{\partial}{\partial r}\left(r^2\frac{\partial R}{\partial r}\right)\right] + \frac{1}{T}\left[\frac{1}{r^2 \sin\theta}\frac{\partial}{\partial\theta}\left(\sin\theta\frac{\partial T}{\partial\theta}\right)\right] - \frac{m^2}{r^2\sin^2\theta} + \frac{2\mu}{\hbar^2}[E - V(r)] = 0 \tag{5.11}$$

If we multiply by r^2 and rearrange, we have

$$\frac{1}{R}\left[\frac{\partial}{\partial r}\left(r^2\frac{\partial R}{\partial r}\right)\right] + \frac{2\mu r^2}{\hbar^2}[E - V(r)] = \frac{m^2}{\sin^2\theta} - \frac{1}{T}\left[\frac{1}{\sin\theta}\frac{\partial}{\partial\theta}\left(\sin\theta\frac{\partial T}{\partial\theta}\right)\right] \tag{5.12}$$

Now the left-hand side of Eq. 5.12 depends only on r and is independent of θ, while the converse is true of the right-hand side. Again, this can hold for all values of the variables only if both sides equal the same constant. Let us call the constant β. We have now completely separated our problem into three equations, each depending upon only one variable.

5.3 THE θ AND ϕ EQUATIONS; SPHERICAL HARMONICS

The three equations to be solved can be written

$$\frac{d^2F(\phi)}{d\phi^2} = -m^2F(\phi) \tag{5.13}$$

$$\frac{1}{\sin\theta}\left\{\frac{d}{d\theta}\left[\sin\theta\frac{dT(\theta)}{d\theta}\right]\right\} - \frac{m^2}{\sin^2\theta}T(\theta) + \beta T(\theta) = 0 \tag{5.14}$$

$$\frac{1}{r^2}\frac{d}{dr}\left[r^2\frac{dR(r)}{dr}\right] + \frac{2\mu}{\hbar^2}[E - V(r)]R(r) - \frac{\beta}{r^2}R(r) = 0 \tag{5.15}$$

We have already seen Eqs. 5.13 and 5.14 in the rigid-rotor problem (as Eqs. 3.22 and 3.29). The solutions are exactly the same as before, except that now we will call

$$\beta = l(l + 1) \tag{5.16}$$

rather than $J(J + 1)$ as in the rigid-rotor problem; i.e., we have new names, m and l, for the quantum numbers relating to the angular variables. The quantum number l is called the *orbital angular-momentum quantum number* or just the *angular-momentum quantum number*. The quantum number m is the *magnetic quantum number*. The *angular-momentum* quantum numbers are frequently given letter symbols (s, p, d, and so on). These familiar designations were originally derived from atomic spectroscopy.

The product of the angular functions,

$$T_{lm}(\theta)F_m(\phi) = Y_{lm}(\theta, \phi) \tag{5.17}$$

again defines a spherical harmonic. The total wave function can be considered to be a product of an r-dependent (radial) function, $R(r)$, and a spherical harmonic

$$\psi(r, \theta, \phi) = R(r)Y_{lm}(\theta, \phi) \tag{5.18}$$

The spherical harmonics are more important when we discuss the symmetry of atomic systems than are the individual $T(\theta)$ and $F(\phi)$.

5.4 THE r EQUATION; THE ENERGY

Equation 5.15 presents us with a new problem, and the solution again involves a series expansion. Rather than go through the details this time, we will outline the procedure. The reader interested in the details is referred to the works listed in the bibliography.

If we substitute $l(l + 1)$ for β and the functional form for $V(r)$ into Eq. 5.15, we obtain

$$\frac{1}{r^2}\left(\frac{d}{dr}r^2\frac{dR}{dr}\right) + \left[-\frac{l(l+1)}{r^2} + \frac{2\mu}{\hbar^2}\left(E + \frac{Ze^2}{r}\right)\right]R = 0 \tag{5.19}$$

Now, to avoid carrying constants and to simplify the form of the equation, let

$$\alpha^2 = -\frac{2\mu E}{\hbar^2} \tag{5.20}$$

$$\lambda = \frac{\mu Z e^2}{\hbar^2 \alpha} \tag{5.21}$$

$$\rho = 2\alpha r \tag{5.22}$$

and

$$S(\rho) = R(r) \tag{5.23}$$

Equation 5.19 becomes

$$\frac{1}{\rho^2}\frac{d}{d\rho}\left[\rho^2\frac{dS(\rho)}{d\rho}\right] + \left[-\frac{l(l+1)}{\rho^2} - \frac{1}{4} + \frac{\lambda}{\rho}\right]S(\rho) = 0 \tag{5.24}$$

The variable r, and consequently ρ, can have any value from zero to infinity. Let us consider Eq. 5.24 as ρ becomes very large. First, carrying out the differentiation, we have

$$\frac{2}{\rho}\frac{dS}{d\rho} + \frac{d^2S}{d\rho^2} + \left[-\frac{l(l+1)}{\rho^2} - \frac{1}{4} + \frac{\lambda}{\rho}\right]S = 0 \tag{5.25}$$

For sufficiently large ρ, all terms with ρ in the denominator become negligibly small, giving

$$\frac{d^2S}{d\rho^2} \cong \frac{1}{4}S \qquad \text{(large } r\text{)} \tag{5.26}$$

The solutions are

$$S(\rho) \sim e^{\pm\rho/2} \qquad \text{(large } r\text{)} \tag{5.27}$$

The positive exponent would go infinite as ρ approached infinity. Since the wave function must remain finite over all space, this is not acceptable; consequently the negative exponent must be chosen. The asymptotic behavior of $S(\rho)$ for large ρ is thus $e^{-\rho/2}$. Over the full range of ρ, we can assume that $S(\rho)$ has the form

$$S(\rho) = P(\rho)e^{-\rho/2} \tag{5.28}$$

The next step is to expand $P(\rho)$ as a series expansion in ρ. Without going into details, this leads to a recursion relation of the form (Pauling and Wilson, Sec. 18d)

$$a_{\nu+1} = \frac{-(\lambda - l - 1 - \nu)}{2(\nu + 1)(l + 1) + \nu(\nu + 1)} a_\nu \tag{5.29}$$

For large ν, the coefficients begin to behave as the coefficients in the series expansion for e^ρ. For large ρ, the product

$$e^\rho e^{-\rho/2} = e^{\rho/2} \tag{5.30}$$

goes infinite. Consequently, the series generated by the recursion formula 5.29 must be terminated after a finite number of terms to keep the wave function finite. Again, as in the previous series expansion, we can accomplish this by requiring that the numerator in the recursion formula be equal to zero for some value of ν. Termination of Eq. 5.29 at the νth term requires that

$$\lambda = l + 1 + \nu \tag{5.31}$$

Since l and ν are both integers, λ will be an integer, which we will call n. Solving for l in terms of n, we find that

$$0 \leq l \leq n - 1 \tag{5.32}$$

There are no limits on n, other than that it be greater than zero. We call n the *principal quantum number.*

If we equate n to λ as defined by Eq. 5.21, we find

$$n = \frac{\mu Z e^2}{\hbar^2 \alpha} \tag{5.33}$$

or, solving for α,

$$\alpha = \frac{\mu Z e^2}{\hbar^2 n} \tag{5.34}$$

But α^2 was defined in terms of energy by Eq. 5.20, so we have

$$\alpha^2 = \frac{\mu^2 Z^2 e^4}{\hbar^4 n^2} = -\frac{2\mu E}{\hbar^2} \tag{5.35}$$

Solving for energy, we obtain

$$E = -\frac{\mu Z^2 e^4}{2n^2\hbar^2} \tag{5.36a}$$

$$= -\frac{2\pi^2 \mu Z^2 e^4}{n^2 h^2} \tag{5.36b}$$

Equation 5.36b is exactly the equation that we obtained from Bohr theory (except that this time we have used the reduced mass instead of the electron mass). It may seem unrewarding to use a much more complicated derivation to arrive at the same answer; however, as we shall see later, modern quantum theory handles

Table 5.1

THE RADIAL PORTION OF THE HYDROGENIC WAVE FUNCTION[a]

n	l	
1	0	$R_{10}(r) = 2(Z/a_0)^{3/2}e^{-\rho/2}$
2	0	$R_{20}(r) = \frac{(Z/a_0)^{3/2}}{2\sqrt{2}}(2 - \rho)e^{-\rho/2}$
2	1	$R_{21}(r) = \frac{(Z/a_0)^{3/2}}{2\sqrt{6}}\rho e^{-\rho/2}$
3	0	$R_{30}(r) = \frac{(Z/a_0)^{3/2}}{9\sqrt{3}}(6 - 6\rho + \rho^2)e^{-\rho/2}$
3	1	$R_{31}(r) = \frac{(Z/a_0)^{3/2}}{9\sqrt{6}}(4 - \rho)\rho e^{-\rho/2}$
3	2	$R_{32}(r) = \frac{(Z/a_0)^{3/2}}{9\sqrt{30}}\rho^2 e^{-\rho/2}$
4	0	$R_{40}(r) = \frac{(Z/a_0)^{3/2}}{96}(24 - 36\rho + 12\rho^2 - \rho^3)e^{-\rho/2}$
4	1	$R_{41}(r) = \frac{(Z/a_0)^{3/2}}{32\sqrt{15}}(20 - 10\rho + \rho^2)\rho e^{-\rho/2}$
4	2	$R_{42}(r) = \frac{(Z/a_0)^{3/2}}{96\sqrt{5}}(6 - \rho)\rho^2 e^{-\rho/2}$
4	3	$R_{43}(r) = \frac{(Z/a_0)^{3/2}}{96\sqrt{35}}\rho^3 e^{-\rho/2}$
		and so on.

[a] a_0 is the Bohr radius and ρ is $2\mu Ze^2 r/n\hbar^2$ (see Eqs. 5.22 and 5.34).

many-electron atoms and chemical bonding in a satisfactory manner, from both a qualitative and a quantitative point of view, while Bohr theory failed on both counts.

The polynomial, $P(\rho)$, introduced in Eq. 5.28 is known as the *associated Laguerre polynomial* and is commonly given the symbol $L_{n+l}^{2l+1}(\rho)$. The radial portion of the hydrogenic wave function thus has the form

$$R_{nl}(r) = \left\{\frac{(n-l-1)!}{2n[(n+l)!]^3}\left(\frac{2Z}{na_0}\right)^3\right\}^{1/2} e^{-\rho/2} L_{n+l}^{2l+1}(\rho) \qquad 5.37$$

where the bracketed term is the normalizing factor (a_0 is the Bohr radius defined in Eq. 1.27). Notice that this function depends on the quantum numbers n and l, even though the energy depends only upon n. The first few radial functions are tabulated in Table 5.1. Several of these are plotted in Figure 5.1. Here we have plotted $R_{nl}(r)$,

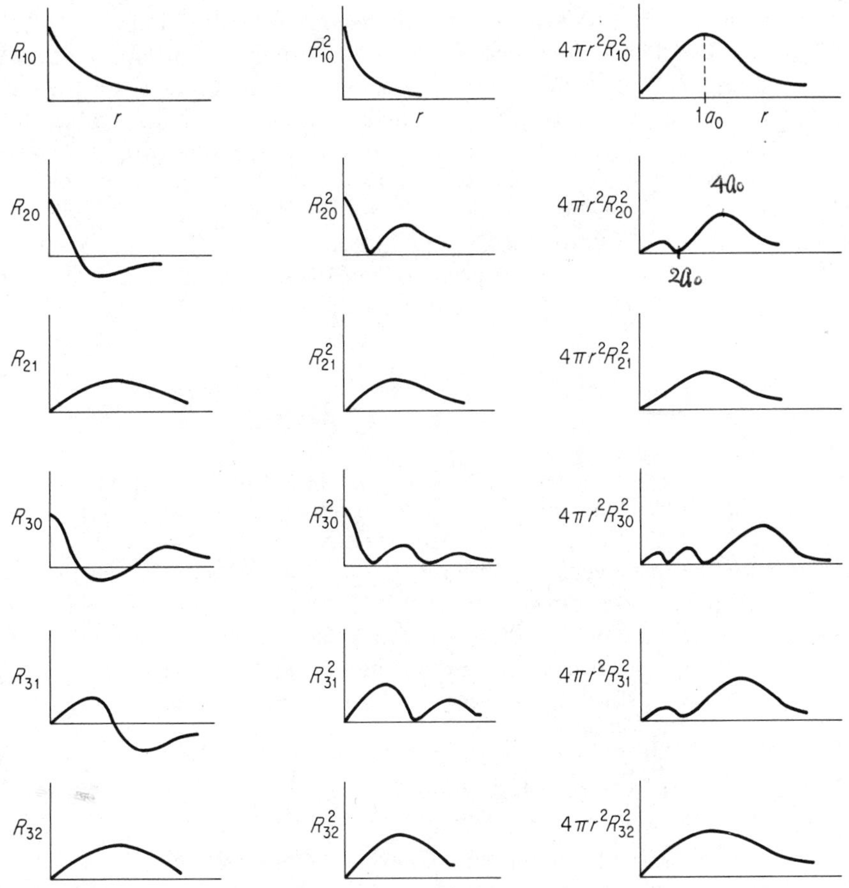

Figure 5.1 The radial behavior of the hydrogenic wave function, $R_{nl}(r)$; the electron density, $R_{nl}^2(r)$; and the radial distribution function, $4\pi r^2 R_{nl}^2(r)$.

$R_{nl}^2(r)$, and $4\pi r^2 R_{nl}^2(r)$. The first two of these have their previously defined meanings. The third gives the probability that the electron will be found in a spherical shell of thickness dr at a radius of r ($4\pi r^2$ is the surface area of a sphere of radius r). $R_{nl}^2(r)$ is an *electron density function*, while $4\pi r^2 R_{nl}^2(r)$ is the *radial distribution function*: it gives the overall probability of finding the electron at r (i.e., between r and $r + dr$).

Two interesting observations can be made about Table 5.1 and Figure 5.1. First, there are one or more nodes (places where the wave function goes to zero) in each radial function, except for those corresponding to the maximum l for a given n. This situation also shows up in the electron density and radial distribution functions. There is zero probability of finding the electron at these positions. Another observation is that, for a Z value of 1 and an n of 1, the maximum in the radial distribution function comes at one Bohr radius. In other words, although we are treating our electron by a wave function and consequently cannot localize it in an orbit, the most probable r value for the electron in the lowest energy state of the hydrogen atom occurs at the value of the radius of the corresponding Bohr orbit.

The nodal properties of a wave function have an important qualitative significance. For a given type of function, the more nodes there are, the higher the energy. For example, comparing the s functions, the one with an n value of 1 has no nodes. The one with $n = 2$ has one. That with $n = 3$ has two, and so forth. Both the number of nodes and the energies increase with increasing n. In the hydrogen atom, all l values for a given n value yield orbitals of the same energy. The $n = 2$, $l = 1$ function has no radial nodes, but all p functions have one angular node (see Figure 3.2). The total number of nodes is thus the same for a $2s$ function and a $2p$ function. The same is true for the $n = 3$ functions and for all other hydrogenic levels.

5.5 THE TOTAL WAVE FUNCTION

The total wave function for the hydrogen atom is the product $R_{nl}(r)T_{lm}(\theta)F_m(\phi)$ (Eq. 5.18). These one-electron functions are frequently called *orbitals*. (The term orbital is commonly used for any one-electron wave function. In the hydrogen-atom case, the one-electron function is the total wave function. This is obviously not the case for many-electron systems.) We have tabulated some of the $R_{nl}(r)$ in Table 5.1. The $T_{lm}(\theta)$ are the functions given in Table 3.1, except that we are now using l and m for the quantum numbers, rather than J and M. $F_m(\phi)$ is just $1/\sqrt{2\pi}e^{im\phi}$. We can write the total wave functions directly. For example, for an n value of 2 and l and m values of 1, we have

$$\psi_{211}(r, \theta, \phi) = \frac{1}{4\sqrt{2\pi}}\left(\frac{Z}{a_0}\right)^{3/2}\frac{Z}{a_0} r \sin\theta e^{-(Z/2a_0)r}e^{i\phi} \qquad 5.38$$

and for the same n and l, but an m value of -1, we have

$$\psi_{21\bar{1}}(r, \theta, \phi) = \frac{1}{4\sqrt{2\pi}}\left(\frac{Z}{a_0}\right)^{3/2}\frac{Z}{a_0} r \sin\theta e^{-(Z/2a_0)r}e^{-i\phi} \qquad 5.39$$

These are the correct eigenfunctions for the appropriate values of n, l, and m. However, they are complex functions, since they contain i.

For qualitative discussions of bonding, it is frequently convenient to express the wave functions in real form. We can do this as we did for the spherical harmonics in Chapter 3, by taking linear combinations of the degenerate $+m$ and $-m$ functions, remembering the identities

$$e^{im\phi} = \cos m\phi + i \sin m\phi \tag{5.40a}$$

and

$$e^{-im\phi} = \cos m\phi - i \sin m\phi \tag{5.40b}$$

Thus, the combinations

$$\psi_{211}^{+} = \frac{1}{2}(\psi_{211} + \psi_{21\bar{1}}) \tag{5.41a}$$

and

$$\psi_{211}^{-} = \frac{i}{2}(\psi_{21\bar{1}} - \psi_{211}) \tag{5.41b}$$

are both real. We have

$$\psi_{211}^{+} = \frac{1}{4\sqrt{2\pi}}\left(\frac{Z}{a_0}\right)^{3/2}\frac{Z}{a_0}re^{-(Z/2a_0)r}\sin\theta\cos\phi \tag{5.42a}$$

and

$$\psi_{211}^{-} = \frac{1}{4\sqrt{2\pi}}\left(\frac{Z}{a_0}\right)^{3/2}\frac{Z}{a_0}re^{-(Z/2a_0)r}\sin\theta\sin\phi \tag{5.42b}$$

From the angular dependence of these (see Eqs. 3.2), we see that ψ_{211}^{+} lies along the x axis and ψ_{211}^{-} along the y axis of a Cartesian coordinate system. However, m is *not* a valid quantum number for these (although $|m|$ is), since each is a combination of the $m = +1$ and the $m = -1$ functions! On the other hand, the $n = 2, l = 1, m = 0$ function

$$\psi_{210} = \frac{1}{4\sqrt{2\pi}}\left(\frac{Z}{a_0}\right)^{3/2}\frac{Z}{a_0}re^{-(Z/2a_0)r}\cos\theta \tag{5.43}$$

is real. The quantum number m (zero, in this case) is a valid quantum number. The function is directed along the z axis of the Cartesian system. In the plots of the angular dependence of wave functions that we usually see, the real forms of the functions are plotted. Only those corresponding to an m value of zero (directed along the z axis, except for an l value of zero) have m as a valid quantum number. The point may appear trivial, since the energy of the hydrogen atom is independent of l or m. As we will see in Section 8.4, however, the energy levels of an atom in a magnetic field or in an electric field depend upon m. The real forms of the rigid-rotor angular functions for several values of J and M were given in Table 3.3 and were plotted in Figure 3.2. These are identical to the real forms of the angular dependence for the hydrogen functions (except with l and m substituted for J and M). In fact, these are the functions describing the three-dimensional angular behavior of any single-particle angular momentum.

In the hydrogen atom, the same angular dependences of the orbitals recur for each principal quantum number. Furthermore, the angular dependences of the orbitals of polyelectron atoms are the same as those of the hydrogen-atom orbitals. Certain spectral properties of atoms depend strongly on the angular dependence of the orbitals from which and to which the electrons are promoted. For example, transitions terminating in orbitals with an l value of zero are characteristically very sharp. Transitions from an l value of zero to an l value of one are characteristically very intense and in fact are the principal lines seen in hydrogen and certain other atoms. For this reason a nomenclature has arisen in which orbitals with an l value of zero are referred to as s (for sharp) orbitals, those with $l = 1$ as p (principal) orbitals, those with $l = 2$ as d (diffuse) orbitals, and those with $l = 3$ as f (fine) orbitals. Orbitals with higher angular momenta are referred to, sequentially, as g, h, i, and so on. In this nomenclature the principal quantum number is listed as a numeral. Thus, the ground state of the hydrogen atom has its electron in a $1s$ orbital.

The hydrogenic energies and wave functions we have obtained are valid for any centrosymmetric, one-electron system, such as the one-electron ions He^+, Li^{2+}, and so on, and the positronium system (a positron and an electron); only the charge and the reduced mass have to be changed to their correct values. They also form the starting point for discussions of many-electron atoms.

BIBLIOGRAPHY

KARPLUS, M., and PORTER, R. N., *Atoms and Molecules*. W. A. Benjamin, Inc., New York, 1970.

PAULING, L., and WILSON, E. B., JR., *Introduction to Quantum Mechanics*. McGraw-Hill Book Company, New York, 1935.

PILAR, F. L., *Elementary Quantum Chemistry*. McGraw-Hill Book Company, New York, 1968.

SLATER, J. C., *Quantum Theory of Atomic Structure*, Vol. 1. McGraw-Hill Book Company, New York, 1960.

PROBLEMS

***5.1** Find the r value for the maximum probability of electron density for the following hydrogenic orbitals as a function of Z (don't forget the $4\pi r^2$ spherical volume element):

(a) $1s$. (b) $2s$. (c) $2p_0$. (d) $2p_{\pm 1}$.

(e) $3d_0$. (f) $3d_{\pm 1}$. (g) $3d_{\pm 2}$.

5.2 What apparent relation is there between the most probable r value and m, for a given n and l?

5.3 Do a contour plot of the electron density of the following hydrogen orbitals in the xz, xy, and yz planes:

Orbital	n	l	m
(a)	2	1	0
(b)	3	1	0
(c)	3	2	0
(d)	4	3	0

***5.4** The first ionization potential for He is .90355 a.u. Assume a model for He in which there are two completely independent electrons in hydrogenic $1s$ orbitals (having an energy $E = -\frac{1}{2}\zeta^2$).

(a) Calculate an effective ζ from the ionization potential.

(b) Calculate the total energy using this effective ζ.

(c) Find the r value for the maximum electron density from this model.

***5.5** Deduce, from group theory, the Δl selection rule for the normal absorption spectrum of the hydrogen atom.

***5.6** In several quantum-mechanical problems in the Schrödinger treatment, series expansions are involved (the hydrogen atom, the harmonic oscillator, the rigid rotor). Explain where the quantization condition comes from in these cases.

5.7 Calculate the ground-state energy of the U^{+91} ion (a uranium atom with only one electron). The virial theorem, which states that $E = -T = \frac{1}{2}V$, is satisfied for quantum-mechanical systems. What is the kinetic energy of an electron in the $1s$ orbital of U^{+91}? What is the average (root-mean-square) velocity of this electron? Relativistic effects can be ignored for particles whose velocity is small compared to that of light. What about the relativistic effects for an electron in the $1s$ orbital of uranium?

chapter six

Approximation methods in quantum chemistry

6.1 INTRODUCTION

The hydrogenic system (the hydrogen atom or any one-electron ion) is the only chemical system for which an exact, closed-form, quantum-mechanical solution is known. Problems relating to many-electron atoms and to molecules must be solved with other methods. The most obvious method is to directly solve the Schrödinger equation by numerical techniques. Many workers have devoted much time and effort to this approach. However, the problem is very complicated. Although results have been obtained for relatively simple systems, using electronic computers, most work on chemically interesting systems employs approximations. The most common methods employed in quantum chemistry are based upon either the *variation principle* or *perturbation theory*.

The *variation principle* employs an approximate wave function that contains some parameters that can be arbitrarily varied. The energy is constructed as a function of these parameters. The parameters are then varied, using the calculus of variations, to minimize the energy. It can be shown that the energy constructed from an exact Hamiltonian and an arbitrary wave function is always greater than or equal to the true energy represented by that Hamiltonian. Thus, the minimization procedure yields the best energy that can be produced from a given form of trial function. If a new trial function can be found that gives a lower energy, it will be a closer approximation to the true energy for that Hamiltonian. In principle, and often in practice, the Hamiltonian can be the exact Hamiltonian for the system, although an approximate Hamiltonian is frequently employed. With an approximate Hamiltonian the true energy is not necessarily a lower bound to the energy from that Hamiltonian.

Perturbation theory starts with an approximate Hamiltonian for the system to which exact solutions may be found. The solution to the desired Hamiltonian is constructed as a linear combination of the exact solutions for the approximate Hamiltonian. The technique involves treating the difference between the desired Hamiltonian and the model Hamiltonian as a *perturbation* on the system. This leads to expressions for the energy and the desired functions in terms of integrals involving the perturbation operator and the unperturbed functions.

In this chapter we will study both the variation principle and perturbation theory. Most of our applications in later chapters will employ one or both of these.

6.2 THE VARIATION PRINCIPLE

Given a Schrödinger equation describing some system

$$\hat{H}\psi_i = E_i\psi_i \tag{6.1}$$

where i is a label indicating the state of the system, there exists a complete set of energy eigenvalues (called an *eigenvalue spectrum*) and the corresponding set of eigenfunctions. We probably cannot calculate or even count these solutions. (For example, there are an infinite number of bound states for the hydrogen atom, since $1 \leq n \leq \infty$, as well as the continuum.) However, if the Schrödinger equation is to have any meaning, they must exist.

The ψ_i that are solutions to Eq. 6.1 are linearly independent. That is,

$$\int \psi_i^* \psi_j \, dv = \delta_{ij} \tag{6.2}$$

where i and j indicate the states, the integration is over all space, and δ_{ij} is the *Kronecker delta function*, a function that equals unity when the indices i and j are the same, and zero when they are different. The functions form a complete *orthonormal* (orthogonal and normalized) function space. (A complete function space is called a *Hilbert space*.) Any other arbitrary function, say u, in the same space can be constructed as a linear combination of these:

$$u = \sum_i a_i \psi_i \tag{6.3}$$

Normalization of u requires that

$$\sum_i a_i^* a_i = 1 \tag{6.4}$$

Consider now the *expectation value* of our Hamiltonian (call it $\langle H \rangle$) with respect to the function u:

$$\begin{aligned} \langle H \rangle &= \int u^* \hat{H} u \, dv \\ &= \sum_i \sum_j \int a_i^* \psi_i^* \hat{H} a_j \psi_j \, dv \\ &= \sum_i \sum_j a_i^* a_j \int \psi_i^* \hat{H} \psi_j \, dv \end{aligned} \tag{6.5}$$

But, because of the Schrödinger relation of Eq. 6.1, $\hat{H}\psi_j$ equals $E_j\psi_j$, giving

$$\begin{aligned}\langle H \rangle &= \sum_i \sum_j a_i^* a_j E_j \int \psi_i^* \psi_j \, dv \\ &= \sum_i \sum_j a_i^* a_j E_j \delta_{ij} \\ &= \sum_i a_i^* a_i E_i \end{aligned} \qquad \mathbf{6.6}$$

Now subtract the lowest (ground-state) energy, E_0, from both sides of Eq. 6.6. This gives

$$\langle H \rangle - E_0 = \sum_i a_i^* a_i (E_i - E_0) \qquad \mathbf{6.7}$$

Each $a_i^* a_i$ is nonnegative, as is $(E_i - E_0)$. Thus,

$$\langle H \rangle - E_0 \geq 0 \qquad \mathbf{6.8}$$

This very important result tells us that the expectation value of a given Hamiltonian with respect to any arbitrary normalized function, $\int u^* \hat{H} u \, dv$, always lies above the true energy of the ground state of the system. (If u is not normalized, the expectation value must be divided by a normalizing factor that equals $\int u^* u \, dv$.) This allows us a way to devise approximate wave functions. If an approximate wave function is constructed with respect to some variational parameters, say λ_k, then the set of variational equations,

$$\frac{\partial \langle H(\lambda_k) \rangle}{\partial \lambda_k} = 0 \qquad \mathbf{6.9}$$

combined with the requirement that the functions remain normalized, gives the best-energy values for the parameters. The $\langle H \rangle$ corresponding to the optimum parameters is the lowest energy that can be obtained from that type of a trial wave function. Furthermore, by Eq. 6.8, it lies above the true ground-state energy of the system unless u is the true wave function. Thus, anything that can be done to the ground-state trial function to cause it to yield a lower energy should be an improvement (at least as far as the energy is concerned). Once the variational wave function has been obtained, it can be used to calculate properties other than the energy as the expectation values of the appropriate operators.

The variation principle is probably the most useful approximation technique in computational quantum chemistry. It suffers from one severe limitation, however. In the general case, it will give only the lowest energy state, of a given spin and symmetry, of the system under consideration. Also, the optimum wave function for energy is not necessarily the best for other properties.

6.3 VARIATIONAL TREATMENT OF THE HELIUM ATOM

To illustrate the use of the variation principle, we will apply it to the ground state of the helium atom. In order to avoid carrying the large number of units we had

A.U

in the hydrogen-atom problem, let us define a new set of units for quantum-chemical calculations. Here, the unit of mass will be the electron rest mass, m_e; that of charge, the charge of the electron, e; that of length, the Bohr radius, a_0; and that of angular momentum, Planck's constant divided by 2π, $\hbar$. With these units, called atomic units, the energy unit is the Hartree energy unit—the potential energy of the ground state of the hydrogen atom (4.3598×10^{-18} J or 27.211652 eV). Using these units, the quantum-mechanical kinetic-energy operator for an electron becomes $-\frac{1}{2}\nabla^2$, while the nuclear attraction becomes $-Z/r$. (Note that these units imply the use of the electron mass, rather than the reduced mass, for the kinetic energy. For very accurate calculations, a correction must be made for this.)

The Hamiltonian for the helium atom contains the kinetic energy for each electron, $-\frac{1}{2}\nabla_i^2$, the nuclear attraction potential for each electron, $-Z/r_i$, and the interelectron repulsion potential, $1/r_{12}$:

$$\hat{H} = -\frac{1}{2}\nabla_1^2 - \frac{Z}{r_1} - \frac{1}{2}\nabla_2^2 - \frac{Z}{r_2} + \frac{1}{r_{12}} \qquad \textbf{6.10}$$

If it were not for the interelectron repulsion term,the variables for the two electrons could be separated, and the problem could be solved in closed form. The Hamiltonian would, in fact, be the sum of two hydrogenic Hamiltonians. The wave function would be the product of two hydrogenic wave functions. This suggests that a suitable trial function might be derived from a product of two hydrogenic functions, but with some variational parameters added. Constructing a many-particle wave function as a product of one-particle functions amounts to using an *independent-particle approximation* for the wave function. The independent-particle wave function for helium has the form

$$\psi(1, 2) = \chi(1)\chi(2) \qquad \textbf{6.11}$$

where the numbers simply tell which electron's motion is being described by the function. The Hamiltonian that we will use, however, is the exact Hamiltonian of Eq. 6.10 (exact, that is, with regard to the effects we are including—relativity and other small effects are not being considered).

We are interested in the ground state of the helium atom, so let us give our trial functions the form of hydrogenic 1*s* orbitals. In order to have a variational parameter, we will replace the nuclear charge in the function with an effective nuclear charge, ζ, which will be our variational parameter. Using $F_0(\phi)$ (Eq. 3.26), $T_{00}(\theta)$ (Table 3.1), and $R_{10}(r)$ (Table 5.1) and atomic units, the one-particle trial function has the form

$$\chi_{1s}(i) = \left(\frac{\zeta_i^3}{\pi}\right)^{1/2} e^{-\zeta_i r_i} \qquad \textbf{6.12}$$

Equation 6.12 is an eigenfunction to a hydrogenic Hamiltonian with a (partially shielded) nuclear charge of ζ_i. The eigenvalue is $-\zeta_i^2/2$. For the helium-atom ground state, we will assume that the two electrons are associated with identical 1*s*-type orbitals. Thus, both ζ_i's will be the same, and we can drop the subscript. We can use the fact that our trial function is hydrogenic in form to save some work in calculating the expectation value of the Hamiltonian. If we subtract and add ζ/r_1 and ζ/r_2 to our Hamiltonian (Eq. 6.10), it can be rearranged to give

$$\hat{H} = \left[-\frac{1}{2}\nabla_1^2 - \frac{\zeta}{r_1}\right] + \left[-\frac{1}{2}\nabla_2^2 - \frac{\zeta}{r_2}\right] + (\zeta - 2)\left[\frac{1}{r_1} + \frac{1}{r_2}\right] + \frac{1}{r_{12}} \qquad \mathbf{6.13}$$

(substituting 2 for Z). The first two bracketed terms are hydrogenic Hamiltonians, of which Eq. 6.12 is an eigenfunction with an eigenvalue of $-\zeta^2/2$. This, in effect, saves us having to evaluate the integral involving the kinetic-energy operator without altering the functional forms of the other operators.

We now have to find the expectation value of the Hamiltonian with respect to the trial function. The trial function is Eq. 6.11 with Eq. 6.12 substituted for the χ's. We will abbreviate it as

$$\psi(1, 2) = 1s(1)1s(2) \qquad \mathbf{6.14}$$

Making use of the eigenvalues of the hydrogenic function, we find the expectation value of the Hamiltonian is (the integration is over the space of both electron 1 and electron 2)

$$\iint 1s^*(1)1s^*(2)\hat{H}1s(1)1s(2)\, dv_1\, dv_2$$

$$= 2\left[-\frac{\zeta^2}{2}\right]\int 1s^*(1)1s(1)\, dv_1 \int 1s^*(2)1s(2)\, dv_2$$

$$+ (\zeta - 2)\left[\int 1s^*(1)\frac{1}{r_1}1s(1)\, dv_1 \int 1s^*(2)1s(2)\, dv_2\right.$$

$$\left. + \int 1s^*(2)\frac{1}{r_2}1s(2)\, dv_2 \int 1s^*(1)1s(1)\, dv_1\right]$$

$$+ \iint 1s^*(1)1s^*(2)\frac{1}{r_{12}}1s(1)1s(2)\, dv_1\, dv_2 \qquad \mathbf{6.15}$$

Note that all constants can be factored out of the integrals and that if there is no operator involving both electrons, the integrals can be separated into integrals over one electron only. Furthermore, the integrals involving the function only, and no operator, are normalization integrals for the functions and therefore equal unity. Also, the terms containing $1/r_i$ are of identical form, except for the label. Consequently, the corresponding integrals are equal. Thus, the expection value of the Hamiltonian is

$$\langle H \rangle = -\zeta^2 + 2(\zeta - 2)\int 1s^*(1)\frac{1}{r_1}1s(1)\, dv_1$$

$$+ \iint 1s^*(1)1s^*(2)\frac{1}{r_{12}}1s(1)1s(2)\, dv_1\, dv_2 \qquad \mathbf{6.16}$$

The trial functions have spherical symmetry. Integrating the spherical volume element, $r^2 \sin\theta\, dr\, d\theta\, d\phi$, over the angular variables, we have

$$dv_i = 4\pi r_i^2\, dr_i \tag{6.17}$$

The first integral of Eq. 6.16 is easy to evaluate:

$$\begin{aligned}\int 1s^*(1)\frac{1}{r_1}1s(1)dv_1 &= \frac{\zeta^3}{\pi}4\pi\int_0^\infty e^{-\zeta r_1}\frac{1}{r_1}e^{-\zeta r_1}r_1^2\, dr_1\\ &= 4\zeta^3\int_0^\infty r_1 e^{-2\zeta r_1}\, dr_1\\ &= \frac{4\zeta^3}{4\zeta^2}\\ &= \zeta\end{aligned} \tag{6.18}$$

The remaining integral is a bit more difficult; however, it can be evaluated in a straightforward manner. We will carry out the integration first over the coordinates of electron 2 and then over those of electron 1. If we consider the r value for electron 1 to be fixed at some value, r_1, we effectively have the charge of that electron evenly distributed over a spherical volume element of radius r_1. In other words, the physical situation is a charged sphere of radius r_1. The classical relationships for the potential arising from a point charge (in this case electron 2) interacting with a charged sphere (electron 1) are well known. If the point charge is outside the sphere, the potential is the same as if the charge on the sphere were localized at the sphere's center. If the point charge is anywhere inside the sphere, the potential has a constant value, the value it would have if it were at the surface.

It is thus convenient to break our integration over r_2 into two regions, $0 \le r_2 \le r_1$ and $r_1 < r_2 < \infty$ (see Figure 6.1). We have the two integrals

$$\int_{r_1}^\infty 1s^*(2)\frac{1}{r_2}1s(2)r_2^2\, dr_2 = 4\zeta^3\int_{r_1}^\infty r_2 e^{-2\zeta r_2}\, dr_2 \tag{6.19}$$

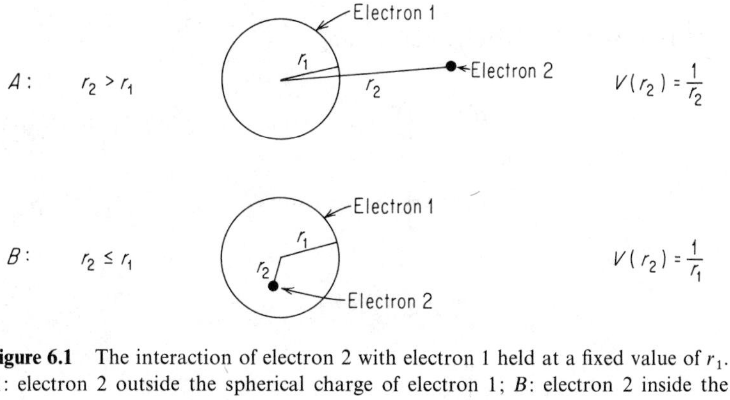

Figure 6.1 The interaction of electron 2 with electron 1 held at a fixed value of r_1. *A*: electron 2 outside the spherical charge of electron 1; *B*: electron 2 inside the spherical charge of electron 1.

and

$$\int_0^{r_1} 1s^*(2)\frac{1}{r_1}1s(2)r_2^2\,dr_2 = \frac{4\zeta^3}{r_1}\int_0^{r_1} r_2^2e^{-2\zeta r_2}\,dr_2 \quad \textbf{6.20}$$

or

$$\int_0^{\infty} 1s^*(2)\frac{1}{r_{12}}1s(2)\,dv_2 = 4\zeta^3\left[\int_{r_1}^{\infty} r_2e^{-2\zeta r_2}\,dr_2 + \frac{1}{r_1}\int_0^{r_1} r_2^2e^{-2\zeta r_2}\,dr_2\right] \quad \textbf{6.21}$$

Now, including the integration over r_1, we have

$$\iint 1s^*(1)1s^*(2)\frac{1}{r_{12}}1s(1)1s(2)\,dv_1\,dv_2$$

$$= 16\zeta^6\int_0^{\infty} r_1^2e^{-2\zeta r_1}\left[\int_{r_1}^{\infty} r_2e^{-2\zeta r_2}\,dr_2 + \frac{1}{r_1}\int_0^{r_1} r_2^2e^{-2\zeta r_2}\,dr_2\right]dr_1 \quad \textbf{6.22}$$

The integrals over r_2 can be evaluated, giving

$$\int_{r_1}^{\infty} r_2e^{-2\zeta r_2}\,dr_2 = \frac{e^{-2\zeta r_2}}{4\zeta^2}(-2\zeta r_2 - 1)\Bigg|_{r_1}^{\infty}$$

$$= \frac{e^{-2\zeta r_1}}{2\zeta}r_1 + \frac{e^{-2\zeta r_1}}{4\zeta^2} \quad \textbf{6.23}$$

$$\int_0^{r_1} r_2^2e^{-2\zeta r_2}\,dr_2 = -r_2^2\frac{e^{-2\zeta r_2}}{2\zeta}\Bigg|_0^{r_1} + \frac{1}{\zeta}\int_0^{r_1} r_2e^{-2\zeta r_2}\,dr_2$$

$$= -r_2^2\frac{e^{-2\zeta r_2}}{2\zeta}\Bigg|_0^{r_1} + \frac{1}{\zeta}\left[\frac{e^{-2\zeta r_2}}{4\zeta^2}(-2\zeta r_2 - 1)\right]\Bigg|_0^{r_1}$$

$$= -r_1^2\frac{e^{-2\zeta r_1}}{2\zeta} - r_1\frac{e^{-2\zeta r_1}}{2\zeta^2} - \frac{e^{-2\zeta r_1}}{4\zeta^3} + \frac{1}{4\zeta^3} \quad \textbf{6.24}$$

Dividing Eq. 6.24 through by r_1 and combining with Eq. 6.23, we get, for Eq. 6.22,

$$\iint 1s^*(1)1s^*(2)\frac{1}{r_{12}}1s(1)1s(2)\,dv_1\,dv_2$$

$$= \frac{16\zeta^6}{4\zeta^3}\int_0^{\infty} r_1e^{-2\zeta r_1}[1 - e^{-2\zeta r_1}(2\zeta r_1 + 1)]\,dr_1$$

$$= 4\zeta^3\left(\int_0^{\infty} r_1e^{-2\zeta r_1}\,dr_1 - 2\zeta\int_0^{\infty} r_1^2e^{-4\zeta r_1}\,dr_1 - \int_0^{\infty} r_1e^{-4\zeta r_1}\,dr_1\right)$$

$$= 4\zeta^3\left(\frac{1}{4\zeta^2} - \frac{2\zeta}{64\zeta^3} - \frac{1}{16\zeta^2}\right)$$

$$= \frac{5}{8}\zeta \quad \textbf{6.25}$$

Substituting the results of Eqs. 6.18 and 6.25 into Eq. 6.16, we find our approximate energy $E(\zeta)$ as a function of the parameter ζ:

$$\begin{aligned} E(\zeta) = \langle H \rangle &= -\zeta^2 + 2(\zeta - 2)\zeta + \tfrac{5}{8}\zeta \\ &= \zeta^2 - \tfrac{27}{8}\zeta \end{aligned} \tag{6.26}$$

Taking the derivative of $E(\zeta)$ with respect to ζ and setting the result equal to zero for the minimization, we have

$$\frac{dE(\zeta)}{d\zeta} = 2\zeta - \frac{27}{8} = 0 \tag{6.27}$$

or

$$\zeta = \tfrac{27}{16} = 1.6875 \tag{6.28}$$

The variational energy is

$$\begin{aligned} E &= (\tfrac{27}{16})^2 - (\tfrac{27}{8})(\tfrac{27}{16}) \\ &= -(\tfrac{27}{16})^2 \\ &= -2.84766 \text{ Hartrees} = -77.48943 \text{ eV} \end{aligned} \tag{6.29}$$

[Note that the virial theorem (see Eq. 1.17) is satisfied, since E equals $-T$ or $\frac{1}{2}V$.] The experimental ground-state electronic energy for helium is -79.02 eV. The absolute error is only ~ 1.9 percent. This is still a relatively large error in chemical terms, however, amounting to more than 35 kcal mol^{-1}. More accurate results can be obtained by using a trial function with more variational parameters. However, even the best independent-particle wave function still gives an energy that is off by about 26 kcal mol^{-1}. This energy error, known as the *correlation energy*, arises because the wave function does not allow for the correlation of the motions of the electrons. To get the exact energy by a variational treatment, we must use a wave function that includes r_{12} in some fashion. (Our comments pointing out the limitations of quantum-chemical results are intended only as warnings, not as discouragement.)

An additional point concerning the two-electron integral in the helium-atom problem is worthy of comment. If either Eq. 6.23 (which arises from the electron arrangement in Figure 6.1A) or Eq. 6.24 (from Figure 6.1B) is used alone in Eq. 6.22 and the integration is carried out over r_1, the same value, $\frac{5}{16}\zeta$, is obtained. In an alternative technique for calculating the two-electron integral, we can consider electron number 1 to be the inner electron and electron 2 the outer, and calculate the repulsion for this. The electrons can then be interchanged and the calculation repeated. The contribution will be the same. The total repulsion energy will be the sum of the two contributions. Since the number labeling of the electrons has no physical significance, this process (known as the Ewald transformation) must give the same result as keeping the number labels constant, as we did.

6.4 PERTURBATION THEORY

In perturbation theory an approximate Hamiltonian, for which the Schrödinger equation can be solved exactly, is employed. Let

$$\hat{H} = \hat{H}^0 + \hat{H}' \tag{6.30}$$

where $\hat{H}$ is the Hamiltonian for the system being studied, $\hat{H}^0$ is an approximate Hamiltonian for which exact solutions can be found, and $\hat{H}'$ is the perturbation. For the helium atom, for example, $\hat{H}^0$ could be chosen as the sum of two hydrogenic Hamiltonians and $\hat{H}'$ as the $1/r_{12}$ term.

Let us express $\hat{H}'$ in terms of an arbitrary parameter, λ, and some other operator, $\hat{V}$:

$$\hat{H}' = \lambda\hat{V} \tag{6.31}$$

where the magnitude of λ is less than unity. The wave function and the energy for any state, N, can be expanded in powers of λ:

$$\psi_N = \psi_N^0 + \lambda\psi'_N + \lambda^2\psi''_N + \ldots \tag{6.32}$$

$$E_N = E_N^0 + \lambda E'_N + \lambda^2 E''_N + \ldots \tag{6.33}$$

where the quantities with a zero superscript represent the zero-order approximation (the result from $\hat{H}^0$), and the primed quantities represent successively higher-order corrections to the quantities. (If λ is less than unity, the series will converge.) The Schrödinger equation, in terms of Eqs. 6.30 through 6.33, becomes

$$\hat{H}\psi_N = E_N\psi_N$$

$$(\hat{H}^0 + \lambda\hat{V})(\psi_N^0 + \lambda\psi'_N + \lambda^2\psi''_N + \ldots)$$
$$= (E_N^0 + \lambda E'_N + \lambda^2 E''_N + \ldots)(\psi_N^0 + \lambda\psi'_N + \lambda^2\psi''_N + \ldots) \tag{6.34}$$

Expanding this and collecting like powers of λ, we have

$$(\hat{H}^0\psi_N^0 - E_N^0\psi_N^0) + (\hat{H}^0\psi'_N + \hat{V}\psi_N^0 - E_N^0\psi'_N - E'_N\psi_N^0)\lambda$$
$$+ (\hat{H}_N^0\psi''_N + \hat{V}\psi'_N - E_N^0\psi''_N - E'_N\psi'_N - E''_N\psi_N^0)\lambda^2 + \ldots = 0 \tag{6.35}$$

Since λ is an arbitrary parameter, the left-hand side of Eq. 6.35 can always equal zero if and only if the coefficient of each power of λ equals zero. We have, for the zero-power term,

$$\hat{H}^0\psi_N^0 = E_N^0\psi_N^0 \tag{6.36}$$

which is the Schrödinger equation for the exactly solvable approximate Hamiltonian. For the first power of λ, we have

$$\hat{H}^0\psi'_N + \hat{V}\psi_N^0 = E_N^0\psi'_N + E'_N\psi_N^0 \tag{6.37}$$

This leads to what we call *first-order perturbation theory*. For λ^2 we have

$$\hat{H}^0\psi''_N + \hat{V}\psi'_N = E_N^0\psi''_N + E'_N\psi'_N + E''_N\psi_N^0 \tag{6.38}$$

which leads to *second-order perturbation theory*. Although this can be continued to any order, second order is as high as is usually employed.

We will adopt here a new abbreviated notation for integration, the *Dirac notation*. The Dirac notation was originally conceived of as a vector notation, but it has been generalized to integrals. The quantity $\langle\psi_i|\hat{O}|\psi_j\rangle$ is defined to mean the same thing as $\int \psi_i^*\hat{O}\psi_j\, dv$, where the integration is over all space.

Let us multiply the first-order equation (6.37) by ψ_N^{0*} and integrate over all space. Using Dirac notation, we have

$$\langle\psi_N^0|\hat{H}^0|\psi'_N\rangle + \langle\psi_N^0|\hat{V}|\psi_N^0\rangle = E_N^0\langle\psi_N^0|\psi'_N\rangle + E'\langle\psi_N^0|\psi_N^0\rangle \qquad \textbf{6.39}$$

The Hamiltonian operator is a Hermitian operator. This means that

$$\begin{aligned}\langle\psi_N^0|\hat{H}^0|\psi'_N\rangle &= \langle\psi'_N|\hat{H}^0|\psi_N^0\rangle \\ &= E_N^0\langle\psi'_N|\psi_N^0\rangle \\ &= E_N^0\langle\psi_N^0|\psi'_N\rangle\end{aligned} \qquad \textbf{6.40}$$

Furthermore, if the zero-order wave functions are normalized, the last integral of Eq. 6.39 equals unity, giving

$$E'_N = \langle\psi_N^0|\hat{V}|\psi_N^0\rangle \qquad \textbf{6.41}$$

or

$$\lambda E'_N = \langle\psi_N^0|\hat{H}'|\psi_N^0\rangle \qquad \textbf{6.42}$$

The first-order correction to the energy is just the expectation value of the perturbation operator with respect to the zero-order wave function. To first order, the energy is

$$E_N \cong E_N^0 + \langle\psi_N^0|\hat{H}'|\psi_N^0\rangle \qquad \textbf{6.43}$$

In order to get the first-order correction to the wave function, we will expand ψ'_N as a linear combination of zero-order functions:

$$\psi'_N = \sum_K c_{KN}\psi_K^0 \qquad \textbf{6.44}$$

Equation 6.37 can be rearranged to give

$$(\hat{H}^0 - E_N^0)\psi'_N = (E'_N - \hat{V})\psi_N^0 \qquad \textbf{6.45}$$

Substituting Eq. 6.44 into this, we have

$$\sum_K c_{KN}(\hat{H}^0 - E_N^0)\psi_K^0 = (E'_N - \hat{V})\psi_N^0 \qquad \textbf{6.46}$$

If we now multiply by ψ_M^0, where M designates one particular eigenfunction, and integrate, we get

$$\sum_K c_{KN}\langle\psi_M^0|\hat{H}^0 - E_N^0|\psi_K^0\rangle = \langle\psi_M^0|E'_N - \hat{V}|\psi_N^0\rangle \qquad \textbf{6.47}$$

All the terms in the summation, except for the one in which K equals M, vanish because of the orthogonality of the functions. $\hat{H}^0$ acting on ψ_M^0 gives E_M^0 times ψ_M^0. The E'_N term on the right vanishes, since E'_N is a scalar and the functions are orthogonal. We are left with

$$c_{MN}(E_M^0 - E_N^0) = -\langle\psi_M^0|\hat{V}|\psi_N^0\rangle \tag{6.48}$$

or

$$c_{MN} = \frac{\langle\psi_M^0|\hat{V}|\psi_N^0\rangle}{E_N^0 - E_M^0} \tag{6.49}$$

Thus

$$\psi'_N = \sum_{M \neq N} \frac{\langle\psi_M^0|\hat{V}|\psi_N^0\rangle}{E_N^0 - E_M^0}\psi_M^0 \tag{6.50}$$

and, since the first-order correction to the wave function is $\lambda\psi'_N$, we have, to first order,

$$\psi_N \cong \psi_N^0 + \sum_{M \neq N} \frac{H'_{MN}}{E_N^0 - E_M^0}\psi_M^0 \tag{6.51}$$

where we have used the abbreviation H'_{MN} for $\langle\psi_M^0|\hat{H}'|\psi_N^0\rangle$.

Let us consider again the second-order equation. If we multiply Eq. 6.38 by ψ_N^{0*} and integrate, we obtain

$$\langle\psi_N^0|\hat{H}|\psi''_N\rangle + \langle\psi_N^0|\hat{V}|\psi'_N\rangle = E_N^0\langle\psi_N^0|\psi''_N\rangle + E'_N\langle\psi_N^0|\psi'_N\rangle + E''_N\langle\psi_N^0|\psi_N^0\rangle \tag{6.52}$$

By the arguments used with Eq. 6.40, the first term on the left equals the first on the right. The second term on the right vanishes because the expansion for ψ'_N (Eq. 6.50) does not include ψ_N^0. The integral in the final term equals unity, and we are left with

$$E''_N = \langle\psi_N^0|\hat{V}|\psi'_N\rangle \tag{6.53}$$

The second-order correction to the energy depends upon the first-order correction to the wave function. Substituting in Eq. 6.50 for ψ'_N and $\lambda^2 E''_N$ for the second-order energy correction, our energy to second order is

$$E_N \cong E_N^0 + H'_{NN} + \sum_{M \neq N} \frac{H'_{NM}H'_{MN}}{E_N^0 - E_M^0} \tag{6.54}$$

The summation in Eq. 6.54 can be infinite (as it will be for atomic problems based upon a hydrogenic zero-order approximation, and for many other cases). This is one reason why perturbation theory often is not carried beyond second order. If the second-order approximation cannot be completely solved, it makes little sense to go to a still higher order.

Group theory is extremely useful in perturbation-theory calculations. Many integrals are found to vanish because of group-theoretical, or symmetry, arguments. Any "observable" of a system must be invariant under any of the system's symmetry operations. In other words, observables are scalars rather than vectors, operators, and so on. From a group-theoretical point of view, this means that any integral, or any other function, that represents an observable must transform as the totally symmetric irreducible representation of the group according to which the system is classified. (The totally symmetric representation is the only real scalar representation of a group and is that representation in which each element of the group is mapped onto the scalar, +1.) Since each function or operator in an integral can be classified according to an irreducible representation—or combination of irreducible representations—of the group, and since the multiplication rules for the representations are known, the representation according to which an integrand transforms can be simply determined. If this representation is not, or does not contain, the totally symmetric representation, we need not evaluate the integral. We know that it will equal zero. In perturbation-theory calculations this can tell us, for example, whether or not the first-order correction to the energy is zero and which coefficients will vanish in the first-order expansion of the wave function or in the second-order expansion of the energy.

6.5 FIRST-ORDER PERTURBATION-THEORY TREATMENT OF HELIUM

Let us go through a perturbation-theory treatment of the ground state of the helium atom to first order. As we mentioned in discussing the variational treatment of helium, if the $1/r_{12}$ term were not present, the Hamiltonian (Eq. 6.10) would be the sum of two hydrogenic Hamiltonians and the problem could be solved exactly. It thus appears to be an ideal candidate for a perturbation-theory treatment. The zero-order Hamiltonian can be expressed as the sum of the hydrogenic Hamiltonians:

$$\hat{H}^0 = \left(-\frac{1}{2}\nabla_1^2 - \frac{Z}{r_1}\right) + \left(-\frac{1}{2}\nabla_2^2 - \frac{Z}{r_2}\right) \tag{6.55}$$

and the perturbation, $1/r_{12}$, as

$$\hat{H}' = \frac{1}{r_{12}} \tag{6.56}$$

The solutions to Eq. 6.55 are simple products of one-electron hydrogenic wave functions. In particular, the zero-order approximation to the ground-state energy will be the sum of two hydrogenic $1s$ energies:

$$\begin{aligned} E_1^0 &= -\frac{2Z^2}{2} \\ &= -4 \text{ Hartrees} = -108.84 \text{ eV} \end{aligned} \tag{6.57}$$

(Remember that, in atomic units, the ground-state hydrogenic energy for one electron is $-Z^2/2$ Hartrees.) This is much below the experimental value of -79.02 eV. The true energy is obviously not a lower bound to the energy derived from perturbation theory, as it is in a variational treatment.

The zero-order ground-state wave function will be a simple product of two hydrogenic 1*s* functions, having the true value of the nuclear charge:

$$\psi_1^0 = 1s(1)1s(2) = \left(\frac{Z^3}{\pi}\right) e^{-Zr_1} e^{-Zr_2} \quad \textbf{6.58}$$

The first-order correction to the energy, $\lambda E_1'$, is

$$\begin{aligned} \lambda E_1' &= \langle \psi_1^0 | \hat{H}' | \psi_1^0 \rangle \\ &= \left\langle 1s(1)1s(2) \left| \frac{1}{r_{12}} \right| 1s(1)1s(2) \right\rangle \end{aligned} \quad \textbf{6.59}$$

This integral is exactly the same as the one we evaluated for the variational treatment of helium, except that now the true value of the nuclear charge (rather than an effective nuclear charge) must be used. Thus, with a Z value of 2,

$$\lambda E_1' = \tfrac{5}{8} Z = 1.25 \text{ Hartrees} \quad \textbf{6.60}$$

The energy to first order is

$$\begin{aligned} E_1 &\cong E_1^0 + \lambda E_1' \\ &= -4 + 1.25 \\ &= -2.75 \text{ Hartrees} = -74.83 \text{ eV} \end{aligned} \quad \textbf{6.61}$$

The error is about 5.3 percent. (Note that the first-order energy lies above the true energy, while the zero-order energy was below the true value.) A second-order treatment would give a more accurate result for the energy, but would require an infinite summation to completely evaluate $\lambda^2 E_1''$:

$$\lambda^2 E_1'' = \sum_{N \neq 1} \frac{H'_{1N} H'_{N1}}{E_1^0 - E_N^0} \quad \textbf{6.62}$$

In practice, only a finite number of terms in the summation need be used, since the sum converges rather rapidly.

Two arguments can be used for deciding on the relative importance of the terms in a second-order perturbation-theory expansion. The most obvious of these arises from a consideration of the denominator. If the terms in the numerator of Eq. 6.62 are of comparable magnitude, those associated with low-lying values of E_N^0—that is, those that would give the smaller denominator—will be of greater importance than those associated with higher energy values. The second argument derives from symmetry and group theory. The perturbation has spherical symmetry and therefore transforms as the totally symmetric irreducible representation of **O**(3). Consequently, only excited states having the same overall symmetry as our zero-order wave function will lead to nonvanishing values for H'_{1N} or H'_{N1}. Our zero-order wave functions have the general form

$$\psi_N^0(1, 2) = nl(1)n'l'(2) \quad \textbf{6.63}$$

where n and n' are principal quantum numbers and l and l' are orbital angular-momentum quantum numbers. The n values do not enter into symmetry arguments; however, the l values are the indices for the irreducible representations, $D^l_{g,u}$, of the **O**(3) group that describes spherical symmetry. A fundamental rule for multiplying irreducible representations is that a product of two irreducible representations contains the totally symmetric irreducible representation if and only if the representations being multiplied are the same. Our zero-order ground-state wave function is

$$\psi_1^0(1, 2) = 1s(1)1s(2) \qquad \textbf{6.64}$$

The s functions are totally symmetric. Thus, in order for the H'_{1N} and H'_{N1} to be nonvanishing, the l and l' from Eq. 6.63 must be the same.

6.6 COMPARISON OF VARIATIONAL AND PERTURBATION METHODS

In the preceding sections, we found that the energy for helium was calculated more accurately (and with only slightly more effort) by the variational treatment than by the first-order perturbation-theory treatment. Furthermore, the variational wave function, which was automatically obtained at the same time as the energy, should be of comparable accuracy. (Note, however, that the "best" wave function for calculating the energy may not be the "best" for calculating some other property.) In addition, the virial theorem was satisfied for the variational treatment, while it was not for the first-order perturbation-theory treatment (here, the kinetic energy is the negative of the zero-order energy). Finally, we saw that in order to obtain the first-order correction to a perturbation theory wave function, an infinite summation is required.

From these considerations, one might wonder if time-independent perturbation-theory has any real utility in quantum chemistry. Actually, it has strengths that, in some situations, make it more useful than the variational principle.

The greatest weakness of the variational principle is that, in its usual form, it will yield only the lowest energy state [having a given spin multiplicity (Section 7.6) and symmetry] of a system. Thus, it is most commonly used only to obtain ground-state energies. Excited-state energies, constructed directly as expectation values from ground-state variational wave functions, have varying accuracy that cannot be reliably estimated. (There are other ways of getting relatively good excitation energies. The best current techniques involve simultaneous optimization of the ground and excited states by a *multiconfiguration self-consistent field* procedure.) On the other hand, there are no state restrictions on perturbation-theory calculations. To a given order, a perturbation-theory calculation of an excited state should be of accuracy comparable to that of a ground state. Furthermore, qualitative answers to questions such as "Will a perturbation of this form affect an energy level?" are often as important as quantitative answers. In such cases, perturbation theory, particularly when used in connection with group theory, can frequently provide quick and simple answers.

In computational problems in quantum chemistry, the variational principle is almost always employed to obtain the ground-state energies and wave functions.

It is frequently used to directly obtain the energies and wave functions for certain excited states, subject to the previous multiplicity and symmetry statement. Perturbation theory is usually simpler to use for problems requiring only qualitative answers. Furthermore, for problems in which a variational solution is impossible, perturbation theory may be the only possible approach. We encounter problems that fit into all these categories. One very important use of perturbation theory is in time-dependent problems, since the form of the variational principle we have presented works only for stationary states.

6.7 TIME-DEPENDENT PERTURBATION THEORY AND SPECTRAL INTENSITIES

Very much of chemistry involves time-dependent processes such as those that are part of chemical kinetics, scattering phenomena, and spectral transitions. The time-dependent Schrödinger equation

$$\hat{H}\Psi = i\hbar \frac{\partial \Psi}{\partial t} \tag{6.65}$$

is the appropriate problem to be solved for such processes. The solution to this was given in Eq. 1.47 as a product of a time-independent part, $\psi(x)$, and a time-dependent part:

$$\begin{aligned} \Psi(x, t) &= \psi(x) \exp(-2\pi i \nu t) \\ &= \psi(x) \exp\left(-\frac{iEt}{\hbar}\right) \end{aligned} \tag{6.66}$$

The direct solution for $\Psi(x, t)$ can be carried out only for limited cases. Usually, time-dependent perturbation theory must be used to solve the problem.

If the system under study has stationary states, the time-independent Schrödinger equation holds for these states:

$$\hat{H}^0 \psi_i = E_i^0 \psi_i \tag{6.67}$$

where $\hat{H}^0$ is the time-independent Hamiltonian, ψ_i^0 is the time-independent wave function, and E_i^0 is the stationary-state energy associated with the ith state. The true Hamiltonian for the time-dependent situation can be expressed in terms of the time-independent $\hat{H}^0$ and a time-dependent $\hat{H}'$:

$$\hat{H} = \hat{H}^0 + \hat{H}' \tag{6.68}$$

We can expand the time-dependent wave function as a sum of terms of the form of Eq. 6.66 by using as the time-independent functions the solutions of Eq. 6.67:

$$\Psi_i = \sum_j c_j \psi_j^0(x) \exp\left(-\frac{iE_j^0 t}{\hbar}\right) \tag{6.69}$$

If we substitute Eqs. 6.68 and 6.69 into Eq. 6.65, the resulting time-dependent Schrödinger equation is

$$(\hat{H}^0 + \hat{H}')\sum_j c_j\psi_j^0 \exp\left(-\frac{iE_j^0 t}{\hbar}\right) = i\hbar\frac{\partial}{\partial t}\sum_j c_j\psi_j^0 \exp\left(-\frac{iE_j^0 t}{\hbar}\right)$$
$$= i\hbar\sum_j\left(\frac{\partial c_j}{\partial t} - \frac{iE_j^0}{\hbar}c_j\right)\psi_j^0 \exp\left(-\frac{iE_j^0 t}{\hbar}\right) \qquad \textbf{6.70}$$

If we multiply Eq. 6.70 from the left by the complex conjugate of one particular time-independent wave function, say ψ_m^0, and integrate, we get

$$\int \psi_m^{0*}(\hat{H}^0 + \hat{H}')\sum_j c_j\psi_j^0 \exp\left(-\frac{iE_j^0 t}{\hbar}\right)dv$$
$$= \int \psi_m^{0*}\sum_j\left(i\hbar\frac{\partial c_j}{\partial t} + E_j^0 c_j\right)\psi_j^0 \exp\left(-\frac{iE_j^0 t}{\hbar}\right)dv \qquad \textbf{6.71}$$

This gives

$$c_m E_m^0 \exp\left(-\frac{iE_m^0 t}{\hbar}\right) + \sum_j c_j \exp\left(-\frac{iE_j^0 t}{\hbar}\right)H'_{mj}$$
$$= c_m E_m^0 \exp\left(-\frac{iE_m^0 t}{\hbar}\right) + i\hbar\frac{dc_m}{dt}\exp\left(-\frac{iE_m^0 t}{\hbar}\right) \qquad \textbf{6.72}$$

or

$$i\hbar\frac{dc_m}{dt} = \sum_j c_j \exp\left[-\frac{i(E_j^0 - E_m^0)t}{\hbar}\right]H'_{mj}$$
$$= \sum_j c_j \exp(2\pi i\nu_{mj}t)H'_{mj} \qquad \textbf{6.73}$$

Equation 6.73 gives the time dependence of the contribution of ψ_m^0 to the total wave function. If the mth stationary state of the system is the initial or final state of some process (such as a spectral transition or a scattering phenomenon), Eq. 6.73 is related to the rate of passage from or to that state. In particular, the process under consideration can occur (i.e., is allowed) if H'_{mj} does not equal zero.

The treatment we have presented to this point is general for any type of perturbation. The only trick is deciding on the perturbation that causes the desired transition and evaluating the integral, H'_{mj}. We will now consider the special case of a spectral transition induced by electromagnetic radiation.

In 1917 Einstein showed that for a spectral transition between two states, say 1 and 2, with populations N_1 and N_2, in an electromagnetic field, the rate of transition between levels is (assuming state 1 has the lower energy)

$$-\frac{dN_1}{dt} = \frac{dN_2}{dt} = N_1 B\rho_\nu - N_2(A + B\rho_\nu) \qquad \textbf{6.74}$$

where ρ_ν is the energy density as a function of frequency (see Eq. 1.4), B is Einstein's coefficient of induced absorption or emission, and A is his coefficient of spontaneous emission. If the sample is left in the field long enough for steady-state conditions (i.e., equilibrium) to be obtained, we have

$$\frac{dN_i}{dt} = 0 \tag{6.75}$$

or

$$\frac{N_2}{N_1} = \frac{B\rho_\nu}{A + B\rho_\nu} \tag{6.76}$$

But from Boltzmann statistics we know that for two states differing in energy by ΔE,

$$\begin{aligned}\frac{N_2}{N_1} &= \exp\left(-\frac{\Delta E}{kT}\right)\\ &= \exp\left(-\frac{h\nu}{kT}\right)\end{aligned} \tag{6.77}$$

Using Planck's equation for the energy density (Eq. 1.8), we see that

$$\frac{A}{B} = \frac{8\pi h\nu^3}{c^3} \tag{6.78}$$

The coefficient, B, can be derived from time-dependent perturbation theory. If needed, the coefficient, A, can be derived from B by the use of Eq. 6.78.

Atoms and molecules are composed of charged particles—electrons and nuclei—that are in continuous motion. A charged particle moving in an electric field interacts with the field. If the charge of the particle is Z, the classical energy of interaction (ignoring the scalar potential) is the scalar product of the vector potential of the field, $\mathbf{A}$, and the particle's velocity, $\mathbf{v}$, multiplied $-Z$:

$$E_{\text{field}} = -Z(\mathbf{A}\cdot\mathbf{v}) \tag{6.79}$$

This can be converted to its quantum-mechanical equivalent and treated directly as the perturbation to a time-independent Hamiltonian. However, a true Hamiltonian for a system in an electric field can be deduced by the use of the Lagrangian and Hamiltonian equations of motion (Slater, Sec. 6-7, App. 4). The result is

$$\begin{aligned}\hat{H} &= \frac{1}{2m}(\hat{p} - Z\mathbf{A})^2 + V\\ &= \frac{1}{2m}(\hat{p}^2 - Z(\hat{p}\mathbf{A} + \mathbf{A}\hat{p}) + Z^2A^2) + V\\ &= \hat{H}^0 + \hat{H}'\end{aligned} \tag{6.80}$$

where V contains only the normal Coulombic interactions. The interaction with an electric field is normally a relatively weak effect; consequently, for most applications the A^2 term is negligible. (The analogous term is important in explaining diamagnetism when the field is magnetic, rather than electric.) Therefore the perturbation is

$$\hat{H}' = -\frac{Z}{2m}(\hat{p}\mathbf{A} + \mathbf{A}\hat{p})$$

$$= -\frac{Z}{2}(\hat{v}\mathbf{A} + \mathbf{A}\hat{v}) \qquad 6.81$$

where we have used the fact that momentum, p, equals mass times velocity, mv. Classically, $\hat{H}'$ would equal $-Z/2\ (\mathbf{v}\cdot\mathbf{A} + \mathbf{A}\cdot\mathbf{v})$, but the scalar products are the same. Therefore, we can write

$$\hat{H}' = -Z\mathbf{A}\hat{v} \qquad 6.82$$

As in time-independent perturbation theory, we will ultimately have to evaluate the expectation value of the perturbation between two states. Although the velocity operator is sometimes used for calculating transition probabilities, it is more usual to convert the perturbation to a form involving coordinates. In order to do this, we use the commutation properties of operators. These properties, in the Schrödinger representation of quantum mechanics, are the same as those for the corresponding matrices in the Heisenberg representation. In particular, Eq. 1.32 relates the time derivative of a property to the commutator of the property with the Hamiltonian. Rewriting this in terms of operators and using the time-independent Hamiltonian, we have

$$[q, \hat{H}^0] = i\hbar\dot{q} \qquad 6.83$$

or

$$\dot{q} = v_q = -\frac{i}{\hbar}[q, \hat{H}^0]$$

$$= -\frac{i}{\hbar}(q\hat{H}^0 - \hat{H}^0 q) \qquad 6.83a$$

Equation 6.83a holds for each component of the velocity. Summing over the components, we obtain

$$\hat{v} = -\frac{i}{\hbar}(\mathbf{r}\hat{H}^0 - \hat{H}^0\mathbf{r}) \qquad 6.83b$$

If we make use of Eq. 6.83b, the time-dependent perturbation becomes, for a single particle,

$$\hat{H}' = i\frac{Z}{\hbar}\mathbf{A}(\mathbf{r}\hat{H}^0 - \hat{H}^0\mathbf{r}) \qquad 6.84$$

or, for a collection of particles,

$$\hat{H}' = \sum_{\mu} i \frac{Z_\mu}{\hbar} \mathbf{A}(\mathbf{r}_\mu \hat{H}^0 - \hat{H}^0 \mathbf{r}_\mu) \qquad \textbf{6.84a}$$

If **A** is assumed to be independent of the spatial coordinates over the extent of the atom or molecule under consideration, H'_{mj} becomes

$$\begin{aligned} H'_{mj} &= \sum_{\mu} i \frac{Z_\mu}{\hbar} \mathbf{A} \int \psi_m^{0*}(\mathbf{r}_\mu \hat{H}^0 - \hat{H}^0 \mathbf{r}_\mu)\psi_j^0 \, dv \\ &= \frac{i(E_j^0 - E_m^0)}{\hbar} \mathbf{A} \int \psi_m^{0*} \left(\sum_\mu Z_\mu \mathbf{r}_\mu \right) \psi_j^0 \, dv \\ &= -i2\pi\nu_{mj} \mathbf{A} \int \psi_m^{0*} \left(\sum_\mu Z_\mu \mathbf{r}_\mu \right) \psi_j^0 \, dv \end{aligned} \qquad \textbf{6.85}$$

where, in the last line, we have used $h\nu_{mj}$ for $(E_m - E_j)$. Note that the operator in the integral in the last line is the dipole moment operator. The operator is averaged between the two states, m and n. The entire integral is frequently referred to as the *transition dipole* and abbreviated $\boldsymbol{\mu}_{mj}$. Equation 6.85 can then be rewritten as

$$H'_{mj} = -i2\pi\nu_{mj} \mathbf{A} \cdot \boldsymbol{\mu}_{mj} \qquad \textbf{6.86}$$

Equation 6.73 can be rewritten as

$$h \frac{dc_m}{dt} = -\sum_j c_j 4\pi^2 \nu_{mj} \mathbf{A} \cdot \boldsymbol{\mu}_m \exp(2\pi i \nu_{mj} t) \qquad \textbf{6.87}$$

If the system is in a particular stationary state, say n, before interaction with the radiation, all the coefficients in the summation will be zero, except for c_n, which will equal unity. Thus, Eq. 6.87 becomes

$$h \frac{dc_m}{dt} = -4\pi^2 \nu_{mn} \mathbf{A} \cdot \boldsymbol{\mu}_{mn} \exp(2\pi i \nu_{mn} t) \qquad \textbf{6.88}$$

The functional form of **A** for a plane-polarized beam of electromagnetic radiation is

$$\begin{aligned} \mathbf{A} &= \mathbf{A}^0 \cos(2\pi\nu t - \alpha - \mathbf{k} \cdot \mathbf{r}) \\ &= \tfrac{1}{2}\mathbf{A}^0 \{\exp[i(2\pi\nu t - \alpha - \mathbf{k} \cdot \mathbf{r})] + \exp[-i(2\pi\nu t - \alpha - \mathbf{k} \cdot \mathbf{r})]\} \end{aligned} \qquad \textbf{6.89}$$

where ν is the frequency, α is the phase angle, and **k** is $2\pi/\lambda$ expressed as a vector. If λ, the wavelength of the radiation, is large compared to the dimensions of the atom or molecule being studied, **A** can be considered to be spatially invariant over the region of space occupied by the system. We have

$$\mathbf{A} \cong \tfrac{1}{2}\mathbf{A}^0 \{\exp[i(2\pi\nu t - \alpha)] + \exp[-i(2\pi\nu t - \alpha)]\} \qquad \textbf{6.90}$$

Substituting this into Eq. 6.88, we get

$$h\frac{dc_m}{dt} = -2\pi^2 \nu_{mn} \mathbf{A}^0 \cdot \boldsymbol{\mu}_{mn} \{\exp(-i\alpha)\exp[2\pi i(\nu + \nu_{mn})t] + \exp(i\alpha)\exp[-2\pi i(\nu - \nu_{mn})t]\} \quad 6.91$$

Integrating with respect to time yields

$$c_m = \frac{\pi i \nu_{mn}}{h}\mathbf{A}^0 \cdot \boldsymbol{\mu}_{mn}\left\{\exp(-i\alpha)\frac{\exp[2\pi i(\nu + \nu_{mn})t]}{\nu + \nu_{mn}} - \exp(i\alpha)\frac{\exp[-2\pi i(\nu - \nu_{mn})t]}{\nu - \nu_{mn}}\right\} + \text{integration constant} \quad 6.92$$

Evaluating this between the limits, zero and t, gives

$$c_m = \frac{\pi i \nu_{mn}}{h}\mathbf{A}^0 \cdot \boldsymbol{\mu}_{mn}\left\{\exp(-i\alpha)\frac{\exp[2\pi i(\nu + \nu_{mn})t] - 1}{\nu + \nu_{mn}} - \exp(i\alpha)\frac{\exp[-2\pi i(\nu - \nu_{mn})t] - 1}{\nu - \nu_{mn}}\right\} \quad 6.93$$

(Note that this removes the time dependence from c_m.) If state m is higher in energy than state n, ν_{mn} is positive and the process is energy absorption. If state m is lower, ν_{mn} is negative and we have emission.

Because of the quantization of energy, energy will be absorbed from or emitted to the electromagnetic radiation only for frequencies (ν) very near ν_{mn}. Thus, the denominator for the first term in brackets will be near zero at frequencies near the emission frequency, while that for the second will be near zero for frequencies near the absorption frequency. When the denominator is near zero, the corresponding complete term will be much larger than the other one. For an absorption process, for example, we can write c_m as

$$c_m = -\frac{\pi i \nu_{mn}}{h}\mathbf{A}^0 \cdot \boldsymbol{\mu}_{mn}\exp(i\alpha)\frac{\{\exp[-2\pi i(\nu - \nu_{mn})t] - 1\}}{\nu - \nu_{mn}} \quad 6.94$$

The probability of the system being transferred from state n to state m by the absorption of radiation will be proportional to $\Psi^*\Psi$, which, with the simplifications we have made, will be just $c_m^* c_m$:

$$\begin{aligned}\Psi^*\Psi &= (c_m^* c_m)_\nu \\ &= \frac{\pi^2 \nu_{mn}^2}{h^2}(\mathbf{A}^0 \cdot \boldsymbol{\mu}_{mn})^2 \frac{\{\exp[-2\pi i(\nu - \nu_{mn})t] - 1\}\{\exp[2\pi i(\nu - \nu_{mn})t] - 1\}}{(\nu - \nu_{mn})^2} \\ &= \frac{4\pi^2 \nu_{mn}^2}{h^2}(\mathbf{A}^0 \cdot \boldsymbol{\mu}_{mn})^2 \frac{\sin^2[\pi(\nu - \nu_{mn})t]}{(\nu - \nu_{mn})^2}\end{aligned} \quad 6.95$$

Equation 6.95 gives the probability of absorption of radiation of a fixed frequency, ν. If we are to find the total transition probability, this must be integrated over all frequencies (from zero to infinity). By choosing $[\pi(\nu - \nu_{mn})t]$ as the integration variable, we can accomplish the integration to give, for the total transition probability,

$$(c_m^* c_m)_{\text{total}} = \frac{4\pi^2 \nu_{mn}^2}{h^2} (\mathbf{A}^0 \cdot \boldsymbol{\mu}_{mn})^2 t \qquad \mathbf{6.96}$$

Equation 6.96 is the appropriate expression for studying the absorption of plane-polarized light by rigidly oriented molecules (such as would be found in a properly alligned single crystal). More commonly, however, spectroscopy involves nonpolarized radiation, a fluid medium, or both. In these cases, the orientations of $\mathbf{A}$ and $\boldsymbol{\mu}_{mn}$ relative to each other must be averaged. The average of $\cos^2 \theta$ [arising from $(\mathbf{A}^0 \cdot \boldsymbol{\mu}_{mn})^2$] is $\frac{1}{3}$. The orientationally averaged transition probability is

$$(c_m^* c_m)_{\text{averaged}} = \frac{4\pi^2 \nu_{mn}^2}{3h^2} \mathbf{A}^{02} \mu_{mn}^2 t \qquad \mathbf{6.97}$$

In terms of the vector potential, the energy density is

$$\rho_\nu = \tfrac{1}{2}\nu^2 \mathbf{A}^{02} \qquad \mathbf{6.98}$$

Therefore,

$$(c_m^* c_m)_{\text{averaged}} = \frac{8\pi^3}{3h^2} \mu_{mn}^2 \rho_\nu t \qquad \mathbf{6.99}$$

This gives the probability that a single molecule will be promoted from state n to state m. An identical expression results from the emission term. The total number of molecules promoted from state n to state m during time t can be found by multiplying Eq. 6.99 by the population of state n, N_n. The total number of molecules being stimulated to emit from state m to state n can be found by multiplying Eq. 6.99 by the population of state m, N_m. The rate of transfer between states can be written

$$\frac{dN_n}{dt} = -\frac{dN_m}{dt} = \frac{d}{dt}(-N_m + N_n)(c_m^* c_m)$$

$$= (-N_m + N_n)\frac{8\pi^3}{3h^2} \mu_{mn}^2 \rho_\nu \qquad \mathbf{6.100}$$

Comparing this with the ρ-dependent part of Eq. 6.74, we see that the Einstein coefficient of stimulated absorption or emission is

$$B = \frac{8\pi^3}{3h^2} \mu_{mn}^2 \qquad \mathbf{6.101}$$

Other mechanisms for the transfer of a system between stationary states are treated analogously. They all depend upon the expectation value of a time-dependent perturbation between the states involved.

BIBLIOGRAPHY

ATKINS, P. W., *Molecular Quantum Mechanics.* Clarendon Press, Oxford, 1970.

PARR, R. G., *Quantum Theory of Molecular Electronic Structure.* W. A. Benjamin, Inc., New York, 1963.

PAULING, L., and WILSON, E. B., JR., *Introduction to Quantum Mechanics.* McGraw-Hill Book Company, New York, 1935.

SLATER, J. C., *Quantum Theory of Atomic Structure*, Vol. 1. McGraw-Hill Book Company, New York, 1960.

PROBLEMS

***6.1** The restoring potential for a molecular vibration is not exactly harmonic. Let

$$V(r) = \frac{k}{2} r^2 + ar^3 + br^4 + cr^5$$

Use first-order perturbation theory to calculate the first three energy levels of an oscillator with this potential. The first three wave functions for the harmonic oscillator are ($\alpha = k/h\nu_0$)

$$\psi_0(x) = \left(\frac{\alpha}{\pi}\right)^{1/4} \exp\left(-\frac{\alpha x^2}{2}\right)$$

$$\psi_1(x) = \left(\frac{4\alpha^3}{\pi}\right)^{1/4} x \exp\left(-\frac{\alpha x^2}{2}\right)$$

$$\psi_2(x) = \left(\frac{\alpha}{4\pi}\right)^{1/4} (2\alpha x^2 - 1) \exp\left(-\frac{\alpha x^2}{2}\right)$$

What generalization can you make about the allowed forms of perturbations for a harmonic oscillator?

***6.2** Assuming a one-electron Gaussian trial function of the form $(2\alpha/\pi)^{3/4} \exp(-\alpha r^2)$, with α a variational parameter, evaluate the variational energy for the helium atom. Note that all integrals will have to be evaluated. They can be found in standard integral tables. [*Hint:* Use the Ewald transformation for the two-electron integral.]

6.3 Find the r value corresponding to the maximum electron probability for the ground state of the helium atom from the results of the previous problem.

6.4 Carry out a second-order perturbation-theory treatment of helium using hydrogenic orbitals and including in the expansion only the terms $1s(1)2s(2)$, $1s(1)3s(2)$, $1s(1)4s(2)$, and $2s(1)2s(2)$ (and the terms involving an interchange of the electrons). Compare the magnitudes of each term in the correction to the energy.

***6.5** The wave function for a particle in a one-dimensional potential well, with the origin in the center and the boundaries at $\pm L/2$, is

$$\psi(x) = \sqrt{\frac{2}{L}} \left\{ \cos\left[\frac{n\pi x}{L} - \frac{(n-1)\pi}{2}\right] \right\}$$

(The energies are independent of where the origin is placed.) Consider a potential well of the following form:

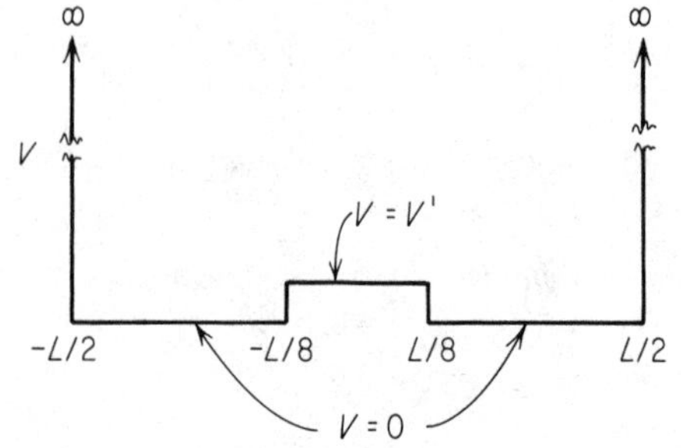

(a) Do a first-order perturbation-theory treatment of the first three energy levels, assuming $V' = h^2/8mL^2$.

(b) Do a second-order perturbation-theory treatment of the first two energy levels, including quantum numbers up to 5 in the expansion.

(c) What apparent symmetry restrictions are there on the integrals appearing in parts (a) and (b)?

***6.6** Repeat Problem 6.5 with a potential well of the following form:

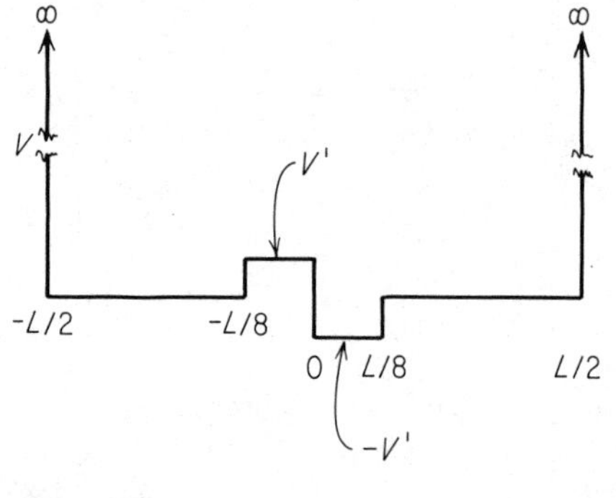

chapter seven

The electronic structure of many-electron atoms

7.1 THE CENTRAL-FIELD APPROXIMATION

Almost all theoretical treatments of many-electron atoms are based upon the *central-field approximation.* In this approximation, it is assumed that each electron moves independently in a spherically averaged field arising from the nucleus and all the other electrons. This was, for example, our implicit assumption in the helium-atom variational problem. With this spherically averaged field, the angular behavior of the one-particle wave functions for the individual electrons will be identical to that of the hydrogen atom. To the extent that the independent-particle wave function is valid, an l and an m quantum number can be assigned to the orbital of each electron. For convenience, an n value is also assigned to each electron, although the hydrogenic n is not a valid quantum number for a many-electron atom. However, the chemical periodicity of the elements (Section 7.2) implies that n is a good approximation to a quantum number.

In the technique most commonly used for actual calculations on many-electron atoms, an assumed spatial distribution of the electrons is chosen. The equations of motion for one electron are solved with the potential calculated from the assumed distribution of the others. The calculated distribution for this electron is then incorporated in the approximate field, another electron is solved for in this new field, and the process is continued until a calculated spatial distribution is obtained for all the electrons. This will most probably be different from the initially assumed distribution. The entire process is then repeated. If the second calculated distribution does not agree with the first, the process is repeated. The cycle is continued until the results at the end of one iteration agree with those from the preceding iteration, to the desired accuracy.

This procedure, called a *self-consistent field* (or S.C.F.) procedure, was developed by D. R. Hartree in 1928. Hartree used direct numerical integration in his work. Most later work employs trial wave functions constructed as linear combinations of some suitable basis functions. Once the integrals involving the various terms in the Hamiltonian have been evaluated with respect to the basis functions, the iteration process involves relatively simple matrix algebra that is ideally suited for electronic computers. Hartree's original energy expression was modified in 1930 by V. Fock to account for the proper interchange symmetry of the electrons. The usual S.C.F. procedure is frequently referred to as the *Hartree-Fock method.*

The Hartree-Fock self-consistent field technique (or some variant thereof) is the procedure most commonly used for numerical quantum-chemical calculations. Very complete Hartree-Fock calculations can be done on modern electronic computers with relative ease for atoms, and with somewhat more difficulty for chemically interesting molecules. However, the results still suffer from the limitations of independent-particle wave functions—wave functions constructed as products of one-electron functions. Other techniques can be added to improve the accuracy of the results, but always at greater computational expense.

7.2 THE PAULI PRINCIPLE AND THE AUFBAU PRINCIPLE

In 1925 (before the papers of Heisenberg and Schrödinger), W. Pauli proposed the well-known *exclusion principle.* In order to explain the periodicity of the elements, he suggested that each electron in a many-electron atom must have four quantum numbers associated with it. Furthermore, "There never exist two or more equivalent electrons in an atom which, in strong fields, agree in all quantum numbers, n, k_1, k_2, m_1." This implies the existence of an additional quantum number not present in the solutions to the hydrogenic problem. This was later identified by Dirac (1933) as the electron *spin* (or *intrinsic angular momentum*) quantum number. (We now designate the quantum numbers as n, l, m, and m_s.) Two electrons can be associated with each orbital corresponding to a given n, l, and m, but these two electrons must have different values of m_s. As we shall soon see, the Pauli principle can be related to the permutational symmetry of the electrons. The four quantum numbers provide a useful mnemonic for determining the orbital occupancy (*configuration*) for many-electron atoms.

For most of the known elements, the orbital occupancy in the neutral atom can be deduced by ordering the orbitals in terms of increasing $(n + l)$, with those of the lower n coming first when there are more than one with a given $(n + l)$ value. The appropriate number of electrons are assigned to each level (each combination of n, l, and m), with two placed in each level, until all the electrons have been assigned. (All

Table 7.1

ORDERING OF ATOMIC ORBITALS IN TERMS OF $(n + l)$ AND OF n

$n + l$	Orbitals			
1	1*s*			
2	2*s*			
3	2*p*	3*s*		
4	3*p*	4*s*		
5	3*d*	4*p*	5*s*	
6	4*d*	5*p*	6*s*	
7	4*f*	5*d*	6*p*	7*s*
8	5*f*	6*d*	7*p*	8*s*

m values corresponding to a given *l* value are energetically degenerate; however, the hydrogenic degeneracy of all *l* values corresponding to a given *n* value has been broken.) The procedure just described is commonly known as the *aufbau principle.*

Table 7.1 shows the level ordering up to an $(n + l)$ value of 8. This is sufficient for all known elements. Each *s* level can accommodate two electrons; each *p*, six; each *d*, ten; and each *f* level, fourteen. Thus, for example, element number 30 (zinc) would have the configuration $(1s)^2(2s)^2(2p)^6(3s)^2(3p)^6(4s)^2(3d)^{10}$. (This is frequently abbreviated as $[\mathrm{Ar}](4s)^2(3d)^{10}$. The configuration of the inert gas, argon, is symbolized by [Ar] and only the valence electrons are specifically listed.) Exceptions to the predicted ordering occur for a few of the transition elements and for a number of the lanthanides and actinides. Some of the exceptions can be explained by an additional rule: when a completely filled or half-filled *d* level can be obtained by transferring one *s* electron from the next lower $(n + l)$, the transfer occurs. For example, Cr (element number 24) has the configuration $[\mathrm{Ar}](4s)^1(3d)^5$, and Cu (element number 29) has the configuration $[\mathrm{Ar}](4s)^1(3d)^{10}$. This gives the correct configuration for Mo (but not W), Ag, and Au. (Actinium and lanthanum violate the simple aufbau principle in that they have d^1, rather than f^1, configurations.) It should be emphasized that the aufbau principle is a bookkeeping aid and does not necessarily give the order of the energy levels (i.e., the ionization potentials) in the atoms.

The chemical periodicity of the elements reflects a periodicity in their electronic configurations. Elements in the same family of the periodic table would have the same numbers of valence electrons associated with a given *l* value, if the $(n + l)$ rule were rigorously obeyed. For example, all inert gases except helium have $(ns)^2(np)^6$ configurations, all members of the oxygen family have $(ns)^2(np)^4$ configurations, all alkali metals have $(ns)^1$ configurations, and so on. In fact, the layout of a modern periodic table reflects the *l* value of the last electron placed by the aufbau principle (Figure 7.1).

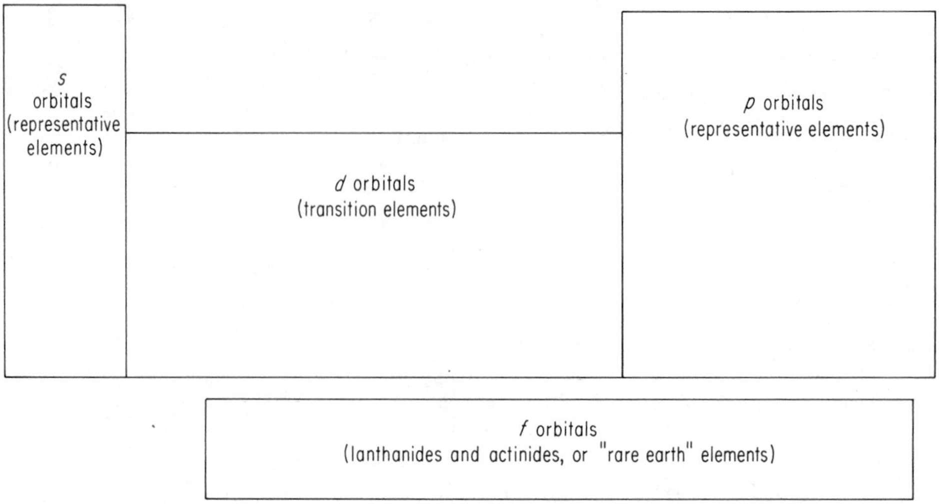

Figure 7.1 Schematic representation of the periodic table, indicating the last orbital type assigned by the aufbau principle. (Hydrogen and helium have been omitted.)

7.3 SYMMETRY GROUPS OF MANY-PARTICLE SYSTEMS

In order to find the symmetry-allowed states arising from a given configuration of a many-electron system, we must know the structure of the complete symmetry group of the system. The overall group structure for describing a many-particle system should include all the symmetries the system can have. The most obvious of these is the spatial symmetry that we have already discussed. Two others are also extremely important: the *intrinsic-angular-momentum symmetry* of the individual particles, and the *permutational symmetry* relating to interchanges of identical particles. The unitary unimodular groups, $\mathbf{SU}(n)$, in which n is $(2s + 1)$ and s is the "spin" of the particle, are used to describe the intrinsic angular momenta of particles. For the electron, this is $\mathbf{SU}(2)$. Although will not explicitly use these groups here (they will be discussed further in Chapter 17), we will use the fact that $\mathbf{SU}(2)$ is *isomorphic* to (has the same group structure as) $\mathbf{R}(3)$, when the double-valued representations are included in $\mathbf{R}(3)$.

The group of a many-particle system is some sort of a product of the groups of the individual particles. The exact form of the product structure depends upon whether or not the independent-particle approximation is being employed. If it is (as will be the case for all our applications), then the product is simply the direct product (see Appendix 2) of the groups for the individual particles, with restrictions on the result imposed by the permutational symmetry. As long as we are considering identical particles (electrons) and spatial symmetry, the groups for the individual particles will be identical, except for the number labels on the particles. These number labels will be permuted by the group representing the permutations of equivalent particles. In an atom, electrons having the same values for their n and l quantum numbers are equivalent. There can be up to two equivalent ns electrons, six np, ten nd, and so on. Application of the methods that we will develop in Section 7.5 shows that when a given level is completely occupied (i.e., the shell is *closed*), any spatial and permutational symmetry requirements for that level are completely satisfied. Consequently, group theory need be applied only to *open shells* (incompletely filled levels).

Let us consider the ground-state electronic configuration of the carbon atom: $(1s)^2(2s)^2(2p)^2$. The $1s$ and $2s$ levels are completely filled; they are closed shells. The $2p$ level, which can accommodate six electrons, contains only two; it is a partially filled, or open, shell. Now let us consider the overall intrinsic (spin) angular momentum of two electrons. This corresponds to the product of two $D^{1/2}$ representations in **R**(3). We have

$$D^{1/2} \times D^{1/2} = D^0 + D^1 \tag{7.1}$$

The resulting values for the total spin, S, are zero and one. The D^0, with an overall S value of zero, corresponds to the nondegenerate *singlet state* and the D^1, with an S of one, to the triply degenerate *triplet state.* (The spin degeneracy, or multiplicity, of a state equals $2S + 1$.) For the overall orbital angular momentum, L, we must take the product

$$D^1 \times D^1 = D^0 + D^1 + D^2 \tag{7.2}$$

since a p orbital has an l value of 1. The state with a total L value of zero is an S state, that with a value of one is a P state, and that with a value of two is a D state. (Note that, in both the spin and orbital angular momenta, capital letters are used to designate total values of a quantum number. Lower-case letters are used for single-particle quantum numbers. The $S, P, D, \ldots$ designation for the total orbital angular momentum is completely analogous to the $s, p, d, \ldots$ designation for the single-particle orbital angular momentum.)

The most general statement of the Pauli principle is based upon interchange (or permutational) symmetry. For systems of *Fermions* (particles with half-integer intrinsic angular momentum—i.e., half-integer spin), the total wave function must be *antisymmetric* with respect to the interchange of two equivalent particles. A single-particle function can be considered to be the product of a function of the spatial coordinates (orbital) and a function of the spin coordinates (intrinsic angular momentum). A many-particle function, then, can be constructed as a product of spin-orbitals—that is, of the space and spin functions of each particle. The final result can be considered to be a product of functions of the overall spin and the overall spatial coordinates.

In Eq. 7.1 we have the representations for the allowed total spin functions for carbon, while in Eq. 7.2 we have the allowed representations for the total orbital functions. If we knew which of these were symmetric and which were antisymmetric under the interchange of the equivalent electrons, we could deduce the overall allowed states for the $(1s)^2(2s)^2(2p)^2$ configuration of carbon. For the simple binary product of a representation of **R**(3) with itself, as is required for the two-electron atomic case, this is easy to determine. As will be seen in Section 7.5, if the original representation has an integer index, the even-indexed representations in the decomposition of the binary product are symmetric and the odd-indexed ones antisymmetric, with respect to the interchange of particles. The converse is true if the original representation has a half-integer index. Thus, for the spin representations of Eq. 7.1, D^0 (the singlet) is antisymmetric while D^1 (the triplet) is symmetric. For the orbital representations, D^0 and D^2 (S and D, respectively) are symmetric, while D^1 (P) is antisymmetric. In order for the total wave function to be antisymmetric, the spin part must be symmetric and the orbital part antisymmetric, or vice versa. Thus the S and D states must be singlet states, while the P state must be a triplet state. The usual designations are ^3P, ^1S, and ^1D. It turns out that the ^3P state is the ground state by Hund's rule (see Section 7.8).

7.4 THE SYMMETRIC GROUP

In order to find Pauli-allowed states for more general systems, we need to use the symmetric permutation group (or, in the language of mathematics, the *symmetric group*). The symmetric group, $\mathbf{S}(N)$, of degree N, is the group that has as its operations all the possible permutations of N objects. For example, if we have two objects, we can arbitrarily label them 1 and 2. Our group, in this case, $\mathbf{S}(2)$, consists of the identity (which we will always call E) and the operation that interchanges the objects. We can write the results

$$\begin{aligned} \{1\ \ 2\} &\xrightarrow{E} \{1\ \ 2\} \\ &\xrightarrow{P} \{2\ \ 1\} \end{aligned} \qquad 7.3$$

The group has only two operations (or two elements) and is said to have an *order* of two. For three objects we can have

$$\begin{aligned} \{1\ \ 2\ \ 3\} &\xrightarrow{E} \{1\ \ 2\ \ 3\} \\ &\xrightarrow{P_1} \{3\ \ 1\ \ 2\} \\ &\xrightarrow{P_2} \{2\ \ 3\ \ 1\} \\ &\xrightarrow{P_3} \{1\ \ 3\ \ 2\} \\ &\xrightarrow{P_4} \{2\ \ 1\ \ 3\} \\ &\xrightarrow{P_5} \{3\ \ 2\ \ 1\} \end{aligned} \qquad 7.4$$

Thus, $\mathbf{S}(3)$ is of order 6. $\mathbf{S}(4)$ is of order 24. In general, the order of $\mathbf{S}(N)$ is $N!$ (Note that $N! = 1 \times 2 \times \ldots \times (N-1) \times N \equiv \prod_{i=1}^{N} i$.) The symmetric group has a finite number of operations, in contrast to the rotation groups, which have an infinite number of infinitesimally different operations. The character table of a symmetric group thus has a finite number of entries.

The character table of a symmetric group, like that of any group, has its rows labeled by the labels of the irreducible representations and its columns by those of the group elements (the permutation operations). For all $\mathbf{S}(N)$ with N greater than two, many of the permutations have the same character in each representation. These

Table 7.2

CHARACTER TABLES FOR $\mathbf{S}(2)$, $\mathbf{S}(3)$ AND $\mathbf{S}(4)$

Y.D.	S(2)	(1^2)	(2)
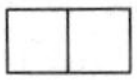	$[2]$	1	1
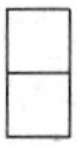	$[1^2]$	1	−1

Y.D.	S(3)	(1^3)	$3(2, 1)$	$2(3)$
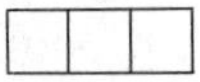	$[3]$	1	1	1
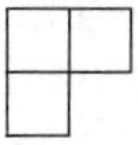	$[2, 1]$	2	0	−1
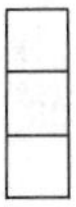	$[1^3]$	1	−1	1

Y.D.	S(4)	(1^4)	$6(2, 1^2)$	$3(2^2)$	$8(3, 1)$	$6(4)$
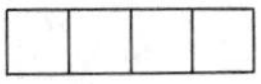	$[4]$	1	1	1	1	1
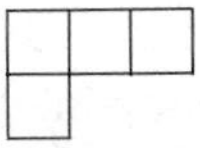	$[3, 1]$	3	−1	−1	0	1
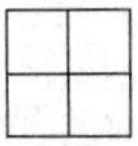	$[2^2]$	2	0	2	−1	0
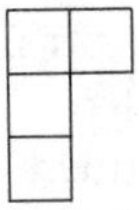	$[2, 1^2]$	3	1	−1	0	−1
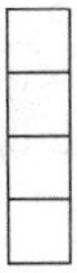	$[1^4]$	1	−1	1	1	−1

characters, and the corresponding permutations, are grouped together into *classes*. The class labels are the actual column labels. The labels used for the classes and for the irreducible representations for the **S**(N) groups are the same. They are the ways that the number, N, can be partitioned. For example, if N is 4, the possible partitions are (4), (3, 1), (2, 2), (2, 1, 1), or (1, 1, 1, 1). These are abbreviated (4), (3, 1), (2^2), (2, 1^2), and (1^4). The individual numbers are called *cycle lengths*. The classes are labeled by the partition in parentheses, while the representations are labeled by the partitions in square brackets. In the character table, the number of permutations in each class is indicated with its label. The significance of the partition notation is explained in the appendix to this chapter.

Table 7.2 presents the character tables for **S**(2), **S**(3), and **S**(4). The symmetric groups up to **S**(7) are given in Appendix 7. The collections of boxes to the left of the character tables in Table 7.2 are the *Young diagrams* for the appropriate representations. Note that these are simply rows of boxes corresponding to the partition structure. Each row corresponds to a cycle length and there is a row for each cycle. The rows are placed in descending numbers of blocks and aligned to the left.

Young diagrams are useful for determining conjugate relations among the representations of the groups. Note that in each **S**(N), for every Young diagram there is another that could be constructed from the first by an interchange of rows and columns, unless such an interchange regenerates the same diagram. Representations bearing this relationship between their Young diagrams are said to be *conjugate* to one another. Thus, in **S**(4), the representations [4] and [1^4] are a conjugate pair, as are [3, 1] and [2, 1^2]. The representation [2^2] is said to be *self-conjugate*. The concept of conjugate representations is important in constructing Pauli-allowed states of many-electron systems. The representation, [1^N], is the totally antisymmetric representation of **S**(N). It turns out that the only way the totally antisymmetric permutational representation can arise, in a product of representations, is by the multiplication of a representation by its conjugate.

7.5 PAULI-ALLOWED STATES

The general form of the Pauli principle is based upon the fact that only one-dimensional permutational states can be physically real. The symmetric group contains only two one-dimensional irreducible representations—the totally symmetric representation and the totally antisymmetric representation. The total wave function for a collection of equivalent particles with half-integer intrinsic angular momentum (*Fermions*) must be totally antisymmetric with respect to the permutations of the equivalent particles. This is related to the double-valued nature of the angular-momentum representations for half-integer angular momentum. The total wave function for a collection of equivalent particles with integer angular momentum (*Bosons*) must be totally symmetric with respect to the permutation of the equivalent particles.

For problems in which the spin does not appear explicitly in the Hamiltonian (this includes most problems we will consider), only the permutational properties of the spin function are required. These can be determined directly from the appropriate symmetric groups, whithout any explicit references to the angular-momentum groups. The spatial functions then have to be adapted to the appropriate permutational symmetry. For Fermions (such as electrons), the spatial and spin functions must transform as conjugate irreducible representations of the appropriate $\mathbf{S}(N)$, so that their product will be totally antisymmetric. For Bosons, the spatial and spin functions must transform as the same irreducible representation, so that their product will be totally symmetric.

The allowed permutational symmetries of spin functions can be obtained directly from Young diagrams. If the intrinsic angular momentum of a particle transforms as D^s within $\mathbf{R}(3)$, then for N particles the representation for the total intrinsic angular momentum will be $(D^s)^N$, with the appropriate permutational restrictions. The only allowed permutational representations within $\mathbf{S}(N)$ are those having no more than $(2s + 1)$ rows in their Young diagrams. For electrons, s is $\frac{1}{2}$, and the allowed Young diagrams can have no more than two rows. Consulting Table 7.2, we see that two equivalent electrons can have permutational symmetry corresponding to $[2]$ and $[1^2]$ within $\mathbf{S}(2)$; three equivalent electrons, corresponding to $[3]$ and $[2, 1]$ of $\mathbf{S}(3)$; and four equivalent electrons, corresponding to $[4]$, $[3, 1]$, and $[2^2]$ of $\mathbf{S}(4)$. The total spin value, S, corresponding to these can be determined by assigning the maximum m_s value to each block in the top row of the Young diagram, and successively lower m_s values to the lower rows, and then summing the values over all blocks of the Young diagram. For electrons, the top row would have $+\frac{1}{2}$ assigned to each block and the second row, $-\frac{1}{2}$. For a particle with an s value of 1, m_s can have the numbers 1, 0, and -1. These would be the numbers assigned to the three rows of the allowed Young diagrams.

To illustrate, let us consider a four-equivalent-electron case. The allowed permutational representations are $[4]$, $[3, 1]$, and $[2^2]$. Assigning the numbers to the blocks in the Young diagrams, we get the following results:

$$[4]: \begin{array}{|c|c|c|c|}\hline \frac{1}{2} & \frac{1}{2} & \frac{1}{2} & \frac{1}{2} \\ \hline \end{array}$$

$$\sum m_s = 4 \times \tfrac{1}{2} = 2$$

$$S = 2 \qquad \textbf{7.5}$$

$$[3, 1]: \begin{array}{|c|c|c|}\hline \frac{1}{2} & \frac{1}{2} & \frac{1}{2} \\ \hline -\frac{1}{2} & \multicolumn{2}{c}{} \\ \cline{1-1} \end{array}$$

$$\sum m_s = 3 \times \tfrac{1}{2} - \tfrac{1}{2} = 1$$

$$S = 1 \qquad \textbf{7.6}$$

$$[2^2]: \begin{array}{|c|c|} \hline \frac{1}{2} & \frac{1}{2} \\ \hline -\frac{1}{2} & -\frac{1}{2} \\ \hline \end{array}$$

$$\sum m_s = 2 \times \tfrac{1}{2} - 2 \times \tfrac{1}{2} = 0$$
$$S = 0 \qquad \textbf{7.7}$$

If the representations within $\mathbf{R}(3)$ of these spin states are required, they are simply D^S. Furthermore, there are as many of each as the dimension of the representation from $\mathbf{S}(N)$ from which it was derived. From the results of Eqs. 7.5 through 7.7 we see that we would have one D^2, three D^1, and two D^0 representations. This is the same thing we would obtain if we were to take the product of four $D^{1/2}$'s.

Once the representations for the spin have been determined, the spatial functions must be adapted to the appropriate permutational representations. There is a systematic way of accomplishing this. Any partition of N numbers, and consequently any class of $\mathbf{S}(N)$, can be expressed in the form

$$(\lambda) = (1^{b_1}, 2^{b_2}, \ldots, N^{b_N}) \qquad \textbf{7.8}$$

where b_1 is the number of cycles of length one, b_2 is the number of cycles of length two, and so on. Obviously, in any partition a number of the b_i will equal zero. With this notation, we can work out the characters for any class in a spatial group that is adapted to the representation $[\lambda]$ of the symmetric group $\mathbf{S}(N)$ by use of the formula

$$\chi_\Gamma(R); [\lambda] = \frac{1}{N!} \sum_{c_P \in \mathbf{S}(N)} h_{c_P} \chi(P)_{[\lambda]} \prod_{i=1}^{N} [\chi(R^i)]^{b_i} \qquad \textbf{7.9}$$

where $\chi_\Gamma(R); [\lambda]$ is the character of operation R in the Γ representation of the spatial group after it has been adapted to representation $[\lambda]$ of $\mathbf{S}(N)$; N is the degree of the $\mathbf{S}(N)$ ($N!$ is its order); the summation is over the classes, c_P, of $\mathbf{S}(N)$; h_{c_P} is the order of the class c_P of $S(N)$; $\chi(P)_{[\lambda]}$ is the character of the permutation P (within the class c_P) for representation $[\lambda]$ of $\mathbf{S}(N)$; the product is over the cycle lengths, i, of the partitions for the class (λ) (see Eq. 7.8); R^i is the operation R of the spatial group raised to the ith power [for example, $C^2(\phi)$ equals $C(2\phi)$]; and $[\chi(R^i)]^{b_i}$ indicates that the resulting character is raised to the b_ith power.

Equation 7.9 is actually more complicated to describe than it is to apply. To illustrate its use, let us first construct the symmetrized and antisymmetrized squares of D^1. This corresponds to adaptations to the representations [2] and $[1^2]$, respectively, of $\mathbf{S}(2)$. The characters for D^1 within $\mathbf{R}(3)$ can be found from Table 3.5, while the character table for $\mathbf{S}(2)$ is found in Table 7.2. Considering first the adaptation to [2], we have for the identity of $\mathbf{R}(3)$

$$\chi_{D^1}(E); [2] = \frac{1}{2!}\{1 \times 1 \times [\chi(E^1)]^2 + 1 \times 1 \times [\chi(E^2)]\}$$

$$= \frac{1}{2!}[1 \times 1 \times (3)^2 + 1 \times 1 \times (3)]$$

$$= 6 \qquad \textbf{7.10}$$

For the $C(\phi)$ of $\mathbf{R}(3)$, we have

$$\begin{aligned}\chi_{D^1}[C(\phi)]; [2] &= \frac{1}{2!}(1 \times 1 \times \{\chi[C(\phi)]\}^2 + 1 \times 1 \times \chi[C^2(\phi)]) \\ &= \frac{1}{2!}[1 \times 1 \times (1 + 2\cos\phi)^2 + 1 \times 1 \times (1 + 2\cos 2\phi)] \\ &= \frac{1}{2}(1 + 4\cos\phi + 4\cos^2\phi + 1 + 2\cos 2\phi) \\ &= \frac{1}{2}(1 + 4\cos\phi + 2 + 2\cos 2\phi + 1 + 2\cos 2\phi) \\ &= \frac{1}{2}(4 + 4\cos\phi + 4\cos 2\phi) \\ &= 2 + 2\cos\phi + 2\cos 2\phi \end{aligned} \qquad \mathbf{7.11}$$

where, in the second line, we have used the fact that $C^2(\phi)$, operating twice with $C(\phi)$, is equivalent to $C(2\phi)$, and in the fourth line we have used the trigonometric identity for $\cos^2\phi$. The resulting reducible representation

$\mathbf{R}(3)$	E	$C(\phi)$
Γ	6	$2 + 2\cos\phi + 2\cos 2\phi$

7.12

reduces to $D^0 + D^2$. The adaptation of D^1 to $[1^2]$ differs only in the character of the class (2) of $\mathbf{S}(2)$. We have

$$\begin{aligned}\chi_{D^1}(E); [1^2] &= \frac{1}{2!}\{1 \times 1 \times [\chi(E)]^2 + 1 \times (-1) \times [\chi(E^2)]\} \\ &= \frac{1}{2}(3^2 - 3) \\ &= 3 \end{aligned} \qquad \mathbf{7.13}$$

and

$$\begin{aligned}\chi_{D^1}[C(\phi)]; [1^2] &= \frac{1}{2!}\{1 \times 1 \times (\chi[C(\phi)])^2 + 1 \times (-1) \times (\chi[C^2(\phi)])\} \\ &= \frac{1}{2}[1 + 4\cos\phi + 4\cos^2\phi - (1 + 2\cos 2\phi)] \\ &= 1 + 2\cos\phi \end{aligned} \qquad \mathbf{7.14}$$

The representation is the irreducible D^1. The spin function associated with [2] is the triplet state. The conjugate $[1^2]$ spatial adaptation goes with this. Thus, the D^1

spatial function goes with the [2] spin function for a ^3P state. Similarly, the [1^2] spin function is the singlet state. This must be associated with a spatial function adapted to [2]. The D^0 and D^2 spatial functions go with the [1^2] spin functions for ^1S and ^1D states.

7.6 TERM SYMBOLS

The symbols we have used, ^1S, ^3P, ^1D, are partial *term symbols*, specifying the total spin angular momentum, S (do not confuse the symbol, S, for the total spin quantum number with the S symbol for a total orbital quantum number of zero), and the total orbital angular momentum, L, for the atom. The complete term symbol includes S, L, and J, the total angular momentum. It has the form $^{(2S+1)}L_J$, where $(2S + 1)$ is the spin multiplicity expressed as a number. L is the orbital angular-momentum quantum number, expressed in the $S, P, D, F, \ldots$ convention. J is the total angular-momentum quantum number (arising from a coupling of the spin and orbital angular momenta), expressed as a number. For many-electron atoms, L and S are only approximate quantum numbers. (L and S cannot be directly observed. The only angular momentum observable is J, the total angular momentum.) The higher the atomic number, the worse the approximation. The total angular momentum, J, is the only true quantum number. For a given electronic configuration of an atom, each different term symbol will correspond to a state of different energy.

Two different coupling schemes are in common use for assigning the J values in term symbols. If the interaction of the spin and orbital angular momenta (a relativistic effect) is negligible compared to electronic repulsion effects (as it is for light atoms), the total L values are found by coupling the one-electron l values, the total S values by coupling the one-electron s values, and the J values by coupling the L and S values. As we have seen, permutational symmetry restricts the allowed combinations of L and S values; however, there are no restrictions on the J values that result from the L-S coupling. A simple product of the representations D^L and D^S gives the allowed D^J representations. For example, for the ^3P state of carbon, we have

$$\begin{aligned} D^S \times D^L &= D^1 \times D^1 \\ &= D^0 + D^1 + D^2 \end{aligned} \tag{7.15}$$

for J values of zero, one, and two, or term symbols $^3P_0$, $^3P_1$, and $^3P_2$. For the singlet states, the spin representation is the totally symmetric D^0; consequently, J will equal L, giving the terms $^1S_0$ and $^1D_2$. The coupling scheme we have just described is frequently referred to as *Russell-Saunders coupling*, or L-S coupling.

The coupling scheme appropriate for heavy atoms, where the spin-orbital interaction is large, is called *j-j coupling*. In this scheme, the l and the s for each electron are coupled to give a one-electron j value. The individual j's are then coupled to give the overall J values. Permutational symmetry comes into play when equivalent electrons with the same j are coupled. The set of states resulting from a given electronic configuration will be the same for either *j-j* coupling or Russell-Saunders coupling. However, S and L are meaningless in *j-j* coupling. Consider a p^2 configuration again. The only j values will be $\frac{1}{2}$ and $\frac{3}{2}$, since

$$D^{1/2} \times D^1 = D^{1/2} + D^{3/2} \tag{7.16}$$

Coupling the equivalent j values of $\frac{1}{2}$ for the two electrons, we have

$$D^{1/2} \times D^{1/2} = D^0 + D^1 \tag{7.17}$$

but only D^0 is antisymmetric. This gives us a J value of zero. Coupling the equivalent j values of $\frac{3}{2}$ gives

$$D^{3/2} \times D^{3/2} = D^0 + D^1 + D^2 + D^3 \tag{7.18}$$

The D^0 and D^2 are antisymmetric, for J values of zero and two. Finally, the non-equivalent j values can be coupled:

$$D^{1/2} \times D^{3/2} = D^1 + D^2 \tag{7.19}$$

Both of these are allowed, owing to the nonequivalence. We see, then, that we have two states with $J = 0$ (from Eqs. 7.17 and 7.18), one with $J = 1$ (from Eq. 7.19), and two with $J = 2$ (from Eqs. 7.18 and 7.19), just as we found in the Russell-Saunders coupling for carbon. The two different schemes make perturbation-theory calculations, and qualitative discussions based upon perturbation arguments, easier at the two extremes. For most of this text we will confine our attention to Russell-Saunders coupling.

7.7 RUSSELL-SAUNDERS COUPLING FOR NITROGEN AND PROTACTINIUM

We will apply Russell-Saunders coupling to find Pauli-allowed states and term symbols for two more systems. As our next example, let us consider the ground-state configuration of nitrogen, N: $(1s)^2(2s)^2(2p)^3$. We now have three electrons in the open p shell. The appropriate symmetric group is **S**(3). Looking at the Young diagrams of **S**(3) (Table 7.2), we see that the allowed spin representations are [3] and [2, 1]. These correspond to S values of $\frac{3}{2}$ and $\frac{1}{2}$, respectively, and are quartet and doublet spin states. The spatial function for the quartet state must be adapted to the $[1^3]$ representation, the conjugate of [3], while that for the doublet must be adapted to [2, 1], since [2, 1] is self-conjugate. Again, the p orbitals transform as D^1. For the adaptation of the identity to $[1^3]$ we have

$$\begin{aligned} \chi_{D^1}(E); [1^3] &= \frac{1}{3!}\{1 \times 1 \times [\chi(E)]^3 + 3 \times (-1) \times [\chi(E^2)][\chi(E)] \\ &\quad + 2 \times 1 \times [\chi(E^3)]\} \\ &= \frac{1}{6}[27 - (3 \times 3 \times 3) + (2 \times 3)] \\ &= 1 \end{aligned} \tag{7.20}$$

In this instance, this is sufficient to tell us that the only result will be D^0, since only D^0 has an identity character of unity. Let us continue, however, and find $\chi[C(\phi)]$. We have

$$\begin{aligned}\chi_{D^1}[C(\phi)]; [1^3] &= \frac{1}{3!}(1 \times 1 \times \{\chi[C(\phi)]\}^3 \\ &\quad + 3 \times (-1) \times \{\chi[C^2(\phi)]\}\{\chi[C(\phi)]\} \\ &\quad + 2 \times 1 \times \{\chi[C^3(\phi)]\}) \\ &= \frac{1}{6}[(7 + 12 \cos \phi + 6 \cos 2\phi + 2 \cos 3\phi) \\ &\quad - 3(1 + 4 \cos \phi + 2 \cos 2\phi + 2 \cos 3\phi) \\ &\quad + 2(1 + 2 \cos 3\phi)] \\ &= 1\end{aligned} \tag{7.21}$$

(In Eq. 7.21 we have used trigonometric identities to obtain simplifications for finding the products of characters under the rotations.) We have, for the quartet state, only the one term, $^4S_{3/2}$.

Now let us adapt D^1 to $[2, 1]$ to obtain the allowed terms for the doublet state. For the identity we have

$$\begin{aligned}\chi_{D^1}(E); [2, 1] &= \frac{1}{3!}(1 \times 2 \times [\chi(E)]^3 + 3 \times 0 \times [\chi(E^2)][\chi(E)] \\ &\quad + 2 \times (-1) \times [\chi(E^3)]) \\ &= \frac{1}{6}(54 + 0 - 6) \\ &= 8\end{aligned} \tag{7.22}$$

and for $C(\phi)$

$$\begin{aligned}\chi_{D^1}[C(\phi)]; [2, 1] &= \frac{1}{3!}(1 \times 2 \times \{\chi[C(\phi)]\}^3 \\ &\quad + 3 \times 0 \times \{\chi[C^2(\phi)]\}\{\chi[C(\phi)]\} \\ &\quad + 2 \times (-1) \times \{\chi[C^3(\phi)]\}) \\ &= \frac{1}{6}[(14 + 24 \cos \phi + 12 \cos 2\phi + 4 \cos 3\phi) \\ &\quad + 0 - (2 + 4 \cos 3\phi)] \\ &= 2 + 4 \cos \phi + 2 \cos 2\phi\end{aligned} \tag{7.23}$$

Our representation, which is reducible, is

R(3)	E	$C(\phi)$
Γ	8	$2 + 4 \cos \phi + 2 \cos 2\phi$

7.24

This reduces to $D^1 + D^2$. We can have doublet P and D states. Coupling L and S to find J, we see that we have, for the P states,

$$D^{1/2} \times D^1 = D^{1/2} + D^{3/2} \tag{7.25}$$

and for the D states,

$$D^{1/2} \times D^2 = D^{3/2} + D^{5/2} \tag{7.26}$$

Our allowed doublet terms are $^2P_{1/2}$, $^2P_{3/2}$, $^2D_{3/2}$, and $^2D_{5/2}$.

The term symbols for any other open-shell atom or ion can be determined analogously to the way we have determined them for carbon and nitrogen. If the system has more than one open shell, as occurs for the ground states of many transition elements and rare earths, and for many excited states of all atoms except hydrogen, each open shell is first treated separately and the results are combined. There are no restrictions on the combination of the different open shells, since the electrons in them are not equivalent. For example, element number 91, protactinium, has the ground-state configuration, Pa: [Rn]$(7s)^2(6d)^1(5f)^2$. The d level and the f level are both open. The single 6d electron leads to a ^2D substate, while the two 5f electrons lead to ^1S, ^1D, ^1G, and ^1I and ^3P, ^3F, and ^3H substates. The ^2D substate can be coupled to all the others in all possible combinations. Table 7.3 shows all the resulting terms from this coupling. Actually, j-j coupling is more appropriate for Pa.

Table 7.3

POSSIBLE TERMS ARISING FROM AN f^2d^1 ELECTRONIC CONFIGURATION

f^2 Substate	Terms Arising from Coupling with d^1 Substate
^1S	$^2D_{3/2}$, $^2D_{5/2}$
^1D	$^2S_{1/2}$, $^2P_{1/2}$, $^2P_{3/2}$, $^2D_{3/2}$, $^2D_{5/2}$, $^2F_{5/2}$, $^2F_{7/2}$, $^2G_{7/2}$, $^2G_{9/2}$
^1G	$^2D_{3/2}$, $^2D_{5/2}$, $^2F_{5/2}$, $^2F_{7/2}$, $^2G_{7/2}$, $^2G_{9/2}$, $^2H_{9/2}$, $^2H_{11/2}$, $^2I_{11/2}$, $^2I_{13/2}$
^1I	$^2G_{7/2}$, $^2G_{9/2}$, $^2H_{9/2}$, $^2H_{11/2}$, $^2I_{11/2}$, $^2I_{13/2}$, $^2J_{13/2}$, $^2J_{15/2}$, $^2K_{15/2}$, $^2K_{17/2}$
^3P	$^2P_{1/2}$, $^2P_{3/2}$, $^2D_{3/2}$, $^2D_{5/2}$, $^2F_{5/2}$, $^2F_{7/2}$, $^4P_{1/2}$, $^4P_{3/2}$, $^4P_{5/2}$, $^4D_{1/2}$, $^4D_{3/2}$, $^4D_{5/2}$, $^4D_{7/2}$, $^4F_{3/2}$, $^4F_{5/2}$, $^4F_{7/2}$, $^4F_{9/2}$
^3F	$^2P_{1/2}$, $^2P_{3/2}$, $^2D_{3/2}$, $^2D_{5/2}$, $^2F_{5/2}$, $^2F_{7/2}$, $^2G_{7/2}$, $^2G_{9/2}$, $^2H_{9/2}$, $^2H_{11/2}$, $^4P_{1/2}$, $^4P_{3/2}$, $^4P_{5/2}$, $^4D_{1/2}$, $^4D_{3/2}$, $^4D_{5/2}$, $^4D_{7/2}$, $^4F_{3/2}$, $^4F_{5/2}$, $^4F_{7/2}$, $^4F_{9/2}$, $^4G_{5/2}$, $^4G_{7/2}$, $^4G_{9/2}$, $^4G_{11/2}$, $^4H_{7/2}$, $^4H_{9/2}$, $^4H_{11/2}$, $^4H_{13/2}$
^3H	$^2F_{5/2}$, $^2F_{7/2}$, $^2G_{7/2}$, $^2G_{9/2}$, $^2H_{9/2}$, $^2H_{11/2}$, $^2I_{11/2}$, $^2I_{13/2}$, $^2J_{13/2}$, $^2J_{15/2}$, $^4F_{3/2}$, $^4F_{5/2}$, $^4F_{7/2}$, $^4F_{9/2}$, $^4G_{5/2}$, $^4G_{7/2}$, $^4G_{9/2}$, $^4G_{11/2}$, $^4H_{7/2}$, $^4H_{9/2}$, $^4H_{11/2}$, $^4H_{13/2}$, $^4I_{9/2}$, $^4I_{11/2}$, $^4I_{13/2}$, $^4I_{15/2}$, $^4J_{11/2}$, $^4J_{13/2}$, $^4J_{15/2}$, $^4J_{17/2}$
Total	$1^2S_{1/2}$, $3^2P_{1/2}$, $3^2P_{3/2}$, $4^2D_{3/2}$, $4^2D_{5/2}$, $5^2F_{5/2}$, $5^2F_{7/2}$, $5^2G_{7/2}$, $5^2G_{9/2}$, $4^2H_{9/2}$, $4^2H_{11/2}$, $3^2I_{11/2}$, $3^2I_{13/2}$, $2^2J_{13/2}$, $2^2J_{15/2}$, $1^2K_{15/2}$, $1^2K_{17/2}$, $2^4P_{1/2}$, $2^4P_{3/2}$, $2^4P_{5/2}$, $2^4D_{1/2}$, $2^4D_{3/2}$, $2^4D_{5/2}$, $2^4D_{7/2}$, $3^4F_{3/2}$, $3^4F_{5/2}$, $3^4F_{7/2}$, $3^4F_{9/2}$, $2^4G_{5/2}$, $2^4G_{7/2}$, $2^4G_{9/2}$, $2^4G_{11/2}$, $2^4H_{7/2}$, $2^4H_{9/2}$, $2^4H_{11/2}$, $2^4H_{13/2}$, $1^4I_{9/2}$, $1^4I_{11/2}$, $1^4I_{13/2}$, $1^4I_{15/2}$, $1^4J_{11/2}$, $1^4J_{13/2}$, $1^4J_{15/2}$, $1^4J_{17/2}$

7.8 RELATIVE STABILITY OF TERMS

There are rules for the relative stability of the various terms arising from a given electronic configuration. First of all, *Hund's rule* (actually, there are several "Hund's rules," although this is the one commonly implied) states that, for a given configuration, the most stable state has the highest spin multiplicity (see Section 7.10). Thus, for the atoms we have considered, the lowest-energy state of carbon will be a triplet state and that of nitrogen and protactinium will be a quartet state. In addition to Hund's rule for spin, there are rules for the orbital angular momentum, L, and the total angular momentum, J, that always hold only for the lowest-energy (ground) state. For a given S value, the ground state always has the maximum L value. For carbon and nitrogen, there is only one L value for the maximum spin multiplicity; however, there are a number of L values for the quartet state of protactinium. The maximum L value for the quartets is 7, which gives a J state. The value of the total angular-momentum quantum number, J, for the lowest-energy state depends upon the occupancy of the open shell. If there is only a single open shell, and if it is less than half filled, the minimum J value corresponds to the lowest-energy state, while if it is more than half filled, the maximum J value gives the lowest-energy state. If the shell is half filled, the L value of zero will be the only L value associated with the maximum S, and there will be only one possible J value. The J rule does not always hold true if there is more than one open shell. For the atoms we have considered, the ground state term for carbon is $^3P_0$, that for nitrogen is the $^4S_{3/2}$, and, if we assume the minimum J rule holds for protactinium, the ground state would correspond to a $^4J_{11/2}$ term.

It is never necessary to construct symmetry-adapted states and terms using an $\mathbf{S}(N)$ larger than an N value corresponding to one-half the maximum occupancy of a level. This is because of the *hole formalism*. The angular momentum (spin, orbital, or total) for a level that can accommodate q electrons is the same if the level contains N electrons or if it contains $(q - N)$ electrons. Thus, the configurations p^1 and p^5 give rise to the same terms. So do the configurations p^2 and p^4. In the d levels, the pairs d^1 and d^9, d^2 and d^8, d^3 and d^7, and d^4 and d^6 each have the same set of term symbols. For example, we have seen that the $(1s)^2(2s)^2(2p)^2$ configuration of carbon leads to $^3P_0$, $^3P_1$, $^3P_2$, $^1D_2$, and $^1S_0$ terms, with the $^3P_0$ term corresponding to the ground state. These are exactly the same terms obtained for the $(1s)^2(2s)^2(2p)^4$ configuration of oxygen. Here, however, the $^3P_2$ term corresponds to the ground state, since the level is more than half filled.

If only the ground-state term symbol for a given configuration is desired, symmetry adaptation by use of Eq. 7.9 can be completely circumvented. The Young diagrams corresponding to the appropriate L value for a given S value can be constructed directly, and a technique analogous to that used in Eqs. 7.5 through 7.7, for determining the total spin, can be employed to find the maximum L value. The Young diagram arising from multiple occupancy of a level having a given l can have no more than $(2l + 1)$ rows. The maximum m corresponding to the l under consideration is assigned to the top row, with progressively smaller m values being assigned to the succeeding rows. The resultant maximum L value will be the sum of the m values so obtained. For example, for the p^n levels, we have the following Young diagrams for the spatial functions corresponding to the maximum S:

	Configuration					
	p^1	p^2	p^3	p^4	p^5	p^6
Y.D.:	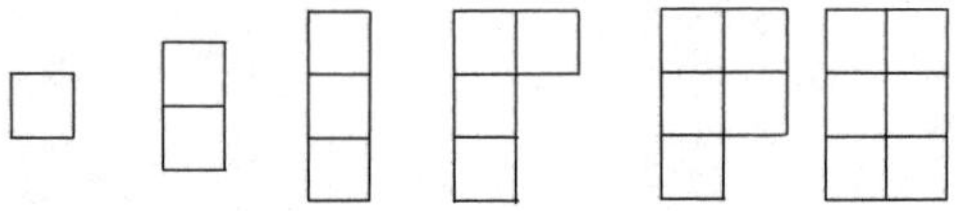					
$\sum m$:	1	1	0	1	1	0
Term:	^2P	^3P	^4S	^3P	^2P	^1S

and for the d^n configurations:

	Configuration									
	d^1	d^2	d^3	d^4	d^5	d^6	d^7	d^8	d^9	d^{10}
Y.D.:	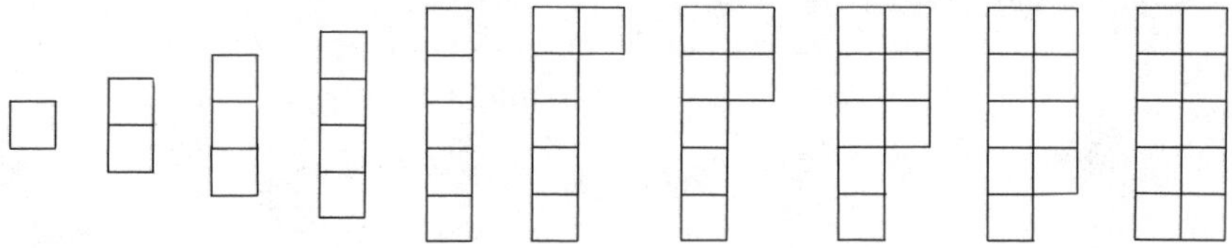									
$\sum m$:	2	3	3	2	0	2	3	3	2	0
Term:	^2D	^3F	^4F	^5D	^6S	^5D	^4F	^3F	^2D	^1S

The energy differences between states having the same S and L but different J values are very small for light atoms, while those involving different S and/or L are considerably larger. For carbon, the relative energies of the states arising from the $(1s)^2(2s)^2(2p)^2$ configuration, with the $^3P_0$ state assigned a value of zero, are 16.4 cm^{-1} for the $^3P_1$ state, 43.5 cm^{-1} for the $^3P_2$, 10,193.7 cm^{-1} for the $^1D_2$, and 21,648.4 cm^{-1} for the $^1S_0$ term. The splittings for the different J values for a given S and L are due to spin-orbit interactions. These are relativistic effects. In order to calculate them, we must explicitly consider spin angular momentum. The splittings arising from different S and L values are due to differences in the interelectronic repulsion effects for the states. Spin angular momentum does not enter into the calculation of these energy differences. Russell-Saunders coupling is appropriate when the interelectronic repulsion effects are much larger than spin-orbit interactions. When the converse is true (as in heavy atoms), j-j coupling is appropriate.

7.9 DETERMINANTAL WAVE FUNCTIONS

Properly antisymmetrized (i.e., symmetry-adapted) wave functions for systems with no degeneracy other than "spin degeneracy" can be constructed as determinants of one-electron functions. If we let $\psi_a(i)$ be a normalized one-electron wave function for electron i, including the spin function, and Ψ be the total wave function, the relation is (Slater, Ch. 12)

$$\Psi = \frac{1}{\sqrt{N!}} \begin{vmatrix} \psi_1(1) & \psi_1(2) & \dots & \psi_1(N) \\ \psi_2(1) & \psi_2(2) & \cdots & \psi_2(N) \\ \dots & \dots & \dots & \dots \\ \psi_N(1) & \psi_N(2) & \dots & \psi_N(N) \end{vmatrix} \qquad \mathbf{7.27}$$

Two general properties of determinants are important in the present context. First, interchanging either two rows or two columns of a determinant causes its value to change signs. Interchanging columns in a determinantal wave function is equivalent to interchanging two electrons; thus, the function is properly antisymmetric. Second, the determinant vanishes if any two rows or any two columns are the same. In the wave function, two identical rows would imply that two of the one-electron functions were identical. Thus, the determinantal function requires that at least one quantum number must be different for any two of the functions, if we are talking about an atomic system.

In order to illustrate the determinantal function, let us specifically consider the two-electron case. The determinantal function is

$$\Psi(1, 2) = \frac{1}{\sqrt{2}} \begin{vmatrix} \psi_1(1) & \psi_1(2) \\ \psi_2(1) & \psi_2(2) \end{vmatrix}$$

$$= \frac{1}{\sqrt{2}} [\psi_1(1)\psi_2(2) - \psi_1(2)\psi_2(1)] \qquad \mathbf{7.28}$$

Note that interchanging the electron labels turns the function into the negative of itself, and that if ψ_1 is the same as ψ_2, the two terms are the same and the function vanishes. In other words, the determinantal function satisfies either statement of the Pauli principle.

Determinantal functions are frequently abbreviated by specifying only the principal diagonal of the determinant, as

$$\Psi = \det \{\psi_1(1)\psi_2(2) \dots \psi_N(N)\} \qquad \mathbf{7.29}$$

or

$$\Psi = |\psi_1 \psi_2 \dots \psi_N| \qquad \mathbf{7.29a}$$

The $(N!)^{-1/2}$ normalizing factor is implied by the notation. For systems with degeneracies, such as partially occupied p, d, f, etc. levels, a linear combination of determinants has to be used. In such cases the appropriate symmetric group can be used to determine the proper linear combination of determinants required. However, it is frequently easier to use the determinantal function for the closed-shell portion of the wave function and to find the open-shell portion from the symmetric group without first including it in the determinant.

7.10 THE HARTREE-FOCK S.C.F. METHOD

In Section 7.1, on the central-field approximation for atomic structure, we gave a brief qualitative discussion of the self-consistent field (S.C.F.) method. In this section we will discuss it in a bit more detail and derive the equations involved. We will use the intermediate equations to show the origins of Hund's rule.

We can write the Schrödinger equation as

$$\hat{H}\Psi = E\Psi \qquad 7.30$$

where Ψ is now the many-electron wave function for the system under consideration. We can express the energy as the expectation value of the Hamiltonian. If Ψ is normalized, we have

$$E = \langle H \rangle = \langle \Psi | \hat{H} | \Psi \rangle \qquad 7.31$$

Let us specifically consider the two-electron case with the wave function expressed as a properly antisymmetrized product of one-electron functions as in Eq. 7.28. This gives

$$\begin{aligned} E &= \frac{1}{2} \langle \psi_1(1)\psi_2(2) - \psi_2(1)\psi_1(2) | \hat{H} | \psi_1(1)\psi_2(2) - \psi_2(1)\psi_1(2) \rangle \\ &= \langle H \rangle \end{aligned} \qquad 7.32$$

For a two-electron atom, the Hamiltonian is

$$\hat{H} = -\frac{1}{2}\nabla_1^2 - \frac{1}{2}\nabla_2^2 - \frac{Z}{r_1} - \frac{Z}{r_2} + \frac{1}{r_{12}} \qquad 7.33$$

The energy can be expressed as the sum of the expectation values of the terms in the Hamiltonian, as

$$\begin{aligned} E &= \langle H \rangle \\ &= -\frac{1}{2}\langle \nabla_1^2 \rangle - \frac{1}{2}\langle \nabla_2^2 \rangle - Z\left\langle \frac{1}{r_1} \right\rangle - Z\left\langle \frac{1}{r_2} \right\rangle + \left\langle \frac{1}{r_{12}} \right\rangle \end{aligned} \qquad 7.34$$

Let us consider each of the terms in Eq. 7.34 individually. For $\langle \nabla_1^2 \rangle$ we have

$$\langle \nabla_1^2 \rangle = \frac{1}{2} \langle \psi_1(1)\psi_2(2) - \psi_2(1)\psi_1(2) | \nabla_1^2 | \psi_1(1)\psi_2(2) - \psi_2(1)\psi_1(2) \rangle \qquad 7.35$$

If we expand this and recognize that the integrals can be separated according to the coordinates of the individual electrons, we obtain

$$\begin{aligned}\langle \nabla_1^2 \rangle = \frac{1}{2}[&\langle \psi_1(1)|\nabla_1^2|\psi_1(1)\rangle\langle \psi_2(2)|\psi_2(2)\rangle \\ &- \langle \psi_1(1)|\nabla_1^2|\psi_2(1)\rangle\langle \psi_2(2)|\psi_1(2)\rangle \\ &- \langle \psi_2(1)|\nabla_1^2|\psi_1(1)\rangle\langle \psi_1(2)|\psi_2(2)\rangle \\ &+ \langle \psi_2(1)|\nabla_1^2|\psi_2(1)\rangle\langle \psi_1(2)|\psi_1(2)\rangle]\end{aligned} \quad \mathbf{7.36}$$

If the individual one-electron functions are members of an orthonormal set, the integrals $\langle \psi_2(2)|\psi_2(2)\rangle$ and $\langle \psi_1(2)|\psi_1(2)\rangle$ equal unity, while $\langle \psi_2(2)|\psi_1(2)\rangle$ and $\langle \psi_1(2)|\psi_2(2)\rangle$ equal zero. Thus, the expectation value of ∇_1^2 reduces to

$$\langle \nabla_1^2 \rangle = \frac{1}{2}[\langle \psi_1(1)|\nabla_1^2|\psi_1(1)\rangle + \langle \psi_2(1)|\nabla_1^2|\psi_2(1)\rangle] \quad \mathbf{7.37}$$

If we repeat the procedure for $\langle \nabla_2^2 \rangle$, we find that

$$\langle \nabla_2^2 \rangle = \frac{1}{2}[\langle \psi_1(2)|\nabla_2^2|\psi_1(2)\rangle + \langle \psi_2(2)|\nabla_2^2|\psi_2(2)\rangle] \quad \mathbf{7.38}$$

The electrons are indistinguishable; consequently, the first term on the right-hand side of Eq. 7.37 is equal to the corresponding term of Eq. 7.38, and the same holds for the second term. This means that

$$\langle \nabla_1^2 \rangle + \langle \nabla_2^2 \rangle = \langle \psi_1|\nabla^2|\psi_1\rangle + \langle \psi_2|\nabla^2|\psi_2\rangle \quad \mathbf{7.39}$$

where we have dropped the electron label on the right-hand side. In a similar fashion, we find that

$$\left\langle \frac{1}{r_1} \right\rangle + \left\langle \frac{1}{r_2} \right\rangle = \left\langle \psi_1 \left| \frac{1}{r} \right| \psi_1 \right\rangle + \left\langle \psi_2 \left| \frac{1}{r} \right| \psi_2 \right\rangle \quad \mathbf{7.40}$$

The $1/r_{12}$ operator involves both electrons. Consequently, the $\langle 1/r_{12} \rangle$ integral cannot be separated into one-electron terms. Expanding the expectation value in terms of the antisymmetric wave function, we find

$$\begin{aligned}\left\langle \frac{1}{r_{12}} \right\rangle = \frac{1}{2}\Bigg[&\left\langle \psi_1(1)\psi_2(2) \left| \frac{1}{r_{12}} \right| \psi_1(1)\psi_2(2) \right\rangle \\ &- \left\langle \psi_1(1)\psi_2(2) \left| \frac{1}{r_{12}} \right| \psi_2(1)\psi_1(2) \right\rangle \\ &- \left\langle \psi_2(1)\psi_1(2) \left| \frac{1}{r_{12}} \right| \psi_1(1)\psi_2(2) \right\rangle \\ &+ \left\langle \psi_2(1)\psi_1(2) \left| \frac{1}{r_{12}} \right| \psi_2(1)\psi_1(2) \right\rangle\Bigg]\end{aligned} \quad \mathbf{7.41}$$

The first and last integrals on the right-hand side of Eq. 7.58 are identical except for an interchange of the labels on the electrons. Thus, they are numerically equal. The same is true for the second and third integrals. Taking these equalities into account, we have

$$\left\langle \frac{1}{r_{12}} \right\rangle = \left\langle \psi_1(1)\psi_2(2) \left| \frac{1}{r_{12}} \right| \psi_1(1)\psi_2(2) \right\rangle - \left\langle \psi_1(1)\psi_2(2) \left| \frac{1}{r_{12}} \right| \psi_2(1)\psi_1(2) \right\rangle \quad \textbf{7.42}$$

Substituting Eqs. 7.39, 7.40, and 7.42 into Eq. 7.34 for the expectation value of the Hamiltonian yields

$$\langle H \rangle = -\frac{1}{2}\langle \psi_1|\nabla^2|\psi_1\rangle - \frac{1}{2}\langle \psi_2|\nabla^2|\psi_2\rangle - Z\left\langle \psi_1 \left| \frac{1}{r} \right| \psi_1 \right\rangle - Z\left\langle \psi_2 \left| \frac{1}{r} \right| \psi_2 \right\rangle + \left\langle \psi_1(1)\psi_2(2) \left| \frac{1}{r_{12}} \right| \psi_1(1)\psi_2(2) \right\rangle - \left\langle \psi_1(1)\psi_2(2) \left| \frac{1}{r_{12}} \right| \psi_2(1)\psi_1(2) \right\rangle \quad \textbf{7.43}$$

The first term is the kinetic energy of an electron in an orbital described by ψ_1, the second is that for an electron in ψ_2. The third and fourth terms are the nuclear attraction for an electron in ψ_1 and ψ_2, respectively. The fifth term is the electrostatic repulsion between two electrons, one in an orbital described by ψ_1 and one in ψ_2. The integral is called the *coulomb integral.* The last term is also an electrostatic-repulsion term; however, its physical interpretation is a bit more complicated. The charge distribution for electron 1 is described by the product $\psi_1^*(1)\psi_2(1)$, and similarly for electron 2. This integral, called the *exchange integral*, is the source of the energy difference between the singlet and triplet states of a two-electron system. The $\psi_\mu(i)$ are functions of the spatial and spin coordinates of electron *i*. The operator is a function of only the spatial coordinates. Thus, owing to the orthogonality of different spin functions, the integral will vanish unless the two spin functions are the same. The integral is intrinsically positive, since it is of the nature of an electrostatic repulsion, but it enters into $\langle H \rangle$ with a negative sign. Consequently, the calculated energy will be lower, for a given spatial behavior of ψ_1 and ψ_2, if the spins are the same than if they are different. The triplet states correspond to the electrons having like spin and thus are of the lower energy of the two spin states that can arise for a two-electron system, in accord with Hund's rule.

Equation 7.43 can be generalized to a many-electron atom. There will be a kinetic-energy and a nuclear-attraction term involving the electron in each occupied spin-orbital (i.e., a function of space and spin), and there will be a coulomb and an exchange term for each pair of electrons. The resulting equation is

$$\langle H \rangle = \sum_{\mu}^{\substack{\text{occ}\\\text{orbitals}}} \left\langle \psi_\mu \left| -\frac{1}{2}\nabla^2 - \frac{Z}{r} \right| \psi_\mu \right\rangle + \sum_{\mu<\nu}\sum \left[\left\langle \psi_\mu\psi_\nu \left| \frac{1}{r_{12}} \right| \psi_\mu\psi_\nu \right\rangle - \left\langle \psi_\mu\psi_\nu \left| \frac{1}{r_{12}} \right| \psi_\nu\psi_\mu \right\rangle \right] \quad \textbf{7.44}$$

where the electron labels are implied by the order of occurrence of the orbitals in the coulomb and exchange integrals. The $\mu < \nu$ restriction on the summation of coulomb and exchange terms is to prevent including each term twice for a pair.

The Hartree-Fock S.C.F. equations are derived from Eq. 7.44. The procedure is to minimize Eq. 7.44 with respect to the orbitals (it is a variational procedure) while enforcing the constraint that the one-electron orbitals must be orthonormal. The technique is the method of *Lagrangian multipliers* (Slater, Sec. 5-1). The function to be varied is the desired function plus all the restraints, each restraint multiplied by an undetermined (constant) multiplier. The variation of this quantity is then set equal to zero. In the present case, the restraints are the normalization of each orbital and the orthogonality of each pair of orbitals. The quantity to be varied is, thus $\left(\langle H \rangle + \sum_\mu \sum_\nu \lambda_{\mu\nu} \langle \psi_\mu | \psi_\nu \rangle\right)$, where the $\lambda_{\mu\nu}$ are the Lagrangian multipliers. Varying this with respect to one of the orbitals, say $\psi_\mu(1)$, we have (reinstating the electron labels)

$$\begin{aligned}\frac{\partial\left(\langle H \rangle + \sum_\mu \sum_\nu \lambda_{\mu\nu} \langle \psi_\mu | \psi_\nu \rangle\right)}{\partial \psi_\mu(1)} &= \left(-\frac{1}{2}\nabla_1^2 - \frac{Z}{r_1}\right)\psi_\mu(1) \\ &+ \sum_\nu \left\langle \psi_\nu(2) \left| \frac{1}{r_{12}} \right| \psi_\nu(2) \right\rangle \psi_\mu(1) \\ &- \sum_\nu \left\langle \psi_\nu(2) \left| \frac{1}{r_{12}} \right| \psi_\mu(2) \right\rangle \psi_\nu(1) + \lambda_{\mu\mu}\psi_\mu(1) \\ &+ \text{complex conjugate}\end{aligned} \qquad \mathbf{7.45}$$

In Eq. 7.45, the terms listed arise from the ψ_μ appearing on the left-hand side of the integral, while their complex conjugates would arise from the ψ_μ appearing on the right-hand side. The summations over μ leave only the one term that contains μ. The summations over ν remain, except for the $\lambda_{\mu\nu}\langle \psi_\mu | \psi_\nu \rangle$ terms. The wave functions can be constructed such that these vanish, except for the case where ν equals μ.

Now, for our function to be minimized, Eq. 7.45 must equal zero. If a function plus its complex conjugate is to equal zero, each must individually equal zero. Thus, we have

$$\begin{aligned}\left(-\frac{1}{2}\nabla_1^2 - \frac{Z}{r_1}\right)\psi_\mu(1) + \sum_\nu \Bigg[\left\langle \psi_\nu(2) \left| \frac{1}{r_{12}} \right| \psi_\nu(2) \right\rangle \psi_\mu(1) \\ - \left\langle \psi_\nu(2) \left| \frac{1}{r_{12}} \right| \psi_\mu(2) \right\rangle \psi_\nu(1) \Bigg] + \lambda_{\mu\mu}\psi_\mu(1) = 0\end{aligned} \qquad \mathbf{7.46}$$

Notice that, except for the second term in the square brackets, each term multiplies the function $\psi_\mu(1)$. Furthermore, the last term is a numerical constant times $\psi_\mu(1)$. If, somehow, the $\psi_\nu(1)$ could be replaced by $\psi_\mu(1)$, the equation would have the form of an eigenvalue equation with $\psi_\mu(1)$ being the eigenfunction and $-\lambda_{\mu\mu}$ being the eigenvalue. Let us write out the offending term in integral notation. We have

$$\left\langle \psi_\nu(2) \left| \frac{1}{r_{12}} \right| \psi_\mu(2) \right\rangle \psi_\nu(1) = \int \psi_\nu^*(2) \frac{1}{r_{12}} \psi_\mu(2)\, dv_2 \psi_\nu(1) \tag{7.47}$$

If we multiply and divide Eq. 7.47 by the product $\psi_\mu^*(1)\psi_\mu(1)$, we have, after some rearranging,

$$\frac{\psi_\mu^*(1)\psi_\mu(1)}{\psi_\mu^*(1)\psi_\mu(1)} \left\langle \psi_\nu(2) \left| \frac{1}{r_{12}} \right| \psi_\mu(2) \right\rangle \psi_\nu(1) = \left[\frac{\int_{v_2} \psi_\mu^*(1)\psi_\nu^*(2) \frac{1}{r_{12}} \psi_\nu(1)\psi_\mu(2)\, dv_2}{\psi_\mu^*(1)\psi_\mu(1)} \right] \psi_\mu(1) \tag{7.48}$$

The integral has the form of the exchange integral, except that the integration has been carried out over only the coordinates of electron 2. The bracketed term is called the *exchange operator* and given the symbol $\hat{K}_{\mu\nu}$. If we call the term $\langle \psi_\nu(2) | 1/r_{12} | \psi_\nu(2) \rangle$ the *coulomb operator*, $\hat{J}_{\mu\nu}$, Eq. 7.46 can be rewritten

$$\left[-\frac{1}{2} \nabla_1^2 - \frac{Z}{r_1} + \sum_\nu (\hat{J}_{\mu\nu} - \hat{K}_{\mu\nu}) \right] \psi_\mu(1) = \varepsilon_\mu \psi_\mu(1) \tag{7.49}$$

where we have defined ε_μ as $-\lambda_{\mu\mu}$. The equation is a pseudo-eigenvalue equation. The operator in the brackets is called the *Fock operator*, $\hat{F}(1)$. The equation can be symbolically written

$$\hat{F}\psi_\mu = \varepsilon_\mu \psi_\mu \tag{7.50}$$

A Fock operator can be constructed for each occupied one-electron orbital of the system. The pseudo-eigenvalue equations can be solved for each orbital. However, the Fock operator contains the operators $\hat{J}_{\mu\nu}$ and $\hat{K}_{\mu\nu}$, which depend upon the electron distributions of all the electrons, other than the particular one involved in the eigenvalue equation. This is why, as we mentioned earlier, the equations must be solved iteratively.

Let us compare the Hartree-Fock expressions to the total energy of the system, in the independent-particle approximation. First of all, from Eq. 7.50, we see that ε_μ is the expectation value of $\hat{F}$ with respect to the one-electron ψ_μ:

$$\langle \psi_\mu | \hat{F} | \psi_\mu \rangle = \varepsilon_\mu \tag{7.51}$$

If we designate the one-electron operator $(-\frac{1}{2}\nabla^2 - Z/r)$ by the symbol, $\hat{h}$, and $(\hat{J}_{\mu\nu} - \hat{K}_{\mu\nu})$ by the symbol, $\hat{G}_{\mu\nu}$, then Eq. 7.44, which is the energy in the independent-particle approximation, becomes

$$E = \sum_\mu^N h_{\mu\mu} + \sum_\mu^{(\nu-1)} \sum_\nu^N G_{\mu\nu} \tag{7.52}$$

where $h_{\mu\mu}$ equals $\langle \psi_\mu | \hat{h} | \psi_\mu \rangle$, $G_{\mu\nu}$ equals the term in brackets in Eq. 7.44, and N is the number of occupied one-electron orbitals. The expectation value of the Fock operator, ε_μ, becomes (from Eq. 7.49, note that $G_{\mu\mu}$ equals zero)

$$\varepsilon_\mu = h_{\mu\mu} + \sum_{\nu \neq \mu}^{N} G_{\mu\nu} \tag{7.53}$$

Consider, now, the sum of ε_μ over all occupied one-electron orbitals for the system

$$\sum_{\mu}^{N} \varepsilon_\mu = \sum_{\mu}^{N} h_{\mu\mu} + \sum_{\mu}^{N} \sum_{\nu \neq \mu}^{N} G_{\mu\nu} \tag{7.54}$$

Comparing Eqs. 7.52 and 7.54, we see, first of all, that the total energy is not the sum of the Fock eigenvalues (which, for reasons we will see momentarily, we will call orbital energies). In effect, the sum of the orbital energies counts the electron interactions twice. The total energy can, however, be expressed in terms of the orbital energies if we subtract out this excess electron interaction:

$$E = \sum_{\mu}^{N} \varepsilon_\mu - \sum_{\mu}^{(\nu-1)} \sum_{\nu}^{N} G_{\mu\nu} \tag{7.55}$$

Let us consider the ionization potential of an atom. This will be the difference in total electronic energy between the ion and the neutral atom. We can assign our number labels to the electrons in such a manner that the electron labeled by N is the one removed from, say, the orbital labeled λ. We have, using Eqs. 7.52 and 7.53 and assuming the other orbitals are unchanged,

$$\begin{aligned} E(\text{ion}) - E(\text{neutral}) &= \sum_{\mu}^{(N-1)} h_{\mu\mu} + \sum_{\mu}^{(\nu-1)} \sum_{\nu}^{(N-1)} G_{\mu\nu} - \sum_{\mu}^{N} h_{\mu\mu} - \sum_{\mu}^{(\nu-1)} \sum_{\nu}^{N} G_{\mu\nu} \\ &= -h_{\lambda\lambda} - \sum_{\nu \neq \lambda}^{N} G_{\nu\lambda} \\ &= -\varepsilon_\lambda \end{aligned} \tag{7.56}$$

Thus, the ionization potential for removing an electron from ψ_λ is the negative of the Hartree-Fock eigenvalue for that orbital *if it is assumed that there is no reorganization of the electron distribution for the other electrons.* This is commonly known as *Koopmans' theorem.* In fact, this is not a valid assumption. The field experienced by the remaining electrons will be altered when an electron is removed. A better approach would require solving the S.C.F. problem for the ion and for the neutral atom and subtracting the resulting energies. However, the Koopmans' theorem ionization potential is a fairly good approximation to the ionization potential, and is frequently used because of its simplicity. As examples, the Koopmans' theorem values for the first ionization potentials of helium, beryllium, and neon are .918 eV, .309 eV, and .850 eV, respectively, from calculations near the Hartree-Fock limit, while the experimental values are .899 eV, .341 eV, and .789 eV.

In early S.C.F. calculations, the equations were solved by starting with a suitable trial function and varying it numerically until charge self-consistency was obtained. Most current Hartree-Fock calculations involve expanding the ψ_μ as linear combinations of some suitable basis functions. The linear expansion coefficients are the variational parameters. We will defer the details of the calculations until a later section. We will discuss S.C.F. calculations on both atoms and molecules after we have had some preliminary discussion of molecules. Most important for our present purposes are the concept of orbital energy and a feeling for the behavior of the two types of interelectronic-repulsion terms that appear in the total-energy expression when it is constructed from a properly antisymmetrized product of one-electron orbitals.

7.A APPENDIX: SOME PROPERTIES OF THE SYMMETRIC GROUP

A natural way of designating a permutation is to list the original labels of the objects in a row and the labels of the *positions to which these objects go* in a second row directly below the first. For $\mathbf{S}(3)$ we have, referring back to Eq. 7.4,

$$\begin{aligned} E &= \begin{pmatrix} 1 & 2 & 3 \\ 1 & 2 & 3 \end{pmatrix} \\ P_1 &= \begin{pmatrix} 1 & 2 & 3 \\ 2 & 3 & 1 \end{pmatrix} \\ P_2 &= \begin{pmatrix} 1 & 2 & 3 \\ 3 & 1 & 2 \end{pmatrix} \\ P_3 &= \begin{pmatrix} 1 & 2 & 3 \\ 1 & 3 & 2 \end{pmatrix} \\ P_4 &= \begin{pmatrix} 1 & 2 & 3 \\ 2 & 1 & 3 \end{pmatrix} \\ P_5 &= \begin{pmatrix} 1 & 2 & 3 \\ 3 & 2 & 1 \end{pmatrix} \end{aligned} \tag{7.A1}$$

(This is the active convention. Some authors use the passive convention in which the second row lists the objects that go to the positions above them in the first row.) With this notation we can define the product of two permutations. The convention for multiplying the operations of any group is that the operation on the right is performed first. We have

$$\begin{aligned} P_1 P_2 &= \begin{pmatrix} 1 & 2 & 3 \\ \downarrow & \downarrow & \downarrow \\ 2 & 3 & 1 \end{pmatrix} \begin{pmatrix} 1 & 2 & 3 \\ \downarrow & \downarrow & \downarrow \\ 3 & 1 & 2 \end{pmatrix} \\ &= \begin{pmatrix} 1 & 2 & 3 \\ 1 & 2 & 3 \end{pmatrix} \end{aligned} \tag{7.A2}$$

Note the method of constructing the product. Starting with the operation on the right, object one goes to position three. This then becomes object three for the other permutation, which then goes to position one. Similarly, starting on the right, two goes to one goes to two and three goes to two goes to three. The arrows have been added in Eq. 7.A2 simply to show the results of the permutations.

There is a more compact notation for permutations than the one we have given. The *cycle-structure* notation uses only one line of numbers. A starting number is chosen. The second number (in the active convention) is the position to which that object goes. The third number is the position to which the object in the resulting position goes, and so on, until the cycle is completed by reaching a position from which the object returns to the starting position for that cycle. If all the numbers are not used in the first cycle, a new starting number outside the first cycle is chosen and the process is repeated until all the numbers are used. The numbers in each cycle are enclosed in parentheses. The cycles can be of any length, containing from 1 to N numbers. For **S**(3) we have

$$\begin{aligned}
E &= \begin{pmatrix} 1 & 2 & 3 \\ 1 & 2 & 3 \end{pmatrix} = (1)(2)(3) \\
P_1 &= \begin{pmatrix} 1 & 2 & 3 \\ 2 & 3 & 1 \end{pmatrix} = (1 \quad 2 \quad 3) \\
P_2 &= \begin{pmatrix} 1 & 2 & 3 \\ 3 & 1 & 2 \end{pmatrix} = (1 \quad 3 \quad 2) \\
P_3 &= \begin{pmatrix} 1 & 2 & 3 \\ 1 & 3 & 2 \end{pmatrix} = (1)(2 \quad 3) \\
P_4 &= \begin{pmatrix} 1 & 2 & 3 \\ 2 & 1 & 3 \end{pmatrix} = (1 \quad 2)(3) \\
P_5 &= \begin{pmatrix} 1 & 2 & 3 \\ 3 & 2 & 1 \end{pmatrix} = (1 \quad 3)(2)
\end{aligned} \qquad \textbf{7.A3}$$

The multiplication is only slightly more complicated using this notation. Again, the permutation on the right is accomplished first. A starting number is chosen from the permutations on the right. This is set down as the first number in the product. The number to which this starting number goes is found on the left, and in the product the second number is the number to which this goes. Next, start with the number just entered in the product and repeat the operations until a cycle is completed in the product. The process is repeated for a second starting number, if necessary, until the product contains all the numbers. For example, for the product $(1 \quad 2 \quad 3 \quad 4 \quad 5) \times (1 \quad 2 \quad 4)(3 \quad 5)$ from **S**(5), we can start with the 1 on the right. This goes to 2 on the right. On the left, 2 goes to 3, so in the product 1 is followed by 3. Next, the 3 on the right is chosen. This goes to 5. On the left, 5 goes to 1. Thus, the first cycle is completed in the product. Next, 2 can be chosen on the right. This goes to 4. On the left 4 goes to 5, so in the second cycle in the product, 2 is followed by 5. Next,

Table 7.A1

THE MULTIPLICATION TABLE FOR THE S(3) GROUP[a]

S(3)	(1)(2)(3)	(1 2 3)	(1 3 2)	(1)(2 3)	(1 2)(3)	(1 3)(2)
(1)(2)(3)	(1)(2)(3)	(1 2 3)	(1 3 2)	(1)(2 3)	(1 2)(3)	(1 3)(2)
(1 2 3)	(1 2 3)	(1 3 2)	(1)(2)(3)	(1 2)(3)	(1 3)(2)	(1)(2 3)
(1 3 2)	(1 3 2)	(1)(2)(3)	(1 2 3)	(1 3)(2)	(1)(2 3)	(1 2)(3)
(1)(2 3)	(1)(2 3)	(1 3)(2)	(1 2)(3)	(1)(2)(3)	(1 3 2)	(1 2 3)
(1 2)(3)	(1 2)(3)	(1)(2 3)	(1 3)(2)	(1 2 3)	(1)(2)(3)	(1 3 2)
(1 3)(2)	(1 3)(2)	(1 2)(3)	(1)(2 3)	(1 3 2)	(1 2 3)	(1)(2)(3)

[a] The convention is (left operation) × (top operation).

5 is chosen on the right. This goes to 3, which, on the left, goes to 4, the next number in this cycle in the product. On the right, 4 goes to 1, which goes to 2 on the left. The cycle is completed and all numbers have been used. We have

$$(1\ \ 2\ \ 3\ \ 4\ \ 5) \times (1\ \ 2\ \ 4)(3\ \ 5) = (1\ \ 3)(2\ \ 5\ \ 4) \qquad \textbf{7.A4}$$

Using either notation and either rule for multiplication, we can set up a multiplication table for a symmetric group. The multiplication table for **S**(3) is shown in Table 7.A1. This exhibits all the characteristics of the multiplication table of any finite group. Note that each row and each column contains all the operations of the group and that each operation occurs only once in each row and each column. In particular, there are always two operations that, when multiplied together, give the identity as the product. These operations are the *inverse* of each other. The operations (1)(2)(3), (1)(2 3), (1 2)(3), and (1 3)(2) are their own inverses, while (1 2 3) and (1 3 2) are the inverse of each other—e.g., $P_1 = P_2^{-1}$, or $P_1P_2 = P_2P_1 = E$.

Operations and their inverses are used to define *conjugates* of operations. If A^{-1} is the inverse of the operation A and if B is some operation, then the product $A^{-1}BA$ is the conjugate of B with respect to A. Sets of operations that are mutually conjugate to each other (i.e., are related by conjugate relationships) are the *classes* of the group. Table 7.A2 gives the table of conjugates for **S**(3). Examining this, we see that (1)(2)(3) is not conjugate to any other operation, so it is in a class by itself. The operations (1 2 3) and (1 3 2) are mutually conjugate and so form a class, as do the operations (1)(2 3), (1 2)(3), and (1 3)(2). The class structure of any finite group can be determined from a table of conjugates. There are, for our purposes, two significant points pertaining to classes. First, for any group, there are the same number of irreducible representations as there are classes. Second, in a character table, all operations of a given class have the same character.

For the symmetric groups there is a simpler way to obtain the class structure. It turns out that all operations having the same *cycle structure* belong to the same class—that is, those operations having the same numbers of cycles of the same length.

Table 7.A2

TABLE OF CONJUGATES FOR **S**(3)

A^{-1}	(1)(2)(3)	(1 2 3)	(1 3 2)	(1)(2 3)	(1 2)(3)	(1 3)(2)	A
(1)(2)(3)	(1)(2)(3)	(1 2 3)	(1 3 2)	(1)(2 3)	(1 2)(3)	(1 3)(2)	(1)(2)(3)
(1 3 2)	(1)(2)(3)	(1 2 3)	(1 3 2)	(1 2)(3)	(1 3)(2)	(1)(2 3)	(1 2 3)
(1 2 3)	(1)(2)(3)	(1 2 3)	(1 3 2)	(1 3)(2)	(1)(2 3)	(1 2)(3)	(1 3 2)
(1)(2 3)	(1)(2)(3)	(1 3 2)	(1 2 3)	(1)(2 3)	(1 3)(2)	(1 2)(3)	(1)(2 3)
(1 2)(3)	(1)(2)(3)	(1 3 2)	(1 2 3)	(1 3)(2)	(1 2)(3)	(1)(2 3)	(1 2)(3)
(1 3)(2)	(1)(2)(3)	(1 3 2)	(1 2 3)	(1 2)(3)	(1)(2 3)	(1 3)(2)	(1 3)(2)

Thus, from **S**(3) we have one operation having three cycles, (1)(2)(3), two having one cycle, (1 2 3) and (1 3 2), and three having two cycles, each of lengths 2 and 1: (1)(2 3), (1 2)(3), and (1 3)(2). The classes of the symmetric group are commonly labeled by the *partitions* indicating the number of cycles of each length. The length of the cycle is indicated by the appropriate number, and the number of cycles of that length by a superscript. The label for the identity of **S**(3) is thus (1^3), the class containing the two three-cycles is labeled (3), and the third class is (2, 1). In character tables, a number is put in front of these to indicate the number of operations in that class.

Let us return briefly to the multiplication table for **S**(3) (Table 7.A1). If we look near the lower right-hand corner, we see that the operations having a cycle length of three can be obtained as a product of two operations that are simply transpositions of two objects. For example

$$(1\quad 2)(3) \times (1)(2\quad 3) = (1\quad 2\quad 3) \qquad \textbf{7.A5}$$

The operation (1 2)(3) is a transposition of objects 1 and 2, while (1)(2 3) is a transposition of 2 and 3. This is general. A cycle of order n can always be resolved into a product of $(n - 1)$ transpositions. This leads to the concept of *even* and *odd* classes—classes that can be resolved into an even or an odd number of transpositions, or interchanges. Looking back at the classes in the character tables of Table 7.2, we see that for **S**(2) the class (1^2) contains zero (an even number) interchanges, while (2) contains one (an odd number). Thus, the classes are even and odd, respectively. For **S**(3), the class (1^3) has only cycles of order one, therefore there are no transpositions and the class is even; the class (2, 1) has one cycle of order two, or one transposition, and is an odd class; the class (3) has one cycle of order three, which can be resolved into two transpositions, so the class is even. For **S**(4) the classes (1^4), (2^2), and (3, 1) are even, while $(2, 1^2)$ and (4) are odd.

Examination of the character tables (Table 7.2) reveals that each symmetric group has two, and only two, one-dimensional irreducible representations. One of these has +1 for all the characters and is the totally symmetric irreducible representation. The other has characters of +1 for the even classes and −1 for the odd classes; this is the totally *antisymmetric* representation. One-dimensional totally symmetric representations occur for all groups, and totally antisymmetric representations occur for all symmetric groups (but not for all other groups). The other representations have mixed behavior for interchanges. The permutational symmetry of a function that is antisymmetric to the interchange of particles is reflected by the totally antisymmetric irreducible representation.

Consider now the product of a representation with its conjugate. We will take the representations from **S**(4) as examples. We have

	(1^4)	$6(2, 1^2)$	$3(2^2)$	$8(3, 1)$	$6(4)$	
$[4] \times [1^4]$:	1×1	$1 \times (-1)$	1×1	1×1	$1 \times (-1)$	
	1	-1	1	1	-1	
$[4] \times [1^4] = [1^4]$						**7.A6**
$[3, 1] \times [2, 1^2]$:	3×3	$(-1) \times 1$	$(-1) \times (-1)$	0×0	$1 \times (-1)$	
	9	-1	1	0	-1	
$[3, 1] \times [2, 1^2] = [1^4] + [2^2] + [3, 1] + [2, 1^2]$						**7.A7**
$[2^2] \times [2^2]$:	2×2	0×0	2×2	$(-1) \times (-1)$	0×0	
	4	0	4	1	0	
$[2^2] \times [2^2] = [1^4] + [4] + [2^2]$						**7.A8**

(The results for the product representations can be verified by summing the indicated irreducible representations.) We see that the product of an irreducible representation with its conjugate always contains the totally antisymmetric irreducible representation. This is true for all symmetric groups. (Equation 7.A8 also illustrates that the product of a representation with itself contains the totally symmetric irreducible representation. This is true for all groups of any type.)

The product representations of Eqs. 7.A7 and 7.A8 are said to be *reducible*, since they can be resolved into a sum of irreducible representations. There is a systematic procedure, which is based upon the orthogonality properties of irreducible representations, for reducing reducible representations of any finite group. It turns out that, if you multiply together the characters, χ, of the operations, R, for two irreducible representations, say Γ_i and Γ_j, and sum the result over all of the operations, the result is δ_{ij} (the Kronecker delta) times the order of the group, g (actually, since characters can be complex, the complex conjugate of one of the characters should be used). We have

$$\sum_R \chi^*(R)_{\Gamma_i}\chi(R)_{\Gamma_j} = g\delta_{ij} \qquad \textbf{7.A9}$$

Since all the operations in a class have the same character, Eq. 7.A9 can be simplified by multiplying the product of the character by the order of the class (the number of operations in the class), h_c, and summing over the classes, c:

$$\sum_c h_c \chi^*(R_c)_{\Gamma_i} \chi(R_c)_{\Gamma_j} = g\delta_{ij} \qquad \textbf{7.A10}$$

(This relation can be verified with the numbers in any of Eqs. 7.A6 through 7.A8.) Any arbitrary representation, Γ, can be expressed as a linear combination of irreducible representations,

$$\Gamma = \sum_i \mathrm{a}_i \Gamma_i \qquad \textbf{7.A11}$$

where a_i is the number of times Γ_i appears in Γ. Similarly, a character in a reducible representation is

$$\chi(R_c)_\Gamma = \sum_i a_i \chi(R_c)_{\Gamma_i} \qquad \textbf{7.A11a}$$

Combining Eqs. 7.A10 and 7.A11a, we obtain

$$a_i = \frac{1}{g} \sum_c h_c \chi^*(R_c)_{\Gamma_i} \chi(R_c)_\Gamma \qquad \textbf{7.A12}$$

Equation 7.A12 is valid for any finite group. To illustrate its use, consider the product representation given in Eq. 7.A7. The order of **S**(4) is 24. The orders of each of the classes are listed with the classes in the character table (Table 7.2). The characters are all real, so we need not worry about the complex conjugates. We have, where each quantity is given in the same order as the corresponding quantity in Eq. 7.A6,

$$\begin{aligned} a_{[4]} &= \frac{1}{24}\{[1 \times 1 \times 9] + [6 \times 1 \times (-1)] + [3 \times 1 \times 1] + [8 \times 1 \times 0] \\ &\quad + [6 \times 1 \times (-1)]\} \\ &= 0 \end{aligned} \qquad \textbf{7.A13}$$

$$\begin{aligned} a_{[3,1]} &= \frac{1}{24}\{[1 \times 3 \times 9] + [6 \times (-1) \times (-1)] + [3 \times (-1) \times 1] \\ &\quad + [8 \times 0 \times 0] + [6 \times 1 \times (-1)]\} \\ &= 1 \end{aligned} \qquad \textbf{7.A14}$$

$$\begin{aligned} a_{[2^2]} &= \frac{1}{24}\{[1 \times 2 \times 9] + [6 \times 0 \times (-1)] + [3 \times 2 \times 1] + [8 \times (-1) \times 0] \\ &\quad + [6 \times 0 \times (-1)]\} \\ &= 1 \end{aligned} \qquad \textbf{7.A15}$$

$$a_{[2,1^2]} = \frac{1}{24}\{[1 \times 3 \times 9] + [6 \times 1 \times (-1)] + [3 \times (-1) \times 1] + [8 \times 0 \times 0] + [6 \times (-1) \times (-1)]\} = 1 \qquad \textbf{7.A16}$$

$$a_{[1^4]} = \frac{1}{24}\{[1 \times 1 \times 9] + [6 \times (-1) \times (-1)] + [3 \times 1 \times 1] + [8 \times 1 \times 0] + [6 \times (-1) \times (-1)]\} = 1 \qquad \textbf{7.A17}$$

BIBLIOGRAPHY

CHISHOLM, C. D. H., *Group Theoretical Techniques in Quantum Chemistry.* Academic Press, New York, 1976.

CONDON E. U., and SHORTLEY, G. H., *The Theory of Atomic Spectra.* Cambridge University Press, New York, 1935.

FLURRY, R. L., JR., *Symmetry Groups: Theory and Chemical Applications.* Prentice-Hall, Inc., Englewood Cliffs, N. J., 1980.

HAMERMESH, M., *Group Theory.* Addison-Wesley Publishing Co., Reading, Mass., 1962.

KAPLAN, I. G., *Symmetry of Many-Electron Systems.* Academic Press, New York, 1975.

KARPLUS, M., and PORTER, R. N., *Atoms and Molecules.* W. A. Benjamin, Inc., New York, 1970.

SLATER, J. C., *Quantum Theory of Atomic Structure*, Vol. 1. McGraw-Hill Book Company, New York, 1960

WIGNER, E. P., *Group Theory.* Academic Press, New York, 1959.

PROBLEMS

7.1 Repeat Problem 6.4 with a properly antisymmetrized perturbational wave function.

***7.2** Give the ground-state electronic configurations for each of the following neutral atoms: C, N, O, V, Si, Ni, Se, Ti, Cl, and Pm.

***7.3** Give the ground-state electronic configurations for the following ions: O^{2-}, Zn^{2+}, Fe^{3+}, and Ag^{+}.

7.4 If there were no permutational restrictions on polyelectron wave functions, what terms would be expected to arise from the open shells of C, N, O, and V?

7.5 Forget everything you know about the Pauli principle, and use the symmetric group to try to symmetry-adapt three electrons to a 1s orbital (i.e., try to adapt D^0 to $[1^3]$ or to $[2, 1]$).

***7.6** (Problems 7.6 and 7.7 are based on the appendix to the chapter.) Reduce the given representations in the indicated S(N) groups into the correct irreducible representations. (Note that the representations called Γ_R are referred to as the *regular representations* of the respective groups. The relation that you will find for these is general for all finite groups.)

S(2)	(1^2)	(2)
Γ	6	2
Γ_R	2	0

S(3)	(1^3)	3(2, 1)	2(3)
Γ	4	−2	1
Γ_R	6	0	0

S(4)	(1^4)	$6(2, 1^2)$	$3(2^2)$	8(3, 1)	6(4)
Γ	14	4	2	−1	−2
Γ_R	24	0	0	0	0

7.7 List the irreducible representations and classes (don't worry about the order of the classes) of S(5), S(6), and S(7). Show the conjugate of each representation. Write out the totally symmetric and the totally antisymmetric representations.

7.8 What representations of S(5), S(6), and S(7) can represent the permutational symmetry of electron spin functions?

***7.9** Find all possible term symbols for the lowest-energy configurations of the atoms Li through Ne. What is the ground-state term symbol of each?

***7.10** Find the term symbols for the ground states of the atoms Sc through Zn.

***7.11** Give the term symbols for all allowed terms arising from the $(4s)^2(3d)^8$ configuration of Ni. Which corresponds to the ground state?

***7.12** Write in detail (in terms of integrals) the total-energy expression for the ground state of Be.

***7.13** Write in detail (in terms of integrals) the Fock operator for a 2s orbital for the ground state of Be.

7.14 What is the independent-particle model in describing many-electron systems? What is the source of error when doing calculations using this model?

7.15 Briefly describe the Hartree-Fock approximation for calculating the energy of a many-electron atom.

7.16 What is the Pauli principle, in terms of permutational symmetry?

7.17 What are the required relations between the permutational symmetry of a spatial function and its corresponding spin function? Why?

7.18 Predict the ground-state electronic configurations and term symbols for elements 116, 119, and 120, should they ever be made.

***7.19** Write the antisymmetrized wave function for the ground state of Li (a) as a determinant and (b) in expanded form. [Call your one-electron spin orbitals $1s(i)$, $1\bar{s}(i)$, and $2s(i)$, where a bar indicates a "spin" of $-\frac{1}{2}$ and the absence of a bar a "spin" of $+\frac{1}{2}$.]

***7.20** (a) Write out, in terms of one- and two-electron integrals, the energy for Li and for Li^+, and the Hartree-Fock orbital energy for the $2s$ orbital of Li.

(b) Show that the orbital energy for this orbital approximates the negative of the ionization potential of Li.

chapter eight

The electronic spectra of many-electron atoms

8.1 INTRODUCTION

The information that led to the development of the quantum theory of the hydrogen atom came entirely from spectroscopy. We have previously defined spectroscopy as the study of the absorption or emission of electromagnetic radiation by a system. Two pieces of information can be obtained from a spectroscopy experiment: energy differences between states (from wavelengths or frequencies) and intensities. Each spectral transition has associated with it a characteristic energy and a characteristic intensity. Every substance has its own characteristic spectrum, thus making spectroscopy an extremely useful analytical tool.

If we neglect nuclear spin states and intranuclear energy levels, all structural features of atoms have to do with the electronic structure. The energy changes associated with spectral transitions are accompanied by changes in electronic structure. In most of the usual experiments, the spectra can be described in the independent-particle model by a change in electronic configuration of the atom. Normally, the two configurations involved in a given transition differ by only one electron. We saw in Chapter 7 that most electronic configurations (except for closed-shell configurations) have a number of terms associated with them. These terms are the labels of the atomic energy states. Spectral transitions represent a transition from a specific term of a given configuration to another specific term, usually of another configuration.

Intensities of spectral transitions are measures of the amount of radiation absorbed or emitted in the transition. They are related to the probability that a transition will be observed when the system is influenced by the radiation (see Section

6.7). Selection rules tell whether or not the probability (and, hence, the intensity) of a given transition should be zero. (The transitions are said to be forbidden or allowed.) They do not give any information about the relative intensities of allowed transitions. In practice, a transition that should be allowed may have such a low intensity that it appears to be forbidden.

Selection rules can be determined by the symmetry of the system being studied. These are derived on the basis of some model for the system. The symmetry principles upon which the selection rules are based are exact. If the model chosen to describe the selection rules is accurate, there will never be a transition observed that is predicted to be forbidden. Frequently, however, selection rules do appear to be violated. When this occurs, the error reflects a deficiency in the model being used rather than in the theory. Such discrepancies can be used to improve the model for the system.

8.2 EXAMPLES

Lithium. Figure 8.1 presents an energy-level diagram for the lithium atom. Note that only the spin-multiplicity and L values are indicated for the states. The lines connecting the states correspond to the observed transitions for direct absorption or emission of electromagnetic radiation. All the indicated states above the ground state correspond to configurations in which the single 2*s* electron has been promoted to a higher level. States are also possible in which one of the 1*s* electrons is promoted; however, these occur at considerably higher energy than the indicated states. Since all these states correspond to only one electron outside a closed shell, only one spin state, the doublet, can occur.

Looking at the transitions, we see that from the ground state (^2S), only transitions to and from ^2P states are observed. The lowest-energy excited state is a ^2P state. Transitions from this state to and from ^2S and ^2D states are observed, but none to ^2P states. This is an indication of a selection rule for changes in L. The apparent L selection rule is $\Delta L = \pm 1$. In this case, since only one electron is in the valence shell, since this one electron is promoted in the transitions, and since the total L equals the one-electron l, Δl also equals ± 1. In fact, Δl is what determines ΔL. The principal quantum number of the valence electron, n, is conventionally used to label different atomic configurations, even though it is not really a valid quantum number. A number of different states having a given L (or l) value can be observed, starting from a given state. Thus, there does not appear to be a selection rule involving n.

If the spectrum of lithium is studied under sufficiently high resolution, many of the levels shown in Figure 8.1 are seen to be split into two levels. These correspond to the two possible J values arising from a doublet state when L is not equal to zero. The transitions from the ground state to the first ^2P state are shown schematically in Figure 8.2. Transitions to and from both $^2P_{1/2}$ and $^2P_{3/2}$ are observed. Thus, apparently ΔJ can be either 0 or ± 1. (As we shall see later, these are the only allowed values of ΔJ for a one-electron promotion.)

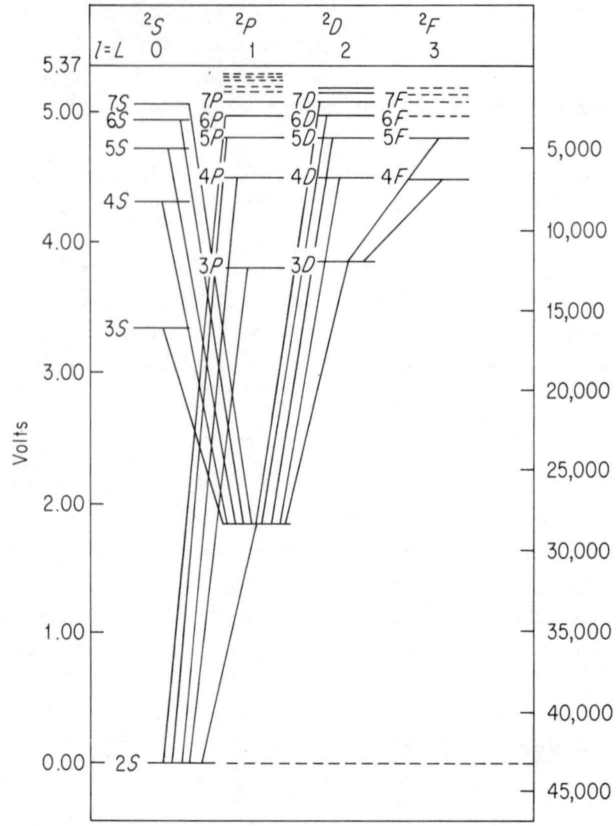

Figure 8.1 Energy level diagram of the Li atom. The principal quantum numbers for the configurations leading to the indicated terms are indicated. The doublet structure is not included. (Adapted from G. Herzberg, *Atomic Spectra and Atomic Structure*, Dover Publications, Inc., New York, 1944. By permission.)

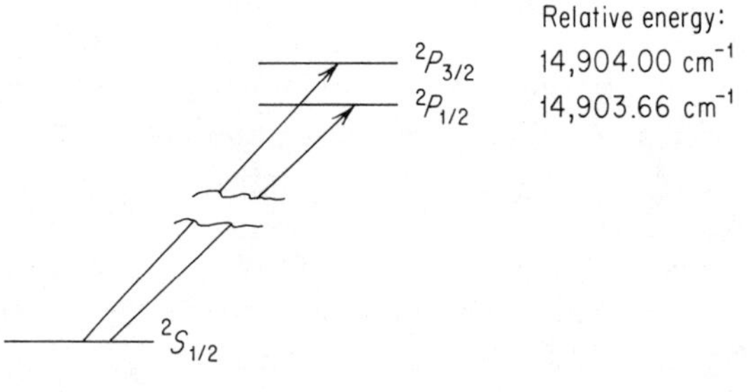

Figure 8.2 Transitions from the $^2S_{1/2}$ ground state of lithium to the first ^2P excited states.

Table 8.1

OBSERVED SPLITTINGS (IN cm^{-1}) BETWEEN THE $J = L - \frac{1}{2}$ AND THE $J = L + \frac{1}{2}$ LEVELS FOR THE VARIOUS TERMS IN THE SPECTRA OF THE ALKALI METALS

	Atom				
Configuration[a]	Li	Na	K	Rb	Cs
np	.34	17.20	57.72	237.60	554.11
nd	—[b]	−.05[c]	−1.10	2.96	42.94
nf	—	—	0	−.01	−.10
$(n+1)p$	0	5.63	18.76	77.50	181.01
$(n+1)d$	.04	−.04	−.51	2.26	20.97
$(n+1)f$	—	0	0	−.01	−.07
$(n+2)p$	0	2.52	8.41	35.09	82.64
$(n+2)d$	.02	−.02	−.24	1.51	11.69
$(n+2)f$	0	0	0	−.01	−.07

[a] Principal quantum number and orbital designation of the outermost level. Since these involve only one electron outside a closed shell, the spin multiplicity is 2 for all levels, and L equals l.

[b] No such level exists.

[c] A negative entry means the $L + \frac{1}{2}$ level lies lower than the $L - \frac{1}{2}$.

The energy splitting between levels having a given S and L but different J values is caused by spin-orbit interaction (a relativistic effect). The splittings between levels having different S values or different L values are electrostatic in nature (see Section 7.10). We see from the spectrum of lithium that the splittings that arise from electrostatic interactions are much the larger of the two. The nonrelativistic Hamiltonian we have been using is completely adequate for most purposes when the spin-orbit interactions are this small. (The magnitude of the spin-orbit coupling gets larger, however, as the atomic number of the elements increases.) All alkali metals (group IA of the periodic table) have energy-level diagrams similar to Figure 8.1. The levels with L not equal to zero are split by spin-orbit interactions. The magnitudes of the splitting for a few of the levels are shown in Table 8.1, indicating the increase with increasing atomic number.

Carbon. The energy-level diagram for carbon, with the observed transitions, is shown in Figure 8.3. Here, because of two electrons outside the closed shell, each configuration can lead to singlet and triplet states. We see that only transitions between different triplet states or between different singlet states are allowed. Transitions in which one state is a triplet and the other is a singlet are not observed. Thus, the spin selection rule is that ΔS must equal zero. Changes in total orbital angular momentum, ΔL, are seen to be zero or ± 1, while the changes in the one-electron orbital angular momentum, Δl, can again only equal ± 1. Here, also, the

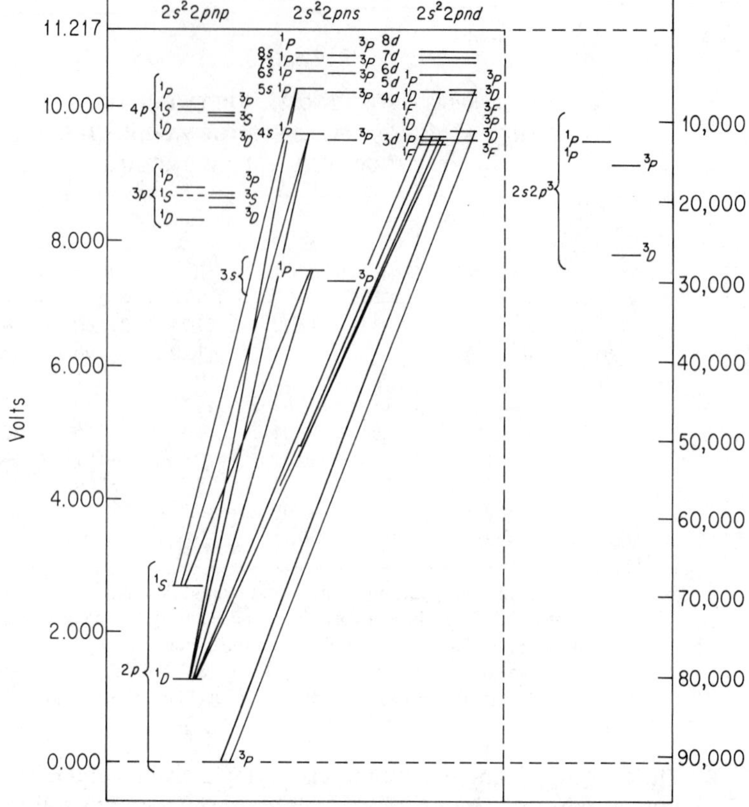

Figure 8.3 Energy level diagram of the C atom, with observed transitions indicated. (Adapted from Herzberg. By permission—see Fig. 8.1.)

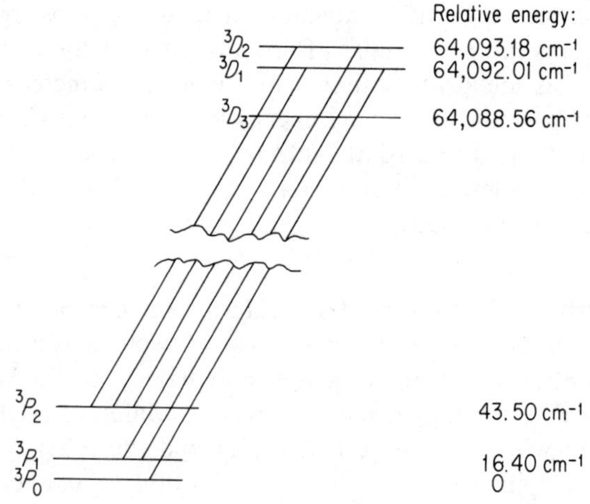

Figure 8.4 Transitions from the ^3P ground states of carbon to the first ^3D excited states.

levels with an L value not equal to zero are split under high resolution. The levels and transitions involving the ^3P ground state and the first excited ^3D state are shown in Figure 8.4. This illustrates that ΔJ may be zero or ± 1.

Beryllium. The energy-level diagram for beryllium, with the observed transitions, is shown in Figure 8.5. The ground-state configuration is a closed shell, $(1s)^2(2s)^2$; consequently there is only the $^1S_0$ term. However one new feature shows up. Observed states arise from configurations in which two electrons have been

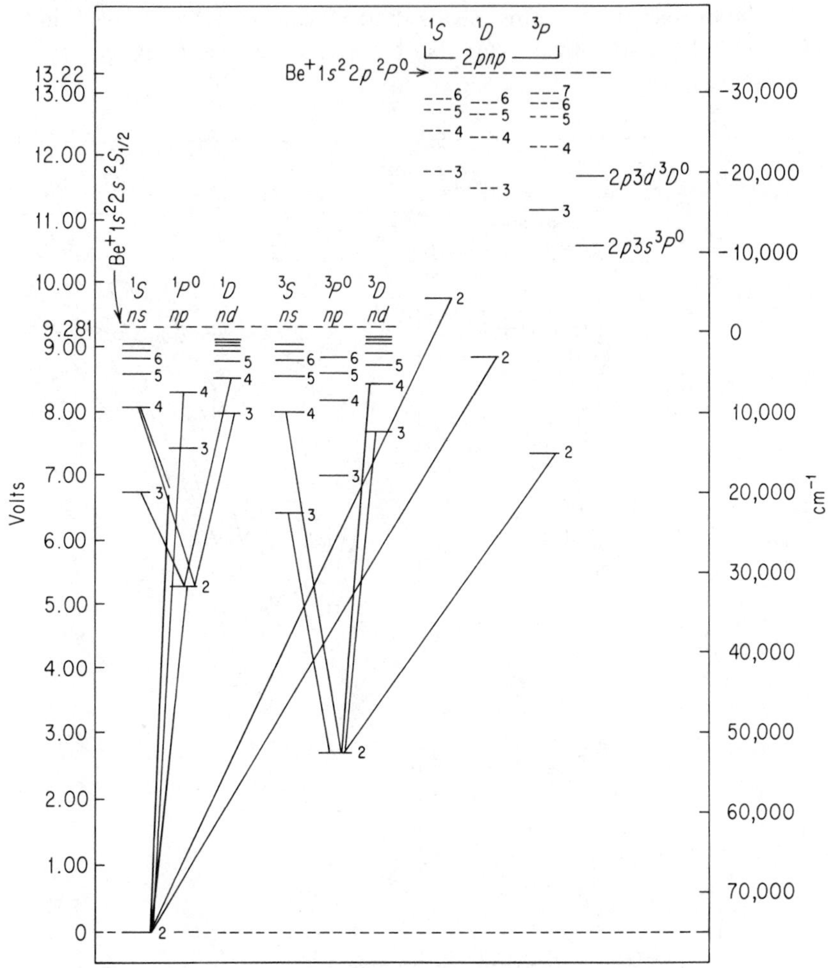

Figure 8.5 Energy level diagram of the Be atom, with normal and anomalous transitions. The normal singlet and triplet transitions are drawn to the left; the anomalous transitions to the right. (Adapted from Herzberg. By permission—see Fig. 8.1)

promoted from the ground-state configuration. These are referred to as *anomalous* states. Among the light elements, these anomalous states are observed only for the alkaline earth elements (group IIA of the periodic table). They occur fairly commonly throughout the periodic table for heavy elements. Their rarity among the light elements has to do primarily with the energies of the states. It usually requires less energy to ionize a light element than it does to reach a configuration in which two electrons have been promoted from the ground-state configuration.

If a transition from the ground state to an anomalous term is to be described within the Russell-Saunders coupling scheme, each promoted electron must obey the ± 1 selection rule for the one-electron Δl. Since the total L is constructed from the one-electron l's, this means that ΔL can be ± 2 when each individual l changes by ± 1. A ΔL of zero can also be observed if one l changes by $+1$ and the other by -1.

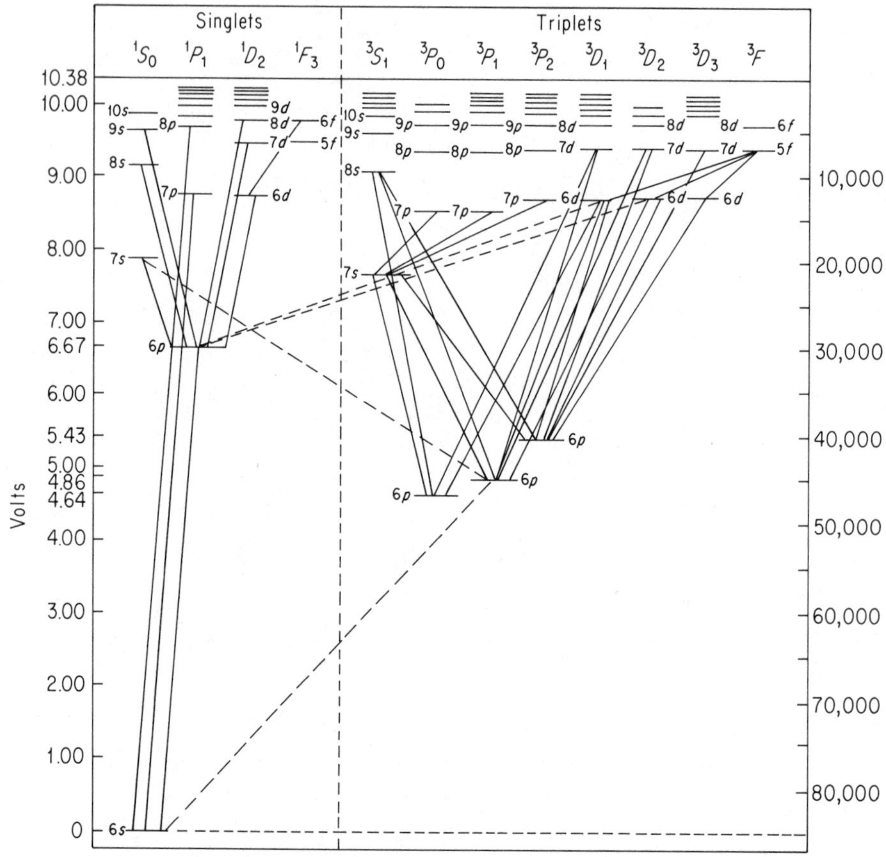

Figure 8.6 Energy level diagram of the Hg atom, with observed transitions indicated. Dashed lines indicate transitions between singlet and triplet states. (Adapted from Herzberg. By permission—see Fig. 8.1.)

Mercury. The energy-level diagram for mercury is shown in Figure 8.6, along with the observed transitions. The new feature for mercury is the fact that singlet-triplet transitions are observed. This is why photochemists often use mercury as a sensitizer for obtaining the population of triplet states of organic molecules. The ΔS selection rule breaks down because, owing to the large magnitude of the spin-orbit interaction terms, S is no longer a valid quantum number. (The terms "singlet" and "triplet" are inappropriate; however, they are still conventionally used.) The only valid quantum number is J. Careful examination of Figure 8.6 reveals that a Δj selection rule of ± 1 (where j is the one-electron total angular momentum) and a ΔJ rule of zero or ± 1 are obeyed.

8.3 SELECTION RULES IN THE ABSENCE OF EXTERNAL FIELDS

We have briefly discussed selection rules for the rigid rotor and the harmonic oscillator. We will now consider the symmetry-related selection rules involved in atomic spectroscopy. The intensity of the absorption of energy for a transition from state i to state j may be defined as the energy absorbed from an incident beam of unit cross section, per unit time. Mathematically, the relation is (see Section 6.7)

$$I_{j \leftarrow i} = (N_i - N_j) h\nu_{ji} B_{ji} \rho(\nu_{ji}) l \qquad \textbf{8.1}$$

where N_i and N_j are the populations of the states (usually expressed as atoms/cm^3), ν_{ji} is the frequency at which the transition occurs, B_{ji} is the coefficient of stimulated absorption, $\rho(\nu_{ji})$ is the energy density, and l is the thickness of the sample. (If conditions are such that an appreciable fraction of the light is absorbed, l has to be replaced by a decaying exponential function.) The reason N_i and N_j both appear in Eq. 8.1 is that radiation stimulates both absorption and emission with equal probability. The relative populations of the levels are determined by the Boltzmann factor. For transitions involving a promotion of an electron, N_j is usually negligible. (In light elements, terms arising from the same ground-state configuration, but differing only in J, are closely enough spaced in energy to have significant populations of terms other than the lowest. However, direct transitions between these are not normally observed.) The *transition probability* per unit time, $w_{j \leftarrow i}$, is defined as

$$w_{j \leftarrow i} = B_{ji} \rho(\nu_{ji}) \qquad \textbf{8.2}$$

For a transition to be observed (i.e., to be allowed), B_{ji} must not equal zero.

An expression for B_{ji} was obtained from time-dependent perturbation theory in Section 6.7. The result is (Eq. 6.101)

$$B_{ji} = 2\pi/3\hbar^2 |\langle \psi_j | \hat{\mu} | \psi_i \rangle|^2 \qquad \textbf{8.3}$$

where $\hat{\mu}$ is the dipole moment operator. (The integral in Eq. 8.3, the transition dipole, is frequently abbreviated as μ_{ji}.) Thus, a transition is allowed if $\langle \psi_j | \hat{\mu} | \psi_i \rangle$ is nonzero. The symmetry imposed selection rule for the transition dipole is the same as for any integral representing an observable. The integral can be nonvanishing only if the product of representations $\Gamma_j \times \Gamma_\mu \times \Gamma_i$ contains the totally symmetric representation of the group describing the system. For an atom, the appropriate group is the group of the sphere, **O**(3). The dipole operator is a vector operator, so it, like any other vector operator, transforms as D_u^1 within **O**(3). Furthermore, it is a function of spatial coordinates only. Thus, if we let our ψ_i, which is a function of spatial and spin coordinates, be represented as the product of a spatial function, say ϕ_i, and a spin function, say σ_i, within the Russell-Saunders scheme, then the transition dipole can be written as

$$\mu_{ji} = \langle\phi_j|\hat{\mu}|\phi_i\rangle\langle\sigma_j|\sigma_i\rangle \qquad \textbf{8.4}$$

The integral $\langle\sigma_j|\sigma_i\rangle$ vanishes unless the representations describing the spin are the same for each state, since the totally symmetric irreducible representation can occur only for the product of a representation with itself. The total spin quantum number, S, is the index for the irreducible representation corresponding to spin in **O**(3). Thus, the selection rule for spin is

$$\Delta S = 0 \qquad \textbf{8.5}$$

The ΔL selection rule can be obtained from $\Gamma_j \times \Gamma_\mu \times \Gamma_i$, where the Γ_j and Γ_i now are the representations only of the spatial portion of the wave function. We can write the requirement as

$$\Gamma_j \times \Gamma_i \times \Gamma_\mu \supset D_g^0 \qquad \textbf{8.6}$$

Since only the product of a representation with itself contains the totally symmetric representation, the requirement is that $\Gamma_j \times \Gamma_i$ must contain Γ_μ. But, letting i and j be the indexes for the representations (i.e., the l), we have

$$D_p^i \times D_{p'}^j = \sum_{k=|i-j|}^{(i+j)} D_{p''}^k \qquad \textbf{8.7}$$

In the summation, k can equal one whenever $|i - j|$ equals one, and also when i does not equal zero, when i equals j. However, the selection rule requires that p'' must be u. If the representations are considered to be for one-electron orbitals, this will not be the case when i equals j, but will be when i equals $j \pm 1$. Thus, we are left with the selection rule

$$\Delta l = \pm 1 \qquad \textbf{8.8}$$

However, states having the same total L value can arise from different electronic configurations. Consequently, ΔL may be zero, even though Δl cannot be.

$$\begin{aligned} \Delta L &= 0, \pm 1 \qquad (L \neq 0) \\ &= \pm 1 \qquad (L = 0) \end{aligned} \qquad \textbf{8.9}$$

To find the ΔJ selection rule, Eq. 8.7 is again applicable; however, the g, u behavior for each term is determined by the L value. Consequently the parity restriction is taken care of by the ΔL selection rule, and a ΔJ of zero is allowed, if J is not equal to zero:

$$\begin{aligned} \Delta J &= 0, \pm 1 \qquad (J \neq 0) \\ &= \pm 1 \qquad (J = 0) \end{aligned} \tag{8.10}$$

In the independent-particle approximation the selection rules must be satisfied for the single electron being promoted. In the Russell-Saunders coupling scheme S and L are constructed from the individual one-electron quantum numbers, so the selection rules for Δs and Δl are the same as for ΔS and ΔL (except that Δl cannot equal zero). In the j-j coupling scheme S and L are no longer valid quantum numbers. The total angular momentum, J, is constructed from one-electron j values. Here, the selection rules for ΔJ are still valid, and the selection rules for the one-electron j's are the same as for ΔJ.

8.4 THE STARK AND ZEEMAN EFFECTS

Thus far, our discussions of spectroscopy have concerned systems uninfluenced by external fields. External electric and magnetic fields have effects on atoms' energy levels and consequently their spectra. The external field destroys the spherical symmetry of the field felt by the electrons in an atom. The effective symmetry is reduced to that of the field itself. The result is a breaking of the degeneracies of degenerate states. The effect caused by an external electric field is called the *Stark effect*, while that caused by a magnetic field is called the *Zeeman effect*.

When an atom is placed in a homogeneous external field, the direction of the field imposes a unique direction on the system. Thus, the group of the sphere, in which all directions are equivalent, is no longer appropriate. Rotation about the axis aligned in the field direction is different from rotation about axes perpendicular to the field. The effective rotational symmetry is reduced to that of the two-dimensional rotation group, **R**(2). The rotation axis is the external field vector. The effect on the other types of symmetry operations possessed by **O**(3) depends upon whether the field is an electric or a magnetic field. An electric field behaves as a vector in the field direction. This vector changes sign on inversion. Consequently the inversion operation is not a symmetry operation for the electric field. On the other hand, the electric-field vector is symmetric with respect to reflection in any plane that contains the vector. In point-group notation (we will use the Schönflies notation for naming point groups) this is called $\mathbf{C}_{\infty v}$ symmetry (the $\mathbf{C}_{\infty}$ group is the pure two-dimensional rotation group in Schönflies notation).

A magnetic field behaves as what is called an *axial vector* or a *pseudovector*. The lines of force associated with a magnetic field describe a circular motion about an axis along the field direction (Figure 8.7). Reversal of the direction of the field reverses the direction of the lines of force. Considering the symmetry of the situation, we see that reflection in a plane perpendicular to the field direction does not change

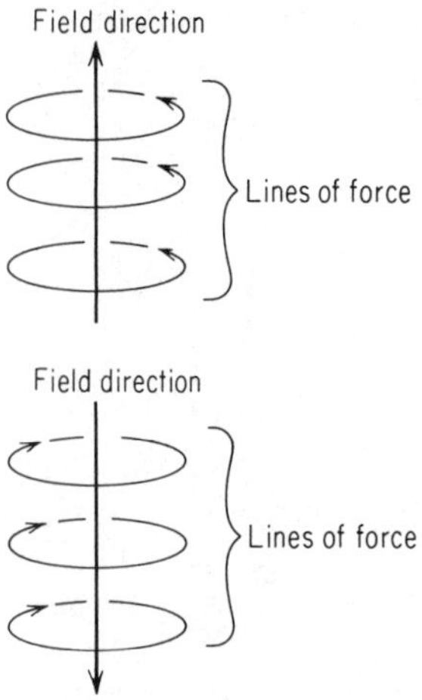

Figure 8.7 Lines of force associated with a magnetic field.

the direction of the lines of force, while reflection in a plane containing the axis does. (The magnetic field behaves as a rotation about a vector, rather than as a vector.) The symmetry of the magnetic field is $\mathbf{C}_{\infty h}$ in the Schönflies notation.

The character tables for $\mathbf{C}_{\infty v}$ (the symmetry of an electric field) and $\mathbf{C}_{\infty h}$ (the symmetry of a magnetic field) are shown in Table 8.2. Notice that, except for those labeled as Σ representations, the character under the rotation operation of the $\mathbf{C}_{\infty v}$ group is $2 \cos \lambda\phi$, where λ is an integer (or half-integer, for double-valued representations). This is of the same form as the last term in a D^j representation from the $\mathbf{R}(3)$ group. The conventional labeling of the representations of $\mathbf{C}_{\infty v}$ is based upon this relationship. The Greek letters used as representation labels are the Greek equivalents of S, P, D, and so on. Just as the spectroscopic notation for atoms uses S for an L value of zero, P for one, and so forth, the notation for the representations of $\mathbf{C}_{\infty v}$ uses Σ for a λ value of zero, Π for a λ of one, and so on. In the $\mathbf{C}_{\infty v}$ group the Σ representations (the $+$ and $-$ superscripts refer to the behavior with respect to the planes of symmetry) are one-dimensional, while all others are two-dimensional. A state that is characterized by one of these two-dimensional representations is doubly degenerate.

In the $\mathbf{C}_{\infty h}$ group, all representations are one-dimensional, even though those corresponding to the same $|\lambda|$ value are paired and given the same representation label. The representation labels use the same Greek letters, corresponding to the various values of λ, as do the representations of $\mathbf{C}_{\infty v}$. The g and u have the usual meaning regarding behavior with respect to inversion. The fact that the representations are all one-dimensional means that there are no symmetry-induced degeneracies when the appropriate group describing a system is $\mathbf{C}_{\infty h}$.

When an atom is placed in an electric field, the effective symmetry is reduced from $\mathbf{O}(3)$ to $\mathbf{C}_{\infty v}$. The states of the field-free atom are classified according to the

Table 8.2

THE $\mathbf{C}_{\infty v}$ AND $\mathbf{C}_{\infty h}$ CHARACTER TABLES[a]

$\mathbf{C}_{\infty v}$	E	$2C(\phi)$	$\infty\sigma_v$
Σ^+	1	1	1
Σ^-	1	1	-1
Π	2	$2\cos\phi$	0
Δ	2	$2\cos 2\phi$	0
Γ^λ	2	$2\cos\lambda\phi$	0

$\mathbf{C}_{\infty h}$	E	$C(\phi)$	$C(-\phi)$	i	$S(-\phi)$	$S(\phi)$	σ_h
Σ_g	1	1	1	1	1	1	1
Π_g	1	$e^{i\phi}$	$e^{-i\phi}$	1	$-e^{i\phi}$	$-e^{-i\phi}$	-1
	1	$e^{-i\phi}$	$e^{i\phi}$	1	$-e^{-i\phi}$	$-e^{i\phi}$	-1
Δ_g	1	$e^{2i\phi}$	$e^{-2i\phi}$	1	$e^{2i\phi}$	$e^{-2i\phi}$	1
	1	$e^{-2i\phi}$	$e^{2i\phi}$	1	$e^{-2i\phi}$	$e^{2i\phi}$	1
Γ_g^λ	1	$e^{\lambda i\phi}$	$e^{-\lambda i\phi}$	1	$(-1)^\lambda e^{\lambda i\phi}$	$(-1)^\lambda e^{-\lambda i\phi}$	$(-1)^\lambda$
	1	$e^{-\lambda i\phi}$	$e^{\lambda i\phi}$	1	$(-1)^\lambda e^{-\lambda i\phi}$	$(-1)^\lambda e^{\lambda i\phi}$	$(-1)^\lambda$
Σ_u	1	1	1	-1	-1	-1	-1
Π_u	1	$e^{i\phi}$	$e^{-i\phi}$	-1	$e^{i\phi}$	$e^{-i\phi}$	1
	1	$e^{-i\phi}$	$e^{i\phi}$	-1	$e^{-i\phi}$	$e^{i\phi}$	1
Δ_u	1	$e^{2i\phi}$	$e^{-2i\phi}$	-1	$-e^{2i\phi}$	$-e^{-2i\phi}$	-1
	1	$e^{-2i\phi}$	$e^{2i\phi}$	-1	$-e^{-2i\phi}$	$-e^{2i\phi}$	-1
Γ_u^λ	1	$e^{\lambda i\phi}$	$e^{-\lambda i\phi}$	-1	$-(-1)^\lambda e^{\lambda i\phi}$	$-(-1)^\lambda e^{-\lambda i\phi}$	$-(-1)^\lambda$
	1	$e^{-\lambda i\phi}$	$e^{\lambda i\phi}$	-1	$-(-1)^\lambda e^{-\lambda i\phi}$	$-(-1)^\lambda e^{\lambda i\phi}$	$-(-1)^\lambda$

[a] The v and h subscripts on the σ indicate, respectively, vertical planes (containing the rotation axis) and horizontal planes (perpendicular to the rotation axis). The half-integer representations are classified as the appropriate Γ^λ.

representations of $\mathbf{O}(3)$. In the field, they can be classified according to the representations of $\mathbf{C}_{\infty v}$. Except for the one- and two-dimensional representations, the representations of $\mathbf{O}(3)$ become reducible when mapped onto $\mathbf{C}_{\infty v}$.

When descending in symmetry, we use the characters from the representations of the higher-symmetry group as the characters in the lower-symmetry group and reduce the resulting representation in the lower-symmetry group, if it is not already an irreducible representation. For example, for the D_g^0 representation of $\mathbf{O}(3)$ mapped onto $\mathbf{C}_{\infty v}$ we have

$\mathbf{O}(3)$ / $\mathbf{C}_{\infty v}$	E / E	$C(\phi)$ / $2C(\phi)$	σ / σ_v
$\Sigma^+ = D_g^0$	1	1	1

8.11

Thus, D_g^0 in $\mathbf{O}(3)$ becomes Σ^+ in $\mathbf{C}_{\infty v}$. For the D_u^1 representation we have

$\mathbf{O}(3)$ / $\mathbf{C}_{\infty v}$	E / E	$C(\phi)$ / $2C(\phi)$	σ / σ_v
D_u^1	3	$1 + 2\cos\phi$	1

8.12

By inspection we see that

$$D_u^1\,[\mathbf{O}(3)] \quad \Rightarrow \quad \Sigma^+ + \Pi\ (\mathbf{C}_{\infty v}) \tag{8.13}$$

For the $D_g^{5/2}$ representation of $\mathbf{O}(3)$ we have

$\mathbf{O}(3)$ / $\mathbf{C}_{\infty v}$	E / E	$C(\phi)$ / $2C(\phi)$	σ / σ_v
$D_g^{5/2}$	6	$2\cos\frac{1}{2}\phi + 2\cos\frac{3}{2}\phi + 2\cos\frac{5}{2}\phi$	0

8.14

The result is

$$D_g^{5/2}\,[\mathbf{O}(3)] \quad \Rightarrow \quad \Gamma^{1/2} + \Gamma^{3/2} + \Gamma^{5/2} \tag{8.15}$$

The general result is

$$D_p^j\,[\mathbf{O}(3)] \quad \Rightarrow \quad \sum_{\lambda=0}^{j} \Gamma^\lambda \qquad \text{(for integer } j\text{)}$$

$$\Rightarrow \quad \sum_{\lambda=1/2}^{j} \Gamma^\lambda \qquad \text{(for half-integer } j\text{)} \tag{8.16}$$

where Γ^0 corresponds to Σ, Γ^1 to Π, and so on, and the + or − on the Σ representation depends upon the sign of the character under the σ_v.

The physical result of placing an atom in an electric field is that the $(2j + 1)$-fold degeneracy of a level in spherical symmetry is partially lifted. The magnitude of the effect can be determined qualitatively from first-order perturbation theory. The perturbation has the form

$$\hat{H}' = \hat{\mu}F_z \tag{8.17}$$

where $\hat{\mu}$ is the dipole operator and F_z is the electric-field strength (the z subscript indicates that the z direction is chosen as the field direction). The first-order perturbation is the integral $\langle\psi_N^0|\hat{H}'|\psi_N^0\rangle$. There is a lowering of the electronic energy by interaction with the field. Since the lower the z component of angular momentum for an angular-momentum function, the more along the z direction it is directed (see Figure 3.2), the larger the λ value, the smaller the shift in energy. Figure 8.8 shows the effect on the ground and first excited levels of sodium.

The selection rules for the Stark effect can be deduced in a manner similar to any other selection rules. The triple product $\Gamma_j \times \Gamma_\mu \times \Gamma_i$ must contain the totally symmetric irreducible representation, or $\Gamma_j \times \Gamma_i$ must contain Γ_μ. In $\mathbf{O}(3)$, Γ_μ transformed as D_u^1. We saw in Eq. 8.13 that, in $\mathbf{C}_{\infty v}$, this reduces to Σ^+ and Π. Thus, if $\Gamma_j \times \Gamma_i$ contains either Σ^+ or Π, the transition from state i to state j will be allowed.

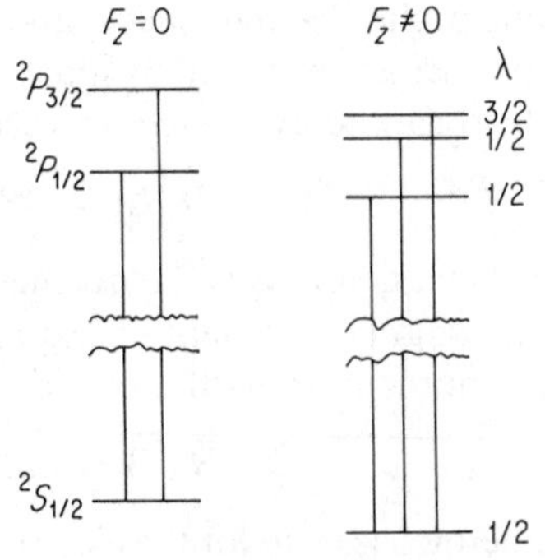

Figure 8.8 The Stark effect for the sodium atom. In the absence of a field, a doublet is seen in the orange region of the spectrum. In an electric field, this is split into a triplet, due to the splitting of the $^2P_{3/2}$ term by the field.

There is a simple rule for multiplying the representations of $\mathbf{C}_{\infty v}$. It is

$$\Gamma^{\lambda} \times \Gamma^{\lambda'} = \Gamma^{|\lambda - \lambda'|} + \Gamma^{(\lambda + \lambda')} \tag{8.18}$$

If λ equals λ', both Σ^+ and Σ^- representations will occur in the product with $\Gamma^{2\lambda}$. The product thus contains Σ^+ if λ equals λ' and contains Π (Γ^1) if λ equals $\lambda' \pm 1$. The selection rule is

$$\Delta\lambda = 0, \pm 1 \tag{8.19}$$

When an atom is placed in a magnetic field, the effective symmetry is reduced from $\mathbf{O}(3)$ to $\mathbf{C}_{\infty h}$. The representations according to which the states transform can be found in a manner analogous to the $\mathbf{C}_{\infty v}$ case. The reduction according to Σ, Π, and so on is exactly the same as it was in $\mathbf{C}_{\infty v}$; however, these are now separable. If we let m equal λ with its sign included, the one-dimensional representations can all be labeled by Γ_p^m. [The parity comes directly from that in $\mathbf{O}(3)$.] The reduction formula, analogous to Eq. 8.16, is

$$D_p^j\,[\mathbf{O}(3)] \Rightarrow \sum_{m=-j}^{j} \Gamma_p^m \tag{8.20}$$

The physical result of placing an atom in a magnetic field is the complete lifting of the $(2j + 1)$-fold degeneracy.

Consider the Zeeman effect on the singlet states arising from the promotion of a single electron in helium. The S value is zero, the L value is the l value of the promoted electron, and the J value is the same as the L value. In a magnetic field, a state with an l value not equal to zero is split into $2l + 1$ levels. The m values labeling these in $\mathbf{C}_{\infty h}$ are the allowed values of the m quantum number for the orbital under consideration. (The same argument could be applied to the low-resolution spectrum of the one-electron hydrogen atom; however, under high resolution, the spectrum would be more complex, owing to splitting of levels classified by various J values.) This is why the m quantum number is called the magnetic quantum number.

The magnitude of the Zeeman effect can be estimated from first-order perturbation theory. The perturbation caused by interaction of the orbital angular momentum with a magnetic field can be expressed, in vector notation, as

$$\hat{H}'(L) = \beta \mathbf{B} \cdot \mathbf{L} \qquad \textbf{8.21}$$

where β is the Bohr magneton, $\mathbf{B}$ is the magnetic-field strength expressed as a vector, and $\mathbf{L}$ is the orbital angular momentum expressed as a vector. For the interaction of the spin angular momentum with the field

$$\hat{H}'(S) = g\beta \mathbf{B} \cdot \mathbf{S} \qquad \textbf{8.22}$$

where g is the magnetogyric ratio for the electron and $\mathbf{S}$ is the spin angular momentum, expressed as a vector. The total perturbation is

$$\hat{H}' = \beta \mathbf{B} \cdot (\mathbf{L} + g\mathbf{S}) \qquad \textbf{8.23}$$

The magnetic field defines the z direction. The components of $\mathbf{L}$ and $\mathbf{S}$ along that direction can be expressed as M_L and M_S. Thus, $\hat{H}'$ can be expressed as

$$\hat{H}' = \beta B(M_L + gM_S) \qquad \textbf{8.24}$$

The first-order correction to the energy increases with both M_L and M_S; consequently, it will increase with M_J, the total magnetic quantum number. The selection rules for ΔM_J within $\mathbf{C}_{\infty h}$ are completely analogous to those for $\Delta\lambda$ within $\mathbf{C}_{\infty v}$.

Several examples of the Zeeman effect are illustrated in Figure 8.9. The first example is the ${}^1P_1 \leftarrow {}^1S_0$ transition of helium. The ground state has a J value of zero and is not split. The ${}^1P_1$ excited state is split into three levels in the magnetic

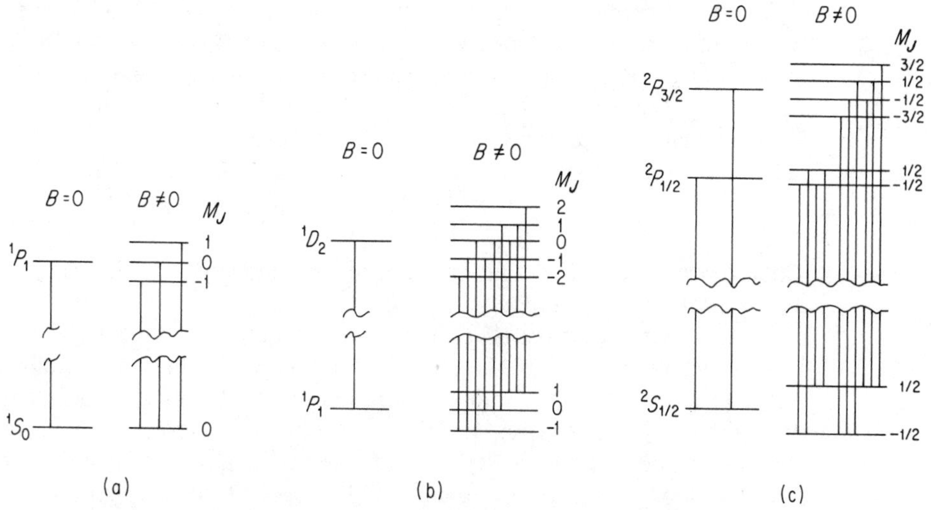

Figure 8.9 The Zeeman effect: (a) Helium; (b) normal Zeeman effect of the ${}^1D_2 \leftarrow {}^1P_1$ transition of cadmium; (c) anomalous Zeeman effect of the ${}^2P \leftarrow {}^2S$ transitions (D line) of sodium.

field. In the absence of a field, a single line is seen, while three lines are seen if the sample is in a magnetic field. When both states involved in a transition have a nonzero J value, both will be split by a field. In the *normal Zeeman effect* the splitting between adjacent levels is the same for both states. This is illustrated for the $^1D_2 \leftarrow {}^1P_1$ transition of cadmium. Even though nine transitions are actually occurring here, only three spectral lines are seen, since the three with ΔM_J of -1 have the same energy difference, as do the three with ΔM_J of 0 and the three with ΔM_J of $+1$. If the g value happens to be different for the different terms, an *anomalous Zeeman effect* occurs. The splittings are different in the different terms. This is the case for the D lines of sodium. Two lines are seen in the absence of a field, but ten are observed under high resolution in a magnetic field.

8.5 LASERS

The laser phenomenon was proposed by Schawlow and Towns in 1958, from considerations based upon the maser phenomenon which, in turn, was discovered by Towns and others in the early 1950s. It presents an interesting departure from the usual behavior in absorption and emission of light. In Section 6.7 (Eq. 6.76) we saw that if a system is in equilibrium with an electromagnetic field, the ratio of the populations of two states, N_2/N_1, (with state 2 lying the higher in energy) is

$$\frac{N_2}{N_1} = \frac{B_{12}\rho_\nu}{A_{21} + B_{12}\rho_\nu} \tag{8.25}$$

where B_{12} is the Einstein coefficient of induced absorption or emission, A_{21} is the coefficient of spontaneous emission, and ρ_ν is the energy density at frequency ν. From the form of the equation we see that the ratio N_2/N_1 approaches unity in the limit of large ρ_ν. N_2 can never be made to exceed N_1 by the direct absorption of electromagnetic radiation.

The rate of energy emission from an excited state is (Eq. 6.75)

$$-\frac{dN_2}{dt} = N_2A_{21} - (N_1 - N_2)B_{12}\rho_\nu \tag{8.26}$$

The maximum emission rate is normally the spontaneous rate that occurs as a rapidly decaying emission after the exciting radiation is shut off (that is, when the energy density is zero). A nonzero radiation field decreases the emission rate, although at normal energy densities and optical frequencies the A_{21} term predominates. If some method other than direct absorption of radiation by state 1 can be found to populate state 2, a population higher than the equilibrium population can be achieved. If the population of state 2 is made to be higher than that of state 1, Eq. 8.26 can be rewritten as

$$-\frac{dN_2}{dt} = N_2A_{21} + (N_2 - N_1)B_{12}\rho_\nu \tag{8.26a}$$

We have a situation where the electromagnetic radiation increases the emission rate. This so-called population inversion is the source of the laser (light amplification through stimulated emission of radiation) phenomenon. [Note that, in the Boltzmann equation

$$\frac{N_2}{N_1} = \exp\left(-\frac{h\nu}{kT}\right) \tag{8.27}$$

a ratio greater than one would imply a negative absolute temperature. The population inversion is sometimes referred to as a negative temperature situation.]

Spontaneous emission always occurs to some extent. If the conditions are right, this can serve as the radiation that stimulates the emission. A spontaneously emitted photon stimulates the emission of other photons, each of which can stimulate the emission of more photons in a cascade effect. In order to get significant amplification, a very long path length is required. This is accomplished by confining the laser substance in a reflective cavity. The distance between the reflectors may be finely adjusted so that only one of a number of frequencies that may be possible from different energy levels is amplified. In effect, the reflectors can form an interferometer.

The light from a laser differs from ordinary light in several respects. Since all the emitted light arises from a chain reaction from the first spontaneously emitted photon, the frequency, the phase angle (see Eq. 6.89), and the direction of propagation are the same for all the photons. The radiation is said to be coherent in time (or monochromatic, the spread in ν is extremely small) and in space (the spread in α is extremely small). In addition, very high power density (power per unit area) can be achieved in these coherent beams. Pulsed lasers concentrate this power in pulses of very short duration. This gives even higher power per unit area per unit time. These properties make lasers useful for a variety of novel applications, ranging from "knives" for microsurgery to surveying the moon from the earth.

For a system to be useful as a laser, it must have at least three, and preferably four or more, accessible energy levels. These are shown in Figure 8.10. In the three-level scheme, the lowest energy state (1) is depopulated by some excitation process,

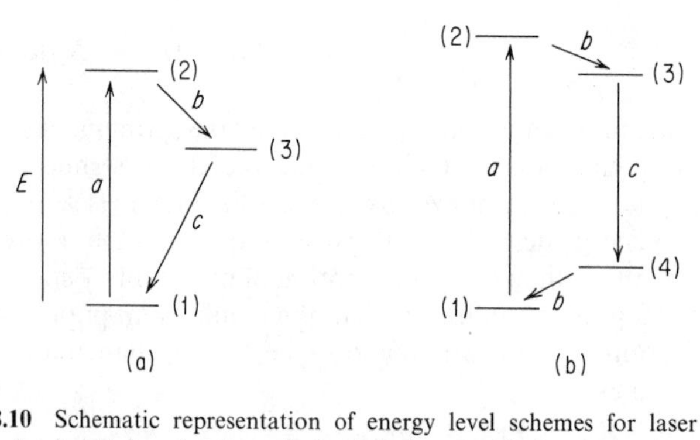

Figure 8.10 Schematic representation of energy level schemes for laser action: (a) three levels; (b) four levels. *a*, Excitation; *b*, nonradiative conversion; *c*, laser emission.

step *a* (absorption of radiation, electric discharge, or some other process). The state (3) is populated from state (2), normally by a nonradiative process, step *b*. If steps *a* and *b* are more rapid than the emission from state (3), a population inversion of N_3/N_1 can occur. Laser emission, step *c*, can be achieved. The three-level scheme requires a large amount of "pumping" power to obtain the population inversion. In the four-level scheme, both states involved in the laser emission are excited states; consequently it is much easier to obtain a high population of the upper laser-active state, relative to the lower.

The first operating laser was a solid-state ruby device, constructed by Maiman in 1960. It was a three-state system involving the Cr^{3+} ions in the ruby. Optical pumping was used. The construction of a helium-neon gas-phase laser was accomplished in 1961 by Jovan and co-workers. Since that time many different materials—gases, liquids, and solids—have been used to construct lasers. The directly available frequencies cover much of the visible and infrared spectral regions. Ultraviolet radiation can be obtained by frequency doubling (utilizing the special properties of nonlinear optics). Continuously—variable—frequency lasers can be built with molecular dyes.

The helium-neon gas laser is particularly interesting within the context of this chapter. The neon is the substance exhibiting the laser action. However, the population inversion arises from a transfer of energy from an excited state of helium to produce an excited state of neon. The helium is excited by an electric discharge (collision with electrons in an electric discharge tube). "Normal" selection rules do not hold for such excitations. Many of the excited helium atoms end up in the lowest excited $^3S_1$ state ($1s^1 2s^1$ configuration), either directly from the excitation or by decay from higher-energy states. The radiative transition from the $^3\mathrm{S}_1$ state to the singlet ground state is spin-forbidden, giving the state a relatively long lifetime. This state lies about 1.6×10^5 cm^{-1} above the ground state of helium. The highest-energy state of the $1s^2 2s^2 2p^5 4s^1$ configuration [which we will abbreviate (Ne^+, $4s$)] of neon lies only 314 cm^{-1} lower in energy than this, relative to the neon ground state. In such a situation, resonant transfer of energy can take place, in which the excitation energy is transferred from the helium to the neon. The states of the (Ne^+, $3p$) and (Ne^+, $3s$) configurations lie between the (Ne^+, $4s$) configuration and the ground state. They are not populated by the excited helium; consequently, there is a population inversion involving various excited states of the neon. The predominant laser activity is from the states of the (Ne^+, $4s$) configuration to those of the (Ne^+, $3p$) configuration. The resulting radiation is in the near-infrared spectral range ($\sim 5.8 - 9.7 \times 10^3$ cm^{-1}). Visible light can be produced when the (Ne^+, $5s$) initial configuration of Ne (which can be populated via the first excited singlet state of helium) decays to the (Ne^+, $3p$) configuration.

The emission behavior of the excited neon is different from that of the other light elements. The selection rules are not those that would be expected from Russell-Saunders coupling. Russell-Saunders coupling predicts four terms, $^1\mathrm{P}_1$, $^3\mathrm{P}_{0,1,2}$, for the (Ne^+, $4s$) configuration and ten terms, $^1\mathrm{S}_0$, $^1\mathrm{P}_1$, $^1\mathrm{D}_2$, $^3\mathrm{S}_1$, $^3\mathrm{P}_{0,1,2}$, $^3\mathrm{D}_{1,2,3}$, for the (Ne^+, $3p$) configuration. The ΔS, ΔL, and ΔJ selection rules predict only eighteen transitions from the higher to the lower configuration. More than this are observed. The apparent selection rules are $\Delta J = 0, \pm 1$ (but only ± 1 for a J of zero) and $\Delta l = \pm 1$. This allows thirty transitions from the (Ne^+, 4s) to the (Ne^+, 3p) configuration. Most of these have been identified in the He-Ne laser.

The excited states of the inert gases are frequently constructed from *pair coupling*, which is intermediate between the Russell-Saunders and the *j-j* couplings. In this scheme the one-electron orbital angular momentum, l, of the promoted electron is coupled to the total angular momentum, J_c, of the unpromoted core (Ne^+ in this case) to give a resultant called K. The one-electron spin, s, is then coupled to K to give the total J. Racah's notation for this lists the orbital of the promoted electron, nl, the K value, and the overall J value as $nl[K]_J$. The l is expressed in the notation s, p, and so on. It also bears a prime if the K value arises from the lower of the two possible J_c values. Consider, for example, the (Ne^+, 4s) and (Ne^+, 3p) configurations of neon. For Ne^+ there are only two term symbols, $^2P_{1/2}$ and $^2P_{3/2}$. For the 4s electron, l is zero and the K values are the same as J, giving $4s'[\frac{1}{2}]$ and $4s[\frac{3}{2}]$. Coupling the spin of the 4s electron gives $4s'[\frac{1}{2}]_0$, $4s'[\frac{1}{2}]_1$, $4s[\frac{3}{2}]_1$, and $4s[\frac{3}{2}]_2$. For the 3p electron, l is one. The K values arising from a J_c of $\frac{3}{2}$ are $\frac{1}{2}$, $\frac{3}{2}$, and $\frac{5}{2}$. Those from a J_c of $\frac{1}{2}$ are $\frac{1}{2}$ and $\frac{3}{2}$. The ten resulting states are $3p[\frac{5}{2}]_{3,2}$, $3p[\frac{3}{2}]_{2,1}$, $3p[\frac{1}{2}]_{1,0}$, $3p'[\frac{3}{2}]_{2,1}$, and $3p'[\frac{1}{2}]_{1,0}$. The states for other configurations and other inert gases can be constructed in a similar manner.

BIBLIOGRAPHY

CONDON, E. U., and SHORTLEY, G. H., *The Theory of Atomic Spectra.* Cambridge University Press, New York, 1935.

HERZBERG, G., *Atomic Spectra and Atomic Structure.* Dover Publications, Inc., New York, 1944.

MOORE, C. E., *Atomic Energy Levels.* NBS Circular 467, Washington, D. C., Vol. I, 1949; Vol. II, 1952; Vol. III, 1958.

PROBLEMS

***8.1** Starting from the indicated states of an atom, what would be the term symbols of states accessible from direct absorption of electromagnetic radiation (assume Russell-Saunders coupling)?

(a) $^1D_2$. (b) $^3P_2$. (c) $^3S_1$. (d) $^2S_{1/2}$.

(e) $^4F_{3/2}$. (f) $^2P_{3/2}$. (g) $^4D_{7/2}$.

***8.2** Schematically, give the energy levels for a $^2P_{3/2} \leftarrow {}^2S_{1/2}$ atomic transition and for a $^3D_3 \leftarrow {}^3P_2$ transition:

(a) In the absence of any external field.

(b) In an external magnetic field with equal splitting in the upper and lower levels.

(c) In an external magnetic field with unequal splitting in the two levels.

(d) In an external electric field.

***8.3** How many spectral lines will be observed in each case in the previous problem?

8.4 Sketch the normal and anomalous Zeeman spectra for a $^2D_{5/2} \leftarrow {}^2P_{3/2}$ transition. Identify the initial and final M values for each line.

8.5 Sketch the expected Stark-effect spectrum for a $^2D_{5/2} \leftarrow {}^2P_{3/2}$ transition, assuming that the Stark splitting is different in the two levels.

***8.6** Which transitions from the indicated initial terms would be forbidden? State the selection rules that are violated. (Assume that the initial and final configurations are different.)

Initial	To				
$^1S_0$	$^1P_1$	$^3P_0$	$^3P_1$	$^3P_2$	$^1D_2$
$^3P_1$	$^3D_1$	$^1D_2$	$^3D_3$	$^1S_0$	$^1P_1$
$^1D_2$	$^1D_2$	$^1P_1$	$^3P_0$	$^1S_0$	$^1F_3$
$^4P_{5/2}$	$^4S_{3/2}$	$^4P_{3/2}$	$^4D_{1/2}$	$^4F_{3/2}$	$^2S_{1/2}$
$^2F_{5/2}$	$^2S_{1/2}$	$^2D_{5/2}$	$^2G_{7/2}$	$^2P_{3/2}$	$^2D_{3/2}$

***8.7** The ground state of nitrogen has the configuration $(1s)^2(2s)^2(2p)^3$ and the term symbol $^4S_{3/2}$. What terms involving the promotion of an electron from a $2p$ orbital to an $n = 3$ orbital can be observed by direct absorption of electromagnetic radiation? Sketch the spectrum (schematically), assuming arbitrary intensities and that (a) the orbital energies are $3s < 3p < 3d$ and (b) the rules for ordering the terms of the excited configurations are the same as for the ground configuration. (This last assumption is not necessarily correct.)

8.8 Sketch the Stark and Zeeman spectra for each allowed transition from the preceding problem. Assume a normal Zeeman effect.

chapter nine

The hydrogen-molecule ion

9.1 GENERAL INTRODUCTION

As interesting as atoms may be, chemistry is concerned primarily with the collections of atoms known as molecules. The failure to predict chemical bonding was one of the serious flaws of Bohr theory. Modern quantum theory does predict the existence of chemical bonding and of molecules. For some simple molecules the available quantum-chemical results are at least as good as the experimental results. For large molecules, however, approximations must be made. Most of our remaining study will be centered on the use of approximate methods to obtain qualitative predictions about the electronic structure and behavior of molecules. We will start with simple systems, so that we can compare the approximate results to more accurate results.

The simplest possible molecular system would require two nuclei (in order to be a molecule) and one electron (to provide the bonding). Such a system is the hydrogen-molecule ion, H_2^+, which has two hydrogen nuclei and an electron. The exact nonrelativistic Hamiltonian for H_2^+ contains the kinetic-energy operators for the electron and each nucleus, the attractive potential for the electron to each nucleus, and the internuclear repulsion potential

$$\hat{H} = -\frac{1}{2}\nabla_1^2 - \frac{m_e}{2M}\nabla_A^2 - \frac{m_e}{2M}\nabla_B^2 - \frac{1}{r_A} - \frac{1}{r_B} + \frac{1}{R_{AB}} \qquad \mathbf{9.1}$$

where the subscript 1 is for the electron, A and B are subscripts for the nuclei, r_A and r_B are the distances from the electron to the nuclei, and R_{AB} is the internuclear separation (see Figure 9.1). Since the mass of the electron, m_e, is our mass unit, the nuclear kinetic-energy terms contain a factor of m_e/M, where M is the nuclear mass, to get them in the same units.

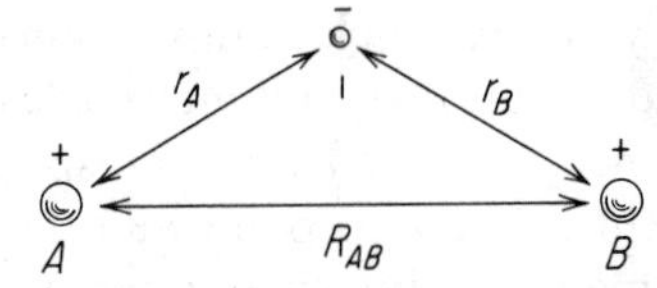

Figure 9.1 The hydrogen molecule ion.

For most of our purposes we will use the *Born-Oppenheimer approximation*—the approximation that nuclear and electronic motions can be treated separately. The qualitative argument is based upon the fact that the nuclear mass is much greater than the electron mass. Thus, the electrons should be able to move much more rapidly than the nuclei, effectively adapting themselves to an optimum distribution for each internuclear separation. The rigorous argument (Slater, App. 2) uses an electronic Hamiltonian, in which the fixed nuclear positions appear as parameters, to obtain a Born-Oppenheimer energy, E_{BO}, as a function of the nuclear coordinates, and an electronic wave function, ψ_{el}. The nuclear Hamiltonian is then set up, using E_{BO} as the potential for the nuclei, to obtain the total energy and a nuclear wave function, ψ_{nuc}. It can then be shown that a total wave function approximated by the simple product, $\psi_{el}\psi_{nuc}$, approximately satisfies the Schrödinger equation for the complete system. The error in this approximation is of the order of the ratio of electron mass to nuclear mass. Thus, the heavier the nuclei, the better the approximation.

Within the Born-Oppenheimer approximation the electronic Hamiltonian is

$$\hat{H}_{el} = -\frac{1}{2}\nabla^2 - \frac{1}{r_A} - \frac{1}{r_B} \tag{9.2}$$

where the subscript for the electron has been dropped. The electronic energy is the eigenvalue of the electronic Schrödinger equation, and the total Born-Oppenheimer energy is

$$E_{BO} = E_{el} + \frac{1}{R_{AB}} \tag{9.3}$$

The H_2^+ problem is simple enough that it can be solved numerically to as much accuracy as is desired, either with or without the Born-Oppenheimer approximation. (It is, of course, easier with the approximation.) With the Born-Oppenheimer approximation, the calculated total ground-state energy is $-.6026342$ Hartree a.u. at an equilibrium separation of 2.0 Bohr a.u. On dissociation, H_2^+ gives a hydrogen atom and a hydrogen ion. A hydrogen ion is a bare proton, so, with our energy conventions that an energy of zero corresponds to the separate particles being at infinite separation, the energy of a hydrogen atom and a hydrogen ion at infinite separation will be just the energy of the hydrogen atom, $-.5$ Hartree. Thus, the calculated dissociation energy is .10263 Hartree. This dissociation energy is called D_e. It is for a fixed internuclear separation and does not account for the zero-point vibrational energy. We can include this within the Born-Oppenheimer approximation by calculating the energy at several internuclear distances near the equilibrium value, considering the energy as a function of displacement, and thereby obtaining a force constant from which the zero-point vibrational energy can be calculated. When this correction is made, the dissociation energy, called D_0 in this case, is .09748 Hartree.

If the Born-Oppenheimer approximation is not used, the calculated total energy is $-.596689$ Hartree at an equilibrium separation of 2.0 Bohrs. Note that this differs by about 1 percent from the Born-Oppenheimer value. The error caused by the approximation is actually worse here than it is in most cases, since the hydrogen nuclei are the lightest of all nuclei. The energy this time already includes the zero-point vibrational energy, since the Hamiltonian includes the nuclear motions. Thus the dissociation energy is D_0 and has a value of .096689 Hartree. The difference between the two calculated values of D_0 is only about .8 percent or, in chemical terms, about half a kilocalorie per mole. Both these values are probably more accurate than the best experimental value.

9.2 *THE UNITED-ATOM AND SEPARATED-ATOM LIMITS*

One of the nice things about computational chemistry is that you can do things with computations that are not likely ever to be done experimentally. For example, Born-Oppenheimer calculations can be done on H_2^+ to find the electronic energy as a function of internuclear separation, R_{AB}, and these calculations can be taken to a separation of zero. This, of course, can never be done experimentally, or even computationally with the complete Hamiltonian, since the energy would go positively infinite as the nuclei were brought together. If the two nuclei, each with a single positive charge, are coalesced, the result is a single point charge, but with a magnitude of two charge units. In other words, as far as the calculation is concerned, it is a helium nucleus. The problem of a helium nucleus with a single electron, He^+, is a hydrogenic problem, and the exact solution is available. If the nuclei are taken to infinite separation, we get a hydrogen atom and a hydrogen ion. The electronic energy levels available are those of the hydrogen atom. The calculations can be performed at all intermediate stages to produce a series of curves for the energy levels as shown in Figure 9.2. The two limits are the *united-atom limit* and the *separated-atom limit*. On the sides of the drawing the principal quantum numbers for the limiting energy levels are given.

Each curve in Figure 9.2 represents an orbital energy level. At the two limits they are the appropriate atomic orbitals. In the intermediate region they correspond to molecular orbitals. Two labels are given to each curve. The first ($1\sigma_g^+$, $1\sigma_u^+$, and so on) is a symmetry label, using the representation labels from the $D_{\infty h}$ point group (which describes the symmetry of H_2^+, or any other homonuclear diatomic molecule or any linear molecule with inversion symmetry) and a running number, which is a pseudo quantum number. The representations of $\mathbf{C}_{\infty v}$ provide labels for the orbitals

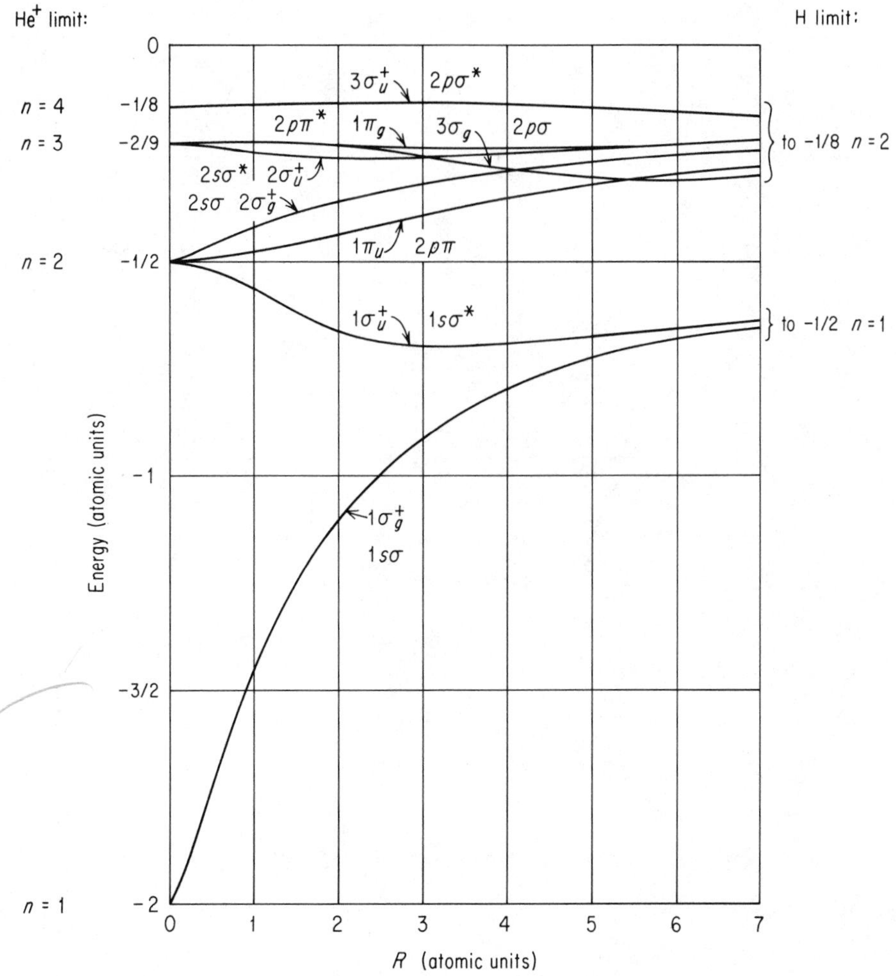

Figure 9.2 Several lowest electronic energy levels of H_2^+ as a function of internuclear distance. (Adapted from J. C. Slater, *Quantum Theory of Molecules and Solids*, Volume 1, McGraw-Hill Book Company, New York, 1963. By permission.)

of heteronuclear diatomic molecules or linear molecules without inversion symmetry. Lower-case symbols are used for the orbitals, since they are one-electron functions. The number labels indicate that a given orbital is the first, second, third, and so on orbital having the indicated symmetry. The second set of labels in Figure 9.2 is based upon the L.C.A.O. description of the molecular orbitals, which we will discuss shortly.

Let us consider a plot of $\psi(z)$ versus z along the z axis (the internuclear axis) for the $1\sigma_g^+$ and $1\sigma_u^+$ orbitals for several internuclear distances. These are shown in Figure 9.3. Consider first the $1\sigma_g^+$ orbital. At a separation of 8 Bohrs the orbital

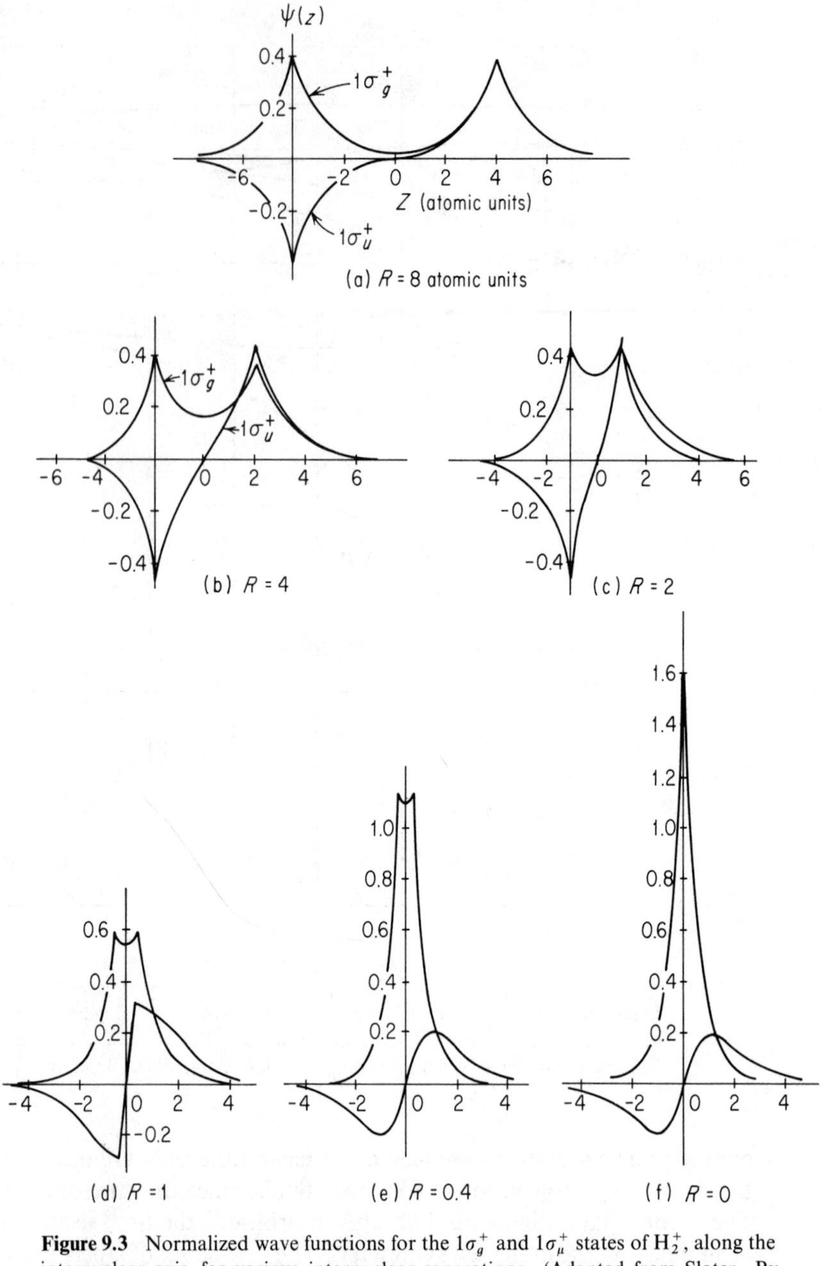

Figure 9.3 Normalized wave functions for the $1\sigma_g^+$ and $1\sigma_\mu^+$ states of H_2^+, along the internuclear axis, for various internuclear separations. (Adapted from Slater. By permission—see Fig. 9.1.)

looks almost like the sum of two hydrogen-atom $1s$ orbitals, one centered on each nucleus:

$$1\sigma_g^+ \sim 1s_A(\mathrm{H}) + 1s_B(\mathrm{H}) \qquad \text{(for large } R\text{)} \qquad \mathbf{9.4}$$

For larger R this becomes an even closer approximation, until at infinite separation it is exact. For an internuclear separation of zero (the united-atom limit), the orbital becomes a $1s$ orbital for He^+:

$$1\sigma_g^+ = 1s(\mathrm{He}^+) \qquad \text{(for } R = 0\text{)} \qquad \mathbf{9.5}$$

At intermediate separations the function goes smoothly from one limit to the other.

At large separation the $1\sigma_u^+$ orbital behaves as the difference of two hydrogenic $1s$ orbitals:

$$1\sigma_u^+ \sim 1s_A(H) - 1s_B(H) \qquad \text{(for large } R\text{)} \qquad \mathbf{9.6}$$

where A is taken to be the atom on the positive z axis. At zero separation, however, this becomes a $2p_0$ orbital of He^+ (directed along the z axis):

$$1\sigma_u^+ = 2p_0(\mathrm{He}^+) \qquad \text{(for } R = 0\text{)} \qquad \mathbf{9.7}$$

This is why the united-atom limit corresponds to a principal quantum number of 2. (Remember that, for a one-electron atom or ion, the $2s$ and $2p$ levels are degenerate with each other. The same is true for the $3s$, $3p$, $3d$, and so on.) If we draw a two-dimensional plot of a contour of constant ψ at large R and at an R of zero, we have the situation in Figure 9.4, for these two orbitals. Note that the symmetry properties of the orbitals, within the $\mathbf{D}_{\infty h}$ symmetry of the molecule, are preserved in going from one limit to the other. A similar situation results for all the other orbitals. For example, Figure 9.5 presents the results for the $1\pi_u$ and $1\pi_g$ orbitals. (Note that, since g and u refer to inversion through the origin, the sum of the two p orbitals is *ungerade*,

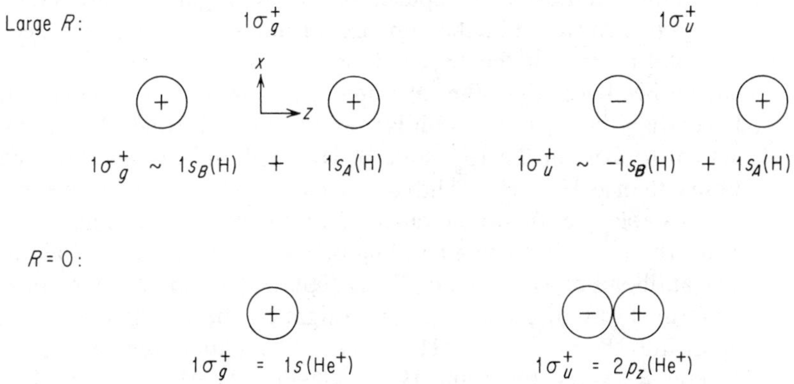

Figure 9.4 Two-dimensional contour of constant ψ for the $1\sigma_g^+$ and $1\sigma_u^+$ orbitals of H_2^+ at large R and at R equals zero. Shapes only are shown. The He^+ functions are more contracted than the H functions.

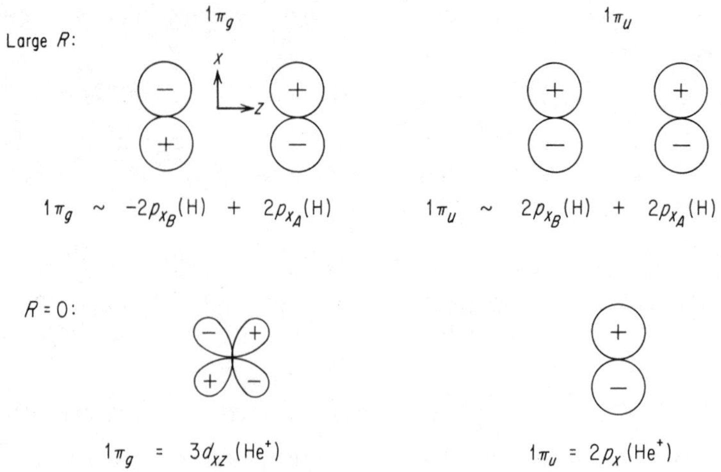

Figure 9.5 Two-dimensional contour of constant ψ for the $1\pi_g$ and $1\pi_u$ orbitals of H_2^+ at large R and at R equals zero. Shapes only are shown. The He^+ functions are more contracted than the H functions.

while the difference is *gerade*.) Considerations such as this can give a qualitative ordering of the molecular-orbital energy levels of H_2^+ without any calculations. The second set of labels for the orbitals given in Figure 9.2 is based upon the separated-atom limit. The notation $1s\sigma$ indicates that the orbital has σ symmetry and goes to the $(1s_A + 1s_B)$ limit, $1s\sigma^*$ goes to the $(1s_A - 1s_B)$ limit, $2p\pi$ has π symmetry and goes to the $(2p_A + 2p_B)$ limit, $2p\pi^*$ has π symmetry and goes to the $(2p_A - 2p_B)$ limit, and so on.

The total energy, in the Born-Oppenheimer approximation, is the sum of the electronic and the nuclear repulsion energies. Figure 9.6 shows the nuclear repulsion energy (which would be the same for all states) along with the electronic energy and the total energy for the $1\sigma_g^+$ and $1\sigma_u^+$ states of H_2^+. Note that only the $1\sigma_g^+$ total energy lies lower than the separated-atom limit (E_H). The fact that this energy is lower than the separated limit is the source of the bonding. It is more stable for the system to exist in the $1\sigma_g^+$ state of H_2^+, with an internuclear separation around 2 Bohrs, than as $H + H^+$. The converse is true, however, for the $1\sigma_u^+$ state. This state is less stable over the whole range of internuclear separations than is $H + H^+$. The $1\sigma_g^+$ orbital is said to be a bonding orbital, while the $1\sigma_u^+$ is said to be antibonding. The antibonding $1\sigma_u^+$ state of H_2^+ is actually a repulsive state. If, somehow, it could be formed (say by the absorption of light by the $1\sigma_g^+$ state), the molecule would fly apart into the more stable H and H^+. The antibonding energy would show up as kinetic energy in the H and H^+ fragments. The $1\sigma_g^+$ state is the only state of H_2^+ that is stable with respect to dissociation to a ground-state hydrogen atom and a hydrogen ion. Certain other states are stable with respect to dissociation to excited states of the hydrogen atom. These are the states in which the orbital having the

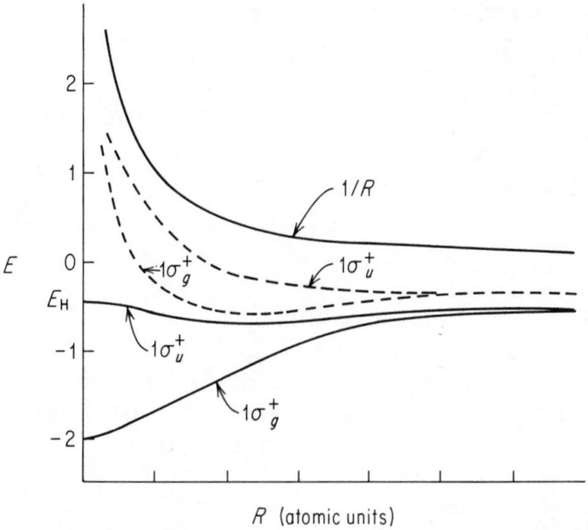

Figure 9.6 The nuclear repulsion energy (upper curve), electronic energy (lower curves), and total energy (dashed curves) for the $1\sigma_g^+$ and $1\sigma_u^+$ states of H_2^+.

plus sign in the separated-atom limiting wave function (the ones without the asterisk in the second notation) is occupied. These orbitals are said to be bonding combinations of the separated-atom functions, while the others are said to be antibonding.

9.3 THE L.C.A.O. APPROXIMATION

We have seen that, in the separated-atom limit, the orbitals of H_2^+ behave as sums or differences of atomic orbitals centered on the two nuclei. Let us assume that the molecular orbital at other separations can be expressed as a linear combination of atomic orbitals. This is the *linear combination of atomic orbitals molecular-orbital* (L.C.A.O.-M.O.) approximation. Let

$$\psi = a\phi_A + b\phi_B \tag{9.8}$$

where ψ is the molecular orbital, ϕ_A and ϕ_B are atomic orbitals centered on nucleus A and nucleus B, and a and b are linear coefficients.

In most L.C.A.O. problems the coefficients are treated as variational parameters and obtained by applying the variational principle. We need to minimize the expectation value of the Hamiltonian, subject to the constraint that the molecular orbital must be normalized. If we allow the Lagrangian multiplier to be the negative of the L.C.A.O.-M.O. energy (see Eq. 7.49), we have

$$\delta\{\langle\psi|\hat{H}|\psi\rangle - E\langle\psi|\psi\rangle\} = 0 \tag{9.9}$$

The bracketed term can be rewritten

$$\{\langle\psi|\hat{H}|\psi\rangle - E\langle\psi|\psi\rangle\} = \langle\psi|\hat{H} - E|\psi\rangle \tag{9.10}$$

If we substitute in Eq. 9.8 for the wave function, we have, remembering that $\langle\phi_A|\hat{H}|\phi_B\rangle$ equals $\langle\phi_B|\hat{H}|\phi_A\rangle$,

$$\begin{aligned}\langle\psi|\hat{H} - E|\psi\rangle &= a^2\langle\phi_A|\hat{H} - E|\phi_A\rangle + b^2\langle\phi_B|\hat{H} - E|\phi_B\rangle + 2ab\langle\phi_A|\hat{H} - E|\phi_B\rangle \\ &= a^2(H_{AA} - E) + b^2(H_{BB} - E) + 2ab(H_{AB} - ES_{AB})\end{aligned} \tag{9.11}$$

where we have made the definitions

$$\begin{aligned}H_{AA} &= \langle\phi_A|\hat{H}|\phi_A\rangle \\ H_{BB} &= \langle\phi_B|\hat{H}|\phi_B\rangle \\ H_{AB} &= \langle\phi_A|\hat{H}|\phi_B\rangle \\ S_{AB} &= \langle\phi_A|\phi_B\rangle\end{aligned} \tag{9.12}$$

The last integral of Eq. 9.12 is called the *overlap integral*. It is a one-electron integral involving atomic orbitals centered on two different atomic centers.

We need now to take the variation of Eq. 9.11 with respect to a and to b and set the results equal to zero. The results are

$$\frac{\partial}{\partial a}\langle\psi|\hat{H} - E|\psi\rangle = 2a(H_{AA} - E) + 2b(H_{AB} - ES_{AB}) = 0 \tag{9.13}$$

$$\frac{\partial}{\partial b}\langle\psi|\hat{H} - E|\psi\rangle = 2a(H_{AB} - ES_{AB}) + 2b(H_{BB} - E) = 0 \tag{9.14}$$

We have two independent equations in the two coefficients (unknowns). The only nontrivial way that these can simultaneously equal zero is for the determinant multiplying the coefficients to equal zero. We have, after dividing through by 2,

$$\begin{vmatrix} H_{AA} - E & H_{AB} - ES_{AB} \\ H_{AB} - ES_{AB} & H_{BB} - E \end{vmatrix} = 0 \tag{9.15}$$

Expanding the determinant, we obtain

$$H_{AA}H_{BB} - (H_{AA} + H_{BB})E + E^2 - H_{AB}^2 + 2H_{AB}S_{AB}E - E^2S_{AB}^2 = 0 \tag{9.16}$$

or, rearranging,

$$E^2(1 - S_{AB}^2) - 2E[\tfrac{1}{2}(H_{AA} + H_{BB}) - H_{AB}S_{AB}] + H_{AA}H_{BB} - H_{AB}^2 = 0 \tag{9.17}$$

This is a quadratic in E. The values of E, in terms of the integrals, can be obtained by use of the quadratic equation. This gives

$$E = \frac{2[\frac{1}{2}(H_{AA}+H_{BB})-H_{AB}S_{AB}] \pm \{4[\frac{1}{2}(H_{AA}+H_{BB})-H_{AB}S_{AB}]^2 - 4(1-S_{AB}^2)(H_{AA}H_{BB}-H_{AB}^2)\}^{1/2}}{2(1 - S_{AB}^2)} \tag{9.18}$$

In the H_2^+ case the orbitals we use on the different nuclei are of the same type; consequently H_{AA} will equal H_{BB}. Making use of this, we can simplify Eq. 9.18 to

$$E = \frac{(H_{AA} - H_{AB}S_{AB}) \pm [H_{AB}^2 - 2H_{AA}H_{AB}S_{AB} + H_{AA}^2S_{AB}^2]^{1/2}}{1 - S_{AB}^2}$$

$$= \frac{(H_{AA} - H_{AB}S_{AB}) \pm (H_{AB} - H_{AA}S_{AB})}{1 - S_{AB}^2} \tag{9.19}$$

The two roots are

$$E_+ = \frac{(H_{AA} + H_{AB})(1 - S_{AB})}{1 - S_{AB}^2}$$

$$= \frac{H_{AA} + H_{AB}}{1 + S_{AB}} \tag{9.20}$$

$$E_- = \frac{(H_{AA} - H_{AB})(1 + S_{AB})}{1 - S_{AB}^2}$$

$$= \frac{H_{AA} - H_{AB}}{1 - S_{AB}} \tag{9.21}$$

We have the energies in terms of three integrals. In order to determine the wave functions, we can substitute these back into Eqs. 9.13 and 9.14. Remembering that H_{AA} equals H_{BB}, we have, substituting for E_+, and dividing through by the factor of two that appears with each term,

$$a\left[H_{AA} - \left(\frac{H_{AA} + H_{AB}}{1 + S_{AB}}\right)\right] + b\left[H_{AB} - \left(\frac{H_{AA} + H_{AB}}{1 + S_{AB}}\right)S_{AB}\right] = 0 \tag{9.22}$$

and

$$a\left[H_{AB} - \left(\frac{H_{AA} + H_{AB}}{1 + S_{AB}}\right)S_{AB}\right] + b\left[H_{AA} - \left(\frac{H_{AA} + H_{AB}}{1 + S_{AB}}\right)\right] = 0 \tag{9.23}$$

Combining these, we find that a equals b. Thus, ψ_+ may be written

$$\psi_+ = a(\phi_A + \phi_B) \tag{9.24}$$

The constant is simply the normalizing constant. It can be evaluated by using the normalization requirement

$$\begin{aligned}\langle\psi_+|\psi_+\rangle &= a^2(\langle\phi_A|\phi_A\rangle + \langle\phi_B|\phi_B\rangle + 2\langle\phi_A|\phi_B\rangle)\\ &= a^2(2 + 2S_{AB})\\ &= 1\end{aligned} \tag{9.25}$$

or

$$a = (2 + 2S_{AB})^{-1/2} \tag{9.26}$$

giving

$$\psi_+ = (2 + 2S_{AB})^{-1/2}(\phi_A + \phi_B) \tag{9.27}$$

Similarly, we find that

$$\psi_- = (2 - 2S_{AB})^{-1/2}(\phi_A - \phi_B) \tag{9.28}$$

The L.C.A.O. treatment of H_2^+ that we have presented has, up to now, been general, since we did not specify which of the separated-atom atomic orbitals were being used. Let us now turn our attention to the specific case of the $1\sigma_g^+$ (or $1s\sigma$) molecular orbital and calculate the ground-state energy. This requires calculating the electronic energy within the Born-Oppenheimer approximation (the orbital energy, in this one-electron case) and adding in the nuclear repulsion energy. For the electronic energy we need the three integrals appearing in Eq. 9.20. The atomic orbitals are 1*s* orbitals. The electronic Hamiltonian is, as in Eq. 9.2,

$$\hat{H}_{\text{el}} = -\frac{1}{2}\nabla^2 - \frac{1}{r_A} - \frac{1}{r_B} \tag{9.2}$$

We can write H_{AA} as

$$\begin{aligned} H_{AA} &= \langle 1s_A|\hat{H}|1s_A\rangle \\ &= \left\langle 1s_A\left|-\frac{1}{2}\nabla^2 - \frac{1}{r_A}\right|1s_A\right\rangle - \left\langle 1s_A\left|\frac{1}{r_B}\right|1s_A\right\rangle \end{aligned} \tag{9.29}$$

The first term on the right is just the energy of a 1*s* hydrogen orbital on atom A, E_H. The second can be evaluated

$$\left\langle 1s_A\left|\frac{1}{r_B}\right|1s_A\right\rangle = N^2 \int e^{-2r_A}\frac{1}{r_B}\,dv \tag{9.30}$$

This is of the same form as the inner part of the integral we used for evaluating the interelectronic repulsion term in the helium-atom problem (Eq. 6.20). Evaluating it between the limits, zero and infinity, gives

$$\left\langle 1s_A\left|\frac{1}{r_B}\right|1s_A\right\rangle = \frac{1}{R}[1 - e^{-2R}(1 + R)] \tag{9.31}$$

where R is the internuclear separation. Thus

$$H_{AA} = E_H - \frac{1}{R}[1 - e^{-2R}(1 + R)] \tag{9.32}$$

For H_{AB} we can write

$$H_{AB} = \left\langle 1s_A\left|-\frac{1}{2}\nabla^2 - \frac{1}{r_B}\right|1s_B\right\rangle - \left\langle 1s_A\left|\frac{1}{r_A}\right|1s_B\right\rangle \tag{9.33}$$

The operator in the first integral is the hydrogenic Hamiltonian of atom B. The function $1s_B$ is an eigenfunction of this, having E_H as the eigenvalue. Thus the first integral equals $E_H S_{AB}$.

The second integral of Eq. 9.33 and S_{AB} are most readily evaluated by the use of elliptical coordinates. The elliptical coordinates are λ, μ, and ϕ, where

$$\lambda = \frac{r_A + r_B}{R} \tag{9.34a}$$

$$\mu = \frac{r_A - r_B}{R} \tag{9.34b}$$

and ϕ is the rotation about the internuclear axis. The volume element in elliptical coordinates is

$$dv = \frac{R^3}{8}(\lambda^2 - \mu^2)\, d\lambda\, d\mu\, d\phi \tag{9.35}$$

The ranges of the coordinates are $1 \le \lambda \le \infty$, $-1 \le \mu \le 1$, and $0 \le \phi \le 2\pi$. For a hydrogenic orbital with a nuclear charge of one (see Eq. 6.12) the overlap integral is

$$\begin{aligned} S_{AB} &= \frac{1}{\pi}\int e^{-r_A} e^{-r_B}\, dv \\ &= \frac{1}{\pi}\int e^{-(r_A+r_B)}\, dv \\ &= \frac{1}{\pi}\int_\lambda \int_\mu \int_\phi e^{-R\lambda}\frac{R^3}{8}(\lambda^2 - \mu^2)\, d\lambda\, d\mu\, d\phi \\ &= \frac{R^3}{8\pi}\left(\int_0^{2\pi} d\phi \int_{-1}^{1} d\mu \int_1^\infty \lambda^2 e^{-R\lambda}\, d\lambda - \int_0^{2\pi} d\phi \int_{-1}^{1} \mu^2\, d\mu \int_1^\infty e^{-R\lambda}\, d\lambda\right) \\ &= \frac{R^3}{8\pi}\left[4\pi \int_1^\infty \lambda^2 e^{-R\lambda}\, d\lambda - \frac{4\pi}{3}\int_1^\infty e^{-R\lambda}\, d\lambda\right] \end{aligned} \tag{9.36}$$

The integrals over λ are of the form

$$\int_y^\infty x^n e^{-ax}\, dx = \frac{n!\,e^{-ay}}{a^{n+1}} \sum_{k=0}^{n} \frac{(ay)^k}{k!} \tag{9.37}$$

Therefore

$$\begin{aligned} S_{AB} &= \frac{R^3}{2}\left[\frac{2}{R^3} e^{-R}\left(1 + R + \frac{R^2}{2}\right) - \frac{e^{-R}}{3R}\right] \\ &= e^{-R}\left(1 + R + \frac{1}{3}R^2\right) \end{aligned} \tag{9.38}$$

Similarly

$$\left\langle 1s_A \left| \frac{1}{r_A} \right| 1s_B \right\rangle = \frac{1}{\pi} \int \frac{1}{r_A} e^{-(r_A + r_B)}\, dv$$

$$= \frac{1}{\pi} \int_\lambda \int_\mu \int_\phi \frac{2}{(\lambda + \mu)R} e^{-R\lambda} \frac{R^3}{8} (\lambda^2 - \mu^2)\, d\lambda\, d\mu\, d\phi$$

$$= \frac{R^2}{4\pi} \int_\lambda \int_\mu \int_\phi e^{-R\lambda} (\lambda - \mu)\, d\lambda\, d\mu\, d\phi$$

$$= \frac{R^2}{4\pi} \left[\int_0^{2\pi} d\phi \int_{-1}^{1} d\mu \int_1^\infty \lambda e^{-R\lambda}\, d\lambda - \int_0^{2\pi} d\phi \int_{-1}^{1} \mu\, d\mu \int_1^\infty e^{-R\lambda}\, d\lambda \right]$$

$$= \frac{R^2}{4\pi} \left[4\pi \frac{e^{-R}}{R^2} (1 + R) \right]$$

$$= e^{-R}(1 + R) \qquad \textbf{9.39}$$

The overall result for H_{AB} is

$$H_{AB} = e^{-R} \left(1 + R + \frac{1}{3} R^2 \right) E_H - e^{-R}(1 + R) \qquad \textbf{9.40}$$

Combining Eqs. 9.32, 9.38, and 9.40 according to Eq. 9.20, we obtain the results shown in Figure 9.7 for the electronic energy. Figure 9.8 shows the calculated total energy.

An examination of Figures 9.7 and 9.8 shows that the results of the simple L.C.A.O. treatment are qualitatively correct. The electronic energy and the total energy go to the correct separated-atom limits. However, the electronic energy does not go the correct united-atom limit; it is off by 25 percent. The total energy does

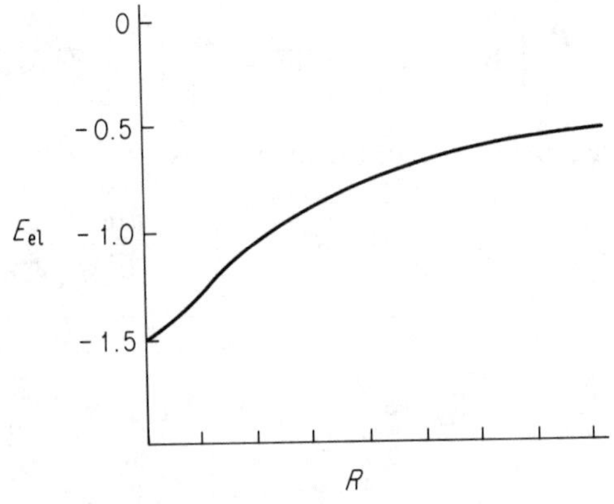

Figure 9.7 Electronic energy for the simple L.C.A.O. $1\sigma_g^+$ orbital of H_2^+. Nuclear charge of unity.

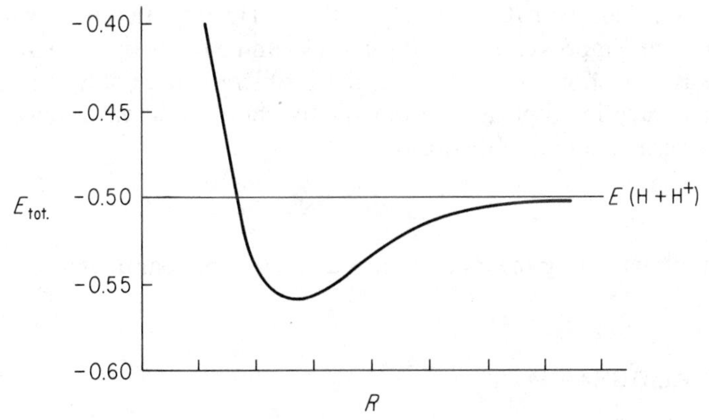

Figure 9.8 The total energy for the L.C.A.O. $1\sigma_g^+$ state of H_2^+. Nuclear charge of unity.

predict bonding. The predicted D_e value is .065 Hartree, about 65 percent of the exact value of .102 Hartree. The predicted equilibrium internuclear separation is 2.5 Bohrs. This is .5 Bohr too long. The results are really not bad, considering the simple model employed.

The simple L.C.A.O. model for H_2^+ can be improved by introducing an effective nuclear charge as a variational parameter in a fashion similar to our variational treatment of the helium atom. This is a reasonable improvement to make, since we know that in the exact treatment, the nuclear charge is 2 in the united-atom limit and 1 in the separated-atom limit. At intermediate separations the effective nuclear charge should lie between these values. Including the effective nuclear charge, ζ, the various terms that enter into the calculations are

$$E_H = -\frac{\zeta^2}{2} \tag{9.41}$$

$$S_{AB} = e^{-\zeta R}\left(1 + \zeta R + \frac{1}{3}\zeta^2 R^2\right) \tag{9.42}$$

$$\left\langle 1s_A \left| \frac{1}{r_B} \right| 1s_A \right\rangle = \frac{1}{\zeta^3 R}[1 - e^{-2\zeta R}(1 + \zeta R)] \tag{9.43}$$

$$\left\langle 1s_A \left| \frac{1}{r_A} \right| 1s_B \right\rangle = \frac{e^{-\zeta R}}{\zeta^2}(1 + \zeta R) \tag{9.44}$$

If ζ is determined variationally at each R value, it goes to the correct united-atom and separated-atom limits. E_{el} also goes to the correct limits. Furthermore, the minimum in E_{tot} occurs at the correct internuclear separation. At this separation of 2.0 Bohrs, ζ has a value of 1.293, the total energy is $-.58651$ Hartree (the exact value is $-.60263$ Hartree), and D_e is .08651 Hartree (the exact value is .10263 Hartree). Further improvements in the energy can be made by allowing more flexibility in the wave function. For example, the problem can be solved to any desired degree of accuracy in elliptical coordinates by choosing a wave function involving a power series in λ and μ of the form

$$\psi = e^{-\alpha\lambda} \sum_m \sum_n c_{mn} \lambda^m \mu^n \qquad \mathbf{9.45}$$

and by treating the c's and the α as variational parameters.

BIBLIOGRAPHY

PARR, R. G., *Quantum Theory of Molecular Electronic Structure.* W. A. Benjamin, Inc., New York, 1963.

PAULING, L., and WILSON, E. B., JR., *Introduction to Quantum Mechanics.* McGraw-Hill Book Company, New York, 1935.

PILAR, F. L., *Elementary Quantum Chemistry.* McGraw-Hill Book Company, New York, 1968.

SLATER, J. C., *Quantum Theory of Molecules and Solids.* Vol. 1. McGraw-Hill Book Company, New York, 1963.

PROBLEMS

9.1 What is the Born-Oppenheimer approximation in molecular quantum mechanics? Why is it useful? Why is it a reasonably good approximation?

***9.2** No excited electronic states have ever been observed for the hydrogen-molecule ion. From a theoretical point of view, would you expect this? Why, or why not?

***9.3** How many bonding and antibonding molecular orbitals can arise in the hydrogen-molecule ion from a set of 3*d* atomic orbitals on the atoms? What is the degeneracy of each? What are the labels of each, within the $\mathbf{D}_{\infty h}$ point group?

***9.4** Find the united-atom limits for the molecular orbitals that can arise from the 3*d* separated-atom limits for H_2^+.

***9.5** "Semiempirical" quantum-mechanical calculations usually involve solving the problem, at some level of approximation, in terms of integrals and then assigning numerical values to the integrals on the basis of experimental data, rather than carrying out the integrations. This approach is based upon the hope that the empirical assignment of values to the integrals will account for things that are difficult or impossible to calculate, and that these effects will carry over to properties other than those used to determine the parameters.

Assume an approximation for Li_2 that involves only the 2*s* valence atomic orbitals in bonding and that neglects electron interactions. The total energy is then the sum of one-electron energies only. Use the following information to estimate the appropriate integrals and evaluate the ionization potential for Li_2.

$$\text{Ionization potential of Li atom} = 5.39 \text{ eV}$$

$$\text{Dissociation energy of } Li_2 = 1.03 \text{ eV}$$

$$\Delta E(2s\sigma_g)(2s\sigma_u) \leftarrow (2s\sigma_g)^2 = 1.74 \text{ eV}$$

9.6 (a) What is the change in electronic energy for the dissociation of the ground state of H_2^+ to $H + H^+$, in terms of E_H, H_{AB}, and S_{AB} (in the L.C.A.O. approximation)?

(b) What about for the first excited state?

(c) Which is greater—the bonding effect of the $1s\sigma$ orbital or the antibonding effect of the $1s\sigma^*$ orbital?

(d) On the basis of the above, would you expect the molecule He_2 to have a stable ground state? Why or why not? What about He_2^+?

chapter ten

The hydrogen molecule

10.1 INTRODUCTION

The simplest neutral molecule is the hydrogen molecule. The complete, nonrelativistic Hamiltonian is

$$\hat{H} = -\frac{1}{2}\nabla_1^2 - \frac{1}{2}\nabla_2^2 - \frac{1}{r_{A1}} - \frac{1}{r_{B1}} - \frac{1}{r_{A2}} - \frac{1}{r_{B2}} + \frac{1}{r_{12}} - \frac{m_e}{2M}\nabla_A^2 - \frac{m_e}{2M}\nabla_B^2 + \frac{1}{R_{AB}} \quad \textbf{10.1}$$

Within the Born-Oppenheimer approximation, which we will be using from now on (unless we explicitly state otherwise), the electronic Hamiltonian is

$$\hat{H}_{\text{el}} = -\frac{1}{2}\nabla_1^2 - \frac{1}{r_{A1}} - \frac{1}{r_{B1}} - \frac{1}{2}\nabla_2^2 - \frac{1}{r_{A2}} - \frac{1}{r_{B2}} + \frac{1}{r_{12}} \quad \textbf{10.2}$$

The total energy at any internuclear separation is again the sum of the electronic energy and the internuclear repulsion. Notice that the electronic Hamiltonian contains the Hamiltonian for an H_2^+ problem for electron number 1 and one for electron number 2 and an interelectronic repulsion term

$$\hat{H}_{\text{el}} = 2\hat{H}_{H_2^+} + \frac{1}{r_{12}} \quad \textbf{10.3}$$

It bears the same relation to the H_2^+ problem as the helium problem does to the hydrogen atom.

One approach to the solution of the H_2 problem might be to use perturbation theory with accurate H_2^+ wave functions. This is not practical, however, since $\langle \psi_i^0 | 1/r_{12} | \psi_i^0 \rangle$ is much more difficult to evaluate in closed form here than it is in the helium problem. Direct numerical calculations have been performed to high accuracy for H_2; however, these are extremely complicated and have not been applied to systems that are very much larger. We are most interested in developing methods that can be applied to a large variety of molecules. The L.C.A.O. treatment for H_2^+ was simple and led to qualitatively correct results. It would thus seem that this would be a logical approach to the H_2 problem.

One question immediately arises, however. In dealing with many-electron atoms we found that, within the independent-particle approximation, our many-electron wave function had to be a product of one-electron functions. The same will be true for many-electron molecules. The question is, if we are to construct our molecular wave function as a linear combination of atomic functions, do we use a product of terms that are themselves a linear combination of one-electron functions, or do we use a linear combination of terms, each of which is a product of one-electron terms? The answer is, both work. They are two different approximations to molecular wave functions. The first leads to *molecular-orbital theory*, while the second leads to *valence-bond theory*. Each approximation has its own advantages and disadvantages.

10.2 THE MOLECULAR-ORBITAL METHOD

We can write the total molecular-orbital wave function for the ground state of the hydrogen molecule as the product of two one-electron $1s\sigma$ molecular orbitals:

$$\Psi_{MO} = 1s\sigma(1)1s\sigma(2) \tag{10.4}$$

If we substitute in the L.C.A.O. form for the $1s\sigma$ molecular orbitals, we have (considering only the spatial functions)

$$\Psi_{MO} = N[1s_A(1) + 1s_B(1)][1s_A(2) + 1s_B(2)] \tag{10.5}$$

The normalizing constant, N, can be found directly:

$$\begin{aligned}\langle \Psi_{MO} | \Psi_{MO} \rangle &= N^2 \langle 1s_A(1)1s_A(2) + 1s_B(1)1s_B(2) + 1s_A(1)1s_B(2) \\ &\quad + 1s_B(1)1s_A(2) | 1s_A(1)1s_A(2) + 1s_B(1)1s_B(2) \\ &\quad + 1s_A(1)1s_B(2) + 1s_B(1)1s_A(2) \rangle \\ &= N^2(4 + 8S + 4S^2) \\ &= 4N^2(1 + S)^2 \end{aligned} \tag{10.6}$$

or

$$N = \frac{1}{2(1 + S)} \tag{10.6a}$$

(For simplicity, we have left the subscripts off S.)

The molecular-orbital electronic energy can be evaluated:

$$\begin{aligned} E_{\text{MO}} &= \langle \Psi_{\text{MO}} | \hat{H} | \Psi_{\text{MO}} \rangle \\ &= 2E_H + \frac{(aa|aa) + (aa|bb) + 2(ab|ab) + 4(aa|ab)}{2(1 + S)^2} \\ &\quad - \frac{2(B|aa) + 2(A|ab)}{1 + S} \end{aligned} \qquad \textbf{10.7}$$

where we have made the abbreviations

$$(aa|aa) = \left\langle 1s_A(1)1s_A(2) \left| \frac{1}{r_{12}} \right| 1s_A(1)1s_A(2) \right\rangle \qquad \textbf{10.8a}$$

$$(aa|bb) = \left\langle 1s_A(1)1s_B(2) \left| \frac{1}{r_{12}} \right| 1s_A(1)1s_B(2) \right\rangle \qquad \textbf{10.8b}$$

$$(ab|ab) = \left\langle 1s_A(1)1s_A(2) \left| \frac{1}{r_{12}} \right| 1s_B(1)1s_B(2) \right\rangle \qquad \textbf{10.8c}$$

$$(aa|ab) = \left\langle 1s_A(1)1s_A(2) \left| \frac{1}{r_{12}} \right| 1s_A(1)1s_B(2) \right\rangle \qquad \textbf{10.8d}$$

$$(B|aa) = \left\langle 1s_A(1) \left| \frac{1}{r_{B1}} \right| 1s_A(1) \right\rangle \qquad \textbf{10.8e}$$

$$(A|ab) = \left\langle 1s_A(1) \left| \frac{1}{r_{A1}} \right| 1s_B(1) \right\rangle \qquad \textbf{10.8f}$$

All these integrals can be evaluated without too much difficulty (Slater, Chap. 3, App. 6). If the calculation is carried out at different values of R_{AB} and hydrogen $1s$ atomic orbitals with a nuclear charge of unity are used, a minimum is found in the total energy at an R value of 1.59 Bohrs. The total energy at this atomic separation is calculated to be -1.0974 Hartrees. The energy of two isolated hydrogen atoms is -1.0 Hartree. Thus, the predicted dissociation energy (based on the true separated-atom limit) is .0974 Hartree. The experimental bond distance is 1.40 Bohrs, while the experimental dissociation energy is .174 Hartree. The calculated results are qualitatively correct, but the numerical values are not very accurate.

As with the hydrogen-molecule ion, the numerical results for the hydrogen molecule can be improved by introducing an effective nuclear charge, ζ, as a variational parameter. Optimizing the energy with respect to ζ at each atomic separation, we find an energy minimum at a separation of 1.38 Bohrs with a total energy of -1.128 Hartrees and a dissociation energy of .128 Hartree. The optimum ζ value at the calculated equilibrium separation is 1.197. The calculated internuclear separation is only .02 Bohr from the correct value. However, the dissociation energy (based on the separated-atom limit) is off by .046 Hartree (about 29 kcal/mole)—a rather large discrepancy in absolute terms.

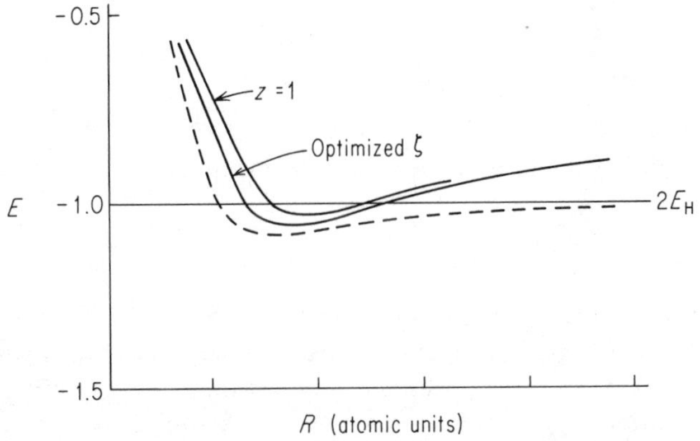

Figure 10.1 Plot of the total ground state energy of the hydrogen molecule as calculated by the molecular orbital method. Top curve, fixed nuclear charge of unity. Middle curve, optimized effective nuclear charge. Dashed curve, exact.

The molecular-orbital total energies are plotted in Figure 10.1 as a function of internuclear separation. The behaviors of both the simple L.C.A.O. molecular-orbital function and the function with ζ optimized are qualitatively correct at separations near the equilibrium internuclear distance. However, both go to the wrong separated-atom limit—about $-.75$ Hartree rather than the correct -1.0 Hartree. This is a common failure of simple molecular-orbital calculations, to which we shall return.

10.3 THE VALENCE-BOND METHOD

The valence-bond method (sometimes called the *Heitler-London* method, after its originators) starts with a product of atomic orbitals, one from each atom. This product is combined with terms resulting from the permutations of the labels of the electrons. For the ground state of the hydrogen molecule we have (again considering only the spatial portion)

$$\Psi_{\text{VB}} = N[1s_A(1)1s_B(2) + 1s_B(1)1s_A(2)] \tag{10.9}$$

Evaluating the normalizing constant, we find

$$\begin{aligned}\langle\Psi_{\text{VB}}|\Psi_{\text{VB}}\rangle &= N^2\langle 1s_A(1)1s_B(2) + 1s_B(1)1s_A(2)|1s_A(1)1s_B(2) + 1s_B(1)1s_A(2)\rangle \\ &= N^2\{\langle 1s_A(1)1s_B(2)|1s_A(1)1s_B(2)\rangle \\ &\quad + \langle 1s_B(1)1s_A(2)|1s_B(1)1s_A(2)\rangle \\ &\quad + 2\langle 1s_A(1)1s_B(2)|1s_B(1)1s_A(2)\rangle\} \\ &= 2N^2(1 + S^2)\end{aligned} \tag{10.10}$$

or

$$N = \frac{1}{\sqrt{2(1 + S^2)}} \qquad \textbf{10.11}$$

The valence-bond energy can be evaluated to give

$$\langle \Psi_{\mathrm{VB}} | \hat{H} | \Psi_{\mathrm{VB}} \rangle = 2E_H + \frac{(aa|bb) - 2(B|aa) + (ab|ab) - 2S(A|ab)}{1 + S^2} \qquad \textbf{10.12}$$

where the various integrals are as defined in Eqs. 10.8. If the total energy is calculated as a function of R_{AB}, using hydrogen 1s atomic orbitals with a unit nuclear charge, a minimum in energy is found at an R value of 1.51 Bohrs. The total energy at this separation is -1.1161 Hartrees for a dissociation energy of .1161 Hartree. The results are a bit better than the simple molecular-orbital values, but not quite as good as those given by the molecular-orbital method with a variational ζ.

Again, the results can be improved by allowing the effective nuclear charge to be variationally determined at each bond length. Doing so gives an energy minimum at an R value of 1.44 Bohrs and a total energy of -1.1389 Hartrees (a bond energy of .1389 Hartree). The ζ value at this R value is 1.166. The energy is slightly better than the molecular-orbital energy with a variational ζ, but the predicted bond length is slightly worse. Plotting the valence-bond total energy as a function of internuclear separation, we obtain Figure 10.2. The most obvious difference between Figures 10.1 and 10.2 is that the valence-bond energies go to the correct separated-atom energy limit of -1.0 Hartree.

In order to find the source of the difference between the molecular-orbital and valence-bond energies at the separated-atom limit, let us compare Eqs. 10.7 and 10.12. The terms S, $(aa|bb)$, $(ab|ab)$, $(aa|ab)$, $(B|aa)$, and $(A|ab)$ all go to zero as R goes to

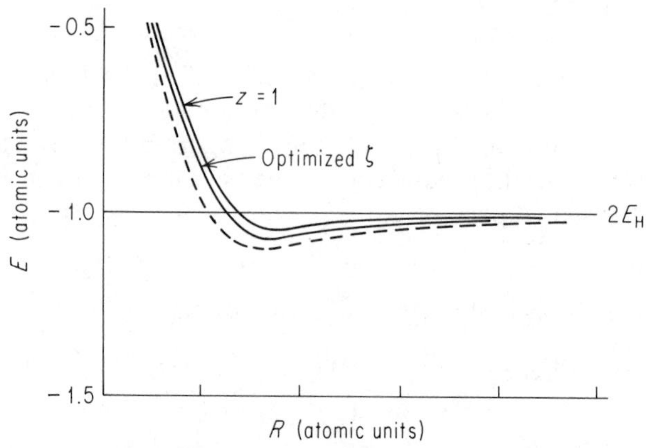

Figure 10.2 Plot of the total ground state energy of the hydrogen molecule as calculated by the valence bond method. Top curve, fixed nuclear charge of unity. Middle curve, optimized effective nuclear charge. Dashed curve, exact.

infinity (see Eqs. 10.8). Thus

$$E_{\text{MO}} \to 2E_H + \tfrac{1}{2}(aa|aa) \qquad (\text{at } R = \infty) \qquad \textbf{10.13}$$

$$E_{\text{VB}} \to 2E_H \qquad (\text{at } R = \infty) \qquad \textbf{10.14}$$

The valence-bond limit is correct, but the molecular-orbital limit is too large by $\frac{1}{2}(aa|aa)$. The integral $(aa|aa)$ is, physically, the interelectron repulsion arising from two electrons on the same atomic center. In order to see why this occurs, we can compare the two wave functions. The valence-bond wave function is

$$\Psi_{\text{VB}} = N[1s_A(1)1s_B(2) + 1s_B(1)1s_A(2)] \qquad \textbf{10.9}$$

while, expanding Eq. 10.5, we have

$$\Psi_{\text{MO}} = N[1s_A(1)1s_B(2) + 1s_B(1)1s_A(2) + 1s_A(1)1s_A(2) + 1s_B(1)1s_B(2)] \qquad \textbf{10.15}$$

The first two terms in Eq. 10.15 (the *covalent* terms) are the two terms that appear in the valence-bond function. The last two terms correspond to two electrons being on the same atom at the same time. While these may be reasonable when the atomic centers are close together, they are physically unreasonable for large values of R. It is the presence of these two *ionic* terms that causes the molecular-orbital energy to go to the wrong limits.

10.4 CONFIGURATION INTERACTION

If some way could be found to remove the ionic terms from the molecular orbital function at large internuclear separations, the molecular orbital energy might be improved. Let us consider a molecular orbital wave function constructed from the product of two one-electron $1s\sigma^*$ orbitals of H_2^+. From the L.C.A.O. construction, we have

$$\Psi'_{\text{MO}} = \frac{1}{2(1-S)}[1s_A(1) - 1s_B(1)][1s_A(2) - 1s_B(2)]$$

$$= \frac{1}{2(1-S)}[1s_A(1)1s_A(2) + 1s_B(1)1s_B(2) - 1s_A(1)1s_B(2) - 1s_B(1)1s_A(2)] \qquad \textbf{10.16}$$

The covalent and ionic terms in this function appear with opposite signs. If Eq. 10.16 were subtracted from Eq. 10.15 at large R values, the ionic terms would be removed. Let an improved wave function Ψ be a linear combination of Eqs. 10.15 and 10.16

$$\Psi = c_1\Psi_{\text{MO}} + c_2\Psi'_{\text{MO}} \qquad \textbf{10.17}$$

and treat the coefficients as variational parameters. In effect, we are mixing the $(1\sigma_g^+)^2$ and the $(1\sigma_u^+)^2$ configurations, thereby performing a *configuration-interaction* calculation. The variational treatment is identical to the one we used for determining the L.C.A.O. function for H_2^+. The resulting determinantal equation is

$$\begin{vmatrix} \langle \Psi_{\text{MO}} | \hat{H} | \Psi_{\text{MO}} \rangle - E & \langle \Psi_{\text{MO}} | \hat{H} | \Psi'_{\text{MO}} \rangle \\ \langle \Psi_{\text{MO}} | \hat{H} | \Psi'_{\text{MO}} \rangle & \langle \Psi'_{\text{MO}} | \hat{H} | \Psi'_{\text{MO}} \rangle - E \end{vmatrix} = 0 \qquad \textbf{10.18}$$

(There is no overlap term in the off-diagonal elements, since the functions $1\sigma_g^+$ and $1\sigma_u^+$ are orthogonal.)

If the configuration-interaction calculation is carried out as a function of R and if a variationally determined effective nuclear charge is included, an equilibrium internuclear separation of 1.45 Bohrs is found. The total energy at this separation is -1.1477 Hartrees and the dissociation energy is .1477 Hartree. The optimum ζ at this separation is 1.193. The energy is better than either the molecular-orbital or the valence-bond energy. Furthermore, the energy goes to the correct separated-atom limit.

The molecular-orbital treatment with configuration interaction gives better results than the valence-bond treatment. The configuration-interaction wave function subtracts out a small part of the ionic contribution of the simple molecular-orbital wave function at internuclear separations near the equilibrium separation and all of it at the separated-atom limit. The valence-bond treatment can be improved if some way can be found to add in some ionic contribution. Let us construct a wave function of the form

$$\Psi = c_1 \Psi_{\text{VB}} + c_2 \Psi'_{\text{VB}} \qquad \textbf{10.19}$$

where

$$\Psi'_{\text{VB}} = \frac{1}{\sqrt{2(1 + S^2)}} [1s_A(1)1s_A(2) + 1s_B(1)1s_B(2)] \qquad \textbf{10.20}$$

Physically, the two terms of Ψ'_{VB} correspond to putting both electrons on atom A and both on atom B, respectively. Equation 10.19 is again a configuration-interaction wave function. Ψ_{VB} corresponds to a covalent valence-bond configuration while Ψ'_{VB} corresponds to the sum of two equally weighted ionic configurations. The coefficients are again determined variationally. The resulting determinant is

$$\begin{vmatrix} \langle \Psi_{\text{VB}} | \hat{H} | \Psi_{\text{VB}} \rangle - E & \langle \Psi_{\text{VB}} | \hat{H} | \Psi'_{\text{VB}} \rangle - \dfrac{2SE}{1 + S^2} \\ \langle \Psi_{\text{VB}} | \hat{H} | \Psi'_{\text{VB}} \rangle - \dfrac{2SE}{1 + S^2} & \langle \Psi'_{\text{VB}} | \hat{H} | \Psi'_{\text{VB}} \rangle - E \end{vmatrix} = 0 \qquad \textbf{10.21}$$

If the valence-bond configuration interaction calculation is carried out as a function of R, and if a variationally determined effective nuclear charge is included, the results turn out to be identical to the molecular-orbital treatment with configuration interaction. If the coefficients are numerically evaluated, the wave functions are also found to be identical. This is a general result: starting from a given basis set, a molecular-orbital treatment with complete configuration interaction gives the same results as a valence-bond treatment with complete configuration interaction.

The results obtained from the configuration-interaction treatments are the best we can obtain for the hydrogen molecule by using a wave function constructed from a single hydrogenic 1*s* orbital on each atom. Adding additional basis functions to the set used to construct the wave function can improve the results, since additional variational parameters are available.

Note that a configuration-interaction wave function actually includes some of the correlation energy. This comes in indirectly through the mixing coefficients, which depend in turn upon the elements of the configuration-interaction determinant. The off-diagonal elements contain electron-repulsion terms corresponding to electronic repulsion between the configurations. Thus, r_{12} is indirectly included in the wave function. (If the configurations differ by two electrons, as in the present case, these are the only terms in the off-diagonal matrix elements.) A complete configuration-interaction treatment, based upon a given basis set, gives all the correlation energy that can be accounted for *within that basis set*. In the present case, the C.I. calculation gives all the correlation energy that can be obtained from a single 1*s* function on each atom. Here, the electrons can avoid each other only by moving back and forth between the two 1*s* atomic orbitals. The results could be improved by adding orbitals of higher principal quantum number to the basis set. Then, the electrons could avoid each other by moving further out from the nuclei. Adding orbitals with higher *l* values would allow angular correlation. A sufficiently large basis set would give the exact energy.

10.5 "EXACT" CALCULATION

In order to improve upon the independent-particle energy, we must include r_{12} in the wave function in some fashion. In 1933, James and Coolidge introduced a wave function of the form

$$\Psi = \frac{e^{-\alpha(\lambda_1+\lambda_2)}}{2\pi} \sum_m \sum_n \sum_j \sum_k \sum_p c_{mnjkp}(\lambda_1^m \lambda_2^n \mu_1^j \mu_2^k u^p + \lambda_1^n \lambda_2^m \mu_1^k \mu_2^j u^p) \qquad \textbf{10.22}$$

where the λ and μ are as described previously (Eqs. 9.34), u depends on the interelectron separation

$$\lambda_i = \frac{r_{Ai} + r_{Bi}}{R_{AB}} \qquad \textbf{10.23a}$$

$$\mu_i = \frac{r_{Ai} - r_{Bi}}{R_{AB}} \qquad \textbf{10.23b}$$

$$u = \frac{2r_{12}}{R_{AB}} \qquad \textbf{10.23c}$$

and α is an arbitrary parameter. Using a thirteen-term expansion of this form, they obtained a dissociation energy that was only about 7×10^{-4} Hartree ($\sim$ .46 kcal mole^{-1}) from the experimental value.

The most accurate calculation to date was performed by Kolos and Wolniewicz in 1965. They used 90 terms of the form of Eq. 10.22 and did not use the Born-Oppenheimer approximation. Their total energy was -1.1744744 Hartrees, for a D_0 value of .1744744 Hartree. This was in slight disagreement with the accepted experimental value. However, when the spectroscopists rechecked their results, they found that the theoretical results were correct, rather than the experimental. In principle, calculations of comparable accuracy can be made on more complicated systems. In practice, it is impractical to carry out such detailed calculations on large systems. Even with the most modern electronic computers, the calculations are prohibitively time-consuming and expensive.

BIBLIOGRAPHY

KARPLUS, M., and PORTER, R. N., *Atoms and Molecules*. W. A. Benjamin, Inc., New York, 1970.

PARR, R. G., *Quantum Theory of Molecular Electronic Structure*. W. A. Benjamin, Inc., New York, 1963.

PAULING, L., and WILSON, E. B., Jr., *Introduction to Quantum Mechanics*. McGraw-Hill Book Company, New York, 1935.

PILAR, F. L., *Elementary Quantum Chemistry*. McGraw-Hill Book Company, New York, 1968.

SLATER, J. C., *Quantum Theory of Molecules and Solids*. Vol. 1. McGraw-Hill Book Company, New York, 1963.

PROBLEMS

***10.1** The He_2^+ molecule is stable (dissociation energy of 3.2 eV). There are two valence-bond structures for He_2^+, each having two electrons on one atom and one on the other. Write a properly antisymmetrized valence-bond wave function for He_2^+.

10.2 Using a semiempirical treatment as in Problem 9.5, the energy of the hydrogen atom, and the following data, estimate the energy of the $(1s\sigma_g)(1s\sigma_u) \leftarrow (1s\sigma_g)^2$ electronic transition of H_2 (the experimental value is 11.37 eV).

$$\text{Dissociation energy} = 4.75 \text{ eV}$$
$$\text{Ionization potential of } H_2 = 15.42 \text{ eV}$$

10.3 Construct a valence-bond wave function for H_2 using $1s$ and $2s$ atomic orbitals. Include only covalent terms. Be sure to include mixing coefficients for terms that are not required by symmetry to be equal.

10.4 Repeat Problem 10.3 for a molecular-orbital wave function.

10.5 Write out the energies obtained from the wave functions from Problems 10.3 and 10.4 in terms of integrals.

***10.6** In general, in configuration-interaction calculations, all possible configurations arising from the available orbitals can be used. (We omitted one of the configurations in the M.O.-C.I. treatment of H_2.) How many configurations can be obtained from three electrons and three molecular orbitals? How many from four electrons and four molecular orbitals? Write a general expression for the number of configurations arising for n electrons and n orbitals.

10.7 Write out the first ten terms in the expansion of the James and Coolidge wave function for H_2 (Eq. 10.22).

chapter eleven

Qualitative treatment of homonuclear diatomic molecules

11.1 INTRODUCTION

We were able to deduce a great deal of information about the electronic structure of many-electron atoms from the quantum-mechanical results for the hydrogen atom. In particular, the angular behavior of the electrons in many-electron atoms was the same as that of the hydrogen-atom wave function. Furthermore, we needed to make only one simple modification in the ordering of energy levels in order to develop a qualitative scheme for deducing the electronic configurations of most many-electron atoms. From these pieces of information we could deduce the periodic behavior of the elements. We could also construct the ground-state term symbols of the atoms.

In a similar manner, we might expect to be able to deduce information about the electronic structure of many-electron molecules, particularly homonuclear diatomic molecules, from the quantum-mechanical results of the hydrogen-molecule ion and the hydrogen molecule. First of all, as in the atomic case, the angular behavior for the orbitals in many-electron diatomic molecules will be the same as in the one-electron molecule. Furthermore, as we shall see, we can use the qualitative energy ordering of many-electron atoms, along with the concept of bonding and antibonding molecular orbitals, to deduce the electronic configurations of the homonuclear diatomic molecules. From this, we can deduce certain properties of the molecules. We can again also construct molecular term symbols.

11.2 BONDING AND ANTIBONDING MOLECULAR ORBITALS

Let us consider again the total energies for a molecule with a single electron in either the bonding (ε_+) or the antibonding (ε_-) one-electron molecular orbital that arises from a linear combination of two 1*s* atomic orbitals. We have

$$\varepsilon_+ = \frac{H_{AA} + H_{AB}}{1 + S} + \frac{1}{R_{AB}} \qquad \textbf{11.1a}$$

$$\varepsilon_- = \frac{H_{AA} - H_{AB}}{1 - S} + \frac{1}{R_{AB}} \qquad \textbf{11.1b}$$

From Eq. 9.29 we saw that H_{AA} includes E_A, the energy of the isolated atom, and $\langle 1s_A|-1/r_B|1s_A\rangle$. This latter integral is attractive, and its magnitude is of the order of $1/R_{AB}$, when R_{AB} is in the bonding range. For total energies, therefore,

$$\varepsilon_+ \sim \frac{E_A + H_{AB}}{1 + S} \qquad \textbf{11.2a}$$

$$\varepsilon_- \sim \frac{E_A - H_{AB}}{1 - S} \qquad \textbf{11.2b}$$

At infinite separation, each of these goes to E_A. The change of energy on bringing the atoms from infinite separation in to the bonding distance is, for the bonding case, ε_+,

$$\varepsilon_+ - E_A \cong \frac{E_A + H_{AB}}{1 + S} - E_A$$

$$= \frac{H_{AB} - E_A S}{1 + S} \qquad \textbf{11.3}$$

and for the antibonding case, ε_-,

$$\varepsilon_- - E_A \cong \frac{E_A - H_{AB}}{1 - S} - E_A$$

$$= -\frac{H_{AB} - E_A S}{1 - S} \qquad \textbf{11.4}$$

The overlap integral, S, always has a value between zero and $+1$, if the orbitals are aligned in phase with each other (i.e., so that like signs of the atomic functions on different centers are overlapping in the bonding region). As a consequence, the antibonding situation is somewhat more antibonding than the bonding is bonding, insofar as the energy is concerned (see Figure 9.6 for the "exact" energies). Both depend directly upon the value of H_{AB}.

Now consider a two-electron molecule. Each molecular orbital can accommodate two electrons, as was the case for atomic orbitals. A two-electron diatomic molecule, arising from two atoms, each contributing a single 1*s* electron, has both electrons in the bonding molecular orbital. This gives rise to a stable molecule (e.g., the hydrogen molecule). If, however, each atom participating were to contribute two 1*s* electrons (as would helium atoms), we would have four electrons in the diatomic molecule. Two of these could be in the bonding molecular orbital, but the other two would have to go into the antibonding molecular orbital. On the basis of this model, the resulting molecular system would be less stable than the two isolated atoms. Thus, the He_2 ground-state molecule does not exist (excited states do exist that are stable with respect to dissociation into excited atoms). Between these two cases the helium molecule ion, He_2^+, has three electrons—two bonding and one antibonding. There is a net bonding effect. The He_2^+ species is quite stable with respect to $He + He^+$, although it is extremely reactive.

Consider now the possibility of 1*s* and 2*s* orbitals from each atom contributing to bonding. The resulting L.C.A.O. molecular orbitals would have the form

$$\psi = c_1 1s_A + c_2 1s_B + c_3 2s_A + c_4 2s_B \tag{11.5}$$

where the *c*'s are again numerical coefficients to be variationally determined. If we label the atomic orbitals with the same numbers as the coefficients in Eq. 11.5, the resulting determinant is

$$\begin{vmatrix} H_{11} - \varepsilon & H_{12} - \varepsilon S_{12} & H_{13} - \varepsilon S_{13} & H_{14} - \varepsilon S_{14} \\ H_{12} - \varepsilon S_{12} & H_{22} - \varepsilon & H_{23} - \varepsilon S_{23} & H_{24} - \varepsilon S_{24} \\ H_{13} - \varepsilon S_{13} & H_{23} - \varepsilon S_{23} & H_{33} - \varepsilon & H_{34} - \varepsilon S_{34} \\ H_{14} - \varepsilon S_{14} & H_{24} - \varepsilon S_{24} & H_{34} - \varepsilon S_{34} & H_{44} - \varepsilon \end{vmatrix} = 0 \tag{11.6}$$

where H_{11} equals $H_{1s_A 1s_A}$, H_{14} equals $H_{1s_A 2s_B}$, and so on, and subscripts are included on the overlap integrals, since several types of overlap occur. The overlap and Hamiltonian integrals involving 1*s* and 2*s* orbitals on the same atom vanish, owing to the orthogonality of atomic orbitals on the same atom. Furthermore, all the terms involving both 1*s* and 2*s* orbitals are small compared to those involving only 1*s* or only 2*s* orbitals. Therefore, to an approximation, Eq. 11.6 can be written

$$\begin{vmatrix} H_{11} - \varepsilon & H_{12} - \varepsilon S_{12} & 0 & 0 \\ H_{12} - \varepsilon S_{12} & H_{22} - \varepsilon & 0 & 0 \\ 0 & 0 & H_{33} - \varepsilon & H_{34} - \varepsilon S_{34} \\ 0 & 0 & H_{34} - \varepsilon S_{34} & H_{44} - \varepsilon \end{vmatrix} = 0 \tag{11.7}$$

Equation 11.7 factors into two 2 × 2 determinants, one involving the $1s_A$ and $1s_B$ atomic orbitals and one involving the $2s_A$ and $2s_B$ atomic orbitals. The approximate energies are

$$\varepsilon_{1s^+} \cong \frac{H_{1s_A 1s_A} + H_{1s_A 1s_B}}{1 + S_{1s_A 1s_B}} \tag{11.8}$$

$$\varepsilon_{1s^-} \cong \frac{H_{1s_A 1s_A} - H_{1s_A 1s_B}}{1 - S_{1s_A 1s_B}} \tag{11.9}$$

$$\varepsilon_{2s^+} \cong \frac{H_{2s_A 2s_A} + H_{2s_A 2s_B}}{1 + S_{2s_A 2s_B}} \tag{11.10}$$

$$\varepsilon_{2s^-} \cong \frac{H_{2s_A 2s_A} - H_{2s_A 2s_B}}{1 - S_{2s_A 2s_B}} \tag{11.11}$$

where we have explicitly used the atomic orbital labels as subscripts. The energy of Eq. 11.8 is bonding with respect to dissociation to the 1*s* atomic orbitals, while that of Eq. 11.9 is antibonding. That of Eq. 11.10 is bonding with respect to the 2*s* atomic orbitals (even though it would lie higher in energy than ε_{1s^-}), while Eq. 11.11 is antibonding.

To this approximation, a similar type of factoring can be carried out for the bonding and antibonding combinations of each type of atomic orbital used in forming the molecular orbitals of a homonuclear diatomic molecule. Each bonding orbital will have the form

$$\varepsilon_+ \cong \frac{H_{AA} + H_{AB}}{1 + S_{AB}} \qquad \textbf{11.12}$$

and each antibonding the form

$$\varepsilon_- \cong \frac{H_{AA} - H_{AB}}{1 - S_{AB}} \qquad \textbf{11.13}$$

Complications arise, however, when degenerate atomic orbitals are involved. Consider, for example, a set of 2*p* atomic orbitals. Because of their nonspherical

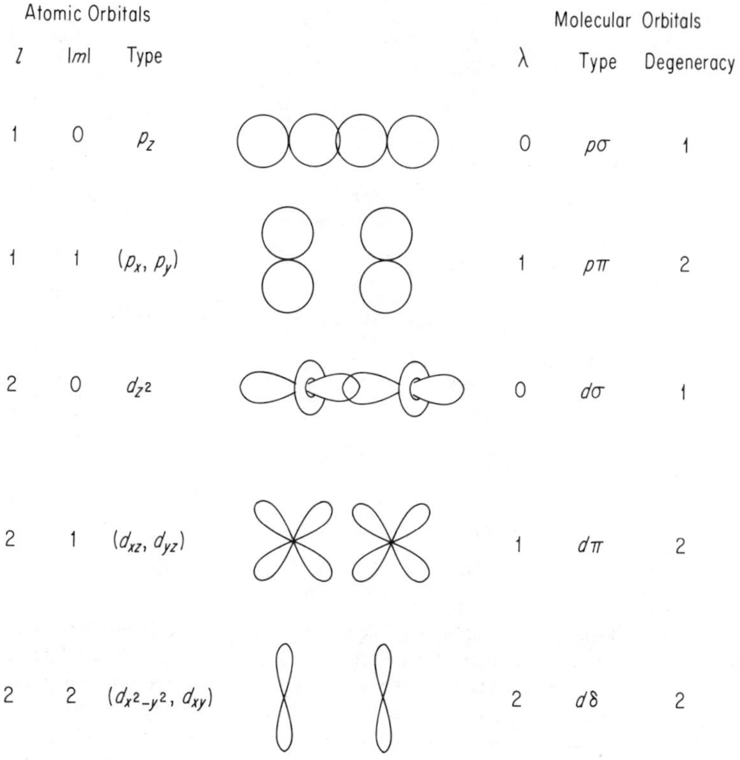

Figure 11.1 The interaction of atomic orbitals having an l value greater than zero. The smaller the value of λ, the greater the interaction.

shape, the interactions depend upon the orientation of the orbitals. The local symmetry (*site symmetry*) of the potential energy experienced by an atom in a diatomic molecule is $\mathbf{C}_{\infty v}$. The degeneracy of the atomic energy levels is partially lifted, as in the Stark effect. Levels with m values of zero are nondegenerate, while those with higher $|m|$ values are doubly degenerate. The molecular axis defines the z axis; consequently the orbitals with an m value of zero are directed along the axis. Successively higher $|m|$ values lie at successively larger angles from this axis. At bonding distances, the overlap, and hence the interactions, of the orbitals that are most directly pointed toward each other (those with an m value of zero) should be the greatest, while those directed successively further away will have successively less interaction. This is illustrated in Figure 11.1.

The magnitude of $|m|$ defines a λ value for a representation from the $\mathbf{D}_{\infty h}$ point group of the molecule. Thus, from a set of atomic orbitals having a given l value, the bonding σ orbital (λ of zero) should be the most bonding and the antibonding σ the most antibonding, the π (λ of one) bonding and antibonding should be next, and so on. All except the σ are doubly degenerate. The overall energy-level scheme arising from 1s, 2s, and 2p atomic orbitals, on the basis of this reasoning, is shown in Figure 11.2. (Orbitals that are neither bonding nor antibonding in behavior show

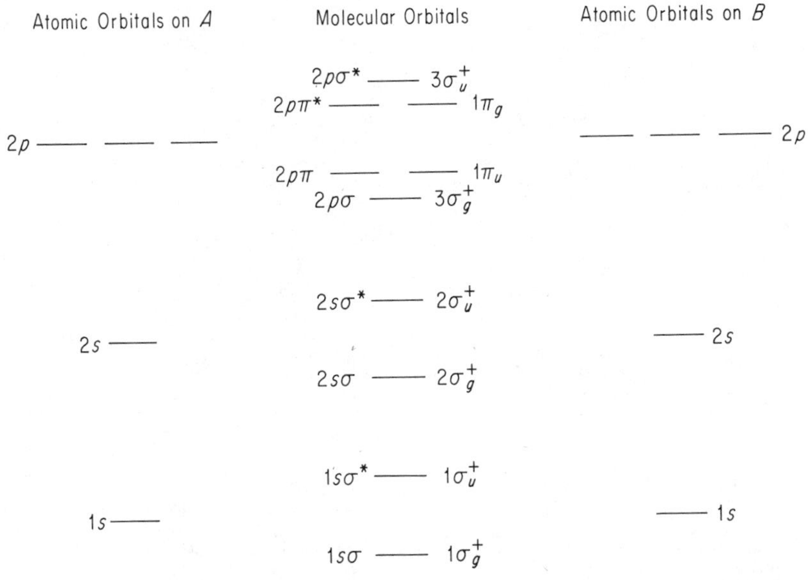

Figure 11.2 Qualitative one-electron molecular orbital energy level scheme for a homonuclear diatomic molecule, using 1s, 2s, and 2p atomic orbitals. Two notations are given for the molecular orbitals. The notation on the left indicates the atomic orbitals from which the molecular orbitals are derived, the λ value (as σ or π) and the bonding or antibonding (with an asterisk) nature. The notation on the right uses the symmetry label from the $\mathbf{D}_{\infty h}$ point group, with a running index to indicate the order of occurrence for an orbital of a given symmetry type.

up in polyatomic molecules. These are called *nonbonding* orbitals.) There are exceptions to this ordering, just as there were to the $(n + l)$ ordering for atomic orbitals.

11.3 ELECTRONIC CONFIGURATIONS

Figure 11.2 can be used to find the electronic configurations of most of the homonuclear diatomic molecules arising from atoms up to an atomic number of 10. The rules are the same as in the aufbau principle for atoms. Each nondegenerate molecular orbital can accommodate two electrons. The doubly degenerate orbitals can accommodate a total of four. Each atom contributes its full complement of electrons to the molecule.

Thus, for example, for Li_2 each lithium atom contributes three electrons, for a total of six. Two electrons go into each of the $1s\sigma$, $1s\sigma^*$, and $2s\sigma$ orbitals. The configuration is written $(1s\sigma)^2(1s\sigma^*)^2(2s\sigma)^2$. The antibonding $1s\sigma^*$ in effect cancels the bonding effect of the $1s\sigma$; however, the $2s\sigma$ orbital is bonding with respect to the $2s$ atomic orbitals. We are left with a two-electron chemical bond. As a consequence, the molecule is predicted to be stable.

In the nitrogen molecule, N_2, each nitrogen atom contributes seven electrons, for a total of 14. The configuration is $(1s\sigma)^2(1s\sigma^*)^2(2s\sigma)^2(2s\sigma^*)^2(2p\sigma)^2(2p\pi)^4$. The bonding effect of the $1s\sigma$ and $2s\sigma$ orbitals is canceled by the corresponding antibonding effect of the $1s\sigma^*$ and $2s\sigma^*$ orbitals. However, there are still six electrons (three electron pairs, three two-electron bonds) in the $2p\sigma$ and $2p\pi$ orbitals that are bonding with respect to the isolated $2p$ atomic orbitals. Since the bonding and antibonding effects of occupied bonding and antibonding pairs of orbitals cancel to give no net bonding, these pairs of orbitals are sometimes treated as nonbonding orbitals. Thus, for example, the valence-shell electrons in N_2 would be classified as two nonbonding (arising from the $2s$ atomic orbitals) and six bonding (from the $2p$), for a net triple bond in the molecule.

The predicted configurations for the first ten homonuclear diatomic molecules are tabulated in Table 11.1. The table also includes the bond dissociation energies for the molecules. Examination of Table 11.1 shows that the molecules that are predicted to be stable are stable, while those that are predicted to be unstable are unstable. Furthermore, there is a qualitative correlation of the bond dissociation energy with the net number of bonding electrons in the molecule. The electronic configuration for any other homonuclear diatomic molecule can be constructed in an analogous fashion.

Even though their ground states are predicted to be unstable, excited states that are stable with respect to dissociation to excited atomic states are observed for many molecules that have no net bonding electrons in their ground-state configurations. An example is the first $^3\Sigma_u^+$ state of He_2. This state arises from a $(1s\sigma)^2(1s\sigma^*)^1(2s\sigma)^1$ configuration, and it dissociates to a ground-state helium atom

Table 11.1

ELECTRONIC CONFIGURATION OF THE FIRST TEN HOMONUCLEAR DIATOMIC MOLECULES, AS PREDICTED FROM FIGURE 11.2, ALONG WITH THE EXPERIMENTAL BOND DISSOCIATION ENERGIES

Molecule	Configuration	Net Bonding Electrons	Bond Energy (kcal mol^{-1})
H_2	$(1s\sigma)^2$	2	104
He_2	$(1s\sigma)^2(1s\sigma^*)^2$	0	0
Li_2	$(1s\sigma)^2(1s\sigma^*)^2(2s\sigma)^2$	2	27
Be_2	$(1s\sigma)^2(1s\sigma^*)^2(2s\sigma)^2(2s\sigma^*)^2$	0	0
B_2[a]	$(1s\sigma)^2(1s\sigma^*)^2(2s\sigma)^2(2s\sigma^*)^2(2p\sigma)^2$	2	67
C_2[a]	$(1s\sigma)^2(1s\sigma^*)^2(2s\sigma)^2(2s\sigma^*)^2(2p\sigma)^2(2p\pi)^2$	4	144
N_2	$(1s\sigma)^2(1s\sigma^*)^2(2s\sigma)^2(2s\sigma^*)^2(2p\sigma)^2(2p\pi)^4$	6	227
O_2	$(1s\sigma)^2(1s\sigma^*)^2(2s\sigma)^2(2s\sigma^*)^2(2p\sigma)^2(2p\pi)^4(2p\pi^*)^2$	4	119
F_2	$(1s\sigma)^2(1s\sigma^*)^2(2s\sigma)^2(2s\sigma^*)^2(2p\sigma)^2(2p\pi)^4(2p\pi^*)^4$	2	37
Ne_2	$(1s\sigma)^2(1s\sigma^*)^2(2s\sigma)^2(2s\sigma^*)^2(2p\sigma)^2(2p\pi)^4(2p\pi^*)^4(2p\sigma^*)^2$	0	0

[a] The qualitative prediction is wrong for these two molecules. The $2p\pi$ level lies lower in energy than the $2p\sigma$.

and a $^3S_1$ excited-state atom, with a $(1s)^1(2s)^1$ configuration. The fact that the $2s\sigma$ molecular orbital is bonding with respect to the $2s$ atomic orbital gives the molecule stability in this state (the dissociation energy is about 60 kcal/mol). The fact that singlet-triplet transitions are forbidden gives it a long enough lifetime to be observed.

The qualitative ordering of the molecular-orbital energies of homonuclear diatomic molecules is possible because the diagonal Hamiltonian matrix elements (which approximate atomic orbital energy levels) for a given atomic orbital have the same value for both atoms. The major interactions are between like orbitals on the two atoms. The energy splitting of the levels of the resulting molecular orbitals is, to a first approximation, symmetrical, relative to their energy value. The relative magnitudes of σ-type and π-type splittings can be deduced from geometric arguments.

Such a qualitative ordering is not generally possible for heteronuclear diatomic molecules. Atomic orbitals of the same type, but on two chemically different atoms, have different energies. Their major interactions may be with other type atomic orbitals on the other atom, rather than with the same type. Even a qualitative discussion of the molecular-orbital energy levels for such molecules usually requires some use of the methods discussed in Chapter 12. In a very few cases, the atoms are sufficiently similar in behavior that their molecular-orbital energy levels are approximated by those in Figure 11.2. Carbon monoxide, CO, is the most notable example. Even though the oxygen atomic orbitals lie lower in energy than those of the carbon, the resulting molecular orbitals behave similarly, in energy, to the homonuclear scheme. The molecular electronic configuration is essentially the same as that of N_2. In fact, many of its properties are near those of N_2. In particular, the dissociation energy is a bit greater than that for N_2 ($\sim$ 257 kcal mol^{-1}), and the molecule has a very small dipole moment.

11.4 MOLECULAR TERM SYMBOLS

The determination of term symbols for molecules is analogous to that for atomic term symbols. If all levels in a molecule are completely filled, the spin state is the totally antisymmetric singlet, while the spatial state must be the totally symmetric state, which carries the label Σ_g^+ in the $\mathbf{D}_{\infty h}$ point group. If, however, any levels are partially occupied, the spatial and spin functions have to be properly adapted to permutational symmetry, so that the total wave function will be antisymmetric.

As in atoms, the spatial and spin portions must be adapted to conjugate representations within the appropriate permutation group. For diatomic molecules, this is relatively easy, since the highest orbital degeneracy that we can have is two. The doubly degenerate orbitals can accommodate one, two, three, or four electrons. With four electrons, the level is completely filled, so no symmetry adaptation is required. With one electron (or with three in the degenerate orbitals, by the hole formalism), the spatial representation of the state is just that of the orbital. Thus, the only adaptation required is for the case when two electrons are in a degenerate level. This can be accomplished by the methods of Section 7.5, using the character table for the $\mathbf{D}_{\infty h}$ group (Table 11.2). However, just as in the atomic case, when only two electrons are involved, we can use the multiplication rules without having to explicitly adapt the representations to permutational symmetry.

Table 11.2

CHARACTER TABLE FOR THE $\mathbf{D}_{\infty h}$ POINT GROUP (INTEGER VALUES OF λ, ONLY)

$\mathbf{D}_{\infty h}$	E	$2C(\phi)$	$\infty\sigma_V$	i	$2S(\phi)$	∞C_2
Σ_g^+	1	1	1	1	1	1
Σ_g^-	1	1	-1	1	1	-1
Π_g	2	$2\cos\phi$	0	2	$-2\cos\phi$	0
Δ_g	2	$2\cos 2\phi$	0	2	$2\cos 2\phi$	0
Γ_g^λ	2	$2\cos\lambda\phi$	0	2	$(-1)^\lambda 2\cos\lambda\phi$	0
Σ_u^+	1	1	1	-1	-1	-1
Σ_u^-	1	1	-1	-1	-1	1
Π_u	2	$2\cos\phi$	0	-2	$2\cos\phi$	0
Δ_u	2	$2\cos 2\phi$	0	-2	$-2\cos 2\phi$	0
Γ_u^λ	2	$2\cos\lambda\phi$	0	-2	$-(-1)^\lambda 2\cos\lambda\phi$	0

The multiplication rules for the $\mathbf{D}_{\infty h}$ point group are

$$\Gamma^{\mu} \times \Gamma^{\nu} = \Gamma^{|\mu-\nu|} + \Gamma^{(\mu+\nu)} \quad \mathbf{11.14}$$

$$\Gamma^{\mu} \times \Gamma^{\mu} = \Sigma^{+} + \Sigma^{-} + \Gamma^{2\mu} \quad \mathbf{11.15}$$

$$g \times g = u \times u = g \quad \mathbf{11.16}$$

$$g \times u = u \quad \mathbf{11.17}$$

$$+ \times + = - \times - = + \quad \mathbf{11.18}$$

$$+ \times - = - \quad \mathbf{11.19}$$

In the direct product of a representation with itself (Eq. 11.15) the representation Σ^- is the antisymmetrized product, while $\Sigma^+ + \Gamma^{2\mu}$ is the symmetrized product. Thus, for the spatial function for any degenerate level occupied by two electrons in a diatomic molecule, the Σ^- function is always antisymmetric with respect to interchange of the electrons, while the Σ^+ and $\Gamma^{2\mu}$ functions are symmetric. This holds for heteronuclear as well as homonuclear systems, since the $\mathbf{D}_{\infty h}$ and $\mathbf{C}_{\infty v}$ representations differ only by the inclusion of the g, u behavior. It also holds for linear polyatomic molecules.

The oxygen molecule has the ground-state configuration

$$(1s\sigma)^2(1s\sigma^*)^2(2s\sigma)^2(2s\sigma^*)^2(2p\sigma)^2(2p\pi)^4(2p\pi^*)^2.$$

Changing the labels to the representation labels from $\mathbf{D}_{\infty h}$ and adding a running number label for the order of occurrence of each symmetry type, this becomes $(1\sigma_g^+)^2(1\sigma_u^+)^2(2\sigma_g^+)^2(2\sigma_u^+)^2(3\sigma_g^+)^2(1\pi_u)^4(1\pi_g)^2$. The partially filled level is a π_g level. Taking the product of this representation with itself, we have

$$\Pi_g \times \Pi_g = \Sigma_g^+ + \Sigma_g^- + \Delta_g \quad \mathbf{11.20}$$

The Σ_g^- level is antisymmetric and must be associated with the symmetric triplet spin state. The other two are symmetric and must be associated with the antisymmetric singlet spin state. The states arising from the lowest-energy configuration of the oxygen molecule are ${}^3\Sigma_g^-$, ${}^1\Delta_g$, and ${}^1\Sigma_g^+$, in order of increasing energy. The ${}^3\Sigma_g^-$ state is the ground state, while the other two states lie 7981.1 cm^{-1} and 13,195.2 cm^{-1}, respectively, above the ground state.

BIBLIOGRAPHY

FLURRY, R. L., JR., *Symmetry Groups: Theory and Chemical Applications*. Prentice-Hall, Inc., Englewood Cliffs, N. J., 1980.

KARPLUS, M., and PORTER, R. N., *Atoms and Molecules*. W. A. Benjamin, Inc., New York, 1970.

PILAR, F. L., *Elementary Quantum Chemistry*. McGraw-Hill Book Company, New York, 1968.

SLATER, J. C., *Quantum Theory of Molecules and Solids*. Vol. 1. McGraw-Hill Book Company, New York, 1963.

PROBLEMS

***11.1** Tell whether molecular orbitals having the following $\mathbf{D}_{\infty h}$ symmetry labels, for homonuclear diatomic molecules, will be bonding or antibonding:

(a) σ_g^+. (b) σ_u^+. (c) π_g. (d) π_u.

(e) δ_g. (f) δ_u. (g) ϕ_g. (h) ϕ_u.

***11.2** Predict the ground-state electronic configurations and the ground-state term symbols for the molecules C_2, P_2, and V_2 on the basis of the simple molecular-orbital scheme for diatomic molecules.

***11.3** Give the ground-state electronic configurations and term symbols for the following molecular ions: C_2^+, C_2^-, N_2^+, N_2^-, O_2^+, O_2^-, F_2^{2+}, F_2^+, F_2^-.

***11.4** Tell whether each of the molecular ions from Problem 11.3 would be expected to have a larger or smaller dissociation energy than the corresponding neutral molecule.

chapter twelve

Electronic structure of polyatomic molecules

12.1 NONEMPIRICAL CALCULATIONS

The generalization of the calculations we discussed for the hydrogen molecule to polyatomic molecules is straightforward. The nonrelativistic electronic Hamiltonian for a molecule has the general form

$$\hat{H} = \sum_{i}^{\text{electrons}} \left(-\frac{1}{2} \nabla_i^2 - \sum_{A}^{\text{atoms}} \frac{Z_A}{r_{Ai}} \right) + \sum_{i<j}^{\text{electrons}}\sum \frac{1}{r_{ij}} \qquad \textbf{12.1}$$

The terms in the parentheses are functions of the coordinates of only one electron, while the double summation represents the two-electron repulsion interactions. The Schrödinger equation can be solved by any of the methods we have mentioned previously. Direct numerical calculations have been carried out on small systems (of the order of ten electrons). However, these are impractical for most chemically interesting systems (for example, benzene contains 42 electrons). Most calculations on systems of any significant size start with the independent-particle wave function. The Hartree-Fock S.C.F. molecular-orbital formalism also is most commonly employed. Valence-bond calculations are used less commonly for large molecules.

The approximations that are used limit the accuracy that can be achieved in the total energy in molecular calculations. The most common limit is that due to the use of independent-particle wave functions. If a wave function were employed that had sufficient flexibility to obtain an exact solution of the Hartree-Fock equations (called the *Hartree-Fock limit*), there would still be an error because the wave function did not include the interelectronic coordinates. The energy difference between the Hartree-Fock limit and the true nonrelativistic energy is the *correlation energy* or *correlation error*. The correlation energy in a molecule is typically .5 to 1 percent of the total energy. (In water, for example, it is about .4 Hartree, or about 250 kcal mol^{-1} out of a total energy of -76.48 Hartrees.) In most cases, however, it is impractical to even try to achieve the Hartree-Fock limit. Consequently the error, when an independent-particle wave function is used, usually is even larger.

Many chemically interesting properties involve energy differences. Spectral transitions involve the difference in energy between different states in the same molecule. Heats of reaction involve the differences in total energy of the reactants and products. Calculations of the structures of a molecule involve calculating the total energy of the molecule as a function of changes in structural parameters (bond lengths and bond angles) and finding the energy minimum. Calculations of activation energies for chemical reactions involve calculating the total energy of the reacting system along the reaction pathway. For any of these applications, the errors in the absolute energies tend to cancel if the individual calculations are performed with the same set of approximations. This can give results for energy differences that have a greater absolute accuracy than do the total energies for the individual calculations. Computational quantum chemistry has produced useful results in all the areas mentioned above, in spite of all the approximations. Appendix 3 gives some examples of results based on nonempirical calculations.

Quantum calculations are probably of the greatest use when applied to systems for which no experimental information is available. Examples where important information has been obtained in such situations include the prediction of the properties of molecular fragments that might be found in outer space or in flames or explosions. There have been recent successes in deducing the mechanisms of metabolic action of drugs from calculations on the drug molecules and postulated receptor sites. Such calculations have led to the synthesis of drugs having specific desired actions.

Research is continually underway to improve the accuracy of molecular calculations. In the 1960s, the introduction of Gaussian functions as basis functions for molecular calculations (implementing a 1950 suggestion of S. F. Boys) greatly reduced the computational time required to obtain good molecular Hartree-Fock solutions, thereby making calculations on larger molecules practical. Various methods for compensating at least partially for the correlation error have been developed and others are being developed. A complete configuration-interaction calculation based on functions from a given basis set would eliminate all the correlation error that could be handled by that basis set; however, the problem quickly becomes unmanageably large. For this reason, C.I. calculations use a limited number of configurations. Multiconfiguration S.C.F. methods have been developed recently that optimize the functions in the excited configuration at the same time as the functions are optimized for the ground configuration. These and many other types of research will continually improve the accuracy of molecular calculations. In the meantime, useful results are continually being obtained by techniques currently available.

12.2 METHODS EMPLOYING MORE APPROXIMATIONS

Because of the complexity of solving the molecular equations to even the Hartree-Fock level, various workers have developed methods that employ more approximations. Some of these involve a systematic examination of the total-energy equations based upon one-electron molecular orbitals, which are linear combinations of basis functions, in order to find types of integrals over basis functions that might be negligibly small. These integrals are then systematically omitted from the calculations. For example, the electronic energy, in terms of molecular orbitals, can be written

$$E = \sum_{i}^{\text{occ}} \langle \psi_i | \hat{h} | \psi_i \rangle + \sum_{i<j}\sum \left(\left\langle \psi_i \psi_j \left| \frac{1}{r_{12}} \right| \psi_i \psi_j \right\rangle - \left\langle \psi_i \psi_j \left| \frac{1}{r_{12}} \right| \psi_j \psi_i \right\rangle \delta_{m_{si} m_{sj}} \right) \qquad \textbf{12.2}$$

where the operator, $\hat{h}$, contains the one-electron terms of Eq. 12.1. If a linear combination of basis functions, χ_μ,

$$\psi_i = \sum_{\mu} c_{i\mu} \chi_\mu \qquad \textbf{12.3}$$

is used for the molecular orbitals, the two-electron integrals have the form

$$\left\langle \psi_i \psi_j \left| \frac{1}{r_{12}} \right| \psi_i \psi_j \right\rangle = \sum_{\mu}\sum_{\nu}\sum_{\lambda}\sum_{\sigma} c_{i\mu}^* c_{j\nu}^* c_{i\lambda} c_{j\sigma} \left\langle \chi_\mu \chi_\nu \left| \frac{1}{r_{12}} \right| \chi_\lambda \chi_\sigma \right\rangle \qquad \textbf{12.4}$$

$$\left\langle \psi_i \psi_j \left| \frac{1}{r_{12}} \right| \psi_j \psi_i \right\rangle = \sum_{\mu}\sum_{\nu}\sum_{\lambda}\sum_{\sigma} c_{i\mu}^* c_{j\nu}^* c_{j\lambda} c_{i\sigma} \left\langle \chi_\mu \chi_\nu \left| \frac{1}{r_{12}} \right| \chi_\lambda \chi_\sigma \right\rangle \qquad \textbf{12.5}$$

For n basis functions, there will be n^4 of the two-electron integrals over the basis functions. Many of these will equal each other; however, the number of unique values will still be greater than $n^4/4$, unless there is a great deal of symmetry in the system. Even then, the number can be reduced by at most only about an order of magnitude.

The general two-electron integral over atomic-centered basis functions, which appears in Eqs. 12.4 and 12.5, can contain as many as four different basis functions, or as few as one, and they can range over from one to four centers. The functions χ_μ and χ_λ are the functions of electron one, while χ_ν and χ_σ are those of electron two. If μ and λ are not the same, then electron one is distributed between the two basis functions, which can be on the same or different atomic centers. (This is referred to in general as *differential overlap* and as *diatomic differential overlap* when the centers are different.) The same is true for electron two, when ν and σ are different. Physically, then, the repulsion is between two overlap distributions. Crudely, it is the repulsion between two "smeared-out" charge distributions, rather than between two localized distributions, as it is when μ equals λ and ν equals σ. Those integrals having the same index for all four χ's have the largest magnitude. Those with the same index for the two functions of electron one and the same for those of electron two have the next largest magnitude, for a given separation of the basis-function centers, and so on. Furthermore, the magnitude of the integrals involving basis functions on different sites falls off exponentially with the distance of separation of the basis functions.

As suggested by Parr, an analysis of these factors allows many of the integrals to be neglected. The PRDDO (partial retention of differential diatomic overlap) method developed by Lipscomb and co-workers, is based directly upon this approach. Many other neglect-of-overlap approximations, such as CNDO (complete neglect of differential overlap), INDO (intermediate neglect of differential overlap), and MINDO (modified intermediate neglect of differential overlap), also are in common use.

Another approach is to absorb the electron-repulsion effects into an effective one-electron Hamiltonian and express the total Hamiltonian as a sum of one-electron effective Hamiltonians:

$$\hat{H} = \sum_i \hat{h}_i^{\text{eff}} \tag{12.6}$$

Only one-electron integrals remain. These are evaluated from experimental data, rather than by an attempt at integration over the functions. (Normally, only valence-shell orbitals are considered in the basis set.) This is the basis for the various Hückel molecular-orbital methods. The simple Hückel methods are not iterative methods. A single solution of the one-electron secular equation is all that is required. Modifications of the Hückel method exist that adjust the effective Hamiltonian integrals to compensate for calculated charge reorganizations. These must be solved iteratively in a manner analogous to true S.C.F. methods.

Still other methods combine the two previous approaches. Some of the two-electron integrals are retained. Some or all of the integrals are evaluated from experiment. The resulting S.C.F.-like equations must be solved iteratively. (Again, only valence-shell orbitals are normally retained.) This is the basis for the CNDO and INDO methods developed by Pople and co-workers, the MINDO method of Dewar and co-workers, the Pariser, Parr, and Pople method for pi-electron systems, and numerous other methods.

These *semiempirical methods*, as they are called, are all much faster and easier than the nonempirical methods. Consequently, they are useful for still larger molecules. Their biggest drawback is that their accuracy is unpredictable. The empirical parameters are evaluated on the basis of certain types of experimental data taken from certain types of molecules. The methods may give poor predictions for unrelated types of experiments or even for the same experiments on unrelated types of molecules. For this reason they are often more useful for studying trends in closely related series of molecules than for computing absolute properties of individual molecules. Even so, they are still very useful techniques for studying relatively large molecular systems.

12.3 THE HÜCKEL MOLECULAR-ORBITAL METHOD

We are interested here in learning the principles behind quantum-chemical calculations and in learning how to apply the results to explain chemical bonding and other chemical phenomena. Any of the techniques for molecular calculations could be employed. Most of them, however, require electronic computers for calculations on systems of chemical interest. Although most practicing quantum chemists are heavy computer users, we will concentrate on problems that can be solved with, at most, a pocket calculator. The reason is that you can best learn something by doing it. Feeding the numbers into a computer and staring at the output is not doing the calculations. Writing the programs for the computer would serve the desired purpose, but progress would then be too slow to allow covering the desired material in a reasonable time. For these reasons we will concentrate on the simplest of the molecular-orbital methods we have mentioned, the Hückel method.

The Hückel method embodies the principles that are employed in more sophisticated calculations. It does not involve the evaluation of complex integrals or the iterative solution of the secular equation. With the aid of group theory, which we will develop for its innate utility, fairly complex problems can be solved, at the Hückel level of approximation, fairly easily.

In the previous section we mentioned that Hückel theory uses only the valence-shell orbitals and an effective Hamiltonian and that the integrals are evaluated empirically. "Classical" Hückel theory is for planar π-electron systems. In such systems, double bonds are considered to be composed of a σ bond lying in the plane of the molecule and a π bond having a node in the plane. (The σ, π, etc. nomenclature is taken over from the theory of diatomic molecules. If a bond between two atoms has cylindrical symmetry it is called a σ bond; if it has the symmetry of a π orbital of a diatomic molecule it is called a π bond.) It is assumed that the π-bond system of a molecule can be treated separately from the σ system. Classical Hückel theory deals exclusively with the π system.

In molecules composed of atoms from the representative elements, each atom has four valence orbitals available. In planar π-electron systems only one of the p orbitals from each atom contributes to the π system. The others are in the σ system. Thus, the basis set for the π molecular orbitals is much smaller than it would be for the complete valence set. The justification for considering only the π system is that these are the highest occupied and lowest vacant molecular orbitals in energy. Furthermore, owing to symmetry, one-electron integrals between the σ- and π-type basis functions vanish. The lowest-energy spectral transitions, the first ionization potentials, and the electron affinities of π systems all involve the energies of π orbitals. Chemical reactions involving such systems usually involve greater changes in the π system than in the σ system. Much useful information about the chemistry of π systems can be obtained from simple Hückel theory.

Since the Hückel Hamiltonian is a sum of one-electron effective Hamiltonians, and since these one-electron Hamiltonians all have the same form (see Eq. 12.1), the

Hückel approximation amounts to solving the L.C.A.O. equations for one electron moving in the field of all the atomic cores (the nuclei and all electrons except those donated to the π system). The result is a set of one-electron molecular orbitals and their corresponding energies. The electrons are assigned to the molecular orbitals to give the proper molecular-orbital configurations. If necessary, permutational symmetry can be applied to form the proper states, although, at the Hückel level, all states from a given electronic configuration have the same energy. If desired, configuration interaction can be used along with electronic repulsion to improve calculated spectroscopic transition energies. If properties other than energy are desired, they can be obtained as the expectation values of the appropriate operators. In certain cases perturbation theory may be useful for improving calculated results. For our initial work we will concentrate on π-electron systems.

12.4 BUTADIENE AND ACROLEIN

The butadiene and acrolein systems each contain four π electrons. They are large enough to fully illustrate the Hückel method, yet simple enough that the equations are easy to solve. Also, the butadiene problem can be simplified by symmetry. The σ structure of butadiene is

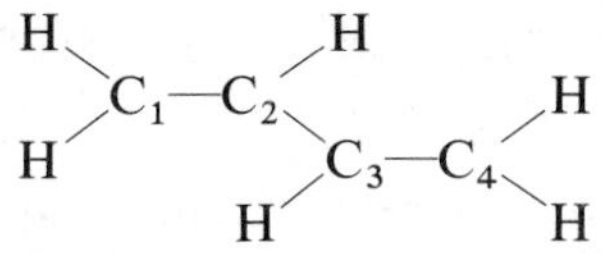

(1)

There is one p-π atomic orbital on each carbon atom, from which the π molecular orbitals will be constructed. Let us label these χ_1 to χ_4 and assume that they are normalized. The molecular orbitals have the form

$$\psi_i = c_{i1}\chi_1 + c_{i2}\chi_2 + c_{i3}\chi_3 + c_{i4}\chi_4 \qquad \textbf{12.7}$$

The secular equation can be derived by varying the quantity $\langle\psi_i|\hat{h} - \varepsilon_i|\psi_i\rangle$, where ε_i is the one-electron molecular orbital energy, with respect to the coefficients and setting the results equal to zero. We have

$$\langle\psi_i|\hat{h} - \varepsilon_i|\psi_i\rangle$$
$$= \langle c_{i1}\chi_1 + c_{i2}\chi_2 + c_{i3}\chi_3 + c_{i4}\chi_4|\hat{h} - \varepsilon_i|c_{i1}\chi_1 + c_{i2}\chi_2 + c_{i3}\chi_3 + c_{i4}\chi_4\rangle \qquad \textbf{12.8}$$

Let us make the definitions

$$\langle\chi_\mu|\hat{h}|\chi_\mu\rangle = \alpha_\mu \qquad \textbf{12.9a}$$

$$\langle\chi_\mu|\hat{h}|\chi_\nu\rangle = \beta_{\mu\nu} \qquad \textbf{12.9b}$$

$$\langle\chi_\mu|\chi_\nu\rangle = S_{\mu\nu} \qquad \textbf{12.9c}$$

Expanding Eq. 12.8, we have

$$\begin{aligned}\langle\psi_i|\hat{h} - \varepsilon_i|\psi_i\rangle = {} & c_{i1}^2(\alpha_1 - \varepsilon_i) + c_{i2}^2(\alpha_2 - \varepsilon_i) + c_{i3}^2(\alpha_3 - \varepsilon_i) + c_{i4}^2(\alpha_4 - \varepsilon_i) \\ & + 2c_{i1}c_{i2}(\beta_{12} - \varepsilon_i S_{12}) + 2c_{i1}c_{i3}(\beta_{13} - \varepsilon_i S_{13}) \\ & + 2c_{i1}c_{i4}(\beta_{14} - \varepsilon_i S_{14}) + 2c_{i2}c_{i3}(\beta_{23} - \varepsilon_i S_{23}) \\ & + 2c_{i2}c_{i4}(\beta_{24} - \varepsilon_i S_{24}) + 2c_{i3}c_{i4}(\beta_{34} - \varepsilon_i S_{34})\end{aligned} \quad \textbf{12.10}$$

Varying this with respect to each of the coefficients and setting the individual results equal to zero, we obtain the set of equations:

$$2c_{i1}(\alpha_1 - \varepsilon_i) + 2c_{i2}(\beta_{12} - \varepsilon_i S_{12}) + 2c_{i3}(\beta_{13} - \varepsilon_i S_{13}) + 2c_{i4}(\beta_{14} - \varepsilon_i S_{14}) = 0 \quad \textbf{12.11a}$$

$$2c_{i1}(\beta_{12} - \varepsilon_i S_{12}) + 2c_{i2}(\alpha_2 - \varepsilon_i) + 2c_{i3}(\beta_{23} - \varepsilon_i S_{23}) + 2c_{i4}(\beta_{24} - \varepsilon_i S_{24}) = 0 \quad \textbf{12.11b}$$

$$2c_{i1}(\beta_{13} - \varepsilon_i S_{13}) + 2c_{i2}(\beta_{23} - \varepsilon_i S_{23}) + 2c_{i3}(\alpha_3 - \varepsilon_i) + 2c_{i4}(\beta_{34} - \varepsilon_i S_{34}) = 0 \quad \textbf{12.11c}$$

$$2c_{i1}(\beta_{14} - \varepsilon_i S_{14}) + 2c_{i2}(\beta_{24} - \varepsilon_i S_{24}) + 2c_{i3}(\beta_{34} - \varepsilon_i S_{34}) + 2c_{i4}(\alpha_4 - \varepsilon_i) = 0 \quad \textbf{12.11d}$$

As usual, the determinant multiplying the coefficients must equal zero for these to be simultaneously satisfied. This leads to the determinantal equation (the *secular equation*)

$$\begin{vmatrix} \alpha_1 - \varepsilon_i & \beta_{12} - \varepsilon_i S_{12} & \beta_{13} - \varepsilon_i S_{13} & \beta_{14} - \varepsilon_i S_{14} \\ \beta_{12} - \varepsilon_i S_{12} & \alpha_2 - \varepsilon_i & \beta_{23} - \varepsilon_i S_{23} & \beta_{24} - \varepsilon_i S_{24} \\ \beta_{13} - \varepsilon_i S_{13} & \beta_{23} - \varepsilon_i S_{23} & \alpha_3 - \varepsilon_i & \beta_{34} - \varepsilon_i S_{34} \\ \beta_{14} - \varepsilon_i S_{14} & \beta_{24} - \varepsilon_i S_{24} & \beta_{34} - \varepsilon_i S_{34} & \alpha_4 - \varepsilon_i \end{vmatrix} = 0 \quad \textbf{12.12}$$

Note the form of the determinant of Eq. 12.12. The order of the determinant is the number of basis functions from which it was constructed. Each element along the principal diagonal is equal to $\alpha_\mu - \varepsilon_i$, where μ is the label of the row and column. The off-diagonal elements have the form $\beta_{\mu\nu} - \varepsilon_i S_{\mu\nu}$ where μ and ν are row or column labels. (Usually the smaller number is written first, since $\beta_{\mu\nu}$ equals $\beta_{\nu\mu}$ and $S_{\mu\nu}$ equals $S_{\nu\mu}$.) The determinant is symmetrical with respect to the principal diagonal. That is, the element in the μth row and the νth column is the same as that in the νth row and the μth column. These are general features of any secular determinant arising from a linear combination of functions. Using this knowledge, we can write a secular determinant without going through the explicit use of the variational equations. Furthermore, the linear equations analogous to Eqs. 12.11 can also be written down directly, since the order of occurrence of the various terms is the same as in the determinantal equation. (In these, the constant factor of 2 is normally omitted.)

A number of simplifying assumptions are made in the butadiene Hückel determinant. First, all the atoms in butadiene are carbon atoms. Consequently, all the α terms are assumed equal. Their indices can be dropped. (If the Hartree-Fock method were being used, the terms corresponding to the α terms would include two-electron terms, which would be dependent on the charge distribution. They would not all be equal.) The $\beta_{\mu\nu}$ integral is essentially a decaying exponential function.

Consequentially, β terms arising from directly bonded atoms should have much greater magnitude than those arising from pairs of atoms that are not directly bonded. Hückel theory ignores β terms involving pairs of atoms not directly bonded. (In more sophisticated calculations these should be included.) Furthermore, there is not a very large variation in the bond lengths involving like atoms in π-electron systems. Hückel theory sets all β terms involving like pairs of atoms to the same value. (In more exact calculations, the variation of β with bond length should be considered.) Finally, Hückel π-electron theory sets all $S_{\mu\nu}$ equal to $\delta_{\mu\nu}$. (The magnitude of S between two directly bonded carbon π orbitals is about .25. Overlap must be retained in exact calculations.) With all these approximations, the Hückel π-electron secular determinant for butadiene becomes

$$\begin{vmatrix} \alpha - \varepsilon_i & \beta & 0 & 0 \\ \beta & \alpha - \varepsilon_i & \beta & 0 \\ 0 & \beta & \alpha - \varepsilon_i & \beta \\ 0 & 0 & \beta & \alpha - \varepsilon_i \end{vmatrix} = 0 \qquad \textbf{12.13}$$

Equation 12.13 has the general form for the Hückel π-electron determinant for any system. Along the diagonal there is an $\alpha - \varepsilon_i$ term corresponding to each basis function (i.e., to each atom). Off the diagonal there is a β term for each directly bonded pair of atoms. There are zeros everywhere else. If the atoms are carbon atoms, no subscripts are used with the α's and β's. Appropriate subscripts are added if other atoms are involved. The same approximations are also made in the linear equations (analogous to Eqs. 12.11). The linear equations for butadiene (omitting the constant factor of 2) become

$$c_{i1}(\alpha - \varepsilon_i) + c_{i2}\beta = 0 \qquad \textbf{12.14a}$$

$$c_{i1}\beta + c_{i2}(\alpha - \varepsilon_i) + c_{i3}\beta = 0 \qquad \textbf{12.14b}$$

$$c_{i2}\beta + c_{i3}(\alpha - \varepsilon_i) + c_{i4}\beta = 0 \qquad \textbf{12.14c}$$

$$c_{i3}\beta + c_{i4}(\alpha - \varepsilon_i) = 0 \qquad \textbf{12.14d}$$

The σ system of acrolein is

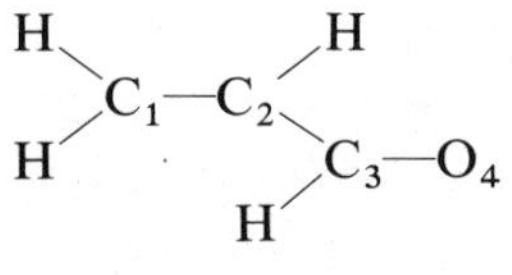

(2)

Again, each carbon or oxygen atom contributes one p-π orbital. The secular determinant can be written immediately as

$$\begin{vmatrix} \alpha - \varepsilon_i & \beta & 0 & 0 \\ \beta & \alpha - \varepsilon_i & \beta & 0 \\ 0 & \beta & \alpha - \varepsilon_i & \beta_{\mathrm{CO}} \\ 0 & 0 & \beta_{\mathrm{CO}} & \alpha_{\mathrm{O}} - \varepsilon_i \end{vmatrix} = 0 \qquad \textbf{12.15}$$

while the linear equations are

$$c_{i1}(\alpha - \varepsilon_i) + c_{i2}\beta = 0 \qquad \textbf{12.16a}$$

$$c_{i1}\beta + c_{i2}(\alpha - \varepsilon_i) + c_{i3}\beta = 0 \qquad \textbf{12.16b}$$

$$c_{i2}\beta + c_{i3}(\alpha - \varepsilon_i) + c_{i4}\beta_{\mathrm{CO}} = 0 \qquad \textbf{12.16c}$$

$$c_{i3}\beta_{\mathrm{CO}} + c_{i4}(\alpha_{\mathrm{O}} - \varepsilon_i) = 0 \qquad \textbf{12.16d}$$

In order to solve the Hückel equations numerically, the carbon-to-carbon β is taken to be the energy unit and the carbon α is frequently taken as the energy zero. For a heteroatom, the parameters are written

$$\alpha_a = \alpha + h_a\beta \qquad \textbf{12.17}$$

$$\beta_{ab} = k_{ab}\beta \qquad \textbf{12.18}$$

where h_a and k_{ab} are empirically determined numerical parameters that depend upon the atom or bond being considered. If we divide through by β and let

$$x_i = \frac{\alpha - \varepsilon_i}{\beta} \qquad \textbf{12.19}$$

the Hückel determinantal equation for butadiene becomes

$$\begin{vmatrix} x_i & 1 & 0 & 0 \\ 1 & x_i & 1 & 0 \\ 0 & 1 & x_i & 1 \\ 0 & 0 & 1 & x_i \end{vmatrix} = 0 \qquad \textbf{12.20}$$

and the linear equations become

$$c_{i1}x_i + c_{i2} = 0 \qquad \textbf{12.21a}$$

$$c_{i1} + c_{i2}x_i + c_{i3} = 0 \qquad \textbf{12.21b}$$

$$c_{i2} + c_{i3}x_i + c_{i4} = 0 \qquad \textbf{12.21c}$$

$$c_{i3} + c_{i4}x_i = 0 \qquad \textbf{12.21d}$$

For acrolein, we have

$$\begin{vmatrix} x_i & 1 & 0 & 0 \\ 1 & x_i & 1 & 0 \\ 0 & 1 & x_i & k_{\mathrm{CO}} \\ 0 & 0 & k_{\mathrm{CO}} & h_{\mathrm{O}} + x_i \end{vmatrix} = 0 \qquad \textbf{12.22}$$

and

$$c_{i1}x_i + c_{i2} = 0 \qquad \textbf{12.23a}$$

$$c_{i1} + c_{i2}x_i + c_{i3} = 0 \qquad \textbf{12.23b}$$

$$c_{i2} + c_{i3}x_i + c_{i4}k_{\mathrm{CO}} = 0 \qquad \textbf{12.23c}$$

$$c_{i3}k_{\mathrm{CO}} + c_{i4}(h_{\mathrm{O}} + x_i) = 0 \qquad \textbf{12.23d}$$

Equations 12.20 and 12.22 each give fourth-order polynomials in x. These lead to four roots from which the orbital energies can be determined by using Eq. 12.19. We can find the coefficients for the L.C.A.O. expansion by substituting the roots into the appropriate linear equations.

If Eq. 12.20 is expanded, we obtain, for butadiene, the polynomial

$$x_i^4 - 3x_i^2 + 1 = 0 \qquad \textbf{12.24}$$

This is a quadratic in x_i^2, giving values of 2.618 and .382 for x_i^2. The values of x_i are thus ± 1.618 and $\pm .618$. By using Eq. 12.19 we find that the energies are, in ascending order (α and β are intrinsically negative),

$$\varepsilon_1 = \alpha + 1.618\beta \qquad \textbf{12.25a}$$

$$\varepsilon_2 = \alpha + .618\beta \qquad \textbf{12.25b}$$

$$\varepsilon_3 = \alpha - .618\beta \qquad \textbf{12.25c}$$

$$\varepsilon_4 = \alpha - 1.618\beta \qquad \textbf{12.25d}$$

Each carbon atom of butadiene contributes one electron to the π system, for a total of four. In the ground state, each of the first two orbitals is doubly occupied, giving a ground-state π electron energy, E_0, of

$$\begin{aligned} E_0 &= 2\varepsilon_1 + 2\varepsilon_2 \\ &= 4\alpha + 4.472\beta \qquad \textbf{12.26} \end{aligned}$$

The first excited state, E_1, would have one electron promoted from ε_2 to ε_3:

$$\begin{aligned} E_1 &= 2\varepsilon_1 + \varepsilon_2 + \varepsilon_3 \\ &= 4\alpha + 3.236\beta \qquad \textbf{12.27} \end{aligned}$$

The first electronic transition, ΔE_1, is the difference between these:

$$\begin{aligned} \Delta E_1 &= E_1 - E_0 \\ &= \varepsilon_3 - \varepsilon_2 \\ &= -1.236\beta \qquad \textbf{12.28} \end{aligned}$$

Experimentally, the first singlet transition occurs at 4.63×10^4 cm^{-1}. (Since Hückel theory ignores electronic repulsion, it cannot predict the singlet-triplet separation.) This can be used to assign a value to β of -3.75×10^4 cm^{-1}. (A better value could be obtained by averaging the results from several molecules.)

The coefficients for the L.C.A.O. equations can be obtained by substituting the roots, one at a time, into the appropriate linear equations and into the normalizing equation. In the Hückel approximation, with $S_{\mu\nu}$ equal to $\delta_{\mu\nu}$, the normalization equation becomes

$$\sum_{\mu} c_{i\mu}^2 = 1 \qquad \textbf{12.29}$$

Let us consider the molecular orbital ψ_1. The value of x that led to ε_1 was -1.618.

Substituting this into Eq. 12.21a, we have

$$-1.618c_{11} + c_{12} = 0$$

$$c_{12} = 1.618c_{11} \qquad \textbf{12.30}$$

Substituting this, along with the value for x, into Eq. 12.21b, we set

$$c_{11} + 1.618c_{11} \times (-1.618) + c_{13} = 0$$

$$c_{13} = 1.618c_{11} \qquad \textbf{12.31}$$

Substituting these into either of Eqs. 12.21c or 12.21d, we find that

$$c_{14} = c_{11} \qquad \textbf{12.32}$$

The normalizing condition

$$c_{11}^2 + c_{12}^2 + c_{13}^2 + c_{14}^2 = 1 \qquad \textbf{12.33}$$

yields

$$c_{11}^2[2 + 2(1.618)^2] = 1$$

$$c_{11} = .3718 \qquad \textbf{12.34a}$$

and, from Eqs. 12.30 through 12.32,

$$c_{12} = .6015 \qquad \textbf{12.34b}$$

$$c_{13} = .6015 \qquad \textbf{12.34c}$$

$$c_{14} = .3718 \qquad \textbf{12.34d}$$

The coefficients for the other molecular orbitals can be found in a similar fashion. The final results are

$$\psi_1 = .3718\chi_1 + .6015\chi_2 + .6015\chi_3 + .3718\chi_4 \qquad \textbf{12.35a}$$

$$\psi_2 = .6015\chi_1 + .3718\chi_2 - .3718\chi_3 - .6015\chi_4 \qquad \textbf{12.35b}$$

$$\psi_3 = .6015\chi_1 - .3718\chi_2 - .3718\chi_3 + .6015\chi_4 \qquad \textbf{12.35c}$$

$$\psi_4 = .3718\chi_1 - .6015\chi_2 + .6015\chi_3 - .3718\chi_4 \qquad \textbf{12.35d}$$

Notice that the number of nodes along the chain goes up as the energy increases. This is general for a wave function constructed from a given basis set; consequently, if the number of nodes in a set of wave functions can be determined by qualitative arguments, a qualitative ordering of their energies can be made. Conversely, if the energies can be ordered qualitatively, the nodal properties of the wave functions can be deduced.

If the determinant for acrolein, Eq. 12.22, is expanded, we have

$$x^4 + h_O x^3 - (2 + k_{CO})x^2 - 2h_O x + k_{CO}^2 = 0 \qquad \textbf{12.36}$$

Reasonable values for the parameters are 1 for both h_O and k_{CO}; i.e.,

$$\alpha_O = \alpha + \beta \qquad \textbf{12.37a}$$

and

$$\beta_{CO} = \beta \qquad \textbf{12.37b}$$

Substituting these into Eq. 12.36 yields

$$x^4 + x^3 - 3x^2 - 2x + 1 = 0 \qquad \textbf{12.38}$$

This can be solved by any of the usual methods for obtaining the roots of a polynomial. The roots are found to be -1.879, -1.000, .347, and 1.523. Using Eq. 12.19, we find the one-electron orbital energies to be

$$\varepsilon_1 = \alpha + 1.879\beta \qquad \textbf{12.39a}$$

$$\varepsilon_2 = \alpha + \beta \qquad \textbf{12.39b}$$

$$\varepsilon_3 = \alpha - .347\beta \qquad \textbf{12.39c}$$

$$\varepsilon_4 = \alpha - 1.523\beta \qquad \textbf{12.39d}$$

The ground-state π-electron energy is

$$\begin{aligned} E_0 &= 2\varepsilon_1 + 2\varepsilon_2 \\ &= 4\alpha + 5.758\beta \end{aligned} \qquad \textbf{12.40}$$

while that for the first excited state is

$$\begin{aligned} E_1 &= 2\varepsilon_1 + \varepsilon_2 + \varepsilon_3 \\ &= 4\alpha + 4.411\beta \end{aligned} \qquad \textbf{12.41}$$

The predicted energy for the first electronic spectral transition is

$$\begin{aligned} \Delta E_1 &= E_1 - E_0 \\ &= \varepsilon_3 - \varepsilon_2 \\ &= -1.347\beta \end{aligned} \qquad \textbf{12.42}$$

If we use the β value calibrated from the butadiene spectrum, this is predicted to be at 5.05×10^4 cm^{-1}. The experimental value for the first singlet $\pi^* \leftarrow \pi$ transition is 4.26×10^4 cm^{-1}—in reasonable agreement for such a simple approximation. (Note, however, that the predicted order of the first transitions for butadiene and for acrolein is wrong. If the Hückel theory is to be used for spectral calculations, the h_a and k_{ab} values should be calibrated against a series of compounds related to those under consideration. Conjugated polyenes and conjugated aldehydes should belong to two different series.)

The L.C.A.O. coefficients for acrolein can be determined in exactly the same manner as those for butadiene. The individual roots of the determinant are substituted into Eqs. 12.23 and used, with the normalization condition (Eq. 12.29), to obtain the desired results:

$$\psi_1 = .2280\chi_1 + .4285\chi_2 + .5774\chi_3 + .6565\chi_4 \qquad \textbf{12.43a}$$

$$\psi_2 = .5774\chi_1 + .5774\chi_2 + 0 - .5774\chi_4 \qquad \textbf{12.43b}$$

$$\psi_3 = .6565\chi_1 - .2280\chi_2 - .5774\chi_3 + .4285\chi_4 \qquad \textbf{12.43c}$$

$$\psi_4 = .4285\chi_1 - .6565\chi_2 + .5774\chi_3 - .2280\chi_4 \qquad \textbf{12.43d}$$

12.5 L.C.A.O. COEFFICIENTS AS VECTORS AND MATRICES

Let us write the L.C.A.O. coefficients for the wave functions of butadiene as column vectors. We have

$$\mathbf{c}_1 = \begin{bmatrix} .3718 \\ .6015 \\ .6015 \\ .3718 \end{bmatrix} \tag{12.44a}$$

$$\mathbf{c}_2 = \begin{bmatrix} .6015 \\ .3718 \\ -.3718 \\ -.6015 \end{bmatrix} \tag{12.44b}$$

$$\mathbf{c}_3 = \begin{bmatrix} .6015 \\ -.3718 \\ -.3718 \\ .6015 \end{bmatrix} \tag{12.44c}$$

$$\mathbf{c}_4 = \begin{bmatrix} .3718 \\ -.6015 \\ .6015 \\ -.3718 \end{bmatrix} \tag{12.44d}$$

Consider now the matrix of the Hückel Hamiltonian, $\mathbf{h}$, multiplied by one of the vectors, say $\mathbf{c}_1$. This gives

$$\begin{aligned} \mathbf{hc}_1 &= \begin{bmatrix} \alpha & \beta & 0 & 0 \\ \beta & \alpha & \beta & 0 \\ 0 & \beta & \alpha & \beta \\ 0 & 0 & \beta & \alpha \end{bmatrix} \begin{bmatrix} .3718 \\ .6015 \\ .6015 \\ .3718 \end{bmatrix} \\ &= \begin{bmatrix} .3718\alpha + .6015\beta \\ .6015\alpha + .9733\beta \\ .6015\alpha + .9733\beta \\ .3718\alpha + .6015\beta \end{bmatrix} \\ &= (\alpha + 1.618\beta) \begin{bmatrix} .3718 \\ .6015 \\ .6015 \\ .3718 \end{bmatrix} \end{aligned} \tag{12.45}$$

or

$$\mathbf{hc}_1 = \varepsilon_1 \mathbf{c}_1 \tag{12.45a}$$

This has the form of an eigenvalue equation. The vector $\mathbf{c}_1$ is said to be an *eigenvector* of $\mathbf{h}$, corresponding to the eigenvalue ε_1. Note also that

$$\mathbf{c}_1^\dagger \mathbf{c}_1 = [.3718 \quad .6015 \quad .6015 \quad .3718]\begin{bmatrix} .3718 \\ .6015 \\ .6015 \\ .3718 \end{bmatrix} \qquad \mathbf{12.46}$$

$$= 1$$

Thus we have

$$\mathbf{c}_1^\dagger \mathbf{h} \mathbf{c}_1 = \varepsilon_1 \qquad \mathbf{12.47}$$

Similar results are obtained for the other vectors. They are each eigenvectors of $\mathbf{h}$ corresponding to the appropriate eigenvalues (energy values). If we construct a square matrix, $\mathbf{C}$, in which the columns are the eigenvector matrices of Eqs. 12.44, we have

$$\mathbf{C} = [\mathbf{c}_1 \quad \mathbf{c}_2 \quad \mathbf{c}_3 \quad \mathbf{c}_4] \qquad \mathbf{12.48}$$

The matrix product $\mathbf{C}^\dagger\mathbf{h}\mathbf{C}$ gives

$$\mathbf{C}^\dagger \mathbf{h} \mathbf{C} = \begin{bmatrix} \alpha + 1.618\beta & 0 & 0 & 0 \\ 0 & \alpha + .618\beta & 0 & 0 \\ 0 & 0 & \alpha - .618\beta & 0 \\ 0 & 0 & 0 & \alpha - 1.618\beta \end{bmatrix} \qquad \mathbf{12.49}$$

The matrix $\mathbf{h}$ is *diagonalized* (put in diagonal form) by a *unitary transformation* with the matrix $\mathbf{C}$. The diagonal elements of the diagonalized matrix are the roots of the secular equation. For computer solution of the problem, the matrix of the operator $\hat{h}$ is constructed. This is of the same form as the determinant, except that the unknown is omitted. The matrix is then diagonalized by a unitary transformation. The diagonal elements of the diagonalized matrix are the roots. The columns of the transformation matrix that accomplishes the diagonalization are the coefficients of the L.C.A.O. molecular orbitals associated with the corresponding energies.

12.6 INTERPRETATION OF THE L.C.A.O. WAVE FUNCTIONS

The normalization requirement for a molecular wave function is

$$\langle \psi_i | \psi_i \rangle = 1 \qquad \mathbf{12.50}$$

If we multiply Eq. 12.50 by the number of electrons in each orbital, f_i, and sum over the occupied orbitals, we have

$$\sum_i^{\text{occ}} f_i \langle \psi_i | \psi_i \rangle = n \qquad \mathbf{12.51}$$

where n is the number of electrons in the system.

We have previously described the square of the wave function $|\psi_i|^2$ as the probability of finding the particle described by ψ_i in various regions of space. For acrolein (or for butadiene), $|\psi_i|^2$ is

$$\begin{aligned}|\psi_i|^2 = {} & c_{i1}^2|\chi_1|^2 + c_{i2}^2|\chi_2|^2 + c_{i3}^2|\chi_3|^2 + c_{i4}^2|\chi_4|^2 \\ & + c_{i1}^*c_{i2}\chi_1^*\chi_2 + c_{i1}^*c_{i3}\chi_1^*\chi_3 + c_{i1}^*c_{i4}\chi_1^*\chi_4 + c_{i2}^*c_{i1}\chi_2^*\chi_1 \\ & + c_{i2}^*c_{i3}\chi_2^*\chi_3 + c_{i2}^*c_{i4}\chi_2^*\chi_4 + c_{i3}^*c_{i1}\chi_3^*\chi_1 + c_{i3}^*c_{i2}\chi_3^*\chi_2 \\ & + c_{i3}^*c_{i4}\chi_3^*\chi_4 + c_{i4}^*c_{i1}\chi_4^*\chi_1 + c_{i4}^*c_{i2}\chi_4^*\chi_2 + c_{i4}^*c_{i3}\chi_4^*\chi_3\end{aligned} \quad \mathbf{12.52}$$

If we integrate this, using the Hückel approximations, we get

$$\langle\psi_i|\psi_i\rangle = c_{i1}^2 + c_{i2}^2 + c_{i3}^2 + c_{i4}^2 \quad \mathbf{12.53}$$

which is just the normalization integral. Thus, $c_{i\mu}^2$ is the integrated probability of finding the single electron, which is in ψ_i, located in χ_μ. The total electron density at μ, q_μ, is the sum of the contributions from each occupied molecular orbital (multiplied, of course, by the appropriate orbital occupancy, f_i):

$$q_\mu = \sum_i^{\text{occ}} f_i c_{i\mu}^2 \quad \mathbf{12.54}$$

The quantity q_μ is sometimes referred to as the total *charge density*, or just the charge density, even though, to be a charge density, it must be multiplied by the charge of the electron. The *net charge* on a center is the difference between q_μ and the number of electrons that the center donated to the molecular orbitals. Note, however, that this definition is for Hückel theory only. If the Hückel overlap neglect were not used, Eq. 12.53 would contain other terms. Various methods are in use for defining the charge on atomic centers when the overlap integral is not neglected.

For butadiene, as for all molecules that can be classified as *alternant hydrocarbons*, q_μ equals 1.00 for all positions, within the Hückel approximation. Since each center donated one electron to the π system, the net charge is zero. This is not the case for acrolein, however. For the ground state of acrolein we have

$$\begin{aligned}q_1 &= 2(.2280)^2 + 2(.5774)^2 \\ &= .771\end{aligned} \quad \mathbf{12.55a}$$

$$\begin{aligned}q_2 &= 2(.4285)^2 + 2(.5774)^2 \\ &= 1.034\end{aligned} \quad \mathbf{12.55b}$$

$$\begin{aligned}q_3 &= 2(.5774)^2 + 2(0)^2 \\ &= .667\end{aligned} \quad \mathbf{12.55c}$$

$$\begin{aligned}q_4 &= 2(.6565)^2 + 2(-.5774)^2 \\ &= 1.529\end{aligned} \quad \mathbf{12.55d}$$

(Note that the sum of the q_μ equals four, as it must, since there are four π electrons in the system.) Each center contributed one electron to the π system. Positions 1 and 3 get back less than one, so they have a net positive charge, while positions 2 and 4 get back more than one and have a net negative charge. The net charges are .229, $-.034$, .333, and $-.529$, respectively. As might be expected, the charge is polarized toward the most electronegative atom, the oxygen.

In its first $\pi^* \leftarrow \pi$ excited state, having the configuration $(\psi_1)^2(\psi_2)(\psi_3)$, the charge in acrolein is reorganized. (It is not in butadiene.) We have

$$q_1 = 2(.2280)^2 + (.5774)^2 + (.6565)^2 = .868 \tag{12.56a}$$

$$q_2 = 2(.4285)^2 + (.5774)^2 + (-.2280)^2 = .753 \tag{12.56b}$$

$$q_3 = 2(.5774)^2 + (0)^2 + (-.5774)^2 = 1.000 \tag{12.56c}$$

$$q_4 = 2(.6565)^2 + (-.5774)^2 + (.4285)^2 = 1.379 \tag{12.56d}$$

The net charges on atoms 1 through 4 are, respectively, .132, .247, 0, and $-.379$. The result of the excitation is that electron density has been lost by the atoms that were the most electron-rich in the ground state and gained by those that were the most electron-deficient. Such charge reorganizations have important consequences in photochemistry.

In 1939, Coulson introduced the concept of a *bond order*. The bond order is defined as

$$p_{\mu\nu} = \sum_{i}^{\text{occ}} f_i c_{i\mu}^* c_{i\nu} \tag{12.57}$$

For directly bonded atoms the bond order is related to the bond length. For butadiene we have, for the ground state,

$$p_{12} = p_{34} = 2(.3718)(.6015) + 2(.6015)(.3718) = .894 \tag{12.58a}$$

$$p_{23} = 2(.6015)(.6015) + 2(.3718)(-.3718) = .447 \tag{12.58b}$$

The bonds between atoms 1 and 2 and between 3 and 4 are 1.35 Å in length, while that between atoms 2 and 3 is 1.46 Å. Thus we see that, for like atoms, the larger the bond order, the shorter the bond. For acrolein the bond orders and bond lengths are

$$p_{12} = .862 \quad (\text{length} = 1.36\ \text{Å}) \tag{12.59a}$$

$$p_{23} = .495 \quad (\text{length} = 1.45\ \text{Å}) \tag{12.59b}$$

$$p_{34} = .758 \quad (\text{length} = 1.22\ \text{Å}) \tag{12.59c}$$

The C—O bond length obviously cannot be directly compared to the C—C bond lengths. However, its bond-length, bond-order correlation should compare favorably with that for other C—O bonds.

12.7 THE FIRST-ORDER DENSITY MATRIX

Bond-order-like quantities can also be calculated for nonbonded atoms. Although these have no direct relation to bond lengths, the resulting matrix is important in quantum chemistry. This matrix is the *first-order density matrix* for a single configuration L.C.A.O. problem. The charge densities are the diagonal elements of this matrix. The first-order density matrix, **P**, for the Hückel ground state of acrolein is

$$\mathbf{P} = \begin{bmatrix} .771 & .862 & .263 & -.367 \\ .862 & 1.034 & .495 & -.104 \\ .263 & .495 & .667 & .758 \\ -.367 & -.104 & .758 & 1.529 \end{bmatrix} \tag{12.60}$$

(Notice that the trace of **P** is the number of electrons in the system.) Any one-electron property of this state of the system can be calculated from the first-order density matrix. If A is some one-electron property of the system, then

$$\begin{aligned} \langle A \rangle &= \mathrm{Tr}\,\mathbf{PA} \\ &= \sum_{\mu}\sum_{\nu} P_{\mu\nu}A_{\mu\nu} \end{aligned} \tag{12.61}$$

where the second line gives the trace of the product in terms of the individual matrix elements. As a specific example, the Hückel energy contains only one-electron operators. Thus the energy is

$$\begin{aligned} E = \langle H \rangle &= \sum_{\mu}\sum_{\nu} P_{\mu\nu}h_{\mu\nu} \\ &= .771\alpha + .862\beta + .862\beta + 1.034\alpha \\ &\quad + .495\beta + .495\beta + .667\alpha + .758\beta_{\mathrm{CO}} \\ &\quad + .758\beta_{\mathrm{CO}} + 1.529\alpha_{\mathrm{O}} \\ &= 4\alpha + 5.759\beta \end{aligned} \tag{12.62}$$

(using Eqs. 12.37 for α_{O} and β_{CO}), which, except for rounding error, is the same as calculated previously. In the independent-particle approximation, properties involving two-electron operators can also be evaluated from the first-order density matrix. Higher-order density matrices can also be defined. All two-electron properties (including the energy of the exact nonrelativistic Hamiltonian) can be determined from the second-order density matrix, and so on. With independent-particle wave functions the higher-order density matrices can be constructed from the first-order matrix. For wave functions including correlation the construction is more complicated. This text will not further discuss the higher-order density matrices.

12.8 MOLECULAR HARTREE-FOCK CALCULATIONS

The Hartree-Fock S.C.F. procedure for molecules is completely analogous to that for atoms, as presented in Section 7.10, when the equations are written in terms of molecular orbitals. The molecular electronic energy equation from which the Hartree-Fock scheme is derived is Eq. 12.2. This differs from the atomic-energy equation, Eq. 7.61, only in that a summation of nuclear attraction terms, rather than a single one as in the atomic case, occurs in the one-electron operator. The Fock operator is now

$$\hat{F}(i) = -\frac{1}{2}\nabla_i^2 - \sum_A \frac{Z_A}{r_{Ai}} + \sum_{j\neq i} (\hat{J}_{ij} - \hat{K}_{ij}) \qquad \textbf{12.63}$$

where the $\hat{J}$ and $\hat{K}$ operators are defined exactly as in the atomic case (Eq. 7.49), but in terms of molecular orbitals instead of atomic orbitals.

When the molecular orbitals are expanded as linear combinations of basis functions, the equations appear more complex. The reason is that the basis functions are not orthogonal. The expansions for the J and K integrals are given in Eqs. 12.4 and 12.5, respectively. If we make the abbreviation

$$\gamma_{\mu\nu\lambda\sigma} = \left\langle \chi_\mu\chi_\nu \left| \frac{1}{r_{12}} \right| \chi_\lambda\chi_\sigma \right\rangle \qquad \textbf{12.64}$$

the Fock matrix can be written

$$F_{\mu\nu} = \left\langle \chi_\mu \left| -\frac{1}{2}\nabla_i^2 - \sum_A \frac{Z_A}{r_{Ai}} \right| \chi_\nu \right\rangle + \sum_{j\neq i}\sum_\mu\sum_\nu\sum_\lambda\sum_\sigma c_{i\mu}^* c_{j\nu}^* c_{i\lambda} c_{j\sigma} (\gamma_{\mu\nu\lambda\sigma} - \gamma_{\mu\nu\sigma\lambda}\,\delta_{m_{si}m_{sj}}) \qquad \textbf{12.65}$$

If we define the matrix $\mathbf{D}$ as the first-order density matrix *without* the orbital occupancy multiplier, this can be written more compactly in terms of the elements of $\mathbf{D}$ as

$$F_{\mu\nu} = \left\langle \chi_\mu \left| -\frac{1}{2}\nabla_i^2 - \sum_A \frac{Z_A}{r_{Ai}} \right| \chi_\nu \right\rangle + \sum_\mu\sum_\nu D_{\mu\nu}(\gamma_{\mu\nu\lambda\sigma} - \gamma_{\mu\nu\sigma\lambda}\delta_{m_{si}m_{sj}}) \qquad \textbf{12.66}$$

The Fock operator is a one-electron operator. Consequently, the L.C.A.O.-type problem to be solved is analogous to the Hückel problem with all off-diagonal and overlap elements included (see Eq. 12.12). However, since the electron-repulsion terms depend upon the charge density, the problem must be solved iteratively. A starting set of coefficients is chosen by some convenient means, usually by solution of a one-electron determinant (either the one-electron portion of the Fock matrix, or the overlap matrix). This is used to construct a starting Fock matrix. The coefficients from this are used to construct the next approximation, and iteration is continued until the L.C.A.O. functions are self-consistent. The convergence can be followed by comparing, in successive iterations, the values of the energy, the elements of the density matrix, the elements of the Fock matrix, or the coefficients. Exactly the same procedure can be used in atomic S.C.F. calculations, if the atomic orbitals are expressed as linear combinations of basis functions.

Many sophisticated computer programs are available for performing S.C.F. calculations on atoms or on polyatomic molecules. (Several of these are available to any interested person from the Quantum Chemistry Program Exchange at Indiana University.) The atom-centered basis functions used in these are almost invariably either gaussian functions (or gaussian-type orbitals, G.T.O.'s), which have an $\exp(-ar^2)$ dependence, or Slater-type orbitals (S.T.O.'s), which have an $\exp(-br)$ dependence, where the a or b are numerical constants depending upon the atom. (While b depends upon an effective nuclear charge and a principal quantum number, a is a totally empirical parameter.). Neither contains the radial nodes of the hydrogenic functions; however, the integrals are easier to evaluate without the nodes, so more basis functions can be used for a given amount of work. Except for the polynomial in r, which gives rise to radial nodes, the functional dependence of the S.T.O.'s is the same as for hydrogenic orbitals. It is different for the G.T.O.'s. On the other hand, the integrals are much easier to evaluate for G.T.O.'s than for S.T.O.'s. Again, more basis functions can be used for a given amount of effort.

Programs using S.T.O.'s are more commonly atomic S.C.F. programs or programs for treating linear molecules (where the integrations are easier). Programs using G.T.O's are common for atomic or general molecular systems.

A word of caution is in order. Because of their rapid fall-off with distance, and their behavior near an r value of zero, wave functions constructed from G.T.O.'s are less accurate in the area very near the nucleus than are wave functions constructed from S.T.O.'s that have comparable (or even less accurate) energies. This frequently shows up in calculations of properties (such as quadrupole coupling constants) that depend upon the charge near the nucleus.

Most of the quantities obtained from the most accurate S.C.F. calculations are the same as those from the simplest Hückel calculations. Basically, these are the orbital energies, the total energies, and the wave functions. Other desired quantities can be obtained as the expectation value of the appropriate operators. Hartree-Fock calculations can be carried out nonempirically to a high degree of accuracy, if sufficient flexibility is allowed in the wave function. These calculations still suffer the limitations of the independent-particle model, however (in particular, the correlation error). Configuration interaction, or other techniques, can be used to correct for the correlation error, with varying degrees of accuracy. As mentioned in Section 12.2, Hartree-Fock-like calculations can also be performed with various additional approximations made to simplify the computations.

12.9 INTENSITIES OF MOLECULAR ELECTRONIC SPECTRAL TRANSITIONS

The intensities of molecular electronic transitions can be computed from calculations of the transition dipole. We can write the wave functions for states m and n, in the molecular-orbital approximation, as a product of one-electron molecular orbitals

$$\Psi_m = \psi_k \prod_{i \neq k}^{\text{occ}} \psi_i \qquad \textbf{12.67a}$$

$$\Psi_n = \psi_l \prod_{j \neq l}^{\text{occ}} \psi_j \qquad \textbf{12.67b}$$

where ψ_k and ψ_l are the orbitals from which and to which the electron is being promoted. The transition dipole is

$$\begin{aligned} \mu_{mn} &= \left\langle \psi_k \prod_{i \neq k} \psi_i | \hat{\mu} | \psi_l \prod_{j \neq l} \psi_j \right\rangle \\ &= \langle \psi_k | \hat{\mu} | \psi_l \rangle \left\langle \prod_i \psi_i \middle| \prod_j \psi_j \right\rangle \\ &= \langle \psi_k | \hat{\mu} | \psi_l \rangle \\ &= \mu_{kl} \end{aligned} \qquad \textbf{12.68}$$

where we have used the fact that the transition dipole operator is a one-electron operator and that the molecular orbitals are normalized. (Note, however, that the dipole moment operator is a many-electron operator, which is a sum over all electrons of one-electron dipole operators. The one-electron operators have the same form in either case.) The dipole operator can be written

$$\hat{\mu} = e \sum_A \mathbf{r}_A \qquad \textbf{12.69}$$

where the $\mathbf{r}_A$ are the position vectors of the sites of each basis function. The summation is over all sites (the atomic centers).

Let us specifically consider the first $\pi^* \leftarrow \pi$ transition of butadiene. The structure was given as **1** in Section 12.4. If the origin of a Cartesian coordinate system is placed in the center of the C_2—C_3 single bond, the x axis placed parallel to the C_1—C_2 and C_3—C_4 bonds, such that C_1 and C_2 have positive x values, the y axis placed in the molecular plane, and directed such that C_1 and C_2 have positive y values, the coordinates are, assuming the bond lengths given in Section 12.6, and 120° bond angles:

	x	y	z
C_1	1.715	.632	0
C_2	.365	.632	0
C_3	−.365	−.632	0
C_4	−1.715	−.632	0

(The transition dipole is actually independent of the axis choice. The present choice was made only for convenience.) The individual position vectors are, in Angstrom units,

$$\mathbf{r}_1 = 1.715\mathbf{i} + .632\mathbf{j} \qquad \textbf{12.71a}$$

$$\mathbf{r}_2 = .365\mathbf{i} + .632\mathbf{j} \qquad \textbf{12.71b}$$

$$\mathbf{r}_3 = -.365\mathbf{i} - .632\mathbf{j} \qquad \textbf{12.71c}$$

$$\mathbf{r}_4 = -1.715\mathbf{i} - .632\mathbf{j} \qquad \textbf{12.71d}$$

where **i** and **j** are the x and y Cartesian unit vectors. The butadiene π-electron wave functions are given in eqs. 12.35. The transition of interest promotes an electron from ψ_2 to ψ_3. The transition dipole can be computed by substituting Eq. 12.69 into 12.68 and using the L.C.A.O. expansion for the wave functions. Within the Hückel approximations, the result is

$$\mu_{kl} = \sum_{\mu} c_{k\mu} c_{l\mu} \mathbf{r}_{\mu} \qquad \textbf{12.72}$$

(Working in atomic units, the electronic charge, e, can be omitted.) For butadiene, we have, using Eqs. 12.35b–c and Eqs. 12.71,

$$\begin{aligned} \mu_{32} &= .6015^2\mathbf{r}_1 - .3718^2\mathbf{r}_2 + .3718^2\mathbf{r}_3 - .6015^2\mathbf{r}_4 \\ &= 2(.6015^2\mathbf{r}_1 - .3718^2\mathbf{r}_2) \\ &= 2[.3718(1.715\mathbf{i} + .632\mathbf{j}) - .1382(.365\mathbf{i} + .632\mathbf{j})] \\ &= .767\mathbf{i} + .283\mathbf{j} \end{aligned} \qquad \textbf{12.73}$$

The transition dipole is a vector having a magnitude of .818 (in units of electronic charge $\times\ 10^{-8}$ cm) and lying 20° off the positive x axis (toward the positive y axis). If the same calculation is repeated for the $\pi^* \leftarrow \pi$ transition of acrolein, using the wave functions of Eqs. 12.43 and the bond lengths given in Eqs. 12.59, the transition dipole is calculated to be 1.066, in the same units, and to lie 22° from the positive x axis. The total intensity of a transition is proportional to the square of the transition dipole (see Eq. 6.101). The relative intensities of the $\pi^* \leftarrow \pi$ transitions of acrolein and butadiene are predicted to be 1.7:1.

BIBLIOGRAPHY

ANDERSON, J. M., *Introduction to Quantum Chemistry*. W. A. Benjamin, Inc., New York, 1969.

ATKINS, P. W., *Molecular Quantum Mechanics*. Clarendon Press, Oxford, 1970.

FLURRY, R. L., JR., *Molecular Orbital Theories of Bonding in Organic Compounds*. Marcel Dekker, Inc., New York, 1968.

LEVINE, I. N., *Quantum Chemistry*. Allyn & Bacon, Boston, 2d ed., 1974.

PARR, R. G., *Quantum Theory of Molecular Electronic Structure*. W. A. Benjamin, Inc., New York, 1963.

STREITWIESER, A., *Molecular Orbital Theory for Organic Chemists*. John Wiley & Sons, New York, 1961.

PROBLEMS

***12.1** Set up the form of the L.C.A.O.-M.O. wave functions for the one-electron molecular orbitals of H_3, using a single $1s$ basis on each hydrogen.

***12.2** Derive the secular equation, within the one-electron approximation for H_3, assuming a linear geometry with equal bond lengths.

***12.3** Repeat Problem 12.2 for an equilateral triangular geometry.

***12.4** If the bond lengths for the linear and triangular H_3 are about the same, which would you expect, on a qualitative basis, to be more stable?

12.5 Write, in detail, the complete, nonrelativistic, Hamiltonian for the B_2 molecule.

***12.6** Write out in detail (in terms of F_{1s1s}, F_{2s1s}, and so on) the L.C.A.O. determinantal equation for an S.C.F. treatment of B_2.

12.7 Write out the Hückel π-electron determinants for the following molecules, in terms of the appropriate α's and β's.

*(a) Ethyleneamine (consider the N as part of the π system)

$$CH_2{=}CH{-}NH_2$$

(b) Cyclopropenyl,

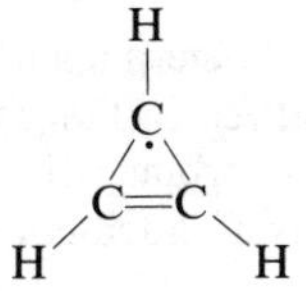

*(c) Benzene,

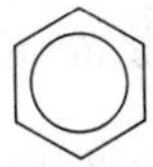

(d) *s*-Triazine,

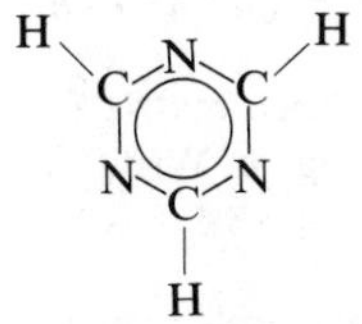

*(e) *p*-Benzoquinone,

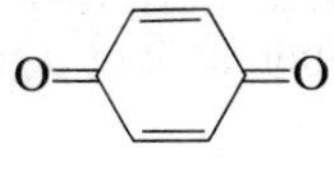

(f) Triphenylene,

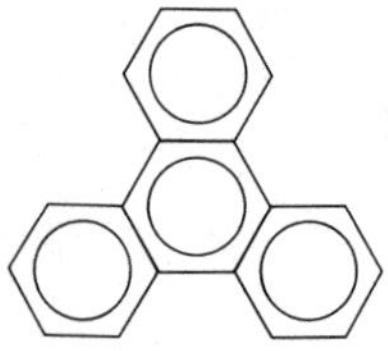

12.8 Given the following wave functions for the π system of ethyleneamine, $CH_2{=}CH{-}NH_2$, calculate the π-electron density, the net charge on each center, and the first-order density matrices, for the ground and first excited states. (Note that the nitrogen contributes two electrons to the π system.)

$$\phi_1 = .2737\chi_1 + .6151\chi_2 + .7394\chi_3$$
$$\phi_2 = .8285\chi_1 + .2398\chi_2 - .5061\chi_3$$
$$\phi_3 = .4886\chi_1 - .7511\chi_2 + .4440\chi_3$$

12.9 Solve for the Hückel π-electron orbital energies, the wave functions, the total ground-state π energy, the energy of the first $\pi^* \leftarrow \pi$ transition, and the charge densities for the following molecules.

(a) Formaldehyde, $CH_2{=}O$.

(b) Cyclopropenyl.

*(c) Glyoxal, $O{=}CH{-}CH{=}O$.

*(d) The carbonate ion, CO_3^{2-} (all O are attached to the C, and the ion has six π electrons).

12.10 Set up the Hückel-level determinant for the valence orbital σ and π systems for ethylene. Assume trigonally (*tr*) hybridized carbon atoms and a single basis, *h*, on the hydrogens. (There will be β terms between the *tr* orbitals on the same atom, as well as between those representing bonds.) Assume that there is no interaction between the σ and π systems.

12.11 Rewrite the determinant from the preceding problem in the *x* convention. Let the π-electron α and β be the reference parameters. Let $\beta_{trtr} = 2.5\beta$, $\beta_{trh} = 2.35\beta$, $\alpha_h = \alpha + .5\beta$, $\alpha_{tr} = \alpha + \beta$, $\beta'_{trtr} = .1\beta$ (on same atom).

12.12 If you have a matrix diagonalization procedure available, find the orbital energy levels, the wave functions, the total ground-state energy, the charge densities, and the energies of the first $\pi^* \leftarrow \pi$, $\pi^* \leftarrow \sigma$, $\sigma^* \leftarrow \pi$, and $\sigma^* \leftarrow \sigma$ transitions for the all-valence-electron ethylene.

12.13 Set up the Hartree-Fock matrix for lithium hydride, LiH. Use a 1*s* and 2*s* basis on Li and a 1*s* basis on the hydrogen. Write out the matrix elements in terms of integrals over the atomic basis set.

12.14 The nonvanishing integrals over the atomic orbitals (Slater-type basis) are given below for LiH at an internuclear separation of $2.6a_0$. Do a few cycles of the S.C.F. calculation. Use only the one-electron integrals for getting the starting coefficients. If you have a computer available, continue the iteration until the agreement between the initial and final density-matrix elements is 1×10^{-5}.

One-electron Integrals:		Two-electron Integrals:	
Basis Functions[a]		Basis Functions	

One-electron		Two-electron	
Overlap:		$\overset{b}{s\ s}$ $\overset{c}{s\ s}$	.64978
s s	1.00000	S s s s	.12216
S s	.00000	S s S s	.01412
S S	1.00000	S S s s	.32298
h h	1.00000	S S S s	.00278
h s	.14903	S S S S	.23448
h S	−.50141	h h h h	.62500
$-\frac{1}{2}\nabla^2$:		h s s s	.15526
s s	3.61037	h s S s	.00762
S s	.69533	h s S S	.04611
S S	.20834	h s h s	.01876
h h	.50000	h S s s	−.17113
h s	−.00084	h S S s	.00002
h S	−.05431	h S S S	−.12728
$1/R_{\text{H}}$:		h S h s	−.02889
h h	1.00000	h S h S	.09722
s s	.38453	h h s s	.37517
S s	.00037	h h S s	.00177
S S	.27859	h h S S	.26305
h s	.18998	h h h s	.06552
h S	−.16662	h h h S	−.23450
$1/R_{\text{Li}}$:			
h h	.37698		
s s	2.68449		
S s	.27275		
S S	.34550		
h s	.07368		
h S	−.30501		

[a] s = $1s$ (Li); S = $2s$ (Li); h = $1s$ (H).
[b] Electron 1.
[c] Electron 2.

chapter thirteen

Point groups

13.1 THE SYMMETRY OF THE HAMILTONIAN

The Hamiltonian of a system is invariant to any change of the coordinate system and to any interchange of equivalent particles. If the individual particles have an intrinsic symmetry (intrinsic angular momentum, or spin), the overall group of the Hamiltonian must also include this. The interrelation between the intrinsic symmetry and the interchange symmetry is the source of the permutational restrictions on a wave function (i.e., the Pauli principle). In this chapter we will concentrate on the symmetry involving changes of the coordinate system, the *spatial symmetry.*

Changes in the coordinate system can be expressed in two ways. Either the positions of the particles can be held fixed and the external coordinate system changed by a rotation, reflection, or inversion (the passive convention), or the external coordinate system can be held fixed and the molecule (with its internal coordinate system) rotated, reflected, or inverted (the active convention). We will adopt the active convention here. (This is much the simpler convention if the phases of the orbitals are to be treated.) Let us consider *trans*-butadiene in two orientations that are rotated with respect to each other by 180° about an axis perpendicular to the molecular plane. In any given coordinate system the Hamiltonian for structure **1** is identical to that for structure **2**, except for labeling. The physical properties of

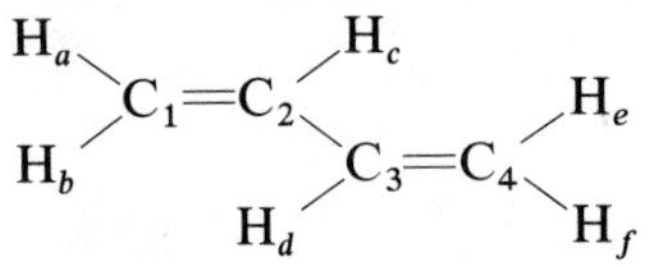

(1)

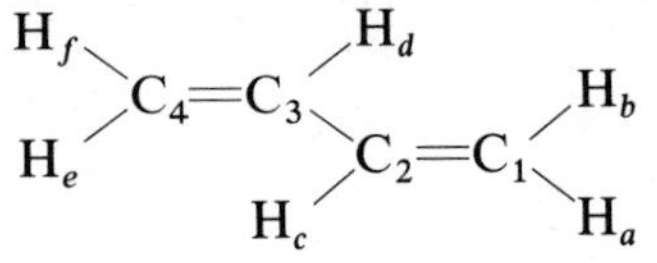

(2)

a molecule cannot depend upon the labels we assign to the atoms. The potential experienced by the electrons and nuclei in **1** is identical to that in **2**. Consequently, the Hamiltonian corresponding to the physical situation is completely identical in orientations **1** and **2**.

Consider now the result of some symmetry operation, $\hat{R}$, acting on the Schrödinger equation for a nondegenerate state

$$\hat{R}\hat{H}\psi = \hat{R}E\psi \qquad \mathbf{13.1}$$

Since the Hamiltonian is invariant to the symmetry operations, it commutes with them. Furthermore, the energy is a constant, so it commutes with any operation. Thus, Eq. 13.1 is completely equivalent to

$$\hat{H}\hat{R}\psi = E\hat{R}\psi \qquad \mathbf{13.2}$$

This means that $\hat{R}\psi$ is an eigenfunction of $\hat{H}$, having the same eigenvalue, E, as does ψ. For this to hold, the most that $\hat{R}$ can do to ψ is to multiply it by a constant, c,

$$\hat{R}\psi = c\psi \qquad \mathbf{13.3}$$

where c can be complex. But normalization requires that

$$\begin{aligned} 1 &= \langle c\psi | c\psi \rangle \\ &= c^*c\langle \psi | \psi \rangle \\ &= c^*c \end{aligned} \qquad \mathbf{13.4}$$

Thus, c may be a complex (or pure real or pure imaginary) number whose magnitude is unity. If we had some way to find the c's corresponding to the various symmetry operations, we would have some limitations placed on the L.C.A.O. functions (or whatever other type of function the wave function might be). This would reduce the number of independent coefficients and, consequently, the size of the secular determinant that would have to be solved.

Let us consider the L.C.A.O. wave functions for the π-electron system of butadiene (Eqs. 12.35) and let R be the operation that carries structure **1** into structure **2**. The results of operating on the various functions are

$$\begin{aligned} \psi_1 &= .3718\chi_1 + .6015\chi_2 + .6015\chi_3 + .3718\chi_4 \\ R\psi_1 &= .3718\chi_4 + .6015\chi_3 + .6015\chi_2 + .3718\chi_1 \\ &= \psi_1 \end{aligned} \qquad \mathbf{13.5a}$$

$$\begin{aligned} \psi_2 &= .6015\chi_1 + .3718\chi_2 - .3718\chi_3 - .6015\chi_4 \\ R\psi_2 &= .6015\chi_4 + .3718\chi_3 - .3718\chi_2 - .6015\chi_1 \\ &= -\psi_2 \end{aligned} \qquad \mathbf{13.5b}$$

$$\begin{aligned} \psi_3 &= .6015\chi_1 - .3718\chi_2 - .3718\chi_3 + .6015\chi_4 \\ R\psi_3 &= .6015\chi_4 - .3718\chi_3 - .3718\chi_2 + .6015\chi_1 \\ &= \psi_3 \end{aligned} \qquad \mathbf{13.5c}$$

$$\psi_4 = .3718\chi_1 - .6015\chi_2 + .6015\chi_3 - .3718\chi_4$$
$$R\psi_4 = .3718\chi_4 - .6015\chi_3 + .6015\chi_2 - .3718\chi_1$$
$$= -\psi_4 \qquad \textbf{13.5d}$$

We see that the constant arising from the operation of R is $+1$ for ψ_1 and ψ_3 and -1 for ψ_2 and ψ_4. That is, ψ_1 and ψ_3 are *symmetric* with respect to a 180° rotation, while ψ_2 and ψ_4 are *antisymmetric* with respect to it. If we had known this beforehand, we could have reduced our 4×4 determinant to two 2×2 determinants.

We have stated that the Hamiltonian must be invariant (i.e., symmetric) with respect to symmetry operations. In fact, the invariances of the Hamiltonian determine the symmetry group of the system. Yet the wave functions may change (perhaps only by a sign) under the symmetry operations. The symmetry group of the wave functions must be the same as the symmetry group of the Hamiltonian. However, different eigenfunctions, which describe the *motions* of the electrons in the system, transform as different irreducible representations of the group. In the example above, ψ_1 and ψ_3 transform as a representation that is symmetric under a 180° rotation, while ψ_2 and ψ_4 transform as a representation that is antisymmetric under it.

13.2 SYMMETRY ELEMENTS AND SYMMETRY OPERATIONS

As we have previously mentioned (Section 3.5), a three-dimensional physical object can have five types of symmetry operations: the identity E; the proper rotation, C_n; the reflection, σ; the inversion, i; and the improper rotation, S_n. For the proper and improper rotations the subscript, n, is the order of the rotation—i.e., 2π divided by the angle of rotation. All objects possess the identity. Objects with any symmetry possess one or more of the others as well. The geometric entities about which the operations are performed are referred to as *symmetry elements*. For example, the axis about which a rotation occurs, the plane through which a reflection occurs, and the point through which an inversion occurs are all symmetry elements. The existence of the operation implies the existence of the elements, and vice versa. The terms symmetry operation and symmetry element are often used interchangeably. A word of caution, however: the symmetry *operations* are the *elements* of the mathematical group.

Let us consider some examples. *Trans*-butadiene has four symmetry operations. The identity is trivial. We have already discussed the 180° rotation, which is denoted C_2. As with any planar molecule, reflection in the plane of the molecule is a symmetry operation. This is denoted as a σ_h, a reflection in a *horizontal* plane (i.e., perpendicular to the rotation axis, which defines the vertical axis). This does not change the positions of any atoms in this case. (Note, however, that it does change the signs on all of the p-π basis functions.) Inversion of all coordinates through an origin at the center of the molecule is a symmetry operation. This produces the same interchange of atom labels as does the C_2. (It changes the signs, as well as the labels, of the p-π

functions.) In this particular case there is one symmetry element (the identity, the axis, the plane, and the point) corresponding to each operation. The group $\{E, C_2, i, \sigma_h\}$ is called the $\mathbf{C}_{2h}$ group. The symmetry elements of butadiene all intersect at the point of inversion. All symmetry elements of any object will intersect at some point; consequently, spatial symmetry groups for individual objects are commonly referred to as *point groups*. The groups used to describe crystals and other systems with repetitive translational symmetry are called *space groups*. We will confine our attention to the point groups of single objects.

Consider now the ammonia molecule, **3**. Figure 13.1 shows the symmetry operations of ammonia. Notice that there is a 120° counterclockwise rotation denoted

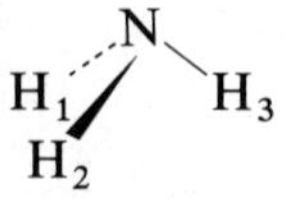

(3)

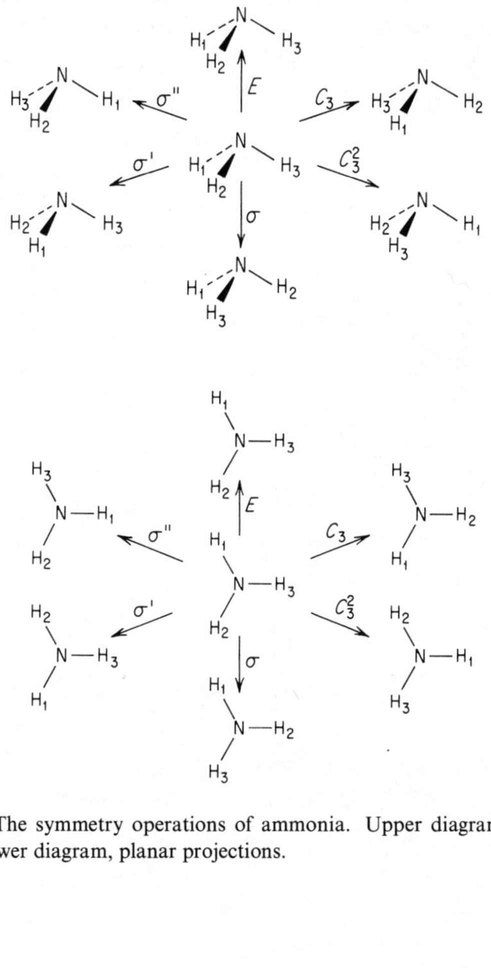

Figure 13.1 The symmetry operations of ammonia. Upper diagram, perspective drawings. Lower diagram, planar projections.

as C_3 and a 240° counterclockwise rotation denoted as C_3^2. This latter operation could also have been considered as a 120° clockwise rotation. There are also three planes of symmetry, each containing the rotation axis and one of the hydrogens. These are denoted σ_v's (v for vertical). The point group $\{E, C_3, C_3^2, \sigma_v, \sigma_v', \sigma_v''\}$ or $\{E, 2C_3, 3\sigma_v\}$ is called the $\mathbf{C}_{3v}$ point group.

Ethane, in its staggered conformation, **4**, has the inversion and the improper

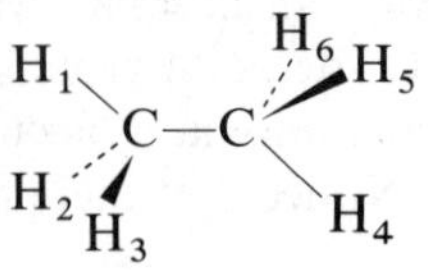

(4)

rotation among its symmetry operations. Overall, it has twelve symmetry operations, $\{E, 2C_3, 3C_2, i, 2S_6, 3\sigma_d\}$. The group is called $\mathbf{D}_{3d}$. One operation of each type is shown in Figure 13.2. The C_3, the inversion, and the σ_d are obvious. The S_6 involves a 60° counterclockwise rotation accompanied by a reflection in a plane perpendicular to the rotation axis. (In this case neither the 60° rotation nor the horizontal reflection is a symmetry operation of the molecule. An alternative description of S_6 is C_3^2 followed by i.) The C_2 axis bisects the pair of planes defined by H_1—C—C—H_4 and H_3—C—C—H_6 and is perpendicular to the C_3 (or S_6)

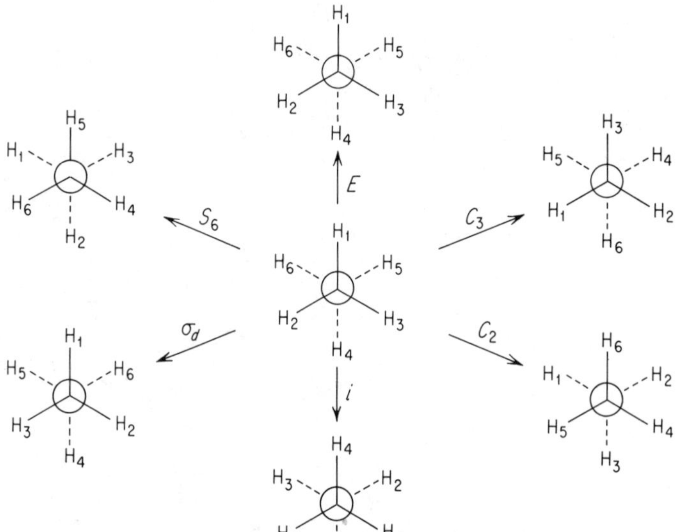

Figure 13.2 The results of one symmetry operation of each class on the staggered conformation of ethane. The Newman projection is used. The view is along the C-C axis, with the C—H bonds on the front atom drawn by a solid line and those on the back by a dashed line.

axis. The planes also contain the principal axis and bisect pairs of C_2 axes. They are given the special name, σ_d (for dihedral).

13.3 GENERATORS AND THE CLASSIFICATION ACCORDING TO POINT GROUPS

Each type of symmetry operation can define a group by itself. For example, consider the operations of $\mathbf{D}_{3d}$. If we have a C_3 operation, we must have all powers of C_3. In particular, we have C_3, C_3^2, and $C_3^3 \equiv E$. This set $\{E, C_3, C_3^2\}$ obeys all the group requirements. Similarly, we have i, $i^2 \equiv E$, for the group $\{E, i\}$; σ, $\sigma^2 \equiv E$ for the group $\{E, \sigma\}$; $S_6, S_6^2 \equiv C_3, S_6^3 = i, S_6^4 = C_3^2, S_6^5, S_6^6 \equiv E$ for the group $\{E, 2C_3, i, 2S_6\}$; and so on. These are all cyclic groups constructed from a single operation. Each is a subgroup of $\mathbf{D}_{3d}$. The individual *independent* operations that any given group can have are called the *generators* of the group. The cyclic groups formed by the generators are subgroups of the full group. The full group is a product of the subgroups. For example, ammonia has C_3 and σ_v as generators. These produce subgroups of orders three and two, respectively. The order of the full $\mathbf{C}_{3v}$ group is 3×2 or 6. Table 13.1 lists generators for all the common types of point groups.

Table 13.1

GENERATORS FOR THE FINITE POINT GROUPS[a]

Group Type	Group	Generators
Axial groups	$\mathbf{C}_1$	None
	$\mathbf{C}_i$	i
	$\mathbf{C}_s$	σ
	$\mathbf{C}_n$	C_n
	$\mathbf{S}_n$	S_n
	$\mathbf{D}_n$	C_n, $\perp C_2$
	$\mathbf{C}_{nv}$	C_n, $\parallel \sigma$
	$\mathbf{C}_{nh}$	C_n, $\perp \sigma$
	$\mathbf{D}_{nh}$	C_n, $\perp C_2$, $\perp \sigma$
	$\mathbf{D}_{nd}$	C_n, $\perp C_2$, $\parallel \sigma$ bisecting pairs of C_2's
Cubic groups	$\mathbf{T}$	C_3, non-$\perp C_2$
	$\mathbf{T}_d$	C_3, non-$\perp C_2$, σ
	$\mathbf{T}_h$	C_3, non-$\perp C_2$, i
	$\mathbf{O}$	C_3, non-$\perp C_4$
	$\mathbf{O}_h$	C_3, non-$\perp C_4$, i
Icosahedral groups	$\mathbf{I}$	C_3, non-$\perp C_5$
	$\mathbf{I}_h$	C_3, non-$\perp C_5$, i

[a] Note that for many of these groups the generators are not unique.

The nomenclature is the Schönflies notation commonly used by spectroscopists and theoreticians.

Assigning the appropriate point group to a system amounts to finding the generators for the group describing the system. Figure 13.3 presents a flow chart for systematically searching out the generators. Let us consider some examples. (Molecular models are helpful in visualizing the operations.) The structure of water

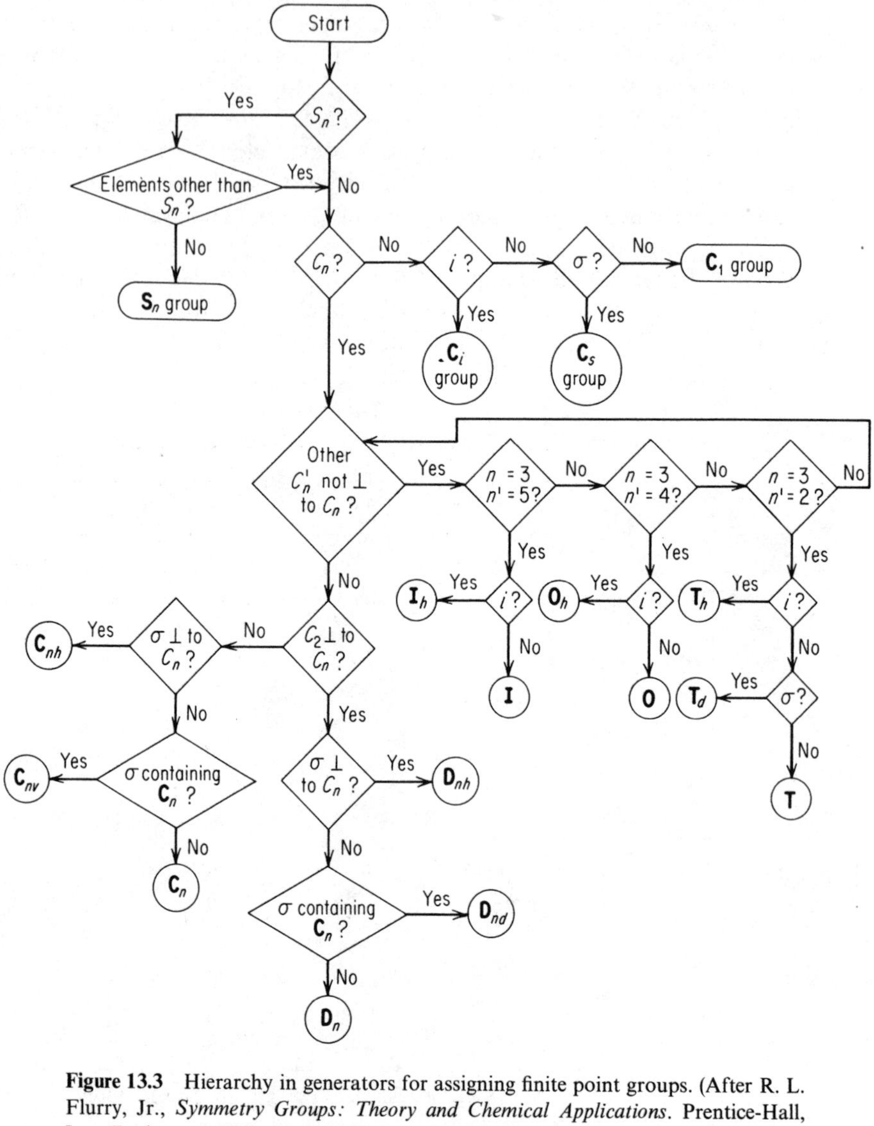

Figure 13.3 Hierarchy in generators for assigning finite point groups. (After R. L. Flurry, Jr., *Symmetry Groups: Theory and Chemical Applications*. Prentice-Hall, Inc., Englewood Cliffs, N. J., 1980. By permission.)

is shown in **5**. The symmetry elements for the two generators for water are fairly

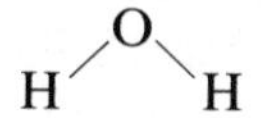

(5)

obvious. They are the C_2 axis, which bisects the two O—H bonds, and the molecular plane, which is a plane of symmetry. Proceeding down the flow chart, there are no S_n operations; there is a C_n with n equal to 2; there are no other C_n operations, either perpendicular or not. There is not a plane perpendicular to the C_2, but there is one containing the C_2. Consequently the group is $\mathbf{C}_{2v}$.

Let us verify the $\mathbf{D}_{3d}$ assignment that we previously made for staggered ethane. There is an S_6 operation, but there are other symmetry elements. There is a C_3, but in this case this is nothing new, since S_6^2 equals C_3. There are no nonperpendicular C_n's, but there are C_2's perpendicular to the C_3. There are no planes of symmetry perpendicular to the C_3, but there are planes of symmetry containing the C_3. Thus, the group is $\mathbf{D}_{3d}$.

As a final example, let us consider methane, **6**. There is an $\mathbf{S}_4$ present, which

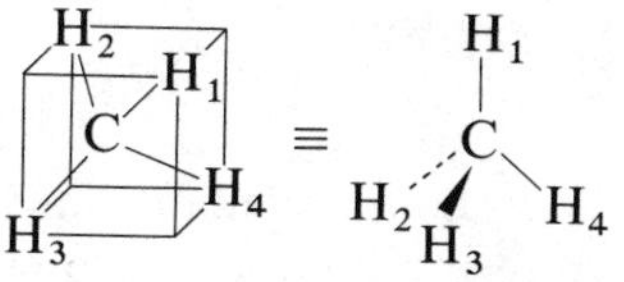

(6)

might be hard to visualize. If the axis bisects the C—H_1 and C—H_3 bonds, the operation takes hydrogen number 1 to number 2, number 2 to number 3, and so on. Other operations are present, however, so missing this one would cause no problems. There is a very obvious C_3 lying along each of the C—H bonds. There are also C_2's bisecting pairs of C—H bonds. These C_2's are not perpendicular to the C_3's. There is not a point of inversion, but there are planes of symmetry containing each pair of C—H bonds. Thus, the point group is $\mathbf{T}_d$. The point groups for other molecules and other physical objects can be assigned in a similar fashion.

13.4 WIGNER'S GRAND ORTHOGONALITY THEOREM

Wigner's *grand orthogonality theorem* provides the starting point for most of the applications we will make of group theory to chemistry. If R is an operation (or group element) of a group, $\mathbf{G}$, which has an order of g; if $\mathbf{R}^\Gamma$ is the Γth irreducible matrix representation, which has a dimension of l, of this group operation; and if $R^\Gamma_{\mu\nu}$ is an element of this matrix; then Wigner's grand orthogonality theorem states that

$$\sum_{R}^{\text{in } G} (R^{\Gamma}_{\mu\nu})^* R^{\Gamma'}_{\mu'\nu'} = \frac{g}{l}\,\delta_{\mu\mu'}\,\delta_{\nu\nu'}\,\delta_{\Gamma\Gamma'} \qquad \textbf{13.6}$$

In words, this says that the irreducible representations form an orthogonal set (from the $\delta_{\Gamma\Gamma'}$). Furthermore, within the same irreducible representation, either the rows (from the $\delta_{\mu\mu'}$) or the columns (from the $\delta_{\nu\nu'}$) of the representation matrices form an orthogonal set.

As an example of the utility of Eq. 13.6, let us set ν and ν' equal to μ and μ' and sum over μ. We get

$$\sum_{\mu}\sum_{R} (R^{\Gamma}_{\mu\mu})^* R^{\Gamma}_{\mu'\mu'} = \frac{g}{l}\sum_{\mu} \delta_{\mu\mu'}\,\delta_{\Gamma\Gamma'} \tag{13.7}$$

But $\sum_{\mu} \delta_{\mu\mu'}$ will equal the dimension of the matrix, l, and $\sum_{\mu} R^{\Gamma}_{\mu\mu}$ equals $\chi(R)_{\Gamma}$. Thus, Eq. 13.7 becomes

$$\sum_{R} \chi^*(R)_{\Gamma}\chi(R)_{\Gamma'} = g\,\delta_{\Gamma\Gamma'} \tag{13.8}$$

This is just the relation we used in Section 7.A to derive the equation for reducing a reducible representation of a finite group (Eq. 7.A12).

13.5 PROJECTION OPERATORS

We have, up to now, used the symbol χ in two contexts—to represent the character of an operation in a representation and to represent a generalized basis function. In the present section, and for some time to come, we will be using both characters and basis functions. To keep from confusing the two, we will adopt a new symbol, u, for basis functions. Thus, an L.C.A.O. molecular orbital will be written

$$\psi_i = \sum_{\mu} c_{i\mu} u_{\mu} \tag{13.9}$$

In Section 12.5 we discussed the coefficients of an L.C.A.O. expansion as vectors and matrices. If we let our basis set be expressed as a row vector, $\mathbf{u}$, we can also express our molecular orbitals as a row vector:

$$\boldsymbol{\psi} = \mathbf{uC} \tag{13.10}$$

Each element of the row vector $\boldsymbol{\psi}$ will be a linear combination of the u_{μ}, with the coefficients coming from the columns of $\mathbf{C}$. If we operate on the $\boldsymbol{\psi}$ vector with an operator, $\hat{R}$, corresponding to some operation, R, of the symmetry group of the system, we can express the result as a product of the $\boldsymbol{\psi}$ vector and the matrix representation, $\mathbf{R}$, of R:

$$\hat{R}\boldsymbol{\psi} = \boldsymbol{\psi}\mathbf{R} \tag{13.11}$$

Notice that $\boldsymbol{\psi}$, transforms *contravariantly* with respect to $\hat{R}$. That is, the matrix result must be written in the opposite order to the operator result. This is always the case when a row vector is being transformed by some operator. Otherwise, the result would not be a row vector.

Let us substitute Eq. 13.10 into the left-hand side of Eq. 13.11. We have

$$\hat{R}\mathbf{u}\mathbf{C} = \boldsymbol{\psi}\mathbf{R} \tag{13.12}$$

Now, multiply Eq. 13.12 from the right by $\mathbf{C}^{-1}$, the inverse of $\mathbf{C}$. This gives

$$\hat{R}\mathbf{u} = \boldsymbol{\psi}\mathbf{R}\mathbf{C}^{-1} \tag{13.13}$$

If we multiply Eq. 13.13 from the left by $(\mathbf{R}^{\Gamma})^*$, the complex conjugate of some irreducible matrix representation, sum over all R, and sum again over the diagonal elements of the matrix, we obtain

$$\sum_{\mu}\sum_{R}(R^{\Gamma}_{\mu\mu})^*\hat{R}\mathbf{u} = \sum_{\mu}\sum_{R}(R^{\Gamma}_{\mu\mu})^*\boldsymbol{\psi}R_{\mu\mu}\mathbf{C}^{-1} \tag{13.14}$$

or

$$\sum_{R}\chi^*(R)_{\Gamma}\hat{R}\mathbf{u} = g\boldsymbol{\psi}\mathbf{C}^{-1} \tag{13.15}$$

where we have used the fact that the $R_{\mu\mu}$ are constants (and thus commute with $\boldsymbol{\psi}$), the definition of a character, and Eq. 13.8. Equation 13.15 tells us that the operator, $\sum_{R}\chi^*(R)_{\Gamma}\hat{R}$, operating on the basis set, $\mathbf{u}$, produces a linear combination of symmetry-adapted wave functions. In terms of the individual basis functions, u_{μ}, and molecular orbitals, ψ_i, Eq. 13.15 can be written as

$$\sum_{R}\chi^*(R)_{\Gamma}\hat{R}u_{\mu} = g\sum_{i}c_i^{\Gamma}\psi_i^{\Gamma} \tag{13.16}$$

The operator is called a *projection operator*, $\hat{P}^{\Gamma}$:

$$\hat{P}^{\Gamma} = \sum_{R}\chi^*(R)_{\Gamma}\hat{R} \tag{13.17}$$

Equation 13.16 can be written in terms of the projection operator as

$$\hat{P}^{\Gamma}u_{\mu} = g\sum_{i}c_i^{\Gamma}\psi_i^{\Gamma} \equiv \lambda^{\Gamma} \tag{13.18}$$

Notice that the function, λ^{Γ}, as defined here is not normalized. It can be normalized with whatever approximations are being employed.

13.6 SYMMETRY-ADAPTED LINEAR COMBINATIONS OF BASIS FUNCTIONS

Equation 13.18 can be used to simplify the construction of linear combinations of basis functions when the system being studied has any symmetry associated with it. Let us again consider the π system of butadiene. The molecule has $\mathbf{C}_{2h}$ point symmetry; however, we will use only a subgroup that interchanges the basis functions,

$\mathbf{C}_2$. (The $\mathbf{C}_i$ subgroup could equally well be used, but the $\mathbf{C}_s$ subgroup could not, since it does not interchange the basis functions.) The character table for the $\mathbf{C}_2$ point group is shown in Table 13.2. We also need the result of operating on the basis functions by the operations of $\mathbf{C}_2$. This is shown in Table 13.3, using the numbering of **7**. Operating on the basis functions, in succession, by the projection operator for

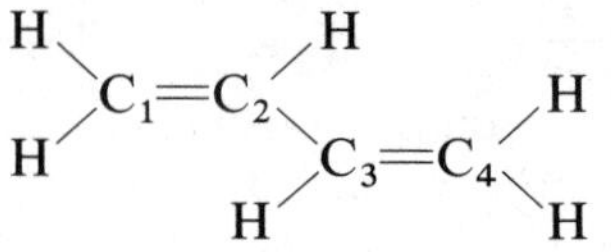

(7)

the A representation of $\mathbf{C}_2$, we obtain

$$\hat{P}^A u_1 = 1 \times Eu_1 + 1 \times C_2 u_1$$
$$= u_1 + u_4 \equiv \lambda_1^A \quad \textbf{13.19a}$$
$$\hat{P}^A u_2 = 1 \times Eu_2 + 1 \times C_2 u_2$$
$$= u_2 + u_3 \equiv \lambda_2^A \quad \textbf{13.19b}$$
$$\hat{P}^A u_3 = 1 \times Eu_3 + 1 \times C_2 u_3$$
$$= u_3 + u_2 = \lambda_2^A \quad \textbf{13.19c}$$
$$\hat{P}^A u_4 = 1 \times Eu_4 + 1 \times C_2 u_4$$
$$= u_4 + u_1 = \lambda_1^A \quad \textbf{13.19d}$$

Table 13.2

CHARACTER TABLE FOR THE C_2 POINT GROUP

$\mathbf{C}_2$	E	C_2
A	1	1
B	1	-1

Table 13.3

RESULTS OF OPERATING ON THE BUTADIENE π BASIS FUNCTIONS WITH THE OPERATIONS OF THE $\mathbf{C}_2$ POINT GROUP

Basis	E	C_2
u_1	u_1	u_4
u_2	u_2	u_3
u_3	u_3	u_2
u_4	u_4	u_1

while the projection operator for the B representation of $\mathbf{C}_2$ gives

$$\hat{P}^B u_1 = 1 \times Eu_1 + (-1) \times C_2 u_1$$
$$= u_1 - u_4 \equiv \lambda_1^B \qquad \textbf{13.20a}$$
$$\hat{P}^B u_2 = 1 \times Eu_2 + (-1) \times C_2 u_2$$
$$= u_2 - u_3 \equiv \lambda_2^B \qquad \textbf{13.20b}$$
$$\hat{P}^B u_3 = 1 \times Eu_3 + (-1) \times C_2 u_3$$
$$= u_3 - u_2 = -\lambda_2^B \qquad \textbf{13.20c}$$
$$\hat{P}^B u_4 = 1 \times Eu_4 + (-1) \times C_2 u_4$$
$$= u_4 - u_1 = -\lambda_1^B \qquad \textbf{13.20d}$$

We can verify that the results are linear combinations of the molecular orbitals by looking at the wave functions given in Eqs. 12.35. We see that

$$\lambda_1^A = .7435\psi_1 + 1.2030\psi_3 \qquad \textbf{13.21a}$$
$$\lambda_2^A = 1.2030\psi_1 - .7435\psi_3 \qquad \textbf{13.21b}$$
$$\lambda_1^B = 1.2030\psi_2 + .7435\psi_4 \qquad \textbf{13.21c}$$
$$\lambda_2^B = .7435\psi_2 - 1.2030\psi_4 \qquad \textbf{13.21d}$$

For our purposes, we are more interested in the fact that if λ_1^A and λ_2^A can be constructed as linear combinations of ψ_1 and ψ_3, then ψ_1 and ψ_3 can also be constructed as linear combinations of λ_1^A and λ_2^A. The same holds for the B functions. Thus, we can write

$$\psi_i^A = c_{i1}^A \lambda_1^A + c_{i2}^A \lambda_2^A \qquad \textbf{13.22a}$$
$$\psi_i^B = c_{i1}^B \lambda_1^B + c_{i2}^B \lambda_2^B \qquad \textbf{13.22b}$$

Since we know the form of λ_1^A, λ_2^A, λ_1^B, and λ_2^B, the variational problem now yields two 2×2 determinantal equations, rather than one 4×4. The 2×2 determinants are much easier to solve. (For comparison, in machine computation, the time required to diagonalize a matrix is roughly proportional to the square of the dimension of the matrix.)

Before constructing the secular determinants from Eqs. 13.22, let us normalize our λ functions. Although in this case this is not necessary, it often is, and usually it is more convenient. If we let

$$\lambda_1^A = N(u_1 + u_4) \qquad \textbf{13.23}$$

we have

$$\langle \lambda_1^A | \lambda_1^A \rangle = N^2 \langle u_1 + u_4 | u_1 + u_4 \rangle$$
$$= N^2 \{ \langle u_1 | u_1 \rangle + \langle u_4 | u_4 \rangle + 2 \langle u_1 | u_4 \rangle \}$$
$$= 2N^2 \qquad \textbf{13.24}$$

within the Hückel approximations. Thus

$$N = \frac{1}{\sqrt{2}} \tag{13.25}$$

Similarly, we find that the normalizing factor is also $1/\sqrt{2}$ for the other λ functions. We have, then,

$$\lambda_1^A = \frac{1}{\sqrt{2}}(u_1 + u_4) \tag{13.26a}$$

$$\lambda_2^A = \frac{1}{\sqrt{2}}(u_2 + u_3) \tag{13.26b}$$

$$\lambda_1^B = \frac{1}{\sqrt{2}}(u_1 - u_4) \tag{13.26c}$$

$$\lambda_2^B = \frac{1}{\sqrt{2}}(u_2 - u_3) \tag{13.26d}$$

The secular determinant arising from Eq. 13.22a is constructed in the same manner as any other secular determinant. The result is

$$\begin{vmatrix} \langle\lambda_1^A|\hat{h}|\lambda_1^A\rangle - \varepsilon_1^A & \langle\lambda_1^A|\hat{h}|\lambda_2^A\rangle - \varepsilon_i^A\langle\lambda_1^A|\lambda_2^A\rangle \\ \langle\lambda_1^A|\hat{h}|\lambda_2^A\rangle - \varepsilon_1^A\langle\lambda_1^A|\lambda_2^A\rangle & \langle\lambda_2^A|\hat{h}|\lambda_2^A\rangle - \varepsilon_i^A \end{vmatrix} = 0 \tag{13.27}$$

The corresponding linear equations are

$$c_{i1}^A(\langle\lambda_1^A|\hat{h}|\lambda_1^A\rangle - \varepsilon_i^A) + c_{i2}^A(\langle\lambda_1^A|\hat{h}|\lambda_2^A\rangle - \varepsilon_i^A\langle\lambda_1^A|\lambda_2^A\rangle) = 0 \tag{13.28a}$$

$$c_{i1}^A(\langle\lambda_1^A|\hat{h}|\lambda_2^A\rangle - \varepsilon_i^A\langle\lambda_1^A|\lambda_2^A\rangle) + c_{i2}^A(\langle\lambda_2^A|\hat{h}|\lambda_2^A\rangle - \varepsilon_i^A) = 0 \tag{13.28b}$$

Evaluating the integrals, we find, using the normalized λ's of Eqs. 13.26 and the Hückel approximations,

$$\begin{aligned} \langle\lambda_1^A|\hat{h}|\lambda_1^A\rangle &= \tfrac{1}{2}\langle u_1 + u_4|\hat{h}|u_1 + u_4\rangle \\ &= \tfrac{1}{2}(\langle u_1|\hat{h}|u_1\rangle + \langle u_4|\hat{h}|u_4\rangle + 2\langle u_1|\hat{h}|u_4\rangle) \\ &= \alpha \end{aligned} \tag{13.29a}$$

$$\begin{aligned} \langle\lambda_1^A|\hat{h}|\lambda_2^A\rangle &= \tfrac{1}{2}\langle u_1 + u_4|\hat{h}|u_2 + u_3\rangle \\ &= \tfrac{1}{2}(\langle u_1|\hat{h}|u_2\rangle + \langle u_1|\hat{h}|u_3\rangle + \langle u_4|\hat{h}|u_2\rangle + \langle u_4|\hat{h}|u_3\rangle) \\ &= \beta \end{aligned} \tag{13.29b}$$

$$\begin{aligned} \langle\lambda_2^A|\hat{h}|\lambda_2^A\rangle &= \tfrac{1}{2}\langle u_2 + u_3|\hat{h}|u_2 + u_3\rangle \\ &= \tfrac{1}{2}(\langle u_2|\hat{h}|u_2\rangle + \langle u_3|\hat{h}|u_3\rangle + 2\langle u_2|\hat{h}|u_3\rangle) \\ &= \alpha + \beta \end{aligned} \tag{13.29c}$$

$$\begin{aligned} \langle\lambda_1^A|\lambda_2^A\rangle &= \tfrac{1}{2}\langle u_1 + u_4|u_2 + u_3\rangle \\ &= \tfrac{1}{2}(\langle u_1|u_2\rangle + \langle u_1|u_3\rangle + \langle u_4|u_2\rangle + \langle u_4|u_3\rangle) \\ &= 0 \end{aligned} \tag{13.29d}$$

(The integrals that do not show up in the final results in each case vanish because of the Hückel approximations.) The secular determinant becomes

$$\begin{vmatrix} \alpha - \varepsilon_i^A & \beta \\ \beta & \alpha + \beta - \varepsilon_i^A \end{vmatrix} = 0 \qquad \textbf{13.30}$$

The corresponding linear equations are

$$c_{i1}^A(\alpha - \varepsilon_i^A) + c_{i2}^A\beta = 0 \qquad \textbf{13.31a}$$

$$c_{i1}^A\beta + c_{i2}^A(\alpha + \beta - \varepsilon_i^A) = 0 \qquad \textbf{13.31b}$$

Dividing through by β and, as before, defining x_i as $(\alpha - \varepsilon_i^A)/\beta$, Eq. 13.30 becomes

$$\begin{vmatrix} x_i & 1 \\ 1 & x_i + 1 \end{vmatrix} = 0 \qquad \textbf{13.32}$$

and Eqs. 13.31 become

$$c_{i1}^A x_i + c_{i2}^A = 0 \qquad \textbf{13.33a}$$

$$c_{i1}^A + c_{i2}^A(x_i + 1) = 0 \qquad \textbf{13.33b}$$

Expanding Eq. 13.32 yields the polynomial

$$x_i^2 + x_i - 1 = 0 \qquad \textbf{13.34}$$

The roots of this are

$$x_1 = -1.618 \qquad \textbf{13.35a}$$

$$x_2 = .618 \qquad \textbf{13.35b}$$

These are the first and third roots of the fourth-order polynomial given in Eq. 12.24. Substituting the first of these into Eqs. 13.33, we find that

$$c_{12}^A = 1.618c_{11}^A \qquad \textbf{13.36}$$

Using the normalization relation, we find that

$$c_{11}^A = .5258 \qquad \textbf{13.37a}$$

$$c_{12}^A = .8506 \qquad \textbf{13.37b}$$

and

$$\psi_1^A = .3718(u_1 + u_4) + .6015(u_2 + u_3) \qquad \textbf{13.38}$$

which is identical to Eq. 12.35a. Substituting in the other value of x yields

$$\psi_2^A = .6015(u_1 + u_4) - .3718(u_2 + u_3) \qquad \textbf{13.39}$$

which is identical to Eq. 12.35c.

The treatment of the B function, Eq. 13.22b, is identical. After evaluating the integrals over the symmetry-adapted functions (the λ's), we get (analogously to Eq. 13.30)

$$\begin{vmatrix} \alpha - \varepsilon_i^B & B \\ \beta & \alpha - \beta - \varepsilon_i^B \end{vmatrix} = 0 \qquad \textbf{13.40}$$

Making the substitution for the x_i gives

$$\begin{vmatrix} x_i & 1 \\ 1 & x_i - 1 \end{vmatrix} = 0 \tag{13.41}$$

or the polynomial

$$x_i^2 - x_i - 1 = 0 \tag{13.42}$$

The roots of this are

$$x_1 = -.618 \tag{13.43a}$$

$$x_2 = 1.618 \tag{13.43b}$$

These are the other two roots we found for Eq. 12.24. The resulting L.C.A.O. wave functions, obtained by substituting these roots into the appropriate linear equations, are

$$\psi_1^B = .6015(u_1 - u_4) + .3718(u_2 - u_3) \tag{13.44a}$$

$$\psi_2^B = .3718(u_1 - u_4) - .6015(u_2 - u_3) \tag{13.44b}$$

These are identical to Eqs. 12.35b and 12.35d, respectively.

13.7 SITE SYMMETRY, INTERCHANGE SYMMETRY, AND CORRELATION DIAGRAMS

In the preceding section we derived symmetry-adapted functions for the butadiene π-electron system by using a subgroup of the complete point-symmetry group. This is a perfectly legitimate way of obtaining this information; often, however, it is important, for any of a number of reasons, to classify symmetry-adapted functions according to the representations of the complete group, rather than just those of a subgroup. One obvious way of doing this is to use the projection operators of the full group and follow the procedure that we used with the subgroup. If we do this with the butadiene π system, we will project out the functions we have called λ_1^A and λ_2^A with the A_u projection operator of $\mathbf{C}_{2h}$ and λ_1^B and λ_2^B with the B_g projection operator. The A_g and B_u projection operators will give a zero when operating on any of the basis functions. Thus, the functions of Eqs. 13.38 and 13.39 should be designated as A_u functions, while those of Eqs. 13.44 should be designated as B_g functions.

With larger groups, it can rapidly become tedious to apply the projection operator for each irreducible representation to each nonequivalent basis function. An easier way of completely classifying the symmetry-adapted functions according to the full group is to use the concepts of *site symmetry* and *interchange symmetry*, along with *correlation diagrams* relating these to the full symmetry group.

In the preceding section we used the $\mathbf{C}_2$ subgroup of the $\mathbf{C}_{2h}$ point group. This subgroup is the simplest subgroup of $\mathbf{C}_{2h}$ that interchanges equivalent basis functions. We say that $\mathbf{C}_2$ is the *interchange-symmetry group* for these functions. Note that the order of the interchange group equals the number of equivalent functions being interchanged. The *site-symmetry group* is the group defined by the symmetry elements passing through the site under consideration. For the butadiene π system, the identity and the plane of symmetry pass through each atom. Thus, the site symmetry of each atom is $\mathbf{C}_s$. The full group is the product of the site-symmetry group and the interchange group. In other molecules there may be different positions that have different site and interchange symmetries. Depending upon the circumstances, either can be as small a group as $\mathbf{C}_1$ or as large as the full point group. In any case, each must be a subgroup of the full group (or the full group), and the product of each site-symmetry group and the corresponding interchange group must give the full group. Frequently, the interchange group may not be unique, as was the case in butadiene where the basis functions could be interchanged by either $\mathbf{C}_2$ or $\mathbf{C}_i$.

Correlation tables are obtained from *mappings* of groups onto subgroups. They tell what representations are obtained in the subgroup from a given representation of the full group, when the symmetry is reduced to that of the subgroup. Alternatively, they tell what representations of the full group can be constructed from specified representations of the subgroup, if the symmetry is increased to that of the full group. (This information comes from the *Frobenius reciprocity theorem.*)

To illustrate the construction of a correlation diagram, let us consider the correlations of $\mathbf{C}_s$ and $\mathbf{C}_2$ to $\mathbf{C}_{2h}$. The appropriate character tables are shown in Table 13.4, along with the mappings of the subgroups onto $\mathbf{C}_{2h}$. If we consider the mapping of $\mathbf{C}_s$, the site symmetry, onto $\mathbf{C}_{2h}$, we see that the identity of $\mathbf{C}_s$ maps onto the identity of $\mathbf{C}_{2h}$ and has a character of $+1$ for all representations of $\mathbf{C}_{2h}$. On the other hand, the σ of $\mathbf{C}_s$ maps onto the σ_h of $\mathbf{C}_{2h}$. Here the character is $+1$ for the A_g and B_u representations of $\mathbf{C}_{2h}$, but -1 for the A_u and B_g representations. The A'

Table 13.4

THE SITE-SYMMETRY ($\mathbf{C}_s$) AND THE INTERCHANGE-SYMMETRY ($\mathbf{C}_2$) GROUPS OF BUTADIENE AND THEIR MAPPINGS ONTO THE FULL GROUP ($\mathbf{C}_{2h}$)

$\mathbf{G}_S = \mathbf{C}_s$	E	σ
A'	1	1
A''	1	-1

$\mathbf{G}_I = \mathbf{C}_2$	E	C_2
A	1	1
B	1	-1

$\mathbf{G}_S = \mathbf{C}_s$	E			σ
$\mathbf{G}_I = \mathbf{C}_2$	E	C_2		
$\mathbf{G} = \mathbf{C}_{2h}$	E	C_2	i	σ
A_g	1	1	1	1
B_g	1	-1	1	-1
A_u	1	1	-1	-1
B_u	1	-1	-1	1

representation of $\mathbf{C}_s$ has a $+1$ under the reflection, while the A'' has a -1. Thus, the A_g and B_u representations of $\mathbf{C}_{2h}$ correlate with the A' of $\mathbf{C}_s$, while the A_u and B_g correlate with the A''. By similar arguments, we see that the A_g and A_u representations of $\mathbf{C}_{2h}$ correlate with the A representation of C_2, while the B_g and B_u correlate with the B. The correlation diagram is shown in Figure 13.4.

In order to use the correlation diagram, we must find the representations according to which the basis functions transform in the site-symmetry group. In the present case we are considering only the p-π basis functions on each center. These are antisymmetric with respect to the plane of symmetry; consequently they transform as A'' within the $\mathbf{C}_s$ group. Consulting Figure 13.4, we see that A'' correlates with the A_u and B_g representations of $\mathbf{C}_{2h}$. This means that these basis functions can lead only to functions of A_u or B_g symmetry in the $\mathbf{C}_{2h}$ point group. Continuing across the table, we see that the A_u representation arises from an A combination of the basis functions under the $\mathbf{C}_2$ interchange symmetry, while the B_g representation arises from a B combination. If we were working systematically, using the correlation diagram from the beginning, we would now apply the $\mathbf{C}_2$ projection operators, as we did in Section 13.6, to obtain the proper symmetry-adapted functions and then the energies and eigenfunctions.

Before we go on to additional examples, two comments on the use of site symmetry and interchange symmetry are in order. First, since the full group can be constructed as a product of the site-symmetry group and the interchange group, there are no generators common to the two groups. Thus, in determining the appropriate site-symmetry and interchange groups, if the generators and one subgroup are known, the other subgroup is automatically determined. For example, for butadiene, in the $\mathbf{C}_{2h}$ point group, the generators are C_2 and σ_h. The site-symmetry group contains the plane of symmetry as its generator; consequently the interchange group contains the C_2 as its generator. For most axial point groups it is as easy to determine one group as it is the other. The generators simply provide a check for the choice. For the cubic and icosahedral groups, however, the site-symmetry group is frequently easier to visualize than is the interchange group. In these cases the generators can be helpful.

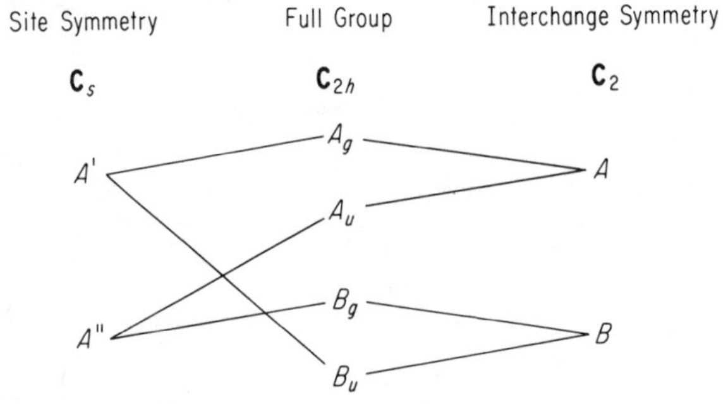

Figure 13.4 Correlation diagram relating the $\mathbf{C}_s$ and $\mathbf{C}_2$ subgroups to the $\mathbf{C}_{2h}$ group.

For example, the molecule, methane, **8**, has $\mathbf{T}_d$ point symmetry. The site sym-

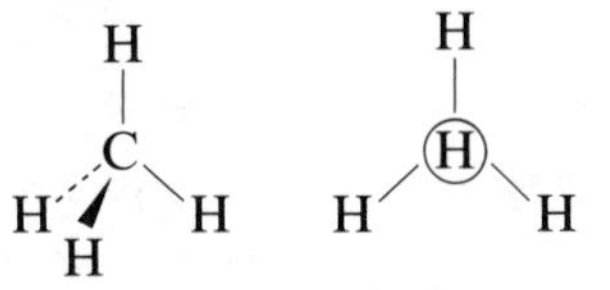

(8)

metry of the hydrogens is obviously $\mathbf{C}_{3v}$, as we see from the second projection. The $\mathbf{C}_{3v}$ group has as generators C_3 and σ. Consulting Table 13.1, we see that $\mathbf{T}_d$ also has a C_2 not perpendicular to the C_3 as a generator. This generator, and the operations generated by it, define the interchange group. In this case the C_3 operation operating on the C_2 produces two more C_2's perpendicular to the first one. Alternatively, the σ_d's that do not contain the C_2 produce S_4's. The appropriate interchange group is either $\mathbf{D}_2$ or $\mathbf{S}_4$. The $\mathbf{S}_4$, which contains only one generator, is usually preferred, since it produces results that are more easily interpreted.

The second comment on the use of site symmetry and interchange groups is actually a warning. Care must be taken, when mapping the site-symmetry and interchange groups onto the full point group, to make sure that the symmetry elements are properly aligned. Confusion frequently arises in the site-symmetry mappings when the full group is one, such as $\mathbf{D}_{4h}$ or $\mathbf{D}_{6h}$, that has more than one class of twofold axes perpendicular to the principal axis and also more than one class of symmetry planes containing the principal axis.

Consider, for example, the benzene molecule. The molecule is a planar hexagon, having $\mathbf{D}_{6h}$ point symmetry. Figure 13.5 shows one symmetry element of each class for a regular hexagon. In benzene, the carbon atoms lie at the vertices of the hexagon. These are intersected by the C_2', the σ_v, and the σ_h. The site-symmetry group is $\mathbf{C}_{2v}$. (Note that the σ_h of $\mathbf{D}_{6h}$ becomes a σ_v with respect to the site-symmetry group.) The generators for $\mathbf{D}_{6h}$ can be chosen as C_6, C_2, and σ. The C_2 and the σ are the generators of $\mathbf{C}_{2v}$; thus, the interchange group has C_6 as its only generator and is the $\mathbf{C}_6$ group.

The correlation diagram for $\mathbf{C}_6$ and $\mathbf{C}_{2v}$ with $\mathbf{D}_{6h}$ is shown in Figure 13.6. There is some arbitrariness in the labeling. This diagram results from the conventions

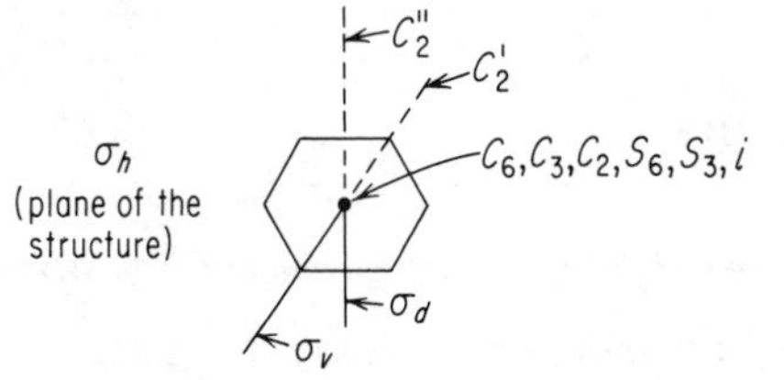

Figure 13.5 One symmetry element of each class for a $\mathbf{D}_{6h}$ structure. The C_2' and σ_v pair could equally well be interchanged with the C_2'' and σ_d pair. The individual C_2' and C_2'' or σ_v and σ_d could not be interchanged.

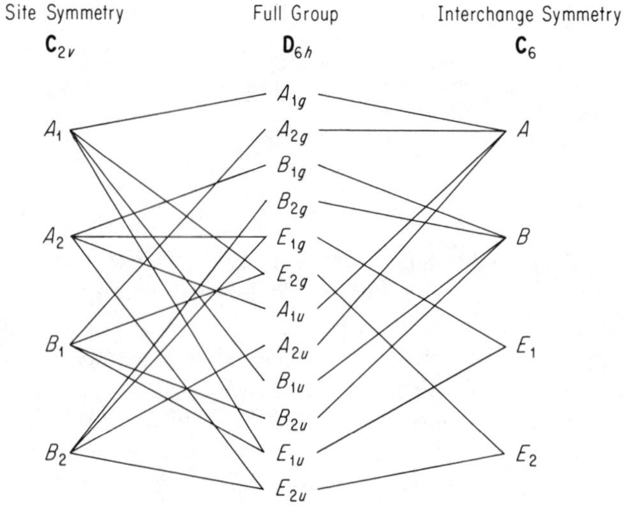

Figure 13.6 The correlation of the $\mathbf{C}_{2v}$ and $\mathbf{C}_6$ subgroups with the $\mathbf{D}_{6h}$ group.

shown in Figure 13.5, and with the σ_h of $\mathbf{D}_{6h}$ being identified with the σ_v of the $\mathbf{C}_{2v}$. The interchange of the orientations of C_2' and σ_v with those of C_2'' and σ_d in $\mathbf{D}_{6h}$ would produce a difference in correlations (the B_1 and B_2 labels of $\mathbf{D}_{6h}$ would be interchanged), as would assigning the σ_h of $\mathbf{D}_{6h}$ to the σ_v' of $\mathbf{C}_{2v}$ (the B_1 and B_2 labels of $\mathbf{C}_{2v}$ would be interchanged, along with the correlations to $\mathbf{D}_{6h}$). Although the labels can vary, the symmetry-adapted functions and any conclusions about the system drawn from symmetry would be unchanged, except for labels.

The final comment concerns the representations of the interchange group. Applying the operations of the interchange group to a member of the basis set produces a function transforming as the *regular representation* of the interchange group. (The regular representation is a reducible representation that contains each irreducible representation, Γ, n_Γ times, where n_Γ is the dimension of the representation Γ.) Thus we know immediately that each projection operator from the interchange group will produce a symmetry-adapted function from each unique basis function.

BIBLIOGRAPHY

CHESTNUT, D. B., *Finite Groups in Quantum Chemistry*. John Wiley & Sons, New York, 1974.

COTTON, F. A., *Chemical Applications of Group Theory*. John Wiley & Sons, New York, 2d ed. 1971.

FLURRY, R. L., JR., *Symmetry Groups: Theory and Chemical Applications*. Prentice-Hall, Inc., Englewood Cliffs, N. J., 1980.

HALL, L. H., *Group Theory and Symmetry in Chemistry*. McGraw-Hill Book Company, New York, 1968.

HOCHSTRASSER, R. M., *Molecular Aspects of Symmetry*. W. A. Benjamin, Inc., New York, 1966.

JAFFÉ, H. H., and ORCHIN, M., *Symmetry in Chemistry*. John Wiley & Sons, New York, 1965.

SCHONLAND, D., *Molecular Symmetry*. Van Nostrand Reinhold Co., New York, 1965.

PROBLEMS

***13.1** Find the point groups of the following molecules (molecular models will be helpful):

(a) Hydrogen peroxide (nonplanar),

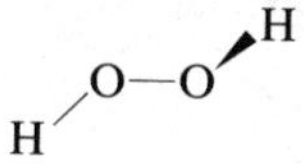

(b) Ethylene,

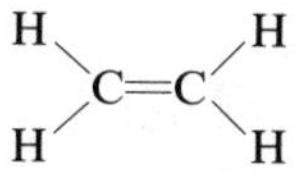

(c) *cis*-Dichloroethylene,

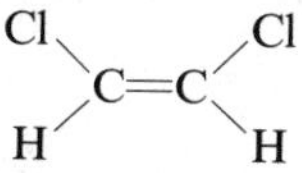

(d) *trans*-Dichloroethylene,

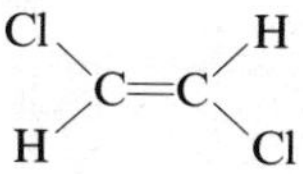

(e) Naphthalene,

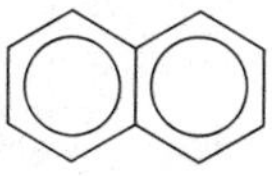

(f) Cyclohexane (chair conformation),

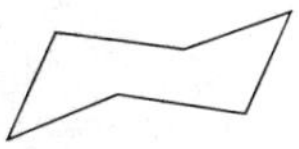

(g) Adamantane,

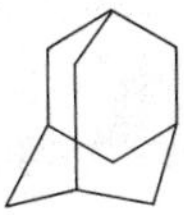

(h) 1-Chloroadamantane,

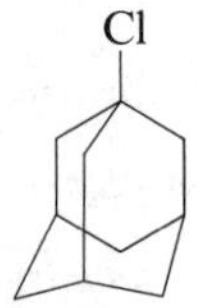

*13.2 Show, on the basis of symmetry arguments, that there are no one-electron integrals between the σ-electron and the π-electron systems of a planar molecule.

*13.3 What symmetry-induced restrictions are there on the two-electron integrals between the σ and π systems?

13.4 Including the lone-pair, n, orbitals, the few highest occupied and lowest unoccupied orbitals in triazine are: a_2'' and e'' (occupied π), a_1' and e' (occupied n), and e'' and a_2'' (vacant π). Which $\pi^* \leftarrow n$ and $\pi^* \leftarrow \pi$ transitions are symmetry-allowed and which are symmetry-forbidden? Give the polarization of each allowed transition.

13.5 Reduce the indicated representations in the indicated point groups. Use the character tables in Appendix 7.

$\mathbf{C}_{3v}$	E	$2C_3$	$3\sigma_v$
Γ_1	8	2	0
Γ_2	6	0	0
Γ_3	3	0	1
Γ_4	3	0	-1

$\mathbf{D}_{3d}$	E	$2C_3$	$3C_2$	i	$2S_6$	$3\sigma_d$
Γ_1	12	0	0	0	0	0
Γ_2	8	2	0	0	0	0
Γ_3	4	1	0	-4	-1	0

13.6 Show, by suitable drawings, that all twelve of the symmetry operations of staggered ethane ($\mathbf{D}_{3d}$ point group) can be generated by successive applications of a C_3, a C_2, and an inversion.

13.7 Construct the correlation diagrams for $\mathbf{C}_s$, $\mathbf{C}_i$, $\mathbf{C}_2$, $\mathbf{C}_3$, $\mathbf{S}_6$, $\mathbf{D}_3$, and $\mathbf{C}_{3v}$ with $\mathbf{D}_{3d}$.

13.8 Find the site symmetry and interchange symmetry for each unique atom in each molecule of Problem 13.1.

13.9 For each molecule in Problem 13.1, assume that each atom has available a set of basis functions transforming as the Cartesian coordinates on that atom. What would be the symmetries of the molecular functions that could be constructed from these?

13.10 Construct the functions of Problem 13.9 for ethylene. (The orientation of the basis functions must be such that equivalent functions are interchanged by the interchange group.)

chapter fourteen

Further examples from Hückel theory

14.1 BENZENE

Since we have already constructed the correlation diagram for benzene (Figure 13.6), let us carry through with the π-electron Hückel problem for benzene. The p-π atomic basis orbitals on each carbon atom are antisymmetric with respect to the σ_h of the $\mathbf{D}_{6h}$ point group of benzene, or to the σ_v of the $\mathbf{C}_{2v}$ site-symmetry group. They are also antisymmetric with respect to the site symmetry C_2, but symmetric with respect to the site symmetry σ_v' (σ_v of $\mathbf{D}_{6h}$). Thus, they transform as the B_2 representation of $\mathbf{C}_{2v}$. From Figure 13.6 we see that the B_2 representation of $\mathbf{C}_{2v}$ correlates with the B_{2g}, E_{1g}, A_{2u}, and E_{2u} representations of $\mathbf{D}_{6h}$. These, then, are the symmetries of the π molecular orbitals of benzene. They, in turn, correlate with the B, E_1, A, and E_2 representations of $\mathbf{C}_6$.

The character table for $\mathbf{C}_6$ is shown in Table 14.1, along with the results of operating on one of the basis functions, u_1, with the operations of $\mathbf{C}_6$. Note that in $\mathbf{C}_6$, as with all other cyclic groups, the E representations are separable into two complex one-dimensional representations. A cyclic group, $\mathbf{C}_n$, thus has n one-dimensional representations. The complex representations can always be combined to give two-dimensional real representations.

The symmetry-adapted combinations of the basis functions can be obtained by applying the projection operators from $\mathbf{C}_6$ to a basis function. For example, the a_{2u} molecular orbital arises from the A projection operator of $\mathbf{C}_6$:

$$\hat{P}^A u_1 = 1 \times u_1 + 1 \times u_2 + 1 \times u_3 + 1 \times u_4 + 1 \times u_5 + 1 \times u_6 \qquad \textbf{14.1}$$

Table 14.1

CHARACTER TABLE FOR THE $\mathbf{C}_6$ POINT GROUP[a,b]

$\mathbf{C}_6$	E	C_6	C_3	C_2	C_3^2	C_6^5
A	1	1	1	1	1	1
B	1	-1	1	-1	1	-1
E_1 {	1	ε	$-\varepsilon^*$	-1	$-\varepsilon$	ε^*
	1	ε^*	$-\varepsilon$	-1	$-\varepsilon^*$	ε
$(E_1^R$	2	1	-1	-2	-1	1)
E_2 {	1	$-\varepsilon^*$	$-\varepsilon$	1	$-\varepsilon^*$	$-\varepsilon$
	1	$-\varepsilon$	$-\varepsilon^*$	1	$-\varepsilon$	$-\varepsilon^*$
$(E_2^R$	2	-1	-1	2	-1	$-1)$
Ru_1	u_1	u_2	u_3	u_4	u_5	u_6

[a] The representations labeled E_1^R and E_2^R are the real forms of the two E representations, which are obtained by adding the two one-dimensional complex forms. A second real form, which is orthogonal to this, can be obtained by subtracting one complex form from the other.

[b] $\varepsilon = e^{2\pi i/6}$.

Normalizing, within the Hückel approximation, we have

$$\psi^{a_{2u}} = \frac{1}{\sqrt{6}}(u_1 + u_2 + u_3 + u_4 + u_5 + u_6) \qquad \textbf{14.2a}$$

Similarly, we find

$$\psi^{b_{2g}} = \frac{1}{\sqrt{6}}(u_1 - u_2 + u_3 - u_4 + u_5 - u_6) \qquad \textbf{14.2b}$$

$$\psi^{e_{1g}} = \frac{1}{\sqrt{12}}(2u_1 + u_2 - u_3 - 2u_4 - u_5 + u_6) \qquad \textbf{14.2c}$$

$$\psi^{e_{2u}} = \frac{1}{\sqrt{12}}(2u_1 - u_2 - u_3 + 2u_4 - u_5 - u_6) \qquad \textbf{14.2d}$$

(The real forms of the e functions are given.) In this case there is only one function of each symmetry (the e functions are each degenerate, and each has two components); consequently the problem is completely solved by the symmetry. The energies are

$$\varepsilon^{a_{2u}} = \langle\psi^{a_{2u}}|\hat{h}|\psi^{a_{2u}}\rangle$$
$$= \alpha + 2\beta \qquad \textbf{14.3a}$$

$$\varepsilon^{b_{2g}} = \langle\psi^{b_{2g}}|\hat{h}|\psi^{b_{2g}}\rangle$$
$$= \alpha - 2\beta \qquad \textbf{14.3b}$$

$$\varepsilon^{e_{1g}} = \langle \psi^{e_{1g}} | \hat{h} | \psi^{e_{1g}} \rangle$$
$$= \alpha + \beta \qquad \textbf{14.3c}$$

$$\varepsilon^{e_{2u}} = \langle \psi^{e_{2u}} | \hat{h} | \psi^{e_{2u}} \rangle$$
$$= \alpha - \beta \qquad \textbf{14.3d}$$

The order of the orbital energy levels is $\varepsilon^{a_{2u}} < \varepsilon^{e_{1g}} < \varepsilon^{e_{2u}} < \varepsilon^{b_{2g}}$. There are six electrons in the π system of benzene. The orbitals labeled by one-dimensional representations can accommodate two electrons, while the two-dimensional labeled (doubly degenerate) orbitals can accommodate four. The π-electron configuration for the ground state of benzene is $(a_{2u})^2(e_{1g})^4$. The ground-state energy, in the Hückel approximation, is

$$E_0 = 2\varepsilon^{a_{2u}} + 4\varepsilon^{e_{1g}}$$
$$= 6\alpha + 8\beta \qquad \textbf{14.4}$$

The energy of the first excited configuration is

$$E_1 = 2\varepsilon^{a_{2u}} + 3\varepsilon^{e_{1g}} + \varepsilon^{e_{2u}}$$
$$= 6\alpha + 6\beta \qquad \textbf{14.5}$$

The predicted first electronic transition is just the energy difference of these, or, more simply, the difference between $\varepsilon^{e_{1g}}$ and $\varepsilon^{e_{2u}}$:

$$\Delta E_1 = \varepsilon^{e_{2u}} - \varepsilon^{e_{1g}}$$
$$= -2\beta \qquad \textbf{14.6}$$

There are actually complications in the first excited state of benzene. There are two partially occupied degenerate orbitals. This leads to more than one state arising from the same configuration, just as was the case for polyelectron atoms with partially filled degenerate levels. In the present case, the representations of the states arising from the $(e_{1g})^3(e_{2u})$ configuration can be found by taking the direct product of the E_{1g} and E_{2u} representations [implicitly using the hole formalism for the $(e_{1g})^3$ substate]. The product can be obtained by multiplying the representations together, character by character, and then reducing the results, as was done in Section 7.4. However, there are rules, based upon the nomenclature, for multiplying the representations of the point groups. These are given in Table 14.2. We have

$$E_{1g} \times E_{2u} = B_{1u} + B_{2u} + E_{1u} \qquad \textbf{14.7}$$

There are three states arising from the $(e_{1g})^3(e_{2u})$ configuration of benzene. Each can have either a singlet or a triplet spin function associated with it, for a grand total of six states. These are all degenerate in the Hückel approximation, since electron repulsion is ignored. However, all are observed in the benzene spectrum. The three singlet transitions occur at $3.81 \times 10^4\ \text{cm}^{-1}$ (B_{2u}), $4.91 \times 10^4\ \text{cm}^{-1}$ (B_{1u}), and $5.59 \times 10^4\ \text{cm}^{-1}$ (E_{1u}).

Table 14.2

MULTIPLICATION PROPERTIES OF IRREDUCIBLE REPRESENTATIONS

General rules:
$A \times A = A$, $B \times B = A$, $A \times B = B$, $A \times E = E$, $B \times E = E$, $A \times T = T$, $B \times T = T$;
$g \times g = g$, $u \times u = g$, $u \times g = u$; $' \times ' = '$, $'' \times '' = '$, $' \times '' = ''$; $A \times E_1 = E_1$, $A \times E_2 = E_2$,
$B \times E_1 = E_2$, $B \times E_2 = E_1$

Subscripts on A or B:
$1 \times 1 = 1$, $2 \times 2 = 1$, $1 \times 2 = 2$, except for $\mathbf{D}_2$ and $\mathbf{D}_{2h}$, where $1 \times 2 = 3$, $2 \times 3 = 1$,
$1 \times 3 = 2$

Doubly degenerate representations:
For $\mathbf{C}_3$, $\mathbf{C}_{3h}$, $\mathbf{C}_{3v}$, $\mathbf{D}_3$, $\mathbf{D}_{3h}$, $\mathbf{D}_{3d}$, $\mathbf{C}_6$, $\mathbf{C}_{6v}$, $\mathbf{D}_6$, $\mathbf{D}_{6h}$, $\mathbf{S}_6$, $\mathbf{O}$, $\mathbf{O}_h$, $\mathbf{T}$, $\mathbf{T}_d$, $\mathbf{T}_h$:

$$E_1 \times E_1 = E_2 \times E_2 = A_1 + A_2 + E_2$$
$$E_1 \times E_2 = B_1 + B_2 + E_1$$

For $\mathbf{C}_4$, $\mathbf{C}_{4v}$, $\mathbf{C}_{4h}$, $\mathbf{D}_{2d}$, $\mathbf{D}_4$, $\mathbf{S}_4$: $E \times E = A_1 + A_2 + B_1 + B_2$
For groups in above lists that have symbols A, B, or E without subscripts; $A_1 = A_2 = A$, and so on.

Triply degenerate representations:
For $\mathbf{T}_d$, $\mathbf{O}$, $\mathbf{O}_h$: $E \times T_1 = E \times T_2 = T_1 + T_2$

$$T_1 \times T_1 = T_2 \times T_2 = A_1 + E + T_1 + T_2$$
$$T_1 \times T_2 = A_2 + E + T_1 + T_2$$

For $\mathbf{T}$, $\mathbf{T}_h$: Drop subscripts 1 and 2 from A and T.

Linear molecules ($\mathbf{C}_{\infty v}$ *and* $\mathbf{D}_{\infty h}$):

$$\Sigma^+ \times \Sigma^+ = \Sigma^- \times \Sigma^- = \Sigma^+;\ \Sigma^+ \times \Sigma^- = \Sigma^-$$
$$\Sigma^+ \times \Pi = \Sigma^- \times \Pi = \Pi;\ \Sigma^+ \times \Delta = \Sigma^- \times \Delta = \Delta; \quad \text{etc.}$$
$$\Pi \times \Pi = \Sigma^+ + \Sigma^- + \Delta$$
$$\Delta \times \Delta = \Sigma^+ + \Sigma^- + \Gamma$$
$$\Pi \times \Delta = \Pi + \Phi$$

or, in general, $\Gamma^{\lambda} \times \Gamma^{\lambda'} = \Gamma^{|\lambda - \lambda'|} + \Gamma^{(\lambda + \lambda')}$

SOURCE: E. B. Wilson, Jr., J. C. Decius, and P. C. Cross, *Molecular Vibration*. Dover Publications, Inc., New York, 1955, p. 331, by permission of E. B. Wilson.

The symmetry-induced spectral selection rules for the finite point groups are the same as they were for the rotation groups (Chapter 3). The product of the representations of the initial and final states must contain the representation of a component of the dipole operator. In the case of benzene, μ_z transforms as A_{2u}, while μ_x and μ_y together transform as E_{1u}. Benzene has a closed-shell, A_{1g}, ground state. Symmetry-allowed transitions can only be to A_{2u} or E_{1u} excited states. The fact that B_{1u} and B_{2u} excited electronic states are observed requires that vibrational excitations accompany the electronic transitions (see Section 16.10) so that the total symmetry of the excited state is either A_{2u} or E_{1u}.

14.2 CYCLOPROPENONE

As another example, let us consider the π system of cyclopropenone, **1**. We have two different kinds of atoms, carbon and oxygen, contributing to the π system

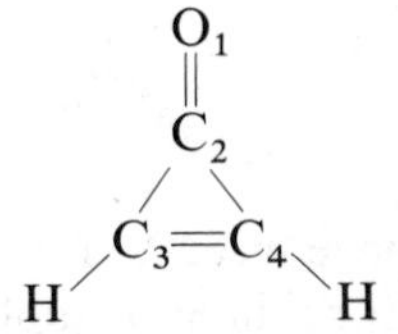

(**1**)

and two different site symmetries for the carbons. The point group of the molecule is $\mathbf{C}_{2v}$. Both generators of $\mathbf{C}_{2v}$ pass through the oxygen atom and carbon atom 2; consequently their site symmetry is the full point symmetry. A p-π orbital on either of these transforms as the B_2 representation, if the molecular plane is chosen as the σ_v. Only a plane of symmetry passes through atoms 3 and 4, so their site symmetry is $\mathbf{C}_s$. Their interchange symmetry is $\mathbf{C}_2$. The correlation diagram relating the $\mathbf{C}_s$ and $\mathbf{C}_2$ groups to $\mathbf{C}_{2v}$ is shown in Figure 14.1. The p-π basis orbitals transform as A'' within $\mathbf{C}_s$; thus, the symmetry-adapted functions arising from atoms 3 and 4 can have A_2 or B_2 symmetry within $\mathbf{C}_{2v}$, and they arise from A and B combinations, respectively, of the $\mathbf{C}_2$ interchange group. In this case the C_2 axis lies in the same plane as the carbon atoms on which the bases are located, and so the C_2 operation takes the basis function u_3 into $-u_4$, and vice versa.

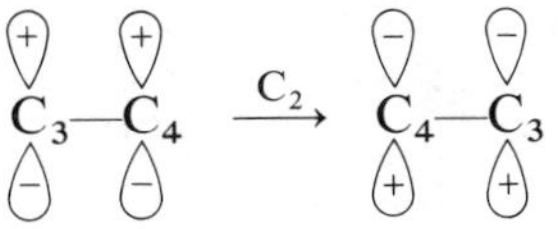

14.8

The A_2 combination of u_3 and u_4 arises from the A projection operator of $\mathbf{C}_2$:

$$\begin{aligned}\lambda^{A_2} &= \hat{P}^A u_3 \\ &= 1 \times u_3 + 1 \times (-u_4) \\ &= u_3 - u_4\end{aligned} \qquad 14.9$$

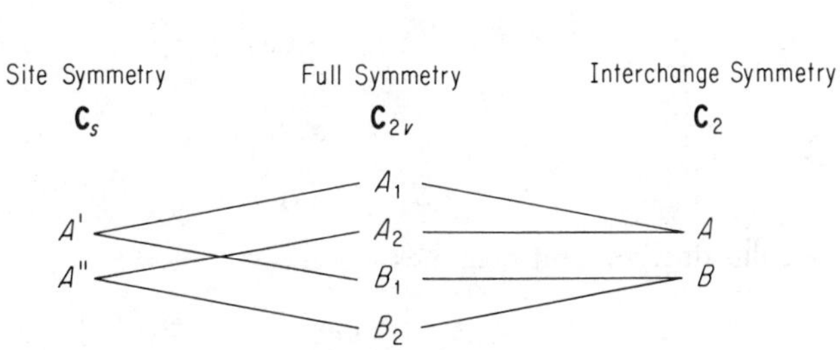

Figure 14.1 Correlation of the $\mathbf{C}_s$ and $\mathbf{C}_2$ subgroups to the $\mathbf{C}_{2v}$ point group. The molecular plane is the plane of the $\mathbf{C}_s$ subgroup and the σ_v of the $\mathbf{C}_{2v}$ group.

or normalizing, within the Hückel approximations,

$$\lambda^{A_2} = \frac{1}{\sqrt{2}}(u_3 - u_4) \qquad \textbf{14.10a}$$

Similarly

$$\lambda^{B_2} = \frac{1}{\sqrt{2}}(u_3 + u_4) \qquad \textbf{14.10b}$$

Overall, we now have three B_2 functions, the p-π orbitals of O_1 and C_2 and λ^{B_2} of Eq. 14.10. There will be three B_2 molecular orbitals, having the form

$$\psi_i^{b_2} = c_{i1}^{b_2}u_1 + c_{i2}^{b_2}u_2 + c_{i3}^{b_2}\lambda^{B_2} \qquad \textbf{14.11}$$

The resulting secular determinant is

$$\begin{vmatrix} \langle u_1|\hat{h}|u_1\rangle - \varepsilon_i & \langle u_1|\hat{h}|u_2\rangle - \varepsilon_i S_{12} & \langle u_1|\hat{h}|\lambda^{B_2}\rangle - \varepsilon_i S_{1\lambda} \\ \langle u_1|\hat{h}|u_2\rangle - \varepsilon_i S_{12} & \langle u_2|\hat{h}|u_2\rangle - \varepsilon_i & \langle u_2|\hat{h}|\lambda^{B_2}\rangle - \varepsilon_i S_{2\lambda} \\ \langle u_1|\hat{h}|\lambda^{B_2}\rangle - \varepsilon_i S_{1\lambda} & \langle u_2|\hat{h}|\lambda^{B_2}\rangle - \varepsilon_i S_{2\lambda} & \langle \lambda^{B_2}|\hat{h}|\lambda^{B_2}\rangle - \varepsilon_i \end{vmatrix} = 0 \qquad \textbf{14.12}$$

Evaluating the integrals, within the Hückel approximations, we have

$$\langle u_1|\hat{h}|u_1\rangle = \alpha_{\mathrm{O}} \qquad \textbf{14.13a}$$

$$\langle u_1|\hat{h}|u_2\rangle = \beta_{\mathrm{CO}} \qquad \textbf{14.13b}$$

$$\langle u_1|\hat{h}|\lambda^{B_2}\rangle = \frac{1}{\sqrt{2}}\langle u_1|\hat{h}|u_3 + u_4\rangle$$

$$= 0 \qquad \textbf{14.13c}$$

$$\langle u_2|\hat{h}|u_2\rangle = \alpha \qquad \textbf{14.13d}$$

$$\langle u_2|\hat{h}|\lambda^{B_2}\rangle = \frac{1}{\sqrt{2}}\langle u_2|\hat{h}|u_3 + u_4\rangle$$

$$= \sqrt{2}\,\beta \qquad \textbf{14.13e}$$

$$\langle \lambda^{B_2}|\hat{h}|\lambda^{B_2}\rangle = \frac{1}{2}\langle u_3 + u_4|\hat{h}|u_3 + u_4\rangle$$

$$= \alpha + \beta \qquad \textbf{14.13f}$$

$$S_{12} = S_{1\lambda} = S_{2\lambda} = 0 \qquad \textbf{14.13g}$$

The secular determinant becomes

$$\begin{vmatrix} \alpha_{\mathrm{O}} - \varepsilon_i & \beta_{\mathrm{CO}} & 0 \\ \beta_{\mathrm{CO}} & \alpha - \varepsilon_i & \sqrt{2}\,\beta \\ 0 & \sqrt{2}\,\beta & \alpha + \beta - \varepsilon_i \end{vmatrix} = 0 \qquad \textbf{14.14}$$

If we let h_{O} and k_{CO} both equal 1, as we did for acrolein, this becomes

$$\begin{vmatrix} \alpha + \beta - \varepsilon_i & \beta & 0 \\ \beta & \alpha - \varepsilon_i & \sqrt{2}\,\beta \\ 0 & \sqrt{2}\,\beta & \alpha + \beta - \varepsilon_i \end{vmatrix} = 0 \tag{14.15}$$

Dividing through by β and letting $(\alpha - \varepsilon_i)/\beta$ equal x_i, we have

$$\begin{vmatrix} x_i + 1 & 1 & 0 \\ 1 & x_i & \sqrt{2} \\ 0 & \sqrt{2} & x_i + 1 \end{vmatrix} = 0 \tag{14.16}$$

Expanding the determinant, we have

$$x(x + 1)^2 - 3(x + 1) = 0 \tag{14.17a}$$

or

$$(x + 1)(x^2 + x - 3) = 0 \tag{14.17b}$$

One root

$$x = -1 \tag{14.18a}$$

is immediately obvious. The other two are the roots of the quadratic equation in the second parentheses of Eq. 14.17b:

$$x = -2.303 \tag{14.18b}$$

$$x = 1.303 \tag{14.18c}$$

The b_2 energy levels are

$$\varepsilon_1^{b_2} = \alpha + 2.303\beta \tag{14.19a}$$

$$\varepsilon_2^{b_2} = \alpha + \beta \tag{14.19b}$$

$$\varepsilon_3^{b_2} = \alpha - 1.303\beta \tag{14.19c}$$

There is only one function having A_2 symmetry; thus the a_2 energy can be determined directly:

$$\begin{aligned} \varepsilon^{a_2} &= \tfrac{1}{2}\langle \lambda^{A_2} | \hat{h} | \lambda^{A_2} \rangle \\ &= \tfrac{1}{2}\langle u_3 - u_4 | \hat{h} | u_3 - u_4 \rangle \\ &= \alpha - \beta \end{aligned} \tag{14.20}$$

This lies between $\varepsilon_2^{b_2}$ and $\varepsilon_3^{b_2}$ in energy. There are four electrons in the π system of cyclopropenone. The first two b_2 levels will be doubly occupied, to give a ground-state energy

$$\begin{aligned} E_0 &= 2\varepsilon_1^{b_2} + 2\varepsilon_2^{b_2} \\ &= 4\alpha + 6.606\beta \end{aligned} \tag{14.21}$$

The first excited state will have one electron promoted to the a_2 orbital:

$$\begin{aligned} E_1 &= 2\varepsilon_1^{b_2} + \varepsilon_2^{b_2} + \varepsilon^{a_2} \\ &= 4\alpha + 4.606\beta \end{aligned} \quad \text{14.22}$$

The transition energy is predicted to be -2β.

The coefficients for the complete molecular orbitals can be found with the aid of the linear equations corresponding to Eq. 14.16. Omitting the superscript symmetry labels, we have

$$c_{i1}(x_i + 1) + c_{i2} = 0 \quad \text{14.23a}$$

$$c_{i1} + c_{i2}x_i + \sqrt{2}c_{i3} = 0 \quad \text{14.23b}$$

$$\sqrt{2}c_{i2} + c_{i3}(x_i + 1) = 0 \quad \text{14.23c}$$

Substituting in the lowest root for x, -2.303, we have

$$-1.303c_{11} + c_{12} = 0 \quad \text{14.24a}$$

$$c_{11} - 2.303c_{12} + \sqrt{2}c_{13} = 0 \quad \text{14.24b}$$

$$\sqrt{2}c_{12} - 1.303c_{13} = 0 \quad \text{14.24c}$$

The first of these gives

$$c_{12} = 1.303c_{11} \quad \text{14.25}$$

Substituting into Eq. 14.24b, we have

$$c_{11} - 3.000c_{11} + \sqrt{2}c_{13} = 0 \quad \text{14.26a}$$

or

$$c_{13} = \sqrt{2}c_{11} \quad \text{14.26b}$$

The Hückel normalizing condition is

$$\sum_{\mu} c_{i\mu}^2 = 1 \quad \text{14.27}$$

Using Eqs. 14.25 and 14.26 gives

$$\begin{aligned} c_{11}^2 + (1.303c_{11})^2 + (\sqrt{2}c_{11})^2 &= 1 \\ 4.698c_{11}^2 &= 1 \\ c_{11} &= .4614 \\ c_{12} &= .6012 \\ c_{13} &= .6525 \end{aligned} \quad \text{14.28}$$

Substituting these values of the coefficients, along with the definition of λ^{B_2} (Eq. 14.10b) into Eq. 14.11, we obtain

$$\psi_1^{b_2} = .4614u_1 + .6012u_2 + .4614(u_3 + u_4) \quad \text{14.29a}$$

Similarly, we find

$$\psi_2^{b_2} = .8165u_1 - .4082(u_3 + u_4) \qquad \textbf{14.29b}$$

$$\psi_3^{b_2} = .3470u_1 - .7991u_2 + .3470(u_3 + u_4) \qquad \textbf{14.29c}$$

The a_2 molecular orbital is, of course, just

$$\psi^{a_2} = .7071(u_3 - u_4) \qquad \textbf{14.29d}$$

14.3 TRIAZINE

As a third example of the application of Hückel theory, let us consider *s*-triazine, **2**.

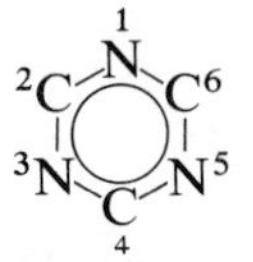

(**2**)

The molecule has $\mathbf{D}_{3h}$ point symmetry. All the nitrogens are equivalent, as are all the carbons. The site symmetry of the nitrogens is the same as that of the carbons: $\mathbf{C}_{2v}$. The interchange symmetry for each is $\mathbf{C}_3$. The correlation diagram is shown in Figure 14.2. The p-π orbitals again transform as the B_2 representation of $\mathbf{C}_{2v}$. Thus, both the carbon basis set and the nitrogen set give rise to an A_2'' and an E'' function, arising, respectively, from the A and E combinations from $\mathbf{C}_3$.

In constructing the symmetry-adapted functions for *s*-triazine there is one additional constraint. When two different sets of atoms have the same site symmetry, the projection operators from the interchange group must operate on basis functions from the two sets that lie on the same symmetry elements. Thus, if u_1 is chosen from

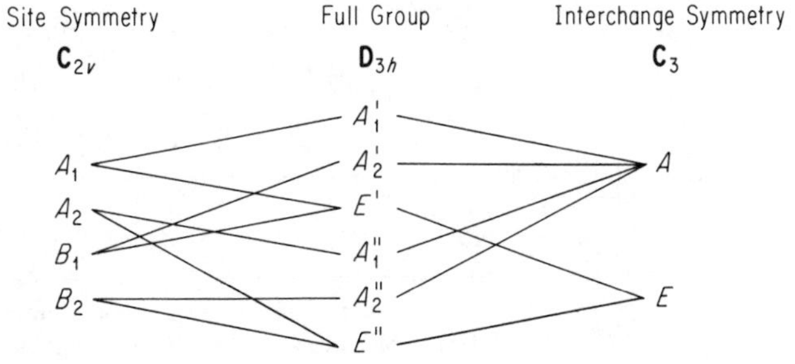

Figure 14.2 Correlation of the $\mathbf{C}_{2v}$ and $\mathbf{C}_3$ subgroups with the $\mathbf{D}_{3h}$ point group. The σ_v of $\mathbf{C}_{2v}$ is the σ_h of $\mathbf{D}_{3h}$.

the nitrogen set, u_4 must be chosen from the carbon set. (In our butadiene and cyclopropenone examples, the symmetry was low enough that this was satisfied automatically.) The $\mathbf{C}_3$ character table, along with the results of operating on u_1 and u_4 by the operations of $\mathbf{C}_3$, is shown in Table 14.3. The resulting normalized symmetry-adapted functions are

$$\lambda_1^{A_2''} = \frac{1}{\sqrt{3}}(u_1 + u_3 + u_5) \quad \textbf{14.30a}$$

$$\lambda_2^{A_2''} = \frac{1}{\sqrt{3}}(u_4 + u_6 + u_2) \quad \textbf{14.30b}$$

$$\lambda_1^{E''} = \frac{1}{\sqrt{6}}(2u_1 - u_3 - u_5) \quad \textbf{14.30c}$$

$$\lambda_2^{E''} = \frac{1}{\sqrt{6}}(2u_4 - u_6 - u_2) \quad \textbf{14.30d}$$

The molecular orbitals will have the form

$$\psi_i^{a_2''} = c_{i1}^{a_2''}\lambda_1^{A_2''} + c_{i2}^{a_2''}\lambda_2^{A_2''} \quad \textbf{14.31a}$$

$$\psi_i^{e''} = c_{i1}^{e''}\lambda_1^{E''} + c_{i2}^{e''}\lambda_2^{E''} \quad \textbf{14.31b}$$

Table 14.3

THE $\mathbf{C}_3$ CHARACTER TABLE, ALONG WITH THE RESULTS OF OPERATING ON u_1 AND u_4 OF *S*-TRIAZINE, **2**, WITH THE OPERATIONS OF THE GROUP[a,b]

$\mathbf{C}_3$	E	C_3	C_3^2
A	1	1	1
E {	1	ε	ε^*
	1	ε^*	ε
(E	2	-1	-1)
Ru_1	u_1	u_3	u_5
Ru_4	u_4	u_6	u_2

[a] The real form of the E representation shown in parentheses is the sum of the seperable complex representations.

[b] $\varepsilon = e^{2\pi i/3}$.

The resulting secular determinants are

$$\begin{vmatrix} H_{NN}^{a_2''} - \varepsilon_i & H_{NC}^{a_2''} \\ H_{NC}^{a_2''} & H_{CC}^{a_2''} - \varepsilon_i \end{vmatrix} = 0 \qquad \textbf{14.32}$$

$$\begin{vmatrix} H_{NN}^{e''} - \varepsilon_i & H_{NC}^{e''} \\ H_{NC}^{e''} & H_{CC}^{e''} - \varepsilon_i \end{vmatrix} = 0 \qquad \textbf{14.33}$$

The matrix elements are

$$\begin{aligned} H_{NN}^{a_2''} &= \langle \lambda_N^{A_2''} | \hat{h} | \lambda_N^{A_2''} \rangle \\ &= \tfrac{1}{3} \langle u_1 + u_3 + u_5 | \hat{h} | u_1 + u_3 + u_5 \rangle \\ &= \alpha_N \end{aligned} \qquad \textbf{14.34a}$$

$$\begin{aligned} H_{NC}^{a_2''} &= \langle \lambda_N^{A_2''} | \hat{h} | \lambda_C^{A_2''} \rangle \\ &= \tfrac{1}{3} \langle u_1 + u_3 + u_5 | \hat{h} | u_4 + u_6 + u_2 \rangle \\ &= 2\beta_{CN} \end{aligned} \qquad \textbf{14.34b}$$

$$\begin{aligned} H_{CC}^{a_2''} &= \langle \lambda_C^{A_2''} | \hat{h} | \lambda_C^{A_2''} \rangle \\ &= \tfrac{1}{3} \langle u_4 + u_6 + u_2 | \hat{h} | u_4 + u_6 + u_2 \rangle \\ &= \alpha \end{aligned} \qquad \textbf{14.34c}$$

$$\begin{aligned} H_{NN}^{e''} &= \langle \lambda_N^{E''} | \hat{h} | \lambda_N^{E''} \rangle \\ &= \tfrac{1}{6} \langle 2u_1 - u_3 - u_5 | \hat{h} | 2u_1 - u_3 - u_5 \rangle \\ &= \alpha_N \end{aligned} \qquad \textbf{14.35a}$$

$$\begin{aligned} H_{NC}^{e''} &= \langle \lambda_N^{E''} | \hat{h} | \lambda_C^{E''} \rangle \\ &= \tfrac{1}{6} \langle 2u_1 - u_3 - u_5 | \hat{h} | 2u_4 - u_6 - u_2 \rangle \\ &= -\beta_{CN} \end{aligned} \qquad \textbf{14.35b}$$

$$\begin{aligned} H_{CC}^{e''} &= \langle \lambda_C^{E''} | \hat{h} | \lambda_C^{E''} \rangle \\ &= \tfrac{1}{6} \langle 2u_4 - u_6 - u_2 | \hat{h} | 2u_4 - u_6 - u_2 \rangle \\ &= \alpha \end{aligned} \qquad \textbf{14.35c}$$

The Hückel parameters for the nitrogens are

$$h_N = .5 \qquad \textbf{14.36a}$$

$$k_{CN} = 1 \qquad \textbf{14.36b}$$

The A_2'' determinant becomes

$$\begin{vmatrix} \alpha + .5\beta - \varepsilon_i & 2\beta \\ 2\beta & \alpha - \varepsilon_i \end{vmatrix} = 0 \qquad \textbf{14.37}$$

while the E'' determinant becomes

$$\begin{vmatrix} \alpha + .5\beta - \varepsilon_i & -\beta \\ -\beta & \alpha - \varepsilon_i \end{vmatrix} = 0 \qquad \textbf{14.38}$$

The roots of Eq. 14.37 are

$$\varepsilon_1^{a_2''} = \alpha + 2.266\beta \qquad \textbf{14.39a}$$

$$\varepsilon_2^{a_2''} = \alpha - 1.766\beta \qquad \textbf{14.39b}$$

Those of Eq. 14.38 are

$$\varepsilon_1^{e''} = \alpha + 1.281\beta \qquad \textbf{14.40a}$$

$$\varepsilon_2^{e''} = \alpha - .781\beta \qquad \textbf{14.40b}$$

Substituting the roots into the linear equations, we obtain the molecular orbitals

$$\begin{aligned} \psi_1^{a_2''} &= .7494\lambda_N^{A_2''} + .6620\lambda_C^{A_2''} \\ &= .4328(u_1 + u_3 + u_5) + .3822(u_2 + u_4 + u_6) \end{aligned} \qquad \textbf{14.41a}$$

$$\begin{aligned} \psi_2^{a_2''} &= .6620\lambda_N^{A_2''} - .7494\lambda_C^{A_2''} \\ &= .3822(u_1 + u_3 + u_5) - .4328(u_2 + u_4 + u_6) \end{aligned} \qquad \textbf{14.41b}$$

$$\begin{aligned} \psi_1^{e''} &= .7880\lambda_N^{E''} - .6156\lambda_C^{E''} \\ &= .3217(2u_1 - u_3 - u_5) + .2513(u_2 - 2u_4 + u_6) \end{aligned} \qquad \textbf{14.42a}$$

$$\begin{aligned} \psi_2^{e''} &= .6156\lambda_N^{E''} + .7880\lambda_C^{E''} \\ &= .2513(2u_1 - u_3 - u_5) - .3217(u_2 - 2u_4 + u_6) \end{aligned} \qquad \textbf{14.42b}$$

In these equations the order of occurrence of the basis functions in the λ_C's has been changed to numerical order. Also, only one component of the doubly degenerate e'' orbitals is listed.

The π-electron system of *s*-triazine contains six electrons. In the ground state, two of these would go into $\psi_1^{a_2''}$ and four into the doubly degenerate $\psi_1^{e''}$:

$$\begin{aligned} E_0 &= 2\varepsilon_1^{a_2''} + 4\varepsilon_1^{e''} \\ &= 6\alpha + 9.656\beta \end{aligned} \qquad \textbf{14.43}$$

The first $\pi^* \leftarrow \pi$ excited state would involve a promotion of an electron from $\psi_1^{e''}$ to $\psi_1^{e''}$. At the Hückel level of approximation the transition energy is

$$\Delta E_1 = -2.062\beta \qquad \textbf{14.44}$$

However, just as in benzene, the first excited configuration involves two partially occupied degenerate orbitals. The resulting states are A_1', A_2', and E'. The A_1' singlet state is observed at 4.40×10^4 cm^{-1} above the ground state. This is a symmetry-forbidden electronic transition, which must be made allowed by vibronic (vibrational-electronic) coupling. There is also an additional complication in the spectrum of *s*-triazine. Each nitrogen has an unshared pair of electrons that is a part of the σ

system, referred to as *nonbonding* (or n) electrons. The first few observed electronic spectral transitions are actually $\pi^* \leftarrow n$ transitions. These cannot be readily calculated by classical Hückel theory, since it considers only π electrons.

14.4 PORPHINE

The porphine dianion, **3**, is large enough to present some additional problems. The

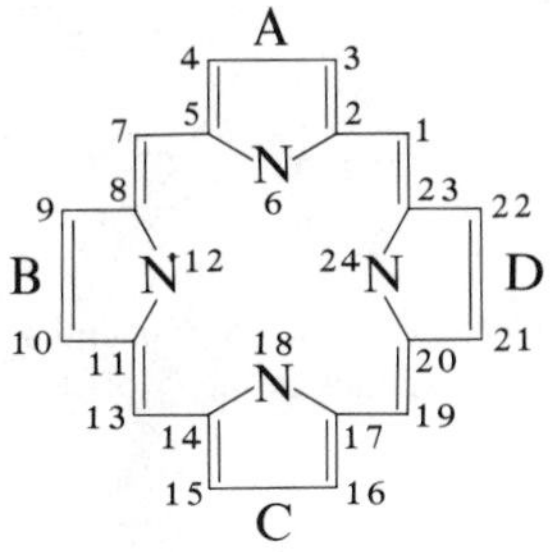

(**3**)

molecule has $\mathbf{D}_{4h}$ point symmetry. (The neutral porphine molecule has protons on two of the nitrogen atoms.) The fragment defined by atoms numbered 1 through 6 is repeated four times in a symmetrical pattern. The labels A, B, C, and D in the drawing indicate the repeating units.

The new complication arises from the fact that there are two sets of atoms with the same site symmetry, which cannot share all the same site-symmetry elements. The sets are carbon atoms 1, 7, 13, and 19 and nitrogen atoms 6, 12, 18, and 24. Each set has $\mathbf{C}_{2v}$ site symmetry, but there is no C_2 axis that passes through both a member of the carbon set and a member of the nitrogen set. This does not present any problems in determining the symmetries of the molecular orbitals, but it does in the construction of the degenerate symmetry-adapted orbitals. There are several ways around the problem. One is to find the site symmetry of the complete repeating unit in the molecule, to adapt all the unique basis functions to this site symmetry, and finally to apply the interchange symmetry to the functions obtained for the repeating unit to find the overall symmetry-adapted functions.

Let the nitrogen set define the first $\mathbf{C}_{2v}$ site symmetry in porphine. The interchange symmetry is $\mathbf{C}_4$. If the C_2 axis of the site symmetry group is aligned with one of the C_2' axes of the $\mathbf{D}_{4h}$ group, the correlation diagram is as in Figure 14.3. The p-π basis function on the nitrogen transforms as B_2 within $\mathbf{C}_{2v}$. We see that the nitrogens can contribute to molecular orbitals of E_g, A_{2u}, or B_{2u} symmetry within $\mathbf{D}_{4h}$. Carbon atom 1 (or any of the same set) also has $\mathbf{C}_{2v}$ site symmetry. However, if the nitrogens lie on the C_2' elements, these carbons must lie on the C_2'' elements. The interchange symmetry can again be $\mathbf{C}_4$. The correlation diagram for this orientation of $\mathbf{C}_{2v}$ and for $\mathbf{C}_4$ with $\mathbf{D}_{4h}$ is given in Figure 14.4. The basis function again

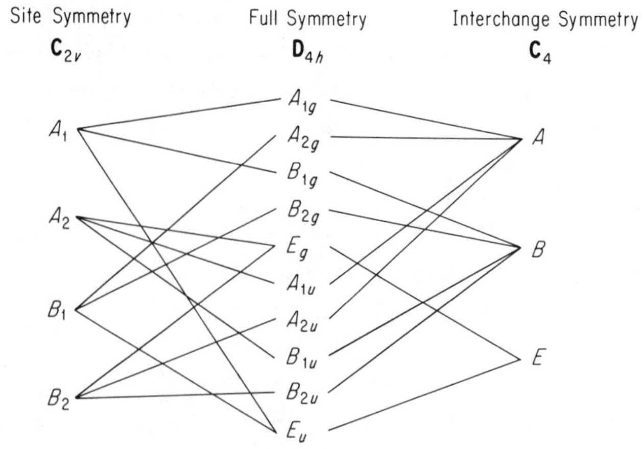

Figure 14.3 Correlation of the $\mathbf{C}_{2v}$ and $\mathbf{C}_4$ subgroups with the $\mathbf{D}_{4h}$ point group. The σ_v of $\mathbf{C}_{2v}$ corresponds to the σ_h of $\mathbf{D}_{4h}$; the C_2 of $\mathbf{C}_{2v}$ to the C_2' of $\mathbf{D}_{4h}$.

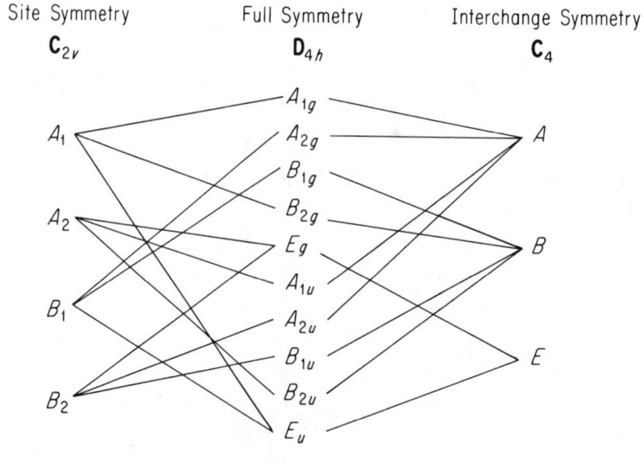

Figure 14.4 Another correlation of the $\mathbf{C}_{2v}$ and $\mathbf{C}_4$ subgroups with the $\mathbf{D}_{4h}$ point group. The C_2 of $\mathbf{C}_{2v}$ is now the C_2'' of $\mathbf{D}_{4h}$. The σ_v of $\mathbf{C}_{2v}$ is the σ_h of $\mathbf{D}_{4h}$.

transforms as B_2 within $\mathbf{C}_{2v}$. This time the correlation is with molecular orbitals of E_g, A_{2u}, and B_{1u} symmetry.

All the other atoms have the plane of the molecule as their only symmetry element. They share $\mathbf{C}_s$ as their common site symmetry. There are two sets—those corresponding to carbon atom 2 and its symmetry-related atoms, and those corresponding to carbon atom 3 and its related set. The interchange group for these is $\mathbf{D}_4$. The correlation diagram is shown in Figure 14.5. The basis functions have A'' symmetry within the $\mathbf{C}_s$ site-symmetry group. From the correlation diagram, we

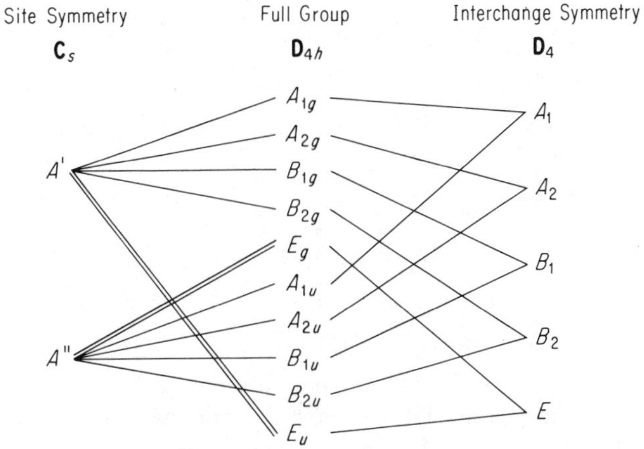

Figure 14.5 Correlation of the $\mathbf{C}_s$ ($\sigma = \sigma_h$ of $\mathbf{D}_{4h}$) and $\mathbf{D}_4$ subgroups with the $\mathbf{D}_{4h}$ point group.

see that each set contributes to two E_g functions (since, from $\mathbf{D}_{4h}$, each E_g representation reduces to two A'', when mapped onto $\mathbf{C}_s$) and one each of A_{1u}, A_{2u}, B_{1u}, and B_{2u}. Overall, there will be six E_g, two A_{1u}, four A_{2u}, three B_{1u}, and three B_{2u} molecular orbitals. There will be a 6×6, a 2×2, a 4×4, and two 3×3 determinants required in the solution of the problem. This will obviously be much simpler than the solution of a 24×24 determinant, as would be required without symmetry adaptation.

Let us work, first, with the degenerate E_g functions. The site symmetry of the repeating unit is $\mathbf{C}_{2v}$. The interchange symmetry for the repeating unit is $\mathbf{C}_4$. The repeating unit must be chosen sufficiently large that all basis functions are generated by the operations of the interchange group acting on it, and such that the operators of the site-symmetry group generate functions outside the set. In the present case we will use the atoms in the A–D quadrant of the molecule. Within the repeating unit, the nitrogen has $\mathbf{C}_{2v}$ site symmetry and $\mathbf{C}_1$ interchange symmetry, while all the carbons have $\mathbf{C}_s$ site symmetry and $\mathbf{C}_2$ interchange symmetry.

We can use the correlation diagram given in Figure 14.1 for our carbon functions. Atoms 1, 2, 3, 22, and 23 (and their symmetry-related partners) each can give rise to A_2 and B_2 functions within $\mathbf{C}_{2v}$. We have, for the A–D unit (the plus signs correspond to the B_2 set),

$$\lambda_1 = \frac{1}{\sqrt{2}}(u_1 \pm u_7) \qquad \textbf{14.45a}$$

$$\lambda_2 = \frac{1}{\sqrt{2}}(u_2 \pm u_5) \qquad \textbf{14.45b}$$

$$\lambda_3 = \frac{1}{\sqrt{2}}(u_3 \pm u_4) \qquad \textbf{14.45c}$$

$$\lambda_4 = \frac{1}{\sqrt{2}}(u_{23} \pm u_8) \qquad \textbf{14.45d}$$

$$\lambda_5 = \frac{1}{\sqrt{2}}(u_{22} \pm u_9) \qquad \textbf{14.45e}$$

$$\lambda_6 = u_6 \quad (B_2 \text{ only}) \qquad \textbf{14.45f}$$

Either the A_2 set or the B_2 set can give rise to E_g molecular functions. The two different sets yield the orthogonal components of the E_g functions. For a given set of molecular functions, the same representation of the site-symmetry group must be used to construct the entire set. The choice of which representation to use can be made on the basis of the function on the nitrogen. The function u_6 on the nitrogen transforms as the B_2 representation of $\mathbf{C}_{2v}$. Thus, the B_2 functions from Eqs. 14.45 (+ signs) must be used.

The correlation diagram for $\mathbf{C}_{2v}$ and $\mathbf{C}_4$ with $\mathbf{D}_{4h}$ is given in Figure 14.3. We see that the B_2 functions of $\mathbf{C}_{2v}$ lead to e_g, a_{2u}, and b_{2u} molecular orbitals. Here, we are interested only in the e_g functions. The character table for the interchange group, $\mathbf{C}_4$, is shown in Table 14.4, along with the results of operating on the basis functions with the operations of $\mathbf{C}_4$. The symmetry-adapted functions, within $\mathbf{D}_{4h}$, can be constructed by operating on the functions of Eqs. 14.45 with the projection operators of $\mathbf{C}_4$. For the e_g functions we have, in general,

$$\Lambda_i^{E_g} = \hat{P}_{\mathbf{C}_4}^{E}\lambda_i \qquad \textbf{14.46}$$

Table 14.4

CHARACTER TABLE FOR $\mathbf{C}_4$, ALONG WITH THE RESULTS OF OPERATING ON THE UNIQUE BASIS FUNCTIONS OF THE STRUCTURE, **3**, WITH THE OPERATIONS OF $\mathbf{C}_4$

$\mathbf{C}_4$	E	C_4	C_2	C_4^3
A	1	1	1	1
B	1	-1	1	-1
E {	1	i	-1	$-i$
	1	$-i$	-1	i
$(E^R$	2	0	-2	$0)$
Ru_1	u_1	u_7	u_{13}	u_{19}
Ru_2	u_2	u_8	u_{14}	u_{20}
Ru_3	u_3	u_9	u_{15}	u_{21}
Ru_6	u_6	u_{12}	u_{18}	u_{24}
Ru_{22}	u_{22}	u_4	u_{10}	u_{16}
Ru_{23}	u_{23}	u_5	u_{11}	u_{17}

The results are

$$\Lambda_1^{E_g} = \frac{1}{2}(u_1 + u_7 - u_{13} - u_{19}) \qquad \textbf{14.47a}$$

$$\Lambda_2^{E_g} = \frac{1}{2}(u_2 + u_5 - u_{14} - u_{17}) \qquad \textbf{14.47b}$$

$$\Lambda_3^{E_g} = \frac{1}{2}(u_3 + u_4 - u_{15} - u_{16}) \qquad \textbf{14.47c}$$

$$\Lambda_4^{E_g} = \frac{1}{2}(u_8 - u_{11} - u_{20} + u_{23}) \qquad \textbf{14.47d}$$

$$\Lambda_5^{E_g} = \frac{1}{2}(u_9 - u_{10} - u_{21} + u_{22}) \qquad \textbf{14.47e}$$

$$\Lambda_6^{E_g} = \frac{1}{\sqrt{2}}(u_6 - u_{18}) \qquad \textbf{14.47f}$$

These are our suitably symmetry-adapted functions, from which our 6×6 determinant can be constructed. Our e_g wave function has the form

$$\psi_i^{e_g} = c_{i1}\Lambda_1^{E_g} + c_{i2}\Lambda_2^{E_g} + c_{i3}\Lambda_3^{E_g} + c_{i4}\Lambda_4^{E_g} + c_{i5}\Lambda_5^{E_g} + c_{i6}\Lambda_6^{E_g} \qquad \textbf{14.48}$$

If the A_2 functions of Eqs. 14.45 had been used, the orthogonal components of Eqs. 14.47a–e would have been obtained. The orthogonal component of Eq. 14.47f would be required with these.

The nondegenerate functions have no arbitrariness associated with them. They can be constructed either from the group functions or from a site-symmetry treatment acting on the individual basis functions. Using the latter approach, we have, for the a_{1u} functions, including normalization:

$$\Lambda_1^{A_{1u}} = \hat{P}_{D_4}^{A_1} u_2$$

$$= \frac{1}{\sqrt{8}}(u_2 - u_5 + u_8 - u_{11} + u_{14} - u_{17} + u_{20} - u_{23}) \qquad \textbf{14.49a}$$

$$\Lambda_2^{A_{1u}} = \hat{P}_{D_4}^{A_1} u_3$$

$$= \frac{1}{\sqrt{8}}(u_3 - u_4 + u_9 - u_{10} + u_{15} - u_{16} + u_{21} - u_{22}) \qquad \textbf{14.49b}$$

For the a_{2u} functions, we get

$$\Lambda_1^{A_{2u}} = \hat{P}_{C_4}^{A} u_1$$

$$= \frac{1}{2}(u_1 + u_7 + u_{13} + u_{19}) \qquad \textbf{14.50a}$$

$$\Lambda_2^{A_{2u}} = \hat{P}_{D_4}^{A_2} u_2$$

$$= \frac{1}{\sqrt{8}}(u_2 + u_5 + u_8 + u_{11} + u_{14} + u_{17} + u_{20} + u_{23}) \quad \textbf{14.50b}$$

$$\Lambda_3^{A_{2u}} = \hat{P}_{D_4}^{A_2} u_3$$

$$= \frac{1}{\sqrt{8}}(u_3 + u_4 + u_9 + u_{10} + u_{15} + u_{16} + u_{21} + u_{22}) \quad \textbf{14.50c}$$

$$\Lambda_4^{A_{2u}} = \hat{P}_{C_4}^{A} u_6$$

$$= \frac{1}{2}(u_6 + u_{12} + u_{18} + u_{24}) \quad \textbf{14.50d}$$

The b_{1u} functions are

$$\Lambda_1^{B_{1u}} = \hat{P}_{C_4}^{B} u_1$$

$$= \frac{1}{2}(u_1 - u_7 + u_{13} - u_{19}) \quad \textbf{14.51a}$$

$$\Lambda_2^{B_{1u}} = \hat{P}_{D_4}^{B_1} u_2$$

$$= \frac{1}{\sqrt{8}}(u_2 - u_5 - u_8 + u_{11} + u_{19} - u_{17} - u_{20} + u_{23}) \quad \textbf{14.51b}$$

$$\Lambda_3^{B_{1u}} = P_{D_4}^{B_1} u_3$$

$$= \frac{1}{\sqrt{8}}(u_3 - u_4 - u_9 + u_{10} + u_{15} - u_{16} - u_{21} + u_{22}) \quad \textbf{14.51c}$$

Finally, the b_{2u} functions are

$$\Lambda_1^{B_{2u}} = \hat{P}_{D_4}^{B_2} u_2$$

$$= \frac{1}{\sqrt{8}}(u_2 + u_5 - u_8 - u_{11} + u_{14} + u_{17} - u_{20} - u_{23}) \quad \textbf{14.52a}$$

$$\Lambda_2^{B_{2u}} = \hat{P}_{D_4}^{B_2} u_3$$

$$= \frac{1}{\sqrt{8}}(u_3 + u_4 - u_9 - u_{10} + u_{15} + u_{16} - u_{21} - u_{22}) \quad \textbf{14.52b}$$

$$\Lambda_3^{B_{2u}} = \hat{P}_{C_4}^{B} u_6$$

$$= \frac{1}{2}(u_6 - u_{12} + u_{18} - u_{24}) \quad \textbf{14.52c}$$

The final energies and wave functions for the π-electron system of porphine are given in Table 14.5. The π system contains 26 electrons (two of the nitrogens donate two

Table 14.5

HÜCKEL π-ELECTRON ENERGIES AND WAVE FUNCTIONS FOR PORPHINE[a]

Symmetry (Γ)	Energy[b]		Coefficients					
			Λ_1^Γ	Λ_2^Γ	Λ_3^Γ	Λ_4^Γ	Λ_5^Γ	Λ_6^Γ
a_{1u}	(1)	+.6180	.8507	.5257				
	(2)	−1.6180	.5257	−.8507				
a_{2u}	(1)	+2.9776	.3041	.6402	.3237	.6268		
	(2)	+1.0232	.0280	.0203	.8736	−.4854		
	(3)	+.3245	.8265	.1896	−.2807	−.4496		
	(4)	−2.2252	.4729	−.7442	.2307	.4114		
b_{1u}	(1)	+1.6180	.6325	.7236	.2764			
	(2)	−.6180	.6325	−.2764	−.7236			
	(3)	−2.0000	.4472	−.6325	.6325			
b_{2u}	(1)	+2.7003	.6125	.3602	.7036			
	(2)	+1.0224	.0198	.8829	−.4692			
	(3)	−1.6227	.7902	−.3013	−.5336			
e_g	(1)	+2.8608	.2520	.0969	.0251	−.6241	−.3354	−.6516
	(2)	+1.2347	.5593	.7105	.3179	.0199	.0848	.2715
	(3)	+1.0220	.0223	.0423	.0209	.0195	.8866	−.4593
	(4)	−.2719	.6387	−.3882	−.5332	−.2145	.1686	.2874
	(5)	−1.6195	.0023	−.4393	.7091	−.4356	.1663	.2945
	(6)	−2.1261	.4639	−.3747	.3327	.6116	−.1956	−.3485

[a] A value of 1.1 was chosen for h_N and 1.3 for k_{CN}. These values were chosen to eliminate accidental degeneracies. Only one component of each degenerate function is given.

[b] In units of β, relative to $\alpha = 0$.

electrons to the π system). The first $\pi^* \leftarrow \pi$ electronic transition is predicted to be from $(3a_{2u})$ to $(4e_g)$. This is predicted to have a transition energy of about $.6\beta$. Porphine itself and substances that contain it all have very low-energy electronic transitions.

14.5 *DENSITY MATRICES INVOLVING DEGENERATE ORBITALS*

Whenever a system has three-fold or higher rotational symmetry, the corresponding point group has degenerate representations and there will be symmetry-induced degeneracies in some of the wave functions and the corresponding energy levels for the system. We have seen these symmetry-induced degeneracies in benzene, *s*-triazine, and porphine. We have so far listed only one real component of the degenerate functions. Use of this one component is sufficient for obtaining the energies. However, if we require charge densities, bond orders, or the density matrix, we must use both components. In fact, if there are partially occupied degenerate levels in the system, the complex form of the wave function may be required.

Let us consider triazine again. The π-electron wave functions are given in Eqs. 14.41 and 14.42. The $\psi_1^{a_2''}$ and $\psi_1^{e''}$ orbitals are completely occupied in the ground state. We can draw an orbital occupancy diagram as

$$\begin{array}{lcc} e'' & \underline{\quad\times\times\quad} & \underline{\quad\times\times\quad} \\ & & \\ a_2'' & \underline{\quad\times\times\quad} & \end{array} \tag{14.53}$$

The a_2'' orbital is (Eq. 14.41a)

$$\psi_1^{a_2''} = .4328(u_1 + u_3 + u_5) + .3822(u_2 + u_4 + u_6) \tag{14.54}$$

Equation 14.42a represents one of the degenerate pair of e'' orbitals; call it $\psi_{1a}^{e''}$:

$$\psi_{1a}^{e''} = .3217(2u_1 - u_3 - u_5) + .2513(u_2 - 2u_4 + u_6) \tag{14.55}$$

It was obtained from symmetry-adapted functions arising from the real E representation of $\mathbf{C}_3$ that resulted from adding the two complex functions (where $\varepsilon = e^{2\pi i/3}$)

$\mathbf{C}_3$	E	C_3	C_3^2
$E\left\{\right.$	1	ε	ε^*
	1	ε^*	ε
E (real) =	2	$\varepsilon + \varepsilon^*$	$\varepsilon + \varepsilon^*$
=	2	2 cos 120°	2 cos 120°
=	2	−1	−1

14.56

If the two complex functions are subtracted, we have

$\mathbf{C}_3$	E	C_3	C_3^2
$E\left\{\right.$	1	ε	ε^*
	1	ε^*	ε
E (imag.) =	0	$\varepsilon - \varepsilon^*$	$\varepsilon^* - \varepsilon$
=	0	$2i$ sin 120°	$-2i$ sin 120°
=	0	$1.7321i$	$-1.7321i$

14.57

By using these in a projection operator, we obtain

$$\hat{P}^E u_1 = 1.7321iu_3 - 1.7321iu_5 \tag{14.58a}$$

$$\hat{P}^E u_4 = 1.7321iu_6 - 1.7321iu_2 \tag{14.58b}$$

or, after normalization,

$$\lambda_{N'}^{E''} = \frac{1}{\sqrt{2}}(u_3 - u_5) \qquad \textbf{14.59a}$$

$$\lambda_{C'}^{E''} = \frac{1}{\sqrt{2}}(u_6 - u_2) \qquad \textbf{14.59b}$$

Since these are degenerate with the functions of Eqs. 14.30c–d, they lead to the same energies as Eqs. 14.40a–b and the same linear equations as in the first lines of Eqs. 14.42a–b. Thus, the resulting wave functions are

$$\psi_{1b}^{e''} = .5572(u_3 - u_5) + .4353(u_2 - u_6) \qquad \textbf{14.60a}$$

$$\psi_{2b}^{e''} = .4353(u_3 - u_5) - .5572(u_2 - u_6) \qquad \textbf{14.60b}$$

Equations 14.55 and 14.60a are the two real components of the first e'' level. They are suitable for constructing the charge densities, bond orders, and density matrix for the ground state of triazine, in which the level is completely occupied. Only the unique elements have to be calculated. Those that are related by symmetry have the same value. For example, for the charge on position 1 (which equals the first diagonal element of the density matrix), we have

$$\begin{aligned} q_1 = P_{11} &= 2 \times .4328^2 + 2 \times (2 \times .3217)^2 + 2 \times 0^2 \\ &= 1.2026 \end{aligned} \qquad \textbf{14.61}$$

This is completely equivalent to that on position 3 (or 5):

$$q_3 = P_{33} = 2 \times .4328^2 + 2 \times (-.3217)^2 + 2 \times .5572^2 \qquad \textbf{14.62}$$

Similarly, P_{13} and P_{35} (and P_{15}) are the same:

$$\begin{aligned} P_{13} &= 2 \times .4328 \times .4328 + 2 \times (2 \times .3217) \times (-.3217) + 2 \times 0 \times .5572 \\ &= -.0393 \end{aligned} \qquad \textbf{14.63a}$$

$$\begin{aligned} P_{35} &= 2 \times .4328 \times .4328 + 2 \times (-.3217) \times (-.3217) \\ &\quad + 2 \times .5572 \times (-.5572) \\ &= -.0393 \end{aligned} \qquad \textbf{14.63b}$$

The complete first-order density matrix for the ground state of the π system of *s*-triazine is

$$\mathbf{P} = \begin{bmatrix} 1.2026 & .6542 & -.0393 & -.3159 & -.0393 & .6542 \\ .6542 & .7974 & .6542 & .0395 & -.3159 & .0395 \\ -.0393 & .6542 & 1.2026 & .6542 & -.0393 & -.3159 \\ -.3159 & .0395 & .6542 & .7974 & .6542 & .0395 \\ -.0393 & -.3159 & -.0393 & .6542 & 1.2026 & .6542 \\ .6542 & .0395 & -.3159 & .0395 & .6542 & .7974 \end{bmatrix} \qquad \textbf{14.64}$$

As we have seen, the first excited $\pi^* \leftarrow \pi$ configuration, $(1a_2'')^2(1e'')^3(2e'')^1$, leads to three states, A_1', A_2', and E'. Constructing the density matrices for these is considerably more complicated than for closed-shell systems. What is required is that the many-electron wave function, rather than the one-electron functions, be symmetry-adapted. The McGlynn book listed in the bibliography describes how to symmetry-adapt a many-electron wave function to the representations for particular states.

BIBLIOGRAPHY

COTTON, F. A., *Chemical Applications of Group Theory*, John Wiley & Sons, New York, 2d ed., 1971.

FLURRY, R. L., JR., *Symmetry Groups: Theory and Chemical Applications*. Prentice-Hall, Inc., Englewood Cliffs, N. J., 1980.

HALL, L. H., *Group Theory and Symmetry in Chemistry*. McGraw-Hill Book Company, New York, 1968.

MCGLYNN, S. P., VANQUICKENBORNE, L. G., KINOSHITA, M., and CARROLL, D. G., *Introduction to Applied Quantum Chemistry*. Holt, Rinehart and Winston, Inc., New York, 1972.

ORCHIN, M., and JAFFÉ, H. H., *Symmetry, Orbitals, and Spectra*. Wiley-Interscience, New York, 1970.

PROBLEMS

Carry out complete Hückel calculations for the indicated molecules. All structures can be found in the problems for Chapters 12 and 13. For all except ethylene (Problem 14.8), use only the π systems. Report the orbital energies, the wave functions, the ground-state energies, and the charge densities on each unique atom. (If properly symmetry-factored, no problem will involve a determinant larger than 3×3.)

***14.1** Cyclopropenyl.

***14.2** Glyoxal. Let $h_O = 1.0$ and $k_{CO} = 1.0$.

14.3 1,2-Dichloroethylene. Include the chlorines, with two π electrons from each. Let $h_{Cl} = 2.0$ and $k_{CCl} = .4$. Can Hückel theory distinguish between *cis* and *trans*?

***14.4** Carbonate. There are six π electrons. Let $h_O = 1.7$ and $k_{CO} = 1.0$.

14.5 Triphenylene.

14.6 *p*-Benzophenone. Use the same parameters as for glyoxal. (Why do you think the carbonate ion should have different parameters?)

14.7 Napthalene.

14.8 Ethylene σ and π systems. See Problems 12.10 and 12.11.

chapter fifteen

Transition-metal compounds

15.1 GENERAL COMMENTS

Transition-metal compounds (also referred to as transition-metal *complexes*) involve a transition-metal atom or ion. In solutions, the groups attached to the metal may be readily exchangeable with other species in solution, or even with the solvent. Except for the number of electrons per atom and the fact that they frequently have open-shell ground states, transition-metal compounds are very much like any other molecular species. Their energies and wave functions can be calculated by any of the basic techniques of molecular quantum mechanics that we have mentioned. However, owing to the large number of electrons in these complexes, the usual types of nonempirical calculations are expensive and time consuming to apply. For this reason, semiempirical methods have been developed. The most common of these are different enough to warrant separate discussion.

The most common transition-metal compounds have only one metal ion, or sometimes a neutral atom, surrounded by a number of groups called *ligands*, with respect to which the metals behave as *Lewis acids* (i.e., electron acceptors). The ligands can be single atoms or monoatomic ions, or they can be polyatomic ions or molecules. The one requirement is that they have unshared pairs of electrons available for sharing with the metal. Such bonding, in which both electrons in an electron-pair bond come from the same species, is often called *coordinate covalent bonding* (or sometimes *dative bonding*). The complexes are frequently referred to as *coordination complexes*. The number of ligands around the metal in a complex is referred to as the metal's *coordination number*.

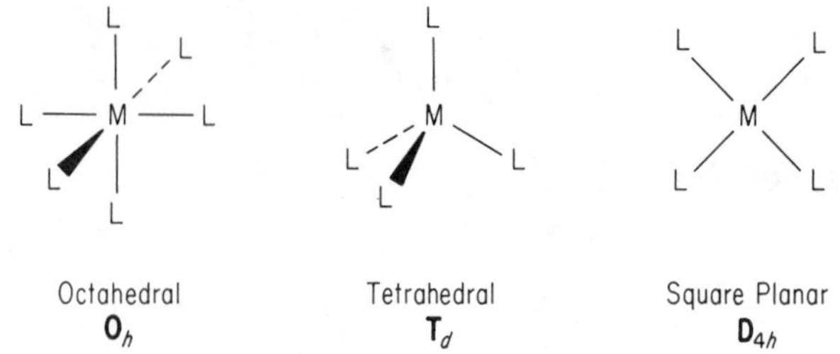

Figure 15.1 The common geometries for transition metal compounds.

The most common coordination number in transition-metal compounds is six. The next most common is four. The orientation of the bonds to the ligands is usually as symmetrical as is consistent with the structure. The sixfold coordination leads to an octahedral orientation of the bonds, while the fourfold leads to either tetrahedral or square-planar (see Figure 15.1). If all the ligands are the same monoatomic species, the point group of the octahedral complexes is $\mathbf{O}_h$, that for the tetrahedral is $\mathbf{T}_d$, and that for the square-planar is $\mathbf{D}_{4h}$. If the ligands are different, the point group will be a subgroup of one of these. The semiempirical theories of the transition-metal complexes make heavy use of the symmetries of the complexes.

Experimentally, the most obviously unique properties of transition-metal compounds are their color and their magnetic behavior. In color, they range from no color at all for zinc compounds to the intense purple of permanganate (MnO_4^-). For most of the metals the color is dependent upon the ligand. In magnetic properties, certain ions always give *diamagnetic* (repelled by a magnetic field) complexes, others always give *paramagnetic* (attracted by a magnetic field) complexes, and still others may give either diamagnetic or paramagnetic complexes, depending upon the ligand. With those ions whose complexes are always paramagnetic, the magnitude of the paramagnetism may strongly depend upon the nature of the ligands. None of this is based on any change in the number of electrons in the bonding system of the complex. Furthermore, there is a correlation between the color and the magnetic properties for different complexes of a given metal.

15.2 CRYSTAL-FIELD THEORY, LIGAND-FIELD THEORY, AND MOLECULAR-ORBITAL THEORY

The simplest of the semiempirical methods for metal complexes, and the one to which we will give the most attention, is *crystal-field theory*. This method, developed in the 1930s by Bethe and Van Vleck, treats the interaction between the metal and the ligands as completely electrostatic. Predictions are based upon the perturbation of the metal *d* orbitals by point charges or point dipoles of the ligands. For octahedral and tetrahedral complexes, the fivefold-degenerate *d* level is split into a twofold-degenerate e level and a threefold t_2 level. The magnitude of the splitting depends upon only one quantity, which is determined empirically. In the square-planar complexes, the d levels are split into four levels, a_{1g}, b_{1g}, b_{2g}, and e_g, of which only one is degenerate. Three splittings must be determined. The magnetic and long-wavelength spectral properties of the complex are assumed to be determined by the behavior of the electrons that came from the free-metal d orbitals but that occupy the split levels in the complex. Since the treatment involves a point charge or dipole approximation for the ligands, complexes with mixed ligands are frequently treated as though all ligands were the same. If desired, an additional perturbation, having the true symmetry, can be applied to these results. Although the crystal-field model is not physically reasonable, it offers a simple explanation for much of the spectral and magnetic behavior of metal complexes.

Ligand-field theory allows for the interaction of the orbitals on the ligand with the metal orbitals, at least implicitly. Molecular-orbital calculations can also be performed on metal complexes, at various levels of approximation. All these more sophisticated calculations predict the existence of levels that are split in the same manner as predicted by crystal-field theory and that are occupied by the same number of electrons as would come from the free-metal d level. The details of the calculated properties can be improved by the more refined calculations, but the primary effects are correctly described (at least qualitatively) in the most primitive crystal-field calculations.

15.3 CRYSTAL-FIELD SPLITTING

The Hamiltonian of a spherically symmetrical transition-metal atom or ion can be schematically written

$$\hat{H} = \hat{T} + \hat{V}_{eN} + \hat{V}_{ee} + \hat{V}_{so} \tag{15.1}$$

where $\hat{T}$ is the kinetic-energy operator, $\hat{V}_{eN}$ is the electron-nucleus contribution to the potential energy, $\hat{V}_{ee}$ is the electron-electron interaction, and $\hat{V}_{so}$ is the spin-orbit interaction (a relativistic term). In the atomic Hamiltonian, the relative magnitudes of $\hat{V}_{ee}$ and $\hat{V}_{so}$ determine whether Russell-Saunders coupling or *j-j* coupling is more appropriate for the atom. Even in the first transiton series, $\hat{V}_{so}$ is far from negligible.

In the crystal-field approximation, the interaction with the nonspherical field of the ligand ions adds to the Hamiltonian one extra term—the crystal-field potential, $\hat{V}_{CF}$:

$$\begin{aligned} \hat{H} &= \hat{T} + \hat{V}_{eN} + \hat{V}_{ee} + \hat{V}_{so} + \hat{V}_{CF} \\ &= \hat{H}_F + \hat{V}_{CF} \end{aligned} \tag{15.2}$$

where $\hat{H}_F$ is the free-atom Hamiltonian. The symmetry of the crystal-field potential is the symmetry of the complex. The perturbation caused by $\hat{V}_{CF}$ partially lifts the degeneracy of the d levels of the metal. As far as the symmetry is concerned, the resulting representations for the metal orbitals, in the point group, can be determined by mapping the D_g^2 representation for the d orbitals from the spherical group, **O**(3), onto the appropriate point group. For octahedral complexes we have, using only the **O** subgroup of $\mathbf{O}_h$,

O	E	$8C_3$	$3C_2$	$6C_4$	$3C_2'$
D_g^2	5	$1 + 2\cos 120° + 2\cos 240°$	$1 + 2\cos 180° + 2\cos 360°$	$1 + 2\cos 90° + 2\cos 180°$	$1 + 2\cos 180° + 2\cos 360°$
=	5	-1	1	-1	1

15.3

This reduces to $E + T_2$ within **O**. The gerade behavior of D_g^2 requires that the orbitals be e_g and t_{2g} within $\mathbf{O}_h$. Thus, the fivefold degeneracy of the d level of the atom is partially lifted in the $\mathbf{O}_h$ symmetry of an octahedral complex. Repeating the process for $\mathbf{T}_d$, the symmetry of the tetrahedral field, we have

$\mathbf{T}_d$	E	$8C_3$	$3C_2$	$6S_4$	$6\sigma_d$
D_g^2	5	$1 + 2\cos 120° + 2\cos 240°$	$1 + 2\cos 180° + 2\cos 360°$	$1 - 2\cos 90° + 2\cos 180°$	1
=	5	-1	1	-1	1

15.4

This reduces to $E + T_2$ within $\mathbf{T}_d$. As far as the degeneracy is concerned, the splitting is the same as in the octahedral field. (The magnitude of the tetrahedral splitting is four-ninths that of the octahedral and the order is reversed.) For the $\mathbf{D}_4$ subgroup of the $\mathbf{D}_{4h}$ group of the square planar case, we have

$\mathbf{D}_4$	E	$2C_4$	C_2	$2C_2'$	$2C_2''$
D_g^2	5	$1 + 2\cos 90° + 2\cos 180°$	$1 + 2\cos 180° + 2\cos 360°$	$1 + 2\cos 180° + 2\cos 360°$	$1 + 2\cos 180° + 2\cos 360°$
=	5	-1	1	1	1

15.5

This reduces to A_1, B_1, B_2, and E. Again the g behavior of the D_g^2 representation carries over for $\mathbf{D}_{4h}$ point symmetry.

The magnitude of the crystal-field splitting can be deduced from qualitative arguments or it can be calculated. We will discuss only the qualitative considerations. The interaction of an electron in a metal d level with a collection of negative ions surrounding the metal has two contributions. There is a uniform shift to higher energy, since the electron is repelled by the negative ions, and there is a splitting of the degeneracy. (Formally, the crystal-field potential has a contribution with spherical symmetry and a contribution that transforms as a fourth-rank tensor.) If the real form of the d orbitals is considered, the relative magnitudes of the splitting can be deduced by noting the distance of the ions from the lobes of the orbitals. The farther the ions are from the lobes, the less the repulsion. The real forms of the d orbitals are designated d_{z^2} (directed mostly along the z axis), $d_{x^2-y^2}$ (directed along the x and y axes), d_{xy} (lying primarily in the xy plane, but with its maximum extension along the xy and $x(-y)$ diagonals of this plane), d_{xz} (along the diagonals of the xz plane), and d_{yz} (along the diagonals of the yz plane). In either the $\mathbf{O}_h$ or the $\mathbf{T}_d$ point group, the first two of these are the e orbitals while the last three are the t_2 orbitals (with a g added in both cases for $\mathbf{O}_h$).

Consider now the positions of the ligands in an octahedral and in a tetrahedral complex, relative to a cube (Figure 15.2). In the octahedral case the ligands are in the centers of the faces of the cube, while in the tetrahedral case they are on alternating vertices of the cube. If the axis system is assumed to lie parallel to the cube's edges, the d_{z^2} and the $d_{x^2-y^2}$ orbitals point toward the centers of the cube's faces, while the d_{xy}, the d_{xz}, and the d_{yz} point toward the centers of the edges. Thus, in the octahedral complex, the e_g orbitals are pointing directly toward the ligands, while the t_{2g} orbitals are directed halfway between them. This means that the e_g orbitals are more destabilized (lie at higher energy) than the t_{2g} orbitals. In the tetrahedral case, none of the orbitals are directed toward any of the ligands. However, the t_2 orbitals are directed more closely to the ligands than are the e orbitals. (It can be seen geometrically that the vertices of the cube are closer to the centers of the edges than they are to the centers of the faces by a factor of $\sqrt{2}$.) Thus, in the tetrahedral case, the t_2 orbitals are more destabilized than are the e orbitals.

With respect to the $\mathbf{D}_{4h}$ complexes, the qualitative argument is a bit more complicated. Here, there are three splitting parameters. Qualitative arguments cannot unambiguously determine the order of the levels. If the axis system is chosen

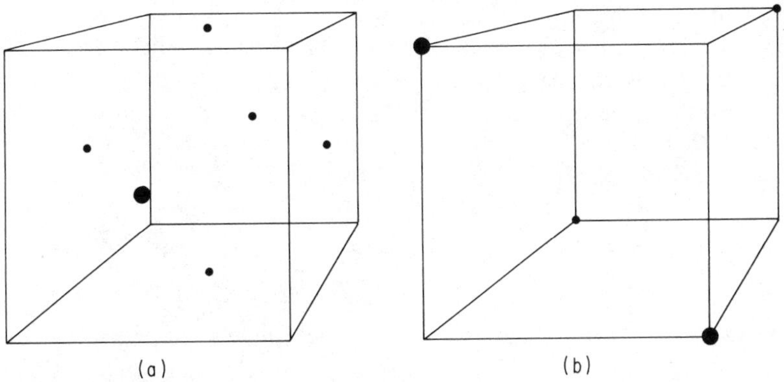

Figure 15.2 Positions of the ligands for (a) an octahedral complex and (b) a tetrahedral complex with reference to a cube. (After R. L. Flurry, Jr., *Symmetry Groups: Theory and Chemical applications.* Prentice-Hall, Inc., Englewood Cliffs, N.J., 1980. By permission.)

so that the x and y coordinates lie along the bonds, the simplest qualitative arguments would place the $d_{x^2-y^2}$ orbital (b_{1g} in $\mathbf{D}_{4h}$ symmetry) as the most destabilized, since it is directed toward the ligands, and the d_{z^2} (a_{1g} in $\mathbf{D}_{4h}$) as the least destabilized, since its maximum extension is perpendicular to the xy plane. The d_{xy} (b_{2g}), which lies in the xy plane but is directed diagonally with respect to the bonds, would be expected to be the second most destabilized. Finally, the d_{xz} and the d_{yz}, which form the e_g set in $\mathbf{D}_{4h}$, would be expected to be the second least destabilized. This logic produces the correct order for the b_{1g} and b_{2g} levels but the reverse of the correct order for the e_g and a_{1g} levels.

An alternative qualitative argument can be based upon the $\mathbf{O}_h$ results. $\mathbf{D}_{4h}$ is a subgroup of $\mathbf{O}_h$; consequently, as the ligands lying on the z axis are removed, the $\mathbf{O}_h$ energy levels should go smoothly to the $\mathbf{D}_{4h}$ results. The e_g level of $\mathbf{O}_h$ correlates with a_{1g} and b_{1g} levels of $\mathbf{D}_{4h}$. By the arguments above, the a_{1g} level should lie lower than the b_{1g}. The t_{2g} level of $\mathbf{O}_h$ correlates with b_{2g} and e_g levels of $\mathbf{D}_{4h}$. Again, by the arguments above, the e_g should be the lower-lying of these two levels. These considerations place e_g as the lowest energy level and b_{1g} as the highest. The relative order of the a_{1g} and b_{2g} levels depends upon the magnitude of the splitting of the e_g and t_{2g} levels of $\mathbf{O}_h$. In most complexes the a_{1g} level lies below the b_{2g}. Figure 15.3 gives the qualitative energy-level scheme for the d orbitals in the $\mathbf{O}_h$, the $\mathbf{T}_d$, and the $\mathbf{D}_{4h}$ complexes.

The actual calculation of the crystal-field splitting requires either some fairly complex geometrical arguments or the use of tensor algebra. Tensor algebra, although beautifully elegant and very useful in many areas of quantum mechanics, has not been developed in this book. Therefore, we will not describe the actual calculation of the splitting. The final results are simple in the octahedral and tetrahedral cases. They are normally expressed in terms of a quantity, Dq, that is the expectation value of a term containing the electron-to-nucleus distance, as a variable, and some constants, including the electronic charge, the effective nuclear charge of the metal, the metal-to-ligand distance, and some numerical constants. The calculated splitting for either the octahedral or the tetrahedral case is $10Dq$. The t_2 levels are $4Dq$ from the unsplit center of gravity; the e levels are $6Dq$ from that point. Experimentally, the electronic transition arising from the d-d promotion is frequently identified with $10Dq$. Alternatively, the splitting may be called Δ and be frankly recognized as an empirical quantity.

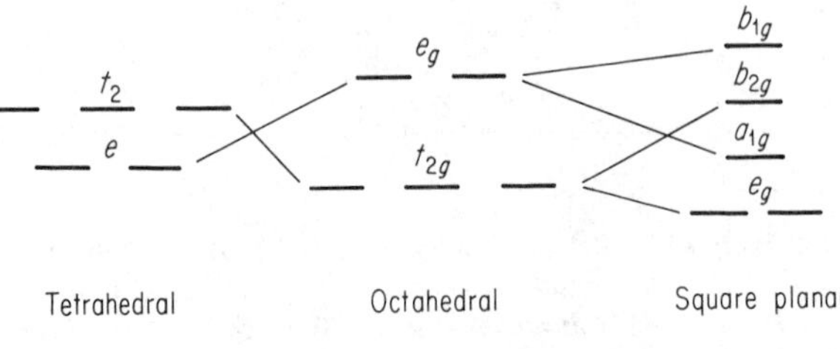

Figure 15.3 Splitting of the d levels in fields of various symmetries.

15.4 MAGNETIC AND OPTICAL PROPERTIES: ONE-ELECTRON APPROXIMATION

The optical and magnetic properties of coordination complexes can be qualitatively explained from a consideration of the splitting of the energy levels available to a single electron in a *d* level, as illustrated in Figure 15.3. The arguments used in Section 15.3 told us that there should be the indicated lifting of the *d*-level degeneracies, but they said nothing about the magnitude of the splittings. In principle, they could be vanishingly small (the weak-field limit) or very large (the strong-field limit). The actual behavior of transition-metal complexes depends upon the nature of the ligand. The stronger the interaction, the nearer the behavior will approach the strong-field limit, and vice versa. The real effects are primarily bonding effects rather than electrostatic. Many uncharged ligands produce a stronger-field effect than many ionic ligands. For example, for some simple ligands, the order is

$$Cl^- < F^- < OH^- < H_2O < NH_3 < CN^- < CO$$

In the weak-field limit, the behavior would be that of the free metal (no crystal-field splitting). In particular, the fivefold degeneracy would remain, and Hund's rule would determine the number of unpaired electrons and hence the magnetic properties of the *d* level. As the number of electrons increased, in a series of metals, the number of unpaired electrons would increase until the *d* level was half filled, then decrease until it was filled. This is just what happens in the ground-state configurations of the neutral atoms from Ti (atomic number 21, one *d* electron) to Zn (atomic number 30, ten *d* electrons). Near the weak-field limit, the degeneracy is broken, but only slightly, and an octahedral complex can have as many as five unpaired electrons. At the weak-field limit, electronic spectral transitions involve the promotion of an electron from one atomic subshell to another. At the strong-field limit, on the other hand, each orbital level of the complex must be filled before any electrons go into the next. An octahedral complex can have no more than three unpaired electrons at the strong-field limit, since that would half fill the t_{2g} level. Furthermore, electronic transitions can take place between the levels into which the *d* level splits. [These electronic transitions are forbidden on symmetry grounds, since the initial and final states are both gerade and the operator is ungerade—they are said to be *Laporte* (parity) *forbidden*. However, they are made weakly allowed by vibronic interactions.]

Real complexes fall between the weak-field and strong-field limits. They are, however, commonly classified as weak-field or strong-field, depending upon which approximation best describes the electron-pairing behavior. Electronic spectral transitions are usually seen between the split levels. To a one-electron first approximation, the energy of such transitions is directly proportional to the crystal-field splitting. When more than one pairing scheme is possible, the magnetic behavior is determined by whether the crystal-field splitting is greater than or less than the *pairing energy* (the excess electronic repulsion energy resulting when two electrons go into the same orbital). Table 15.1 shows the electron pairing in the weak-field and strong-field limits for octahedral and tetrahedral complexes. Table 15.2 shows some representative spectral and magnetic data for some octahedral complexes.

The behavior of the square-planar complexes is not as easy to describe. There are three splitting parameters. Furthermore, each level can have a different pairing

Table 15.1

ELECTRON PAIRING IN OCTAHEDRAL AND TETRAHEDRAL TRANSITION-METAL COMPLEXES

Symmetry	Number of *d* Electrons	Weak-field Configuration	Unpaired Electrons	Strong-field Configuration	Unpaired Electrons
$\mathbf{O}_h$	1	$(t_{2g})^1$	1	$(t_{2g})^1$	1
	2	$(t_{2g})^2$	2	$(t_{2g})^2$	2
	3	$(t_{2g})^3$	3	$(t_{2g})^3$	3
	4	$(t_{2g})^3(e_g)^1$	4	$(t_{2g})^4$	2
	5	$(t_{2g})^3(e_g)^2$	5	$(t_{2g})^5$	1
	6	$(t_{2g})^4(e_g)^2$	4	$(t_{2g})^6$	0
	7	$(t_{2g})^5(e_g)^2$	3	$(t_{2g})^6(e_g)^1$	1
	8	$(t_{2g})^6(e_g)^2$	2	$(t_{2g})^6(e_g)^2$	2
	9	$(t_{2g})^6(e_g)^3$	1	$(t_{2g})^6(e_g)^3$	1
	10	$(t_{2g})^6(e_g)^4$	0	$(t_{2g})^6(e_g)^4$	0
$\mathbf{T}_d$[a]	1	$(e)^1$	1	$(e)^1$	1
	2	$(e)^2$	2	$(e)^2$	2
	3	$(e)^2(t_2)^1$	3	$(e)^3$	1
	4	$(e)^2(t_2)^2$	4	$(e)^4$	0
	5	$(e)^2(t_2)^3$	5	$(e)^4(t_2)^1$	1
	6	$(e)^3(t_2)^3$	4	$(e)^4(t_2)^2$	2
	7	$(e)^4(t_2)^3$	3	$(e)^4(t_2)^3$	3
	8	$(e)^4(t_2)^4$	2	$(e)^4(t_2)^4$	2
	9	$(e)^4(t_2)^5$	1	$(e)^4(t_2)^5$	1
	10	$(e)^4(t_2)^6$	0	$(e)^4(t_2)^6$	0

[a] Very few weak-field tetrahedral complexes are known.

Table 15.2

SPECTRAL AND MAGNETIC PROPERTIES OF d^6 TRANSITION-METAL COMPLEXES

Compound	Color	ΔE (cm^{-1})[a]	Magnetic Behavior[b]
$[Fe(CN)_6]^{4-}$	Yellow	31,000	Diamagnetic
$[Fe(H_2O)_6]^{2+}$	Pale green	10,400	Paramagnetic
$[Co(NH_3)_6]^{3+}$	Yellow	21,000	Diamagnetic
$[Co(H_2O)_6]^{3+}$	Red	16,600	Diamagnetic
$[CoF_6]^{3-}$	Green	11,800	Paramagnetic

[a] The first transition that does not involve a change of multiplicity.
[b] The paramagnetic species have four unpaired electrons.

energy. In order to predict the number of unpaired electrons, one must know whether the pairing is greater than the first crystal-field splitting, the first plus the second, or the sum of the three.

15.5 TERM SYMBOLS

Since degenerate levels may be partially occupied in coordination complexes, different terms can arise from the same electronic configuration. These can be found by use of the methods previously given. The ordering of the terms with respect to energy cannot be completely accomplished without some sort of actual calculation on the many-electron states. Some qualitative ordering can be accomplished by considering the atomic terms at the weak-field limit and the configurations at the strong-field limit.

Consider, for example, a d^3 system in an octahedral field. The quartet state can be expected to be the ground state. In the free metal there are two L values that can be associated with an S value of $\frac{3}{2}$. The allowed terms (neglecting the J values) are ^4F and ^4P, with the ^4F being the ground-state term. We can reasonably expect the quartet state to be the ground state in the complex. Further, the orbital symmetry should arise from the F orbital symmetry of the free metal. Mapping the D_g^3 representation of the spherical group, $\mathbf{O}(3)$, onto the $\mathbf{O}_h$ point group of the complex, we see that we obtain A_{2g}, T_{1g}, and T_{2g}. The ground state of the complex should be one of these. (The P level gives T_{1g}, when mapped onto $\mathbf{O}_h$.)

One can obtain further information by considering the orbital configurations that can arise in the complex. These are $(t_{2g})^3$, $(t_{2g})^2(e_g)$, $(t_{2g})(e_g)^2$, and $(e_g)^3$, in order of increasing energy. (The last of these cannot give rise to the quartet state.) The $(t_{2g})^3$ configuration is the lowest in energy. Adapting this to the $[1^3]$ representation from $\mathbf{S}(3)$, the spatial representation required by a quartet spin state, we see that it leads to the A_{2g} representation of $\mathbf{O}_h$. Thus, the ground state of the complex is the $^4A_{2g}$ state.

The other terms that can be constructed from the various configurations are shown in Table 15.3. In constructing the table, one must adapt the $(t_{2g})^2$ partial configuration to both $[2]$ (for the singlet) and $[1^2]$ (for the triplet) from $\mathbf{S}(2)$. Only the triplet can give rise to the quartet state when coupled to the single e_g electron, but both can give rise to the doublet spin state. Similar considerations apply for the $(e_g)^2$ substate.

At the strong-field limit, the energies can be ordered by configuration and by spin multiplicity within a configuration. The ordering of the different states (terms) within a given configuration and multiplicity cannot be accomplished without the actual computation of the energies of the states. Between the two extremes the ordering depends upon the values of the crystal-field splitting and the electron repulsion. Extensive diagrams of the energy levels, as a function of the parameters, for the possible d level occupancies have been published by Orgel, by Tanabe and Sugano, and by others. These diagrams are useful for estimating the values of the parameters from experimental spectral data.

Table 15.3

TERMS THAT CAN ARISE FROM A d^3 CONFIGURATION IN AN OCTAHEDRAL COMPLEX

Configuration	Spin Multiplicity	Terms
$(t_{2g})^3$	4	$^4A_{2g}$
	2	$^2E_g + {}^2T_{1g} + {}^2T_{2g}$
$(t_{2g})^2(e_g)^1$	4	$^4T_{1g} + {}^4T_{2g}$
	2	$^2A_{1g} + {}^2A_{2g} + 2\,{}^2E_g + 2\,{}^2T_{1g} + 2\,{}^2T_{2g}$
$(t_{2g})^1(e_g)^2$	4	$^4T_{1g}$
	2	$2\,{}^2T_{1g} + 2\,{}^2T_{2g}$
$(e_g)^3$	2	^2E_g

BIBLIOGRAPHY

BALLHAUSEN, C. J., *Introduction to Ligand Field Theory.* McGraw-Hill Book Company, New York, 1962.

COTTON, F. A., *Chemical Applications of Group Theory.* John Wiley & Sons, New York, 2d ed., 1971.

FIGGIS, B. N., *Introduction to Ligand Fields.* Wiley-Interscience, New York, 1966.

HATFIELD, W. E., and PARKER, W. E., *Symmetry in Chemical Bonding and Structure.* Charles E. Merrill Publishing Co., Columbus, O., 1974.

JORGENSEN, C. K., *Absorption Spectra and Chemical Bonding in Complexes.* Addison-Wesley Publishing Co., Inc., Reading, Mass., 1962.

KETTLE, S. F. A., *Coordination Compounds.* Appleton-Century-Crofts, New York, 1969.

PROBLEMS

***15.1** Find the possible term symbols for a triplet $(t_2)^2$ electronic configuration of a tetrahedral molecule.

***15.2** Find the crystal-field splitting of a set of f orbitals in a field of octahedral symmetry.

***15.3** Some typical crystal-field parameters for a square-planar complex are, for the splitting, $\Delta_1 \sim 26{,}500\ \text{cm}^{-1}$, $\Delta_2 \sim 4300\ \text{cm}^{-1}$, $\Delta_3 \sim 7100\ \text{cm}^{-1}$, and for the pairing energy, $\gamma_{ii} \sim 4500\ \text{cm}^{-1}$. Using these parameters, deduce the crystal-field orbital occupancy for the d^1 to d^9 configurations. Find the number of unpaired electrons for each.

***15.4** What is the energy of the first electronic transition for each case from the preceding problem? Note that any difference in electron repulsion for the ground and excited states must be taken into account.

***15.5** Find the ground-state term symbols for each case from Problem 15.3.

15.6 Do a Hückel-level molecular-orbital calculation on the tetrahedral $CoCl_4^{2-}$ complex. Treat the Co as tetrahedrally hybridized, with four equivalent hybridized orbitals. As the basis on the metal, use the tetrahedrally hybridized orbitals (T) and the set of d orbitals. Let each ligand contribute a single σ-type basis, L. Let $\langle T|\hat{h}|T\rangle$ and $\langle T|\hat{h}|L\rangle$ be the reference α and β. Let $\langle d|\hat{h}|d\rangle = \alpha + 1.5\beta$, $\langle L|\hat{h}|L\rangle = \alpha - \beta$, $\langle T|\hat{h}|T'\rangle = .34\beta$, and $\langle T|\hat{h}|d\rangle = .05\beta$. Find the ground-state configuration (assume the strong-field occupancy), the ground-state term symbol, and the energy of the first transition. (Notice that this model does not contain any Hamiltonian matrix elements between the d orbitals and the ligand orbitals.)

chapter sixteen

Molecular vibrations

16.1 NORMAL MODES OF VIBRATION AND NORMAL COORDINATES

In principle, the infrared and Raman spectra of a molecule could be obtained from a direct solution of the Schrödinger equation without the Born-Oppenheimer approximation. This method, however, is usually considered too difficult to be practical, although work is being done along these lines. The vibrational problem is most commonly handled in terms of internal displacement coordinates, representing the displacements of the atoms from their equilibrium positions, and empirically determined force constants. The most detailed vibrational studies determine the force constants from the experimental vibrational spectra and then from these constants calculate the spectrum. The success of the study is judged by how well the calculated and experimental spectra agree with each other.

Studies of molecular vibrations provide another area where a basis-set expansion for a quantum-mechanical problem is of interest to chemists. We saw in Chapter 4 that the quantum-mechanical harmonic oscillator was a good approximation to a vibrating diatomic molecule. With this approximation, the vibrational energy levels for the molecule were simply the eigenvalues of the harmonic-oscillator system. In Section 4.3 we briefly discussed the vibrations of many-atomic molecules in terms of coupled harmonic oscillators. Such a treatment is general, but difficult.

In a *normal mode of vibration* all atoms are moving in such a fashion that they pass through their equilibrium positions simultaneously and reach their maximum displacements simultaneously. The *normal coordinates* are defined as a coordinate system that describes this motion and in which both the kinetic energy and the potential energy can be described as a sum of terms, each dependent upon only one normal coordinate. There are as many normal coordinates for a molecule as there are degrees of vibrational freedom ($3N - 6$ for a general many-atomic molecule, or $3N - 5$ for a linear molecule, where N is the number of atoms). The Hamiltonian for the system is separable into a sum of terms, each dependent upon only one normal coordinate. The vibrational wave function can be expressed as a simple product of functions, each dependent upon only one normal coordinate. Formally, the problem resembles simple Hückel theory, where the Hamiltonian is expressed as a sum of one-electron terms and the many-electron wave function is a simple product of one-electron functions. There is, however, one big difference.

The normal coordinate represents the displacement of many particles (the nuclei). Consequently, the Hamiltonian is not a simple sum of single-particle Hamiltonians, and the wave function is not the product of single-particle functions. Rather, the Hamiltonian is the sum of single-vibration terms, and the wave function is the product of single-vibration functions. The quantum mechanics is that of a property, the vibration, rather than that of a particle. Although this may seem a subtle point, it has important consequences. The vibration transforms as a *Boson*, unlike electrons, which transform as Fermions. Consequently, the many-vibration wave function must be totally symmetric, rather than totally antisymmetric (as must Fermion functions). This means that in the ground vibrational state, all vibrations are associated with the lowest-energy vibrational function. The Boson behavior also shows up in overtones of degenerate vibrations.

16.2 BASIS FUNCTIONS AND THE HAMILTONIAN

The basis functions used to construct normal-coordinate wave functions usually represent localized internal vibrations, such as bond stretches and angle deformations. The normal coordinates can be expressed as linear combinations of the displacements corresponding to these vibrations. Usually mass-weighted coordinates are used. The desired form of the Hamiltonian, in terms of the normal coordinates, Q, is (see Eq. 4.4)

$$\hat{H} = \sum_{a=1}^{3N-L} \frac{1}{2}(\dot{Q}_a^2 + K_a Q_a^2) \qquad \textbf{16.1}$$

where L is 5 or 6, whichever is appropriate, and $\dot{Q}$ is the time derivative of Q. The normal coordinates can be expressed in terms of either mass-weighted Cartesian displacement coordinates ($q_i = m_i^{1/2}x_i$) or mass-weighted internal displacement coordinates (S_m):

$$Q_a = \sum_i c_{ai} q_i \qquad \textbf{16.2}$$

or

$$Q_a = \sum_m c_{am} S_m \qquad \textbf{16.3}$$

(The internal coordinates are not unique. There are an infinite number of possible choices.) It is convenient to express the force constants in terms of the localized internal vibrations; however, the kinetic energy is most easily expressed in terms of Cartesian coordinates. The internal coordinates, on the other hand, can be expressed as linear combinations of the Cartesian coordinates. For solving the problem, everything must be expressed in the same coordinate system. The usual procedure is to first express the kinetic energy in terms of the Cartesian coordinate system and then transform this to the internal system.

The vibrational wave functions can be expressed as linear combinations of the basis functions, χ_s, the internal or Cartesian-coordinate harmonic-oscillator vibrational functions:

$$\psi_v = \sum_s c_s \chi_s \qquad \mathbf{16.4}$$

A logical way to proceed would be to find the expectation value of the Hamiltonian with respect to this expression and minimize it, subject to the normalization constraint, just as was done in the L.C.A.O.-M.O. case. This approach has been taken in a limited number of cases; usually, however, it is not practical. To construct the overlap integrals between the members of the basis set is extremely difficult. It is even more difficult to calculate the interaction integrals. On the other hand, the internal force constants and consequently the matrix elements of the potential energy can be evaluated empirically. So, if a determinantal equation containing only the potential and kinetic energies can be obtained, it can be solved by using available information.

A word of caution is in order about either the Cartesian displacement coordinates or the internal displacement coordinates. There are always more Cartesian coordinates, and there can be more internal coordinates, then there are vibrations. This leads to the so-called "nongenuine" or "zero-frequency" modes in the solution of the problem.

16.3 THE SECULAR DETERMINANT

If we let the set of S_i be a generalized mass-weighted, but nonorthogonal, displacement-coordinate system (our internal coordinates), and $\dot{S}_i$ be the time derivative of S_i, the kinetic energy, T, and the potential energy, V, can be expressed (Wilson, Decius, and Cross, Sec. 4-3)

$$2T = \sum_i \sum_j t_{ij} \dot{S}_i \dot{S}_j \qquad \mathbf{16.5}$$

$$2V = \sum_i \sum_j f_{ij} S_i S_j \qquad \mathbf{16.6}$$

where f_{ij} is a force constant expressing the change of potential energy as a function of the displacements S_i and S_j, and t_{ij} is a constant related to the nonorthogonality of the coordinates. (Remember the diatomic molecule. S_i and S_j give the Δq. In this case there is only one coordinate. The t_{ij} equals unity, and the energy is completely equivalent to that in Eq. 4.4.) From Newton's second law of motion we can write

$$\frac{d}{dt}\frac{\partial T}{\partial \dot{S}_j} + \frac{\partial V}{\partial S_j} = 0 \qquad \textbf{16.7}$$

But, from Eq. 16.5,

$$\frac{\partial T}{\partial \dot{S}_j} = \sum_i \frac{t_{ij}}{2} \dot{S}_i$$

$$\frac{d}{dt}\frac{\partial T}{\partial \dot{S}_j} = \sum_i \frac{t_{ij}}{2} \ddot{S}_i \qquad \textbf{16.8}$$

and from Eq. 16.6

$$\frac{\partial V}{\partial S_j} = \sum_i \frac{f_{ij}}{2} S_i \qquad \textbf{16.9}$$

giving, for Eq. 16.7 (after multiplying through by 2),

$$\sum_i (t_{ij}\ddot{S}_i + f_{ij}S_i) = 0 \qquad \textbf{16.10}$$

This is similar to Eq. 4.14.

A possible solution to Eq. 16.10 for the S_i is

$$S_i = A_i e^{i\omega t} \qquad \textbf{16.11}$$

where ω is $2\pi\nu$. Substituting this into Eq. 16.10, we get

$$\sum_i (-t_{ij}\omega^2 A_i e^{i\omega t} + f_{ij}A_i e^{i\omega t}) = 0 \qquad \textbf{16.12}$$

or

$$\sum_i (f_{ij}A_i - t_{ij}\omega^2 A_i) = 0 \qquad \textbf{16.13}$$

The only way for this to hold in general is for the determinant multiplying the coefficients, A_i, to vanish; i.e.,

$$\det |f_{ij} - t_{ij}\omega^2| = 0 \qquad \textbf{16.14}$$

or in matrix notation

$$|\mathbf{F} - \mathbf{T}\omega^2| = 0 \qquad \textbf{16.14a}$$

The matrix **F** is just the matrix of force constants relating the internal coordinates. **T** is a matrix relating to the nonorthogonality of the internal coordinates (do not confuse it with the kinetic energy). In terms of mass-weighted Cartesian coordinates, q_i, the kinetic energy is

$$2T = \sum_i \dot{q}_i^2$$
$$= \sum_i p_i^2 \qquad 16.15$$

where p_i is the momentum conjugate to q_i—i.e., depending on q_i:

$$p_i = \frac{\partial T}{\partial \dot{q}_i} = \dot{q}_i \qquad 16.16$$

In general, the summation of Eq. 16.15 must be over the $3N$ Cartesian coordinates of the N atoms. Now let the internal coordinate S_t be expressed as a linear combination of the q_i:

$$S_t = \sum_i D_{ti} q_i \qquad 16.17$$

In terms of this, p_i is

$$p_i = \frac{\partial T}{\partial \dot{q}_i}$$
$$= \sum_t \frac{\partial T}{\partial \dot{S}_t} \frac{\partial \dot{S}_t}{\partial \dot{q}_i} \qquad 16.18$$

But $\partial T/\partial \dot{S}_t$ can be defined as the momentum, P_t, and $\partial \dot{S}_t/\partial \dot{q}_i$ is D_{ti} (from Eq. 16.17). Thus, p_i can be expressed

$$p_i = \sum_t P_t D_{ti} \qquad 16.19$$

Substituting into Eq. 16.15 gives

$$2T = \sum_i \sum_t \sum_s P_t D_{ti} D_{is} P_s \qquad 16.20$$

The order of the indexes is chosen to be suggestive of vector-matrix notation. If **P** is a column vector, and **D** is a square matrix, Eq. 16.20 can be rewritten as

$$2T = \mathbf{P}^\dagger \mathbf{D} \mathbf{D}^\dagger \mathbf{P} \qquad 16.21$$

In terms of the unweighted Cartesian coordinate system

$$q_i = m_i^{1/2} x_i$$
$$S_t = \sum_i D_{ti} q_i$$
$$= \sum_i D_{ti} m_i^{1/2} x_i \qquad 16.22$$

But S_t can also be expressed as a linear combination of the unweighted Cartesian coordinates

$$S_t = \sum_i B_{ti} x_i \qquad 16.23$$

Comparing Eqs. 16.22 and 16.23, we see that

$$D_{ti} = m_i^{-1/2} B_{ti} \tag{16.24}$$

or

$$\sum_i D_{ti} D_{is} = \sum_i m_i^{-1} B_{ti} B_{is} \equiv G_{ts} \tag{16.25}$$

If we define the matrix **G** as

$$\mathbf{G} = \mathbf{B}\mathbf{M}^{-1}\mathbf{B}^{\dagger} \tag{16.26}$$

where $\mathbf{M}^{-1}$ is a diagonal matrix having the inverse masses along the diagonal, and **B** is the matrix relating the **S**-coordinate vector to the **x** vector, then we can rewrite Eq. 16.21 as

$$2T = \mathbf{P}^{\dagger}\mathbf{G}\mathbf{P} \tag{16.27}$$

or in terms of individual elements

$$2T = \sum_t \sum_s P_t G_{ts} P_s \tag{16.27a}$$

However, Eq. 16.5 is in terms of $\dot{S}$, not P. But

$$\begin{aligned} \dot{S}_t &= \frac{\partial T}{\partial P_t} \\ &= \sum_s G_{ts} P_s \end{aligned} \tag{16.28}$$

or

$$\begin{aligned} \dot{\mathbf{S}} &= \mathbf{G}\mathbf{P} \\ \mathbf{P} &= \mathbf{G}^{-1}\dot{\mathbf{S}} \end{aligned} \tag{16.29}$$

and

$$\begin{aligned} 2T &= \dot{\mathbf{S}}^{\dagger}(\mathbf{G}^{-1})^{\dagger}\mathbf{G}\mathbf{G}^{-1}\dot{\mathbf{S}} \\ &= \dot{\mathbf{S}}^{\dagger}\mathbf{G}^{-1}\dot{\mathbf{S}} \end{aligned} \tag{16.30}$$

or, comparing Eq. 16.30 with 16.5, we see that, in terms of individual elements,

$$t_{ij} = (G^{-1})_{ij} \tag{16.31}$$

Equation 16.14a becomes

$$|\mathbf{F} - \mathbf{G}^{-1}\omega^2| = 0 \tag{16.32}$$

This can be multiplied through by **G** to give

$$|\mathbf{F}\mathbf{G} - \omega^2\mathbf{E}| = 0 \tag{16.33}$$

where **E** is the identity matrix, having ones along the diagonal and zeros everywhere else. We have succeeded in constructing a determinantal equation that depends on the force constants and the **G** matrix.

16.4 FORCE FIELDS

In terms of the individual elements, Eq. 16.33 becomes

$$\begin{vmatrix} \sum_j f_{1j}G_{j1} - \omega^2 & \sum_j f_{1j}G_{j2} & \sum_j f_{1j}G_{j3} & \cdots \\ \sum_j f_{2j}G_{j2} & \sum_j f_{2j}G_{j2} - \omega^2 & \sum_j f_{2j}G_{j3} & \cdots \\ \cdots & \cdots & \cdots & \cdots \end{vmatrix} = 0 \qquad 16.34$$

As previously mentioned, the f_{ij} are the force constants and the G_{ij} are defined by Eq. 16.25. The G_{ij} are determined by the definition of the internal coordinates, in terms of the Cartesian displacement coordinates. Although their determination can be rather complicated (see the books listed in the bibliography), they are unique for a given choice of internal coordinates. On the other hand, there is no unanimity of opinion on how to choose the force constants. In general, if there are m internal coordinates, there will be $m(m + 1)/2$ possible independent force constants (since f_{ij} equals f_{ji}). Some of these may be equal because of symmetry, but there can still be many more than there are fundamental vibrations for the system.

Various approximations have been applied to limit the number of terms in Eq. 16.6. The most intuitively obvious of these is the *valence-force field*. The potential energy is written only in terms of changes in bond lengths and bond angles:

$$2V_{\text{VFF}} = \sum_b^{\text{bonds}} f_b(\Delta d_b)^2 + \sum_a^{\text{angles}} f_a(d\,\Delta\alpha_a)^2 \qquad 16.35$$

where α_a is half the bond angle and d is the length of the moving bond. This works well for sufficiently simple molecules, but in its simplest form it does not account for torsional motions and other internal motions requiring more than three atoms for their description. (It can be expanded to include such motions.) Its main drawback is that, owing to the lack of off-diagonal f_{ij} terms, it does not allow for interactions between the different potential-energy functions. With the valence-force field the **F** matrix is in diagonal form.

The *generalized valence-force field* is basically the same as the valence-force field but with off-diagonal elements of the **F** matrix included:

$$2V_{\text{GVFF}} = \sum_b^{\text{bonds}} f_b(\Delta d_b)^2 + \sum_a^{\text{angles}} f_a(d\,\Delta\alpha_a)^2 + \sum_b\sum_{b'} f_{bb'}\,\Delta d_b\,\Delta d_{b'}$$
$$+ \sum_a\sum_b f_{ab}(d\,\Delta\alpha_a)(\Delta d_b) + \sum_a\sum_{a'} f_{aa'}(d\,\Delta\alpha_a)(d'\,\Delta\alpha_{a'}) \qquad 16.36$$

Torsional angles are included. This can give more force-constant elements than fundamental vibrations. (Table 16.1 shows the effect of the off-diagonal terms for a few small molecules.) Various other forms have been suggested, some including anharmonic contributions and contributions for nonbonded interactions. Much of the work of a normal coordinate analysis of a vibrational spectrum lies in choosing the type of force field to use.

Table 16.1

COMPARISON OF THE VALENCE-FORCE FIELD[a] AND THE GENERALIZED VALENCE-FORCE FIELD[b] FOR SOME SIMPLE MOLECULES

Molecule	f_b^c	$f_{b'}$	f_a^d	$f_{bb'}^e$	f_{ab}^f
CO_2	16.8	14.2[g]	.57		
	15.5		.57	1.3	
HCN	5.8	17.9	.20		
	5.7	18.6	.20	−.22	
H_2O	7.9		.70		
	8.43		.77	−.10	.25

[a] First line for each molecule.
[b] Second line for each molecule.
[c] Bond stretching force constant (10^2 N m^{-1}).
[d] Bending force constant.
[e] Off-diagonal bond-bond interaction.
[f] Off-diagonal bond-angle interaction.
[g] The first value is calculated from the symmetric stretch, the second from the antisymmetric.

16.5 *SYMMETRY CONSIDERATIONS*

Group theory and symmetry are very useful in the study of molecular vibrations. The determination of the allowed symmetries of the normal modes of vibration and the vibrational selection rules can be completely accomplished by group theory. Furthermore, symmetry-adapted linear combinations of the internal coordinates, called *symmetry coordinates*, can be constructed to reduce the size of the secular determinant. These applications are completely analogous to the corresponding applications involving L.C.A.O. molecular orbitals. We will consider water and methane as examples.

The determination of the symmetries of the vibrations can be accomplished by using site symmetry and correlation diagrams. Consider water, **1**, with the indicated

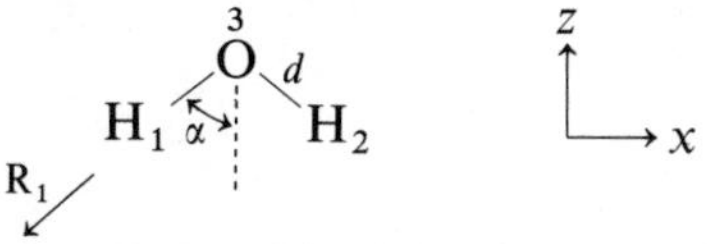

(**1**)

Cartesian coordinate orientation shown to the side (the y axis is into the plane of the paper) and the internal valence coordinates drawn on the molecule. The molecular symmetry is $\mathbf{C}_{2v}$. This is also the site symmetry of the oxygen. The site symmetry of the hydrogens is $\mathbf{C}_s$. For the vibrations, the basis set is the set of Cartesian displacement coordinates on each atom. In the $\mathbf{C}_{2v}$ site symmetry of the oxygen, the

z coordinate transforms as A_1, while x transforms as B_1 and y as B_2. In the $\mathbf{C}_s$ site symmetry of the hydrogens, the local y coordinates transform as A'' while x and z each transform as A'.

The correlation table for $\mathbf{C}_s$ with $\mathbf{C}_{2v}$ was given in Figure 14.1. We see that A' of $\mathbf{C}_s$ correlates with A_1 and B_1 of $\mathbf{C}_{2v}$, while A'' correlates with A_2 and B_2. Thus, the molecular functions that can arise from our basis set are three A_1, three B_1 (one of each from the oxygen and two from the hydrogens), one A_2 (from the hydrogens), and two B_2 (one from the oxygen and one from the hydrogens). These represent all the motions that are possible for the collection of two hydrogens and one oxygen. They include the translations and rotations of the entire molecule (sometimes called nongenuine or zero-frequency vibrations). If these are subtracted out, what remains are the vibrational representations. The translations and rotations can be found from the character tables. The translations transform as the three vectors x, y, and z (in this case they transform as the contributions of the oxygen to the motions)—i.e., the representations A_1, B_1, and B_2. The rotations are also indicated in the character tables, usually as R_x, R_y, and R_z. In the $\mathbf{C}_{2v}$ case these are A_2, B_1, and B_2. Subtracting these out leaves two A_1 and B_1 as the vibrational representations.

The requirement for the selection rules is the usual one. The product of the representations of the initial and final states must contain the representation of the operator. In the case of vibrations, the initial state is normally the ground state, which has the symmetry of the ground-state Hamiltonian. It must thus be totally symmetric. The selection rule is that an allowed vibrational transition must be to an excited vibrational state that transforms as some component of the operator. For normal absorption or emission of radiation (infrared spectroscopy) these are the components of the dipole operator. For $\mathbf{C}_{2v}$ the dipole components transform as A_1, B_1, or B_2. Both symmetries of the water vibrations represent infrared allowed vibrations. For Raman spectroscopy the operator is the polarizability operator, which transforms as the square of the dipole operator. The components of this are x^2, y^2, z^2, xy, xz, and yx. The representations according to which these transform are usually also indicated in the character tables. In the case of $\mathbf{C}_{2v}$, there are components of the polarizability that transform according to each representation. Consequently, any vibration is Raman-allowed in the $\mathbf{C}_{2v}$ point group.

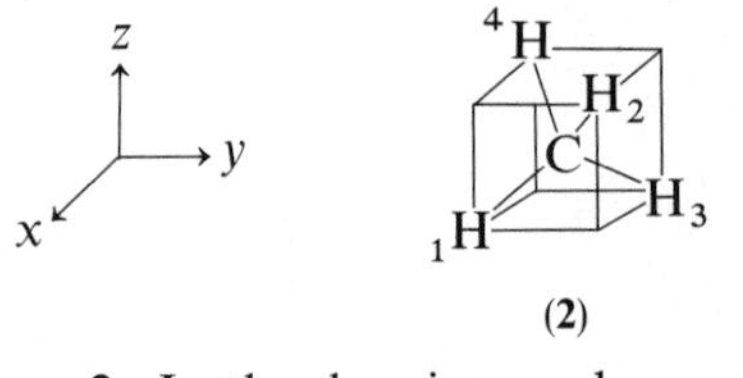

(2)

Now consider methane, **2**. In the drawing we have put the molecule inside a cube. The axis system is aligned parallel to the sides of the cube. The point group is $\mathbf{T}_d$. The site symmetry of the carbon is the full $\mathbf{T}_d$ symmetry. The site symmetry of the hydrogens is $\mathbf{C}_{3v}$. All three coordinates transform together as T_2 in the $\mathbf{T}_d$ group of the carbon. In the $\mathbf{C}_{3v}$ site symmetry the local z axis of the site group, the $\mathbf{C}_3$ axis, transforms as A_1. The local x and y coordinates together transform as E.

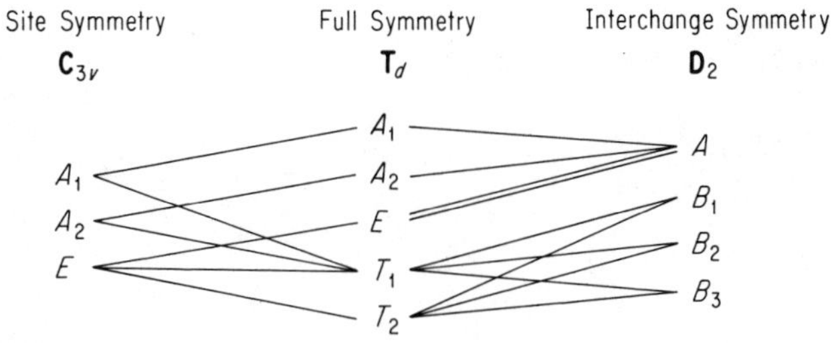

Figure 16.1 Correlations of the $\mathbf{C}_{3v}$ and $\mathbf{D}_2$ subgroups with the $\mathbf{T}_d$ point group.

Figure 16.1 shows the correlation table for the representations of $\mathbf{C}_{3v}$ with $\mathbf{T}_d$. From the figure, we see that the A_1 representation of $\mathbf{C}_{3v}$ correlates with A_1 and T_2 within $\mathbf{T}_d$. The E of $\mathbf{C}_{3v}$ correlates with E, T_1, and T_2 of $\mathbf{T}_d$. Thus, the total motion of the atoms transforms as A_1, E, T_1 (all from the hydrogens) and three T_2 (one from the carbon and two from the hydrogens). The translations transform as T_2 and the rotations as T_1. This leaves $A_1 + E +$ two T_2 for the vibrational representations. The dipole operator has only T_2 components; consequently, only the T_2 vibrations are infrared allowed. The Raman operator has A_1, E, and T_2 components. All vibrations are Raman-allowed.

16.6 WATER

As a first detailed example of a normal coordinate treatment, we will consider water. Let us call the displacements of the Cartesian coordinates X_i, Y_i, and Z_i, and the displacements of the internal coordinates R_i and $2A$. From a consideration of the geometry of structure **1** we see that, allowing the oxygen to move, as well as the hydrogens,

$$R_1 = -X_1 \sin\alpha - Z_1 \cos\alpha + X_3 \sin\alpha + Z_3 \cos\alpha \qquad \mathbf{16.37a}$$

$$R_2 = X_2 \sin\alpha - Z_2 \cos\alpha - X_3 \sin\alpha + Z_3 \cos\alpha \qquad \mathbf{16.37b}$$

$$2A = -\frac{X_1}{d}\cos\alpha + \frac{Z_1}{d}\sin\alpha + \frac{X_2}{d}\cos\alpha + \frac{Z_2}{d}\sin\alpha - \frac{2Z_3}{d}\sin\alpha \qquad \mathbf{16.37c}$$

where d is the bond length. Or, if $\mathbf{R}$ and $\mathbf{X}$ are the column vectors,

$$\mathbf{R} = \begin{bmatrix} R_1 \\ R_2 \\ 2A \end{bmatrix} \qquad \mathbf{X} = \begin{bmatrix} X_1 \\ Z_1 \\ X_2 \\ Z_2 \\ X_3 \\ Z_3 \end{bmatrix} \qquad \mathbf{16.38}$$

we can write

$$\mathbf{R} = \mathbf{BX} \tag{16.39}$$

where $\mathbf{B}$ is the 3×6 rectangular matrix

$$\mathbf{B} = \begin{bmatrix} -\sin\alpha & -\cos\alpha & 0 & 0 & \sin\alpha & \cos\alpha \\ 0 & 0 & \sin\alpha & -\cos\alpha & -\sin\alpha & \cos\alpha \\ -\frac{1}{d}\cos\alpha & \frac{1}{d}\sin\alpha & \frac{1}{d}\cos\alpha & \frac{1}{d}\sin\alpha & 0 & -\frac{2}{d}\sin\alpha \end{bmatrix} \tag{16.40}$$

The matrix $\mathbf{G}$ is

$$\mathbf{G} = \mathbf{BM}^{-1}\mathbf{B}^{\dagger} \tag{16.41}$$

where $\mathbf{M}^{-1}$ is the reciprocal mass matrix

$$\mathbf{M}^{-1} = \begin{bmatrix} m_1^{-1} & 0 & 0 & 0 & 0 & 0 \\ 0 & m_1^{-1} & 0 & 0 & 0 & 0 \\ 0 & 0 & m_2^{-1} & 0 & 0 & 0 \\ 0 & 0 & 0 & m_2^{-1} & 0 & 0 \\ 0 & 0 & 0 & 0 & m_3^{-1} & 0 \\ 0 & 0 & 0 & 0 & 0 & m_3^{-1} \end{bmatrix} \tag{16.42}$$

The resulting $\mathbf{G}$ matrix has the form

$$\mathbf{G} = \begin{bmatrix} g_{11} & g_{12} & g_{13} \\ g_{21} & g_{22} & g_{23} \\ g_{31} & g_{32} & g_{33} \end{bmatrix} \tag{16.43}$$

where the individual matrix elements are

$$g_{11} = m_1^{-1} + m_3^{-1} \tag{16.44a}$$

$$g_{12} = g_{21} = m_3^{-1}\cos 2\alpha \tag{16.44b}$$

$$g_{13} = g_{23} = g_{31} = g_{32} = -\frac{m_3^{-1}}{d^2}\sin 2\alpha \tag{16.44c}$$

$$g_{22} = m_2^{-1} + m_3^{-1} \tag{16.44d}$$

$$g_{33} = \frac{m_1^{-1}}{d^2} + \frac{m_2^{-1}}{d^2} + \frac{2m_3^{-1}}{d^2}(1 - \cos 2\alpha) \tag{16.44e}$$

(Note that, since m_1 equals m_2, g_{22} equals g_{11}.) In the generalized force-field approximation, the $\mathbf{F}$ matrix has the form

$$\mathbf{F} = \begin{bmatrix} f_R & f_{RR} & df_{RA} \\ f_{RR} & f_R & df_{RA} \\ df_{RA} & df_{RA} & d^2 f_A \end{bmatrix} \tag{16.45}$$

Equations 16.43 and 16.45 could be combined according to Eq. 16.33 to obtain the secular determinant and the eigenvalues; however, symmetry adaptation is possible. The $2A$ coordinate has A_1 symmetry within $\mathbf{C}_{2v}$. The R's have $\mathbf{C}_s$ site symmetry, and A' symmetry within it. They contribute to normal coordinates of A_1 and B_1 symmetry. Again consulting Figure 14.1, we see that the symmetric combination of R_1 and R_2 leads to the A_1 function, while the antisymmetric combination leads to the B_1 function. We thus have, after normalization, the symmetry-adapted internal coordinates

$$S_1 = \frac{1}{\sqrt{2}}(R_1 + R_2) \qquad (A_1) \tag{16.46a}$$

$$S_2 = 2A \qquad (A_1) \tag{16.46b}$$

$$S_3 = \frac{1}{\sqrt{2}}(R_1 - R_2) \qquad (B_1) \tag{16.46c}$$

This transformation can be represented in matrix form:

$$\mathbf{S} = \mathbf{UR} \tag{16.47}$$

where $\mathbf{S}$ is the column vector of the S_i, $\mathbf{R}$ is as defined in Eq. 16.38, and $\mathbf{U}$ is the unitary transformation matrix

$$\mathbf{U} = \begin{bmatrix} \frac{1}{\sqrt{2}} & \frac{1}{\sqrt{2}} & 0 \\ 0 & 0 & 1 \\ \frac{1}{\sqrt{2}} & -\frac{1}{\sqrt{2}} & 0 \end{bmatrix} \tag{16.48}$$

We can symmetry-adapt the $\mathbf{F}$ and $\mathbf{G}$ matrices (Eqs. 16.45 and 16.43) by subjecting them to a unitary transformation with $\mathbf{U}$:

$$\mathbf{F}' = \mathbf{UFU}^\dagger$$

$$= \begin{bmatrix} \frac{1}{\sqrt{2}} & \frac{1}{\sqrt{2}} & 0 \\ 0 & 0 & 1 \\ \frac{1}{\sqrt{2}} & -\frac{1}{\sqrt{2}} & 0 \end{bmatrix} \begin{bmatrix} f_{11} & f_{12} & f_{13} \\ f_{12} & f_{11} & f_{13} \\ f_{13} & f_{13} & f_{33} \end{bmatrix} \begin{bmatrix} \frac{1}{\sqrt{2}} & 0 & \frac{1}{\sqrt{2}} \\ \frac{1}{\sqrt{2}} & 0 & -\frac{1}{\sqrt{2}} \\ 0 & 1 & 0 \end{bmatrix}$$

$$= \begin{bmatrix} f_{11} + f_{12} & \sqrt{2}f_{13} & 0 \\ \sqrt{2}f_{13} & f_{33} & 0 \\ 0 & 0 & f_{11} - f_{12} \end{bmatrix} \tag{16.49}$$

$$\mathbf{G}' = \mathbf{U}\mathbf{G}\mathbf{U}^{\dagger}$$

$$= \begin{bmatrix} \frac{1}{\sqrt{2}} & \frac{1}{\sqrt{2}} & 0 \\ 0 & 0 & 1 \\ \frac{1}{\sqrt{2}} & -\frac{1}{\sqrt{2}} & 0 \end{bmatrix} \begin{bmatrix} g_{11} & g_{12} & g_{13} \\ g_{12} & g_{11} & g_{13} \\ g_{13} & g_{13} & g_{33} \end{bmatrix} \begin{bmatrix} \frac{1}{\sqrt{2}} & 0 & \frac{1}{\sqrt{2}} \\ \frac{1}{\sqrt{2}} & 0 & -\frac{1}{\sqrt{2}} \\ 0 & 1 & 0 \end{bmatrix}$$

$$= \begin{bmatrix} g_{11} + g_{12} & \sqrt{2}g_{13} & 0 \\ \sqrt{2}g_{13} & g_{33} & 0 \\ 0 & 0 & g_{11} - g_{12} \end{bmatrix} \qquad \mathbf{16.50}$$

Note that both of these factor into a 2 × 2 matrix corresponding to the A_1 functions and a 1 × 1 corresponding to the B_1. Their product will have the same structure.

The water molecule has three fundamental vibrations. Including the off-diagonal elements (see Eq. 16.45), the force-constant matrix has four unique elements; consequently there is not enough experimental information to completely determine the force constants. We can circumvent the problem by studying isotopically substituted water molecules, assuming that the force constants are not affected by the isotopic substitution. In the vapor phase, the A_1 stretching vibration is observed at 3651.7 cm^{-1}, the B_1 stretch at 3755.8 cm^{-1}, and the A_1 bend at 1595.0 cm^{-1}. Some general qualitative features of vibrational spectra are apparent in as simple a system as water. The asymmetric vibration, from a given set of basis vibrations, generally occurs at a slightly higher energy than does the symmetric vibration. Bends involving a given set of atoms normally occur at significantly lower energies than stretches. In addition to the fundamental transitions, a number of overtone and combination bands are observed in the vibrational spectrum of water. A total of nineteen bands are reported in the infrared spectrum by Herzberg.

16.7 METHANE

The structure of methane was given as **2**. Inscribed in the cube, as in the structure, it is reasonably easy to express the bond stretches and angle bends in terms of the localized Cartesian coordinates. We have

$$R_1 = \tfrac{1}{3}\Big[X_1 - Y_1 - Z_1 + \sum_{i \neq 1}^{5} (-X_i + Y_i + Z_i)\Big] \qquad \mathbf{16.51a}$$

$$R_2 = \tfrac{1}{3}\Big[X_2 + Y_2 + Z_2 + \sum_{i \neq 2}^{5} (-X_i - Y_i - Z_i)\Big] \qquad \mathbf{16.51b}$$

$$R_3 = \tfrac{1}{3}[-X_3 + Y_3 - Z_3 + \sum_{i \neq 3}^{5} (X_i - Y_i + Z_i)] \qquad \textbf{16.51c}$$

$$R_4 = \tfrac{1}{3}[-X_4 - Y_4 + Z_4 + \sum_{i \neq 4}^{5} (X_i + Y_i - Z_i)] \qquad \textbf{16.51d}$$

and, adopting the notation R_5 through R_{10} for the angle deformations,

$$R_5 = 2A_{12} = \frac{1}{\sqrt{6}d}(-Y_1 - Z_1 + Y_2 + Z_2) + \frac{2\sqrt{2}}{\sqrt{3}d}(X_5 + X_3 + X_4) \qquad \textbf{16.51e}$$

$$R_6 = 2A_{13} = \frac{1}{\sqrt{6}d}(X_1 - Y_1 - X_3 + Y_3) - \frac{2\sqrt{2}}{\sqrt{3}d}(Z_5 + Z_2 + Z_4) \qquad \textbf{16.51f}$$

$$R_7 = 2A_{14} = \frac{1}{\sqrt{6}d}(X_1 - Z_1 - X_4 + Z_4) + \frac{2\sqrt{2}}{\sqrt{3}d}(Y_5 + Y_2 + Y_3) \qquad \textbf{16.51g}$$

$$R_8 = 2A_{23} = \frac{1}{\sqrt{6}d}(-X_2 + Z_2 + X_3 - Z_3) - \frac{2\sqrt{2}}{\sqrt{3}d}(Y_5 + Y_1 + Y_4) \qquad \textbf{16.51h}$$

$$R_9 = 2A_{24} = \frac{1}{\sqrt{6}d}(X_2 + Y_2 - X_4 - Y_4) - \frac{2\sqrt{2}}{\sqrt{3}d}(Z_5 + Z_1 + Z_3) \qquad \textbf{16.51i}$$

$$R_{10} = 2A_{34} = \frac{1}{\sqrt{6}d}(Y_3 - X_3 - Y_4 + Z_4) - \frac{2\sqrt{2}}{\sqrt{3}d}(X_5 + X_1 + X_2) \qquad \textbf{16.51j}$$

Notice that all fifteen of the Cartesian coordinates are required to define these; however, we have defined only ten internal coordinates. The **B** matrix is thus a 10×15 matrix. In methane there are nine normal modes of vibration ($3N - 6$). We have more internal coordinates than are required. If we left the problem in this situation, one of the calculated vibrational frequencies would come out to be zero. (Vibrational problems are frequently treated by using more internal coordinates than there are normal modes.)

The **G** matrix can be constructed with the **B** matrix, and the **F** matrix from the R_i coordinates. This would give rise to 10×10 matrices and to a 10×10 secular determinant. We could factor these according to symmetry as we did for water. However, it is easier to define symmetry coordinates (do the symmetry factoring) first, so that the large **F** and **G** matrices will not have to be constructed. The procedure is essentially the same as in constructing matrices with respect to symmetry-adapted functions for molecular-orbital problems. The symmetry-adapted coordinates can be obtained from the site and interchange symmetries. The R_i have a $\mathbf{C}_{3v}$ site symmetry and $\mathbf{D}_2$ interchange symmetry. These coordinates transform as A_1 in $\mathbf{C}_{3v}$. Thus, from Figure 16.1, we see that the stretching coordinates can lead to A_1 or T_2 vibrations. The A combination from the $\mathbf{D}_2$ interchange group gives the A_1 vibration, while B_1, B_2, and B_3 each give one component of the T_2 vibration.

Table 16.2

CHARACTER TABLE FOR $\mathbf{D}_2$ WITH THE RESULTS OF OPERATING ON THE COORDINATES R_1, R_5, R_6, AND R_7 WITH THE OPERATIONS OF $\mathbf{D}_2$

$\mathbf{D}_2$	E	$C_2(z)$	$C_2(y)$	$C_2(x)$
A	1	1	1	1
B_1	1	1	-1	-1
B_2	1	-1	1	-1
B_3	1	-1	-1	1
RR_1	R_1	R_3	R_4	R_2
RR_5	R_5	R_{10}	R_{10}	R_5
RR_6	R_6	R_6	R_9	R_9
RR_7	R_7	R_8	R_7	R_8

We can write these down immediately, using the character table and operation results shown in Table 16.2:

$$S(A_1) = \hat{P}^A R_1 = R_1 + R_2 + R_3 + R_4 \tag{16.52}$$

or, normalizing,

$$S(A_1) = \tfrac{1}{2}(R_1 + R_2 + R_3 + R_4) \tag{16.52a}$$

One component of the T_2 coordinate is

$$S_1(T_2) = \hat{P}^{B_1} R_1 = R_1 + R_3 - R_4 - R_2 \tag{16.53}$$

or, normalizing,

$$S_1(T_2) = \tfrac{1}{2}(R_1 - R_2 + R_3 - R_4) \tag{16.53a}$$

The bending coordinates do not have the same site symmetry as the stretches. However, linear combinations of R_5, R_6, and R_7 can be constructed that have the same $\mathbf{C}_{3v}$ site symmetry as R_1. The two independent linear combinations that can be constructed are

$$R' = \frac{1}{\sqrt{3}}(R_5 + R_6 + R_7) \tag{16.54a}$$

which transforms as A_1 within $\mathbf{C}_{3v}$, and

$$R'' = \frac{1}{\sqrt{6}}(2R_5 - R_6 - R_7) \tag{16.54b}$$

which transforms as E. From Figure 16.1 we see that an A_1 site symmetry contributes to A_1 and T_2 functions within $\mathbf{T}_d$. The totally symmetric combination of the bends is impossible. All the angles cannot expand simultaneously. This eliminates the A_1 vibration arising from R'. The T_2 vibrations from R' are the same as those from

R''. Thus, R' can be ignored in the vibrations. On the other hand, E site symmetry from $\mathbf{C}_{3v}$ contributes to normal coordinates of E, T_1 and T_2 symmetry. Again we can ignore the T_1. We see that our chosen internal coordinates, R_1 through R_7 and R'' (and the coordinates generated from it by the interchange group), span the required A_1, E, and two T_2 representations. Applying the A projection operator from $\mathbf{D}_2$ to Eq. 16.54b, we obtain one component of the E vibration:

$$S(E) = \hat{P}^A R'' = \frac{1}{\sqrt{12}}(2R_5 + 2R_{10} - R_6 - R_7 - R_8 - R_9) \qquad \mathbf{16.55}$$

(after normalization), while the B_1 projection operator yields one component of the T_2 vibration:

$$S_2(T_2) = \hat{P}^{B_1} R'' = \frac{1}{\sqrt{2}}(R_6 - R_9) \qquad \mathbf{16.56}$$

There are only one A_1 and one E vibration; consequently Eqs. 16.52a and 16.55 are two of the final vibrational coordinates. The T_2 coordinates lead to a 2×2 secular determinant, constructed from Eqs. 16.53a and 16.56.

We can construct the symmetry-adapted $\mathbf{F}$ and $\mathbf{G}$ matrices by finding the matrix elements of the operators $\hat{f}$ and $\hat{g}$ with respect to the symmetry-adapted coordinates. For the A_1 coordinate, we have for f^{A_1}:

$$\begin{aligned} f^{A_1} &= \langle S(A_1)|\hat{f}|S(A_1)\rangle \\ &= \tfrac{1}{4}\langle R_1 + R_2 + R_3 + R_4|\hat{f}|R_1 + R_2 + R_3 + R_4\rangle \\ &= \tfrac{1}{4}(f_{11} + f_{22} + f_{33} + f_{44} + 2f_{12} + 2f_{13} + 2f_{14} + 2f_{23} + 2f_{24} + 2f_{34}) \\ &= f_{11} + 3f_{12} \\ &= f_R + 3f_{RR} \end{aligned} \qquad \mathbf{16.57}$$

where like elements have been set equal. Similarly, for g^{A_1}:

$$\begin{aligned} g^{A_1} &= \langle S(A_1)|\hat{g}|S(A_1)\rangle \\ &= g_{11} + 3g_{12} \\ &= g_R + 3g_{RR} \end{aligned} \qquad \mathbf{16.58}$$

We can determine the form of the g elements either by constructing the $\mathbf{G}$ matrix from the $\mathbf{B}$ and $\mathbf{M}^{-1}$ matrices, or from the tables given in the books in the bibliography. For the E coordinate, f^E is

$$\begin{aligned} f^E &= \langle S(E)|\hat{f}|S(E)\rangle \\ &= \tfrac{1}{12}\langle 2R_5 + 2R_{10} - R_6 - R_7 - R_8 - R_9|\hat{f}|2R_5 + 2R_{10} - R_6 - R_7 - R_8 - R_9\rangle \\ &= \tfrac{1}{12}(4f_{5,5} + 4f_{10,10} + f_{6,6} + f_{7,7} + f_{8,8} + f_{9,9} + 8f_{5,10} - 4f_{5,6} - 4f_{5,7} \\ &\quad - 4f_{5,8} - 4f_{5,9} - 4f_{6,10} - 4f_{7,10} - 4f_{8,10} - 4f_{9,10} + 2f_{6,7} + 2f_{6,8} + 2f_{6,9} \\ &\quad + 2f_{7,8} + 2f_{7,9} + 2f_{8,9}) \\ &= f_{5,5} - 2f_{5,6} + f_{5,10} \\ &= d^2(f_A - 2f_{AA} + f_{AA'}) \end{aligned} \qquad \mathbf{16.59}$$

where f_{AA} is the interaction of two angles that share a common bond and $f_{AA'}$ is that for two that do not. Similarly

$$\begin{aligned} g^E &= \langle S(E)|\hat{g}|S(E)\rangle \\ &= g_A - 2g_{AA} + g_{AA'} \end{aligned} \tag{16.60}$$

The T_2 coordinates lead to a 2 × 2 secular determinant. We have, for the **F** matrix elements:

$$\begin{aligned} f_{11}^{T_2} &= \langle S_1(T_2)|\hat{f}|S_1(T_1)\rangle \\ &= \tfrac{1}{4}\langle R_1 - R_2 + R_3 - R_4|\hat{f}|R_1 - R_2 + R_3 - R_4\rangle \\ &= f_{11} - f_{12} \\ &= f_R - f_{RR} \end{aligned} \tag{16.61a}$$

$$\begin{aligned} f_{22}^{T_2} &= \langle S_2(T_2)|\hat{f}|S_2(T_2)\rangle \\ &= \tfrac{1}{2}\langle R_6 - R_9|\hat{f}|R_6 - R_9\rangle \\ &= f_{66} - f_{69} \\ &= d^2(f_A - f_{AA'}) \end{aligned} \tag{16.61b}$$

$$\begin{aligned} f_{12}^{T_2} &= \langle S_1(T_2)|\hat{f}|S_2(T_2)\rangle \\ &= \frac{1}{2\sqrt{2}}\langle R_1 - R_2 + R_3 - R_4|\hat{f}|R_6 - R_9\rangle \\ &= \sqrt{2}(f_{16} - f_{19}) \\ &= \sqrt{2}d(f_{RA} - f_{RA'}) \end{aligned} \tag{16.61c}$$

The f_{RA} is the interaction of a bond stretch with an angle involving that bond, while the $f_{RA'}$ is with one that does not. The elements of the **G** matrix are completely analogous to those of the **F** matrix. From an examination of Eqs. 16.57, 16.59, and 16.61 we see that seven force constants are required for a generalized valence-force-field treatment of methane. There are only four fundamental vibrations, of which only two are infrared active. Determining the others again requires isotopic substitution or the neglect of some of the terms. Experimentally, the two T_2 vibrations are observed at 3020.3 cm^{-1} (stretch) and 1306.2 cm^{-1} (bend) in the gas-phase spectrum. The A_1 vibration is observed at 2914.6 cm^{-1}, and the E probably occurs at 1536 cm^{-1}. Although theoretically allowed in the Raman spectrum, it is not directly observed.

16.8 GENERAL CONSIDERATIONS

We have treated the vibrations of two simple molecules, where the forms of the vibrations were almost totally determined by the symmetry. In both cases more data than could be provided from the spectrum was required for determining all the force constants for a generalized valence-force field. The situation becomes even worse for more complicated molecules. Detailed vibrational analysis of complicated molecules requires the use of fairly sophisticated computer programs and often extensive isotopic substitution. Even so, consideration must be given to the type of force field to be used and to the approximations to be made when applying the force field.

The computer programs used are basically fitting programs. Initial guesses are made for the force constants. The program computes a spectrum based upon these, compares it with the experimental spectrum, and then modifies the initial constants until a best fit is obtained. The initial guesses are usually based upon the fact that force constants for similar motions of a given set of atoms, in similar molecules, have nearly the same values.

Various types of information can be deduced from vibrational studies. If the molecule is simple enough, structural information can be deduced from the **G** matrix. For slightly more complicated molecules, structural information can be verified by checking the consistency of the structural information through the use of a **G** matrix calculated from it. Stretching force constants are proportional to bond strengths. Other force constants depend upon other types of interactions. Consequently, experimentally determined force constants provide checks on ideas concerning chemical bonding.

16.9 OVERTONE AND COMBINATION BANDS

In general, a normal mode can absorb more than one quantum of vibrational energy to produce overtones of the various vibrations. This is forbidden by the harmonic-oscillator selection rule; however, the anharmonicity of real vibrations makes it weakly allowed. Combinations can also be produced in which each of two or more modes receives one or more quanta of energy. The resulting vibrational states in these cases are constructed in a manner somewhat analogous to the construction of electronic states from orbital occupancy. Here, however, the vibrational functions are assumed to be "occupied" by a quantum of vibrational energy. With this one difference, the construction of combination states is completely analogous to that of the electronic states. For example, for a vibrational state in which there is a simultaneous e and t_2 excitation, the resulting states are

$$E \times T_2 = T_1 + T_2 \qquad \textbf{16.62}$$

For overtones, the treatment is again analogous to the electronic case, if the vibration is nondegenerate. If the vibration is degenerate, permutational symmetry adaptation is required. In the vibrational case the vibrational Hamiltonian is a Boson operator, as already mentioned. This means that the degenerate representation must be raised to a symmetrized power, rather than to an antisymmetrized power, as was the case in electronic states. For the nth power we need to adapt to the $[n]$ representation of $\mathbf{S}(n)$. This can be accomplished by use of Eq. 7.9. For the triple excitation of an e vibration of methane we have

$$\chi(R); [3] = \tfrac{1}{6}\{1 \times 1 \times [\chi(R)]^3 + 3 \times 1 \times [\chi(R^2)\chi(R)] + 2 \times 1 \times [\chi(R^3)]\} \qquad 16.63$$

This gives the representation

$\mathbf{T}_d$	E	$8C_3$	$3C_2$	$6S_4$	$6\sigma_d$
Γ	4	1	4	0	0

16.64

This reduces to $A_1 + A_2 + E$ for the vibrational states arising from this overtone.

16.10 VIBRONIC, ROTOVIBRATIONAL, AND ROTOVIBRONIC COUPLING

The complete specification of the state of a molecule in the gas phase requires the specification of its rotational state, its vibrational state, and its electronic state. Spectroscopists study the energy differences between states. In the Born-Oppenheimer and independent-particle approximations, the total wave function is a simple product of the electronic, the vibrational, and the rotational wave functions. The symmetry of the state is that resulting from the product of the representations of the electronic, vibrational, and rotational functions. Spectral selection rules depend on the total symmetry of the initial and final states, not on the individual symmetries of the different types of wave function. Rotational spectroscopy involves smaller energy differences than either electronic or vibrational spectroscopy. Normally a molecule is in its ground electronic and vibrational state when a rotational spectrum is being studied. Thus, only changes in rotational state are occurring, and the symmetry-induced selection rules arise from the representations of the rotational states only. These were discussed in Chapter 3.

Since vibrational transitions occur at higher energies than rotational, a vibrational transition has considerably more energy associated with it than is required for a rotational transition. This means that a vibrational transition is likely to be accompanied by rotational transitions. This has profound effects on the spectrum, showing up in low-pressure gas-phase spectra as rotational fine structure superimposed on the vibrational spectrum. In the liquid phase, the rotational levels are perturbed by molecular interactions and collisions, so what appears in the vapor phase as fine structure becomes in the liquid phase merely a broadening of the vibrational bands. Often the shape of the broadened band is similar to that of the rotational "envelope" (a smooth curve connecting the maxima in the fine structure) in the gas phase. At other times the liquid interactions can completely obscure the band shape. In the crystalline solid state at low temperatures, molecular rotations cannot occur. The fine structure completely disappears and vibrational bands become very sharp. We can obtain selection rules for rotovibrational states by mapping the representations for the rotational states, from **R**(3) and **R**(2), onto the point group of the molecule, and then combining them with the vibrational representations. Figure 16.2 shows the rotational fine structure on the 3020-cm^{-1} vibrational band of methane.

Electronic transitions normally occur at significantly higher energies than vibrational. Consequently, both vibrational and rotational excitations can accompany an electronic transition. Often both of these can be resolved in low-pressure vapor-phase spectra. In liquid-phase spectra the rotational fine structure is lost and the vibrational structure is broadened, owing to collisions and interactions. The vibrational structure may not be resolvable, but the characteristic shape of the band envelope remains. In the crystalline state at low temperatures the vibrational fine structure can show up again as resolvable lines. The representations of the states

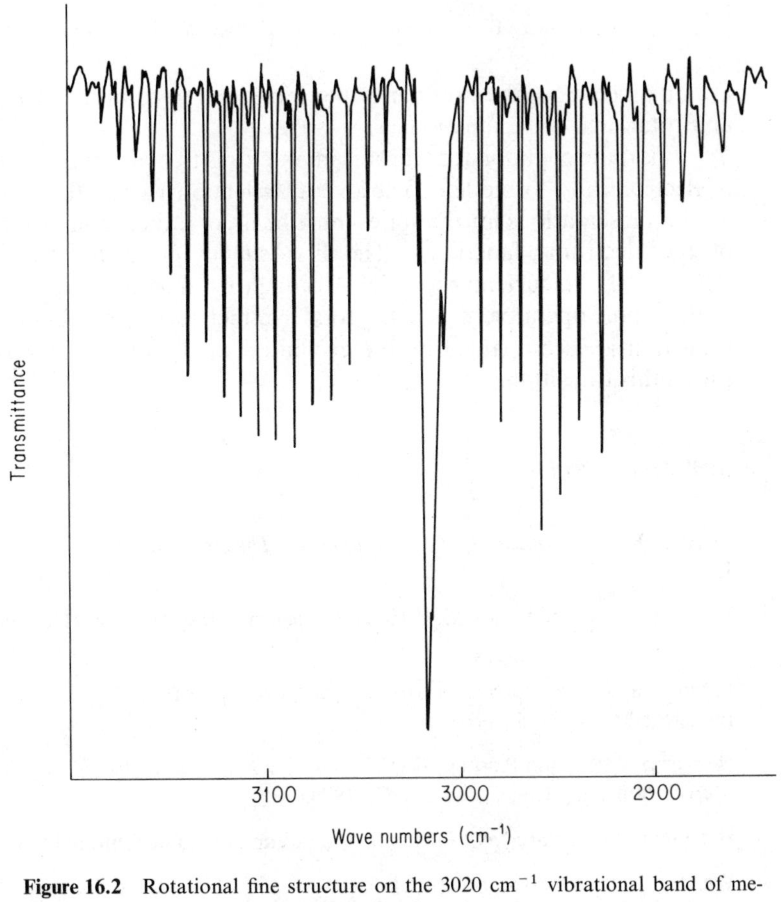

Figure 16.2 Rotational fine structure on the 3020 cm^{-1} vibrational band of methane. Gas phase. (Courtesy of M. A. Goodman.)

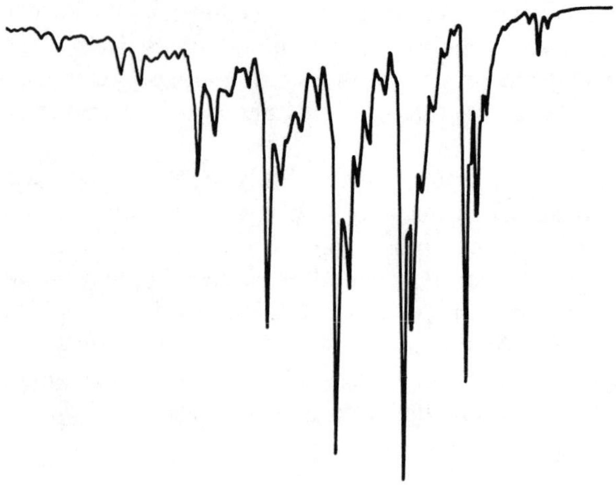

Figure 16.3 Fine structure on the $^1B_{2u}$ electronic band of benzene. Gas phase.

are obtained from the product of the representations of the electronic, vibrational, and rotational wave function.

It is quite common for vibronic interactions to make observable a transition that is electronically forbidden. The overall selection rule is the normal one, but the total representations for the states must be used, rather than just the representations of the electronic functions. The first singlet electronic transition of benzene ($^1B_{2u} \leftarrow {}^1A_{1g}$) is such an example. Although B_{2u} does not transform as a component of the dipole operator, A_{2u} or E_{1u} total symmetries can be obtained if the electronic transition is accompanied by the excitation of B_{1g} or E_{2g} vibrations. Figure 16.3 shows this transition.

BIBLIOGRAPHY

COTTON, F. A., *Chemical Applications of Group Theory*. John Wiley & Sons, New York, 2d ed., 1971.

FERRARO, J. R., and ZIOMEK, J. S., *Introductory Group Theory*. Plenum Press, New York, 2d ed., 1975.

FLURRY, R. L., JR., *Symmetry Groups: Theory and Chemical Applications*. Prentice-Hall, Inc., Englewood Cliffs, N. J., 1980.

HATFIELD, W. E., and PARKER, W. E., *Symmetry in Chemical Bonding and Structure*. Charles E. Merrill Publishing Co., Columbus, O., 1974.

HERZBERG, G., *Infrared and Raman Spectra*. Van Nostrand Reinhold Co., New York, 1945.

WILSON, E. B., DECIUS, J. C., and CROSS, P. C., *Molecular Vibrations*. McGraw-Hill Book Company, New York, 1955.

PROBLEMS

16.1 Determine the possible symmetries of the normal modes of vibration for the following molecules. Tell which are infrared-allowed, which are Raman-allowed, and which are forbidden.

*(a) PF_5 ($\mathbf{D}_{3h}$ point symmetry).

(b) Ethylene ($\mathbf{D}_{2h}$).

(c) $XeOF_4$ (C_{4v}).

(d) Benzene ($\mathbf{D}_{6h}$).

*(e) $PtCl_4^{2-}$ ($\mathbf{D}_{4h}$). (Compare these results to the $\mathbf{T}_d$ results. Infrared spectroscopy is useful for determining the structure of inorganic complexes.)

16.2 Sketch and label (with symmetry labels) the normal modes for BF_3.

***16.3** Assuming a valence-force field (i.e., assuming f_{RR} and f_{RA} to be negligible), calculate f_R and f_A for water from the vibrational data in the last paragraph of Section 16.6.

16.4 Using the valence-force-field data from Problem 16.3, calculate the fundamental vibrational frequencies for D_2O (the experimental values are 1178.7, 2666, and 2789 cm^{-1}).

16.5 Using the generalized valence-force-field data from Table 16.1, calculate the fundamental vibrational frequencies for D_2O.

16.6 Sketch and explain what you would expect to observe at very high resolution in the following spectra:

(a) The microwave spectrum of carbon monoxide (CO).

(b) The infrared spectrum of carbon monoxide.

(c) The ultraviolet spectrum of carbon monoxide.

***16.7** Give the overtone or combination vibrational state or states that can arise from the following excitations of ammonia (the vibrational quantum numbers for the various normal vibrations and their symmetries within $\mathbf{C}_{3v}$ are given):

	$v_1(a_1)$	$v_2(a_1)$	$v_3(e)$	$v_4(e)$
(a)	1	1	0	0
(b)	1	0	1	0
(c)	2	0	0	0
(d)	0	0	2	0
(e)	0	0	1	1
(f)	0	0	2	1
(g)	0	0	0	3

16.8 Construct the symmetry-adapted **F** matrices for the vibrations of ammonia. Assume a generalized valence-force field. Express the results in terms of f_{ij}, and so on.

chapter seventeen

Magnetic phenomena

17.1 INTRODUCTION

A number of magnetic phenomena are of interest to chemists. The most familiar are magnetic susceptibility and related phenomena, and the various types of magnetic resonance. *Magnetic susceptibility* is a bulk property. It and the molecular property, the *magnetic moment*, indicate the interaction of a substance with a magnetic field. There are two types of susceptibility: diamagnetic and paramagnetic (ferromagnetism and antiferromagnetism are special cases of these). A substance that is *diamagnetic* is repelled by a magnetic field. The effect is small and is due to the motion of the electrical charges in the system. A substance that is *paramagnetic* is attracted by a magnetic field. The effect depends upon the existence of a magnetic moment in the atoms or molecules of the substance. The magnetic moment of an atom or molecule, on the other hand, comes primarily from the intrinsic magnetic moments of the fundamental particles of the system (i.e., the electrons and the nuclei) and their interactions. There are also orbital contributions to atomic and molecular magnetic moments, but these are usually small.

Magnetic resonance (or paramagnetic resonance) is a spectral study of the energy-level differences that result between the magnetic states when a substance having a magnetic moment is placed in a magnetic field. Normally the studies concern the states of the nuclei (*nuclear magnetic resonance, N.M.R.*) or of the unpaired electrons (*electron paramagnetic resonance, E.P.R.*, or *electron spin resonance, E.S.R.*) in the system. A great deal of structural information can be obtained from magnetic-resonance experiments, particularly from N.M.R.

The fundamental particles commonly considered by chemists—the electron, the proton, and the neutron—all have nonzero intrinsic angular momentum (or "spin") and consequently have magnetic dipole moments. (We will use the common term "spin," rather than the more unwieldy "intrinsic angular momentum," even though the popular concept of the electron and other fundamental particles as spinning masses leads to many erroneous ideas.) The spins of the electron, the proton, and the neutron have the same magnitude, $\frac{1}{2}$. Atomic nuclei are, according to the chemist, made up of protons and neutrons. The spins of the protons and the neutrons couple to give a net spin to the nucleus, which can then be considered to be a single particle having the appropriate intrinsic angular momentum. The resulting spin for the nucleus corresponds to one of the spins that result from the Clebsch-Gordan series for the appropriate number of spin-$\frac{1}{2}$ particles, but there is no simple general way of predicting which one. However, if there is an odd number of either protons, neutrons, or both, the spin is nonzero.

By far the most commonly studied nucleus in N.M.R. experiments is the proton (spin-$\frac{1}{2}$). Most of the other commonly studied nuclei also have a spin of $\frac{1}{2}$—nuclei such as ^{13}C, ^{15}N, and ^{31}P. These have only a magnetic dipole moment. Nuclear-magnetic-resonance studies are also carried out on higher-spin nuclei; however, all nuclei with spins greater than $\frac{1}{2}$ also have quadrupole, and possibly higher, moments. The quadrupole moment usually causes extensive spectral broadening, which obscures the fine details of the spectra. Electron paramagnetic resonance studies the spin-$\frac{1}{2}$ electron. For these reasons, most of the theory of magnetic resonance has centered around spin-$\frac{1}{2}$ particles. We will treat in detail only spin-$\frac{1}{2}$ particles; however, the treatment will be sufficiently general that it can be extended to higher-spin particles.

17.2 *THE MAGNETIC HAMILTONIAN*

Magnetism is a relativistic phenomenon related to the spin of the fundamental particles. If the relativistic effects are sufficiently small, they can be treated as a perturbation to the nonrelativistic Schrödinger equation. When this approximation is valid, the wave function for the system can be factored into the simple product of a spatial function and a spin function. Normally, magnetic energy levels are more closely spaced than even rotational levels; consequently the spatial wave function is unchanged by changes in the spin function, within the limitations of the Pauli principle, which we will not consider in this connection.

The energy contributions from the spatial functions, the only terms we have discussed in previous chapters, are constant. We can effectively ignore them by assuming that they set the zero of the energy scale. (Note also that, owing to the small energy differences, the different magnetic states of a system have nearly equal Boltzmann populations at normal temperatures.) The magnetic properties depend only upon the spin function. This leads to the commonly used spin-only approximation for describing magnetic phenomena. For most magnetic properties in chemically interesting systems, this is a good approximation. For magnetic effects involving electrons in heavy elements, however, the relativistic effects begin to get larger and the approximation begins to break down. It can also begin to break down in extremely high magnetic fields.

We will concentrate on magnetic effects arising from the spin of the particles being studied. The magnetic moment of a particle is a relativistic effect that we will make no attempt to derive (Corio, Sec. 1.1.C). In terms of the intrinsic angular momentum (spin), **I**, of a particle, expressed as a vector, the magnetic dipole moment is

$$\boldsymbol{\mu} = g\beta\mathbf{I} \qquad \textbf{17.1}$$

where g is the Landé g factor, a dimensionless quantity that equals approximately 2 for the electron and must be experimentally determined for nuclei. β is known as the *magneton*

$$\beta = \frac{q\hbar}{2mc} \qquad \textbf{17.2}$$

where c is the speed of light and q and m are the charge and the mass of the electron, if the electron is the particle under consideration, or that of the proton if nuclei are under consideration. More commonly, if the electron is the particle, Eq. 17.1 is written with a negative sign on the right, so that the magnitude of the electronic charge, e, can be used (as a positive quantity) in Eq. 17.2. We will use the conventions of Eqs. 17.1 and 17.2 so that the equations will be general. Sometimes the magnetic moment is written instead in terms of the magnetogyric ratio, γ, for the particle:

$$\boldsymbol{\mu} = \gamma\hbar\mathbf{I} \qquad \textbf{17.3}$$

We can identify γ in terms of other quantities by comparing Eqs. 17.1 through 17.3.

Within the approximations we have listed, the only contributions to the Hamiltonian are the magnetic terms. These can be written in terms of the classical expressions. The energy of interaction between a magnetic dipole and a magnetic field is the scalar product of the dipole and the field, **H**:

$$\begin{aligned} Z &= \boldsymbol{\mu}\cdot\mathbf{H} \\ &= g\beta\mathbf{I}\cdot\mathbf{H} \end{aligned} \qquad \textbf{17.4}$$

For a many-particle system there is a sum of terms of the form of Eq. 17.4:

$$Z = \sum_i Z_i \qquad \textbf{17.5}$$

Where the Z_i are the single-particle Z terms. There are also the interactions between the particles. The energy of interaction of two magnetic dipoles (on two particles) is proportional to their scalar product

$$V_{ij} = J_{ij}\mathbf{I}_i\cdot\mathbf{I}_j \qquad \textbf{17.6}$$

where J_{ij} is the proportionality constant, which depends upon the separation of the particles and a number of other things. J_{ij} is called the *coupling constant*. Theories for calculating it are not very exact, so we will treat it as an empirical parameter. The total interaction energy for a collection of particles is

$$V = \sum_{i<j}\sum V_{ij} \qquad \textbf{17.7}$$

Both Eqs. 17.4 and 17.6 represent stabilizing interactions. This means that we have the wrong signs for inclusion in the Hamiltonian. The magnetic Hamiltonian can be written

$$\hat{H} = -(Z + V) \qquad \textbf{17.8}$$

The two terms are frequently called the *Zeeman* term and the *coupling* term, respectively.

In magnetic-resonance experiments, resonance frequencies relative to some standard are usually of more interest than are the absolute resonance frequencies. If the resonance frequency of the standard is chosen as the energy zero, the Zeeman term can be rewritten as

$$Z_i^0 - Z_i = g_i\beta(1 - \sigma_i)\mathbf{I}_i\cdot\mathbf{H} \qquad \textbf{17.9}$$

where σ_i is a *shielding constant* relating the Zeeman term to the standard, Z_i^0. In magnetic-resonance experiments, σ_i times the magnitude of the applied field, $\sigma_i|H|$, is referred to as the *chemical shift*.

17.3 THE SPIN-HAMILTONIAN SECULAR EQUATION

Physicists describe the intrinsic angular-momentum behavior of fundamental particles by use of the special unitary groups, **SU**(n), where n equals $(2I + 1)$. A *special unitary group* is the group of all unitary matrices (matrices having inverses equal to their conjugate transposes) of dimension n, with determinants equal to $+1$. In this group the intrinsic angular momentum (spin) of a single particle transforms as the group's first nonscalar irreducible representation (i.e., the first with a dimension greater than unity). Properly symmetry-adapted collections of identical particles transform as the higher-dimensional representations. [The three-dimensional rotation group, **R**(3), is a subgroup of all **SU**(n).] There are two valid labeling schemes for the representations of **SU**(n): the labels from the symmetric group, **S**(N), and angular-momentum labels. These considerations, as well as the fact that the group algebra of the **SU**(n) is well developed, are what make it convenient to use the **SU**(n) for describing spin properties.

When a particle is placed in a magnetic field, the effective symmetry is reduced to that of the field. We discussed this with respect to the atomic Zeeman effect in Chapter 8. For our present purposes, the two-dimensional **R**(2) rotation group is sufficient for the field. The subgroup chain

$$\mathbf{SU}(n) \supset \mathbf{R}(3) \supset \mathbf{R}(2) \qquad \textbf{17.10}$$

provides a set of valid quantum numbers for the magnetic behavior of a collection of identical particles in a field. The **SU**(n) provides the permutational label, the **R**(3) the total-angular-momentum label, and **R**(2) the z component of angular momentum (the m quantum number). If there is more than one set of equivalent magnetic particles in the system, there may also be spatial or permutational symmetries relating the different sets; however, we will not consider the symmetry aspects of such a situation here.

For spin-$\frac{1}{2}$ particles, the appropriate head group for the chain in Eq. 17.10 is **SU**(2). This group is locally isomorphic to (has the same generating function as) **R**(3), if the even-dimensional (half-integer indexed) representations are included in **R**(3). Thus, the angular-momentum information from **SU**(2) is the same as that from **R**(3). There cannot be more than one total spin value for a given permutational symmetry. (We saw this from the Young diagrams presented in Chapter 7.) This is not the case, however, for higher-spin systems.

In constructing a secular determinant, it is convenient to choose a basis set that incorporates as much of the symmetry of the system as is convenient. This reduces the number of matrix elements that must be calculated. In the present case, an optimum basis would be simultaneously symmetry adapted to **SU**(n), **R**(3) and **R**(2) (see Eq. 17.10), or for spin-$\frac{1}{2}$ particles, just **SU**(2) or **R**(3) and **R**(2). An extremely simple basis to use is a *spin-product basis* in which each single-particle function is an eigenfunction of the operations of **R**(2)—i.e., the z component of the angular momentum. (We will call the operator $\hat{I}_z$.) For the spin-$\frac{1}{2}$ case, these are the functions associated with the m_s of $\frac{1}{2}$ and the m_s of $-\frac{1}{2}$; that is, the α and the β spin functions. The simple product functions are not necessarily eigenfunctions of the **R**(3) operations (i.e., of the square of the total angular momentum, which we will call $\hat{I}^2$), but linear combinations that are eigenfunctions can easily be constructed. For two equivalent spin-$\frac{1}{2}$ particles, such as the two protons in H_2, the simple product functions are

$$\sigma_1 = \alpha(1)\alpha(2) \qquad \textbf{17.11a}$$

$$\sigma_2 = \alpha(1)\beta(2) \qquad \textbf{17.11b}$$

$$\sigma_3 = \beta(1)\alpha(2) \qquad \textbf{17.11c}$$

$$\sigma_4 = \beta(1)\beta(2) \qquad \textbf{17.11d}$$

There are two possible values for the total angular momentum for a two spin-$\frac{1}{2}$ system: $I = 1$, arising from the $[2]$ representation of **S**(2), and $I = 0$, arising from the $[1^2]$ representation. Applying the appropriate **S**(2) projection operators to Eqs. 17.11, we obtain the symmetry-adapted functions

$$\sigma_1^{[2]} = \alpha(1)\alpha(2) \qquad \textbf{17.12a}$$

$$\sigma_2^{[2]} = \beta(1)\beta(2) \qquad \textbf{17.12b}$$

$$\sigma_3^{[2]} = \frac{1}{\sqrt{2}}\left[\alpha(1)\beta(2) + \beta(1)\alpha(2)\right] \tag{17.12c}$$

$$\sigma^{[1^2]} = \frac{1}{\sqrt{2}}\left[\alpha(1)\beta(2) - \beta(1)\alpha(2)\right] \tag{17.12d}$$

The first three are the three components of the spin-one triplet state, while the last is the spin-zero singlet. Notice that these are still eigenfunctions of $\hat{I}_z$, having eigenvalues of 1, -1, 0, and 0, respectively, in units of $\hbar$.

Once the Hamiltonian and the basis have been defined, the construction of the matrix elements for the secular determinant is straightforward. In the present case, we can simplify the process by making full use of the properties of angular momentum. Let us rewrite the Hamiltonian of Eq. 17.8 as

$$\hat{H} = -\left\{\sum_i \omega_i \mathbf{n}\cdot\mathbf{I}_i + \sum_{i<j}\sum J_{ij}\mathbf{I}_i\cdot\mathbf{I}_j\right\} \tag{17.13}$$

where $\mathbf{n}$ is a unit vector lying along the field direction, ω_i is the *Larmor frequency*

$$\omega_i = g_i\beta|H| \tag{17.14}$$

and $|H|$ is the magnitude of the field. Since the field defines the z direction, $\mathbf{n}\cdot\mathbf{I}_i$ is just the projection of $\mathbf{I}_i$ on the z axis, I_{zi}. The spin functions we are using are eigenfunctions of $\hat{I}_z$. Consequently, the Zeeman contribution to the matrix elements is easy to calculate.

In order to calculate the coupling terms, it is convenient to define a pair of operators, $\hat{I}^+$ and $\hat{I}^-$, known variously as *ladder* operators, *raising* and *lowering* operators, or *step-up* and *step-down* operators ($i = \sqrt{-1}$):

$$\hat{I}^+ = \hat{I}_x + i\hat{I}_y \tag{17.15a}$$

$$\hat{I}^- = \hat{I}_x - i\hat{I}_y \tag{17.15b}$$

The ladder operators raise or lower the m value of the angular-momentum functions on which they act. If we schematically express an angular-momentum function as a Dirac ket (see Section 6.4), $|I, m\rangle$, in which the quantum number for the total angular momentum, I, and the m value imply the function, the results are (Corio, Sec. 2.2):

$$\hat{I}^+|I, m\rangle = (I - m)^{1/2}(I + m + 1)^{1/2}|I, m + 1\rangle \tag{17.16a}$$

$$\hat{I}^-|I, m\rangle = (I + m)^{1/2}(I - m + 1)^{1/2}|I, m - 1\rangle \tag{17.16b}$$

If the raising operator acts on a function having the maximum m value, the result is zero. The same is true if the lowering operator acts on a function having the minimum m. Specifically, for spin-$\frac{1}{2}$ particles, we have

$$\hat{I}^+\alpha = 0 \tag{17.17a}$$

$$\hat{I}^+\beta = \alpha \tag{17.17b}$$

$$\hat{I}^-\alpha = \beta \tag{17.17c}$$

$$\hat{I}^-\beta = 0 \tag{17.17d}$$

In terms of the ladder operators, the scalar product, $\mathbf{I}_i \cdot \mathbf{I}_j$, is

$$\begin{aligned}\mathbf{I}_i \cdot \mathbf{I}_j &= \hat{I}_{xi}\hat{I}_{xj} + \hat{I}_{yi}\hat{I}_{yj} + \hat{I}_{zi}\hat{I}_{zj} \\ &= \hat{I}_{zi}\hat{I}_{zj} + \tfrac{1}{2}(\hat{I}_i^+\hat{I}_j^- + \hat{I}_i^-\hat{I}_j^+)\end{aligned} \qquad \textbf{17.18}$$

In operator form, the complete magnetic Hamiltonian can now be written

$$H = -\left\{\sum_i \omega_i \hat{I}_{zi} + \sum_{i<j}\sum J_{ij}[\hat{I}_{zi}\hat{I}_{zj} + \tfrac{1}{2}(\hat{I}_i^+\hat{I}_j^- + \hat{I}_i^-\hat{I}_j^+)]\right\} \qquad \textbf{17.19}$$

All the matrix elements of this can be evaluated in terms of the $\hat{I}_{zi}$ and the ladder operators, without any numerical computation. Since the basis functions are eigenfunctions of the $\hat{I}_{zi}$, there can be no off-diagonal terms involving the $\hat{I}_{zi}$. The diagonal terms are a simple sum of single-particle terms from the $\sum_i \omega_i \hat{I}_{zi}$ and of two-particle products from the $\sum_{i<j}\sum J_{ij}\hat{I}_{zi}\hat{I}_{zj}$. The ladder operators contribute only to off-diagonal terms, since they change the basis functions. In the integral, $\langle\psi_A|\hat{H}|\psi_B\rangle$, ψ_A and ψ_B must differ in two and only two m values. One of these must be higher and the other lower, by one unit.

In order to illustrate the construction of the secular determinant, let us construct the determinant for the two-proton N.M.R. problem, using the simple product functions of Eqs. 17.11 as our basis. This gives a 4 × 4 determinant. We have

$$\begin{aligned}H_{11} = \langle\sigma_1|\hat{H}|\sigma_1\rangle &= \langle\alpha(1)\alpha(2)|\hat{H}|\alpha(1)\alpha(2)\rangle \\ &= -(\tfrac{1}{2}\omega_1 + \tfrac{1}{2}\omega_2 + \tfrac{1}{4}J_{12})\end{aligned} \qquad \textbf{17.20a}$$

$$\begin{aligned}H_{12} = \langle\sigma_1|\hat{H}|\sigma_2\rangle &= \langle\alpha(1)\alpha(2)|\hat{H}|\alpha(1)\beta(2)\rangle \\ &= 0\end{aligned} \qquad \textbf{17.20b}$$

$$H_{13} = \langle\sigma_1|\hat{H}|\sigma_3\rangle = 0 \qquad \textbf{17.20c}$$

$$H_{14} = \langle\sigma_1|\hat{H}|\sigma_4\rangle = 0 \qquad \textbf{17.20d}$$

$$\begin{aligned}H_{22} = \langle\sigma_2|\hat{H}|\sigma_2\rangle &= \langle\alpha(1)\beta(2)|\hat{H}|\alpha(1)\beta(2)\rangle \\ &= -(\tfrac{1}{2}\omega_1 - \tfrac{1}{2}\omega_2 - \tfrac{1}{4}J_{12})\end{aligned} \qquad \textbf{17.20e}$$

$$\begin{aligned}H_{23} = \langle\sigma_2|\hat{H}|\sigma_3\rangle &= \langle\alpha(1)\beta(2)|\hat{H}|\beta(1)\alpha(2)\rangle \\ &= \tfrac{1}{2}\langle\alpha(1)\beta(2)|J_{12}\hat{I}_1^+\hat{I}_2^-|\beta(1)\alpha(2)\rangle \\ &= -\tfrac{1}{2}J_{12}\end{aligned} \qquad \textbf{17.20f}$$

$$H_{24} = \langle\sigma_2|\hat{H}|\sigma_4\rangle = 0 \qquad \textbf{17.20g}$$

$$\begin{aligned}H_{33} = \langle\sigma_3|\hat{H}|\sigma_3\rangle &= \langle\beta(1)\alpha(2)|\hat{H}|\beta(1)\alpha(2)\rangle \\ &= -(-\tfrac{1}{2}\omega_1 + \tfrac{1}{2}\omega_2 - \tfrac{1}{4}J_{12})\end{aligned} \qquad \textbf{17.20h}$$

$$H_{34} = \langle\sigma_3|\hat{H}|\sigma_4\rangle = 0 \qquad \textbf{17.20i}$$

$$\begin{aligned}H_{44} = \langle\sigma_4|\hat{H}|\sigma_4\rangle &= \langle\beta(1)\beta(2)|\hat{H}|\beta(1)\beta(2)\rangle \\ &= -(-\tfrac{1}{2}\omega_1 - \tfrac{1}{2}\omega_2 + \tfrac{1}{4}J_{12})\end{aligned} \qquad \textbf{17.20j}$$

The energies of the system can be found from the roots of the resulting 4×4 determinant:

$$\begin{vmatrix} H_{11} - E & 0 & 0 & 0 \\ 0 & H_{22} - E & H_{23} & 0 \\ 0 & H_{23} & H_{33} - E & 0 \\ 0 & 0 & 0 & H_{44} - E \end{vmatrix} = 0 \qquad \textbf{17.21}$$

Because of the zeros, the determinant factors into two 1×1 determinants and a 2×2. The determinant can be solved and the wave functions constructed from the corresponding linear equations, just as in any other determinantal problem. Obviously, H_{11} and H_{44} are two of the eigenvalues, with σ_1 and σ_4 as the corresponding eigenfunctions. The other two roots arise from the 2×2 and give eigenfunctions that are linear combinations of σ_2 and σ_3.

If the nuclei are equivalent (as, for example, in H_2), the determinant need not be solved. The functions of Eqs. 17.12 are completely symmetry-adapted. The energies constructed from these are the energies of the proton states of the H_2 system. The terms given in Eqs. 17.20a and 17.20j are the energies associated with the functions of Eqs. 17.12a and 17.12b, respectively. Equations 17.12c and 17.12d are the wave functions for the two roots of the 2×2 determinant resulting from Eqs. 17.11b and 17.11c. These are

$$\begin{aligned} \langle \sigma_3^{[2]} | \hat{H} | \sigma_3^{[2]} \rangle &= \tfrac{1}{2} \langle \alpha(1)\beta(2) + \beta(1)\alpha(2) | \hat{H} | \alpha(1)\beta(2) + \beta(1)\alpha(2) \rangle \\ &= \tfrac{1}{2} (\langle \alpha(1)\beta(2) | \hat{H} | \alpha(1)\beta(2) \rangle \\ &\quad + \langle \beta(1)\alpha(2) | \hat{H} | \beta(1)\alpha(2) \rangle \\ &\quad + 2 \langle \alpha(1)\beta(2) | \hat{H} | \beta(1)\alpha(2) \rangle) \\ &= \tfrac{1}{2} \{ -(\tfrac{1}{2}\omega_1 - \tfrac{1}{2}\omega_2 - \tfrac{1}{4}J_{12}) \\ &\quad -(-\tfrac{1}{2}\omega_1 + \tfrac{1}{2}\omega_2 - \tfrac{1}{4}J_{12}) - J_{12} \} \\ &= -\tfrac{1}{4}J_{12} \end{aligned} \qquad \textbf{17.22}$$

$$\begin{aligned} \langle \sigma^{[1^2]} | \hat{H} | \sigma^{[1^2]} \rangle &= \tfrac{1}{2} \langle \alpha(1)\beta(2) - \beta(1)\alpha(2) | \hat{H} | \alpha(1)\beta(2) - \beta(1)\alpha(2) \rangle \\ &= \tfrac{3}{4} J_{12} \end{aligned} \qquad \textbf{17.23}$$

Since both nuclei are the same, the subscripts on the ω's can be dropped. We can also drop those on the J, since there is only one J value. With these simplifications, the energy-level scheme is as in Figure 17.1. Notice that in the absence of a field, ω equals zero, and the three [2] levels become degenerate. They represent the triplet spin state of *ortho* hydrogen. The $[1^2]$ level is the singlet spin state of *para* hydrogen. The determinantal procedure we presented is general for systems of spin-$\frac{1}{2}$ particles, although the symmetry adaptation is applicable only to identical particles. Once the energy levels and wave functions are determined, they can be used to calculate the magnetic-resonance spectrum or the susceptibilities, and related phenomena.

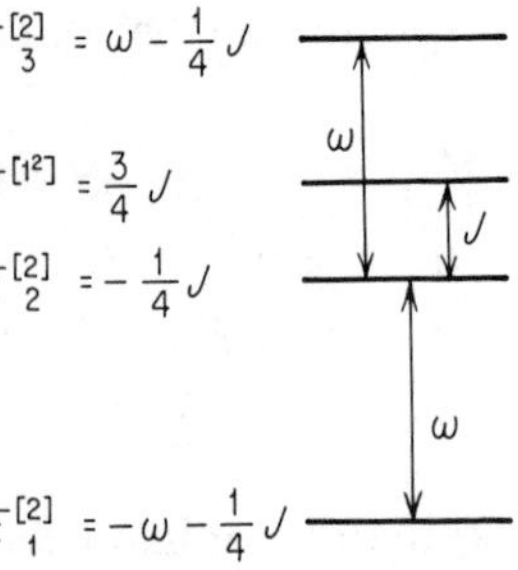

Figure 17.1 Energy levels of the nuclear spin states of H_2 in a magnetic field.

17.4 NUCLEAR MAGNETIC RESONANCE

General. There are two common types of magnetic-resonance experiments, N.M.R. and E.P.R. In order for us to detect an E.P.R. signal, there must be one or more unpaired electrons in the system; i.e., no spectrum is possible for singlet states. (This is obvious from Figure 17.1. The singlet state gives rise to only one energy level.) For this reason E.P.R. is useful, in a qualitative sense, for detecting unpaired electrons.

Usually systems with only one unpaired electron are studied. The single electron gives rise to a doublet state. There is only one transition involving an isolated electron in a magnetic field. However, in a molecule that has paramagnetic nuclei, the energy levels of the electron are split by coupling with the nuclei. The coupling constants (which are called *hyperfine constants* and usually given the symbol a) are proportional to the probability of finding the electron near a specific nucleus. Thus, E.P.R. gives a way of experimentally determining the electron density of the orbital of the unpaired electron. The Landé g value for the electron can be anisotropic (vary with angular orientation) if it is not in a spherical environment. In a fluid medium this averages out, owing to molecular motion; however, it can be observed in the solid state. Analysis of the resulting *g tensor* gives information about the symmetry of the environment of the unpaired electron. We will concentrate here on the isotropic case.

Nuclear magnetic resonance is extremely useful for structure determinations. Each equivalent set of magnetic nuclei has its own characteristic resonance frequency. Furthermore, there is a characteristic splitting pattern caused by nearby groups of magnetic nuclei. If the differences in the ω_i are large compared to the J_{ij} between the sets, the signals from each set do not overlap, and the intensity of the signals is directly proportional to the number of nuclei in the set. (Such a situation is commonly called a *first-order spectrum.*) From a first-order spectrum we can easily determine the number of different equivalent sets (from the number of signals), the relative numbers of nuclei in the sets (from the intensities), and the number of nuclei in adjacent sets (from the splitting of the signals). Much information about the chemical environment of the sets can also be determined from the frequencies. In general, the higher the electron density around a nucleus, the more it is shielded from the magnetic field and consequently the less the splitting between spin states of different m values and the lower the resonance frequency. The same information can be obtained from spectra that are not first-order as from those that are; to do so, however, requires a complete analysis of the spectrum.

N.M.R. of the Hydrogen Molecule. Let us use Figure 17.1 to deduce the N.M.R. spectrum of the hydrogen molecule. Since the **S**(2) labels are valid symmetry labels, and since the transition dipole operator has no permutational components (it is a single-particle operator), the only transitions observed are those between states having the same permutational labels, states $E_1^{[2]}$, $E_2^{[2]}$, and $E_3^{[2]}$. (This is just a more formal way of saying that the total spin quantum number does not change.) Furthermore, a transition between $E_1^{[2]}$ and $E_3^{[2]}$ would require that two particles simultaneously change their quantum states (which is, in the first approximation, forbidden). Thus, the only observed transitions are from $E_1^{[2]}$ to $E_2^{[2]}$ and from $E_2^{[2]}$ to $E_3^{[2]}$. Both of these occur at the same frequency, ω. Note that the transition does not depend upon the coupling constant, J.

These are general features for the magnetic resonance spectra of systems containing only one set of any number of equivalent, spin-$\frac{1}{2}$, magnetic particles. Only one frequency is seen in the spectrum, even though it may correspond to a number of different transitions. This frequency is independent of the coupling within the set. In treatments of such systems this coupling is frequently absorbed in the chosen energy zero. If the particles have spins greater than $\frac{1}{2}$, the coupling within equivalent sets can affect the spectrum.

First-order Spectra. If a system has more than one set of equivalent magnetic particles, the magnetic resonance spectrum becomes more interesting. Each set has its own characteristic ω value associated with it. There are coupling constants expressing interactions between sets as well as those within the sets. The coupling constants between the sets figure prominently in the frequencies of the transitions, although for spin-$\frac{1}{2}$ particles the coupling within the sets still does not affect the spectra. The energy levels can be calculated from the Hamiltonian of Eq. 17.19 and a spin-product basis. Usually, however, if a detailed analysis of the spectrum is undertaken, the ω_i and the J_{ij} are the information that is wanted from an experiment. In this case the problem is solved like the normal coordinate analysis problem. Initial guesses are made for the ω_i and the J_{ij}, and these are refined, by comparison of the calculated spectrum with the observed spectrum, until satisfactory agreement is obtained. In the magnetic resonance spectrum there is usually more experimental data than there are unknowns, so a unique solution is possible.

Consider the N.M.R. problem for a system of two nonequivalent protons. The determinantal equation is again Eq. 17.21. Two of the roots arise directly from Eqs. 17.20a and 17.20j; however, the ω_i terms are no longer equal. The roots of the remaining quadratic are

$$E^{\pm} = \tfrac{1}{4}J \pm \left[\left(\frac{\omega_1}{2} - \frac{\omega_2}{2}\right)^2 + \tfrac{1}{4}J^2\right]^{1/2} \qquad \textbf{17.24}$$

The resulting spin functions have the form

$$\sigma^{\pm} = c_2^{\pm}\alpha(1)\beta(2) + c_3^{\pm}\beta(1)\alpha(2) \qquad \textbf{17.25}$$

Let us make the assumption that $(\omega_1/2 - \omega_2/2)$ is much larger than J, in order to get a first-order spectrum (the system is called, in N.M.R. terminology, an AX spectrum), and call the two energies from Eq. 17.24 E_2 and E_3. We have, for the four energies,

$$E_1 = -\tfrac{1}{2}(\omega_1 + \omega_2) - \tfrac{1}{4}J \qquad \textbf{17.26a}$$

$$E_2 = \tfrac{1}{2}(\omega_2 - \omega_1) + \tfrac{1}{4}J \qquad \textbf{17.26b}$$

$$E_3 = \tfrac{1}{2}(\omega_1 - \omega_2) + \tfrac{1}{4}J \qquad \textbf{17.26c}$$

$$E_4 = \tfrac{1}{2}(\omega_1 + \omega_2) - \tfrac{1}{4}J \qquad \textbf{17.26d}$$

The symmetry labels that we had in Figure 17.1 are no longer valid, since the protons are no longer equivalent. Therefore, transitions are allowed from E_1 to both E_2 and E_3, and from both E_2 and E_3 to E_4. The transition energies are

$$\Delta E_{12} = \omega_2 + \tfrac{1}{2}J \qquad \textbf{17.27a}$$

$$\Delta E_{13} = \omega_1 + \tfrac{1}{2}J \qquad \textbf{17.27b}$$

$$\Delta E_{24} = \omega_1 - \tfrac{1}{2}J \qquad \textbf{17.27c}$$

$$\Delta E_{34} = \omega_2 - \tfrac{1}{2}J \qquad \textbf{17.27d}$$

Schematically, the so-called AX, spin-$\tfrac{1}{2}$, spectrum should look as shown in Figure 17.2. Notice that there are two doublets, centered at ω_1 and ω_2. The splitting in each doublet is J. In effect, each proton has its own characteristic resonance frequency, but the signal of each is split by the magnetic states of the other. The total intensities of the signals for the two protons are the same, since there is only one proton per set, and since the Boltzmann population of the states is essentially equal. This is the characteristic behavior of any first-order spectrum. If there is more than one proton in a set, the pattern the set induces on the signals of the protons that are split by it has relative intensities that follow the binomial coefficients, the coefficients of the successive terms in the expansion, $(a + b)^n$.

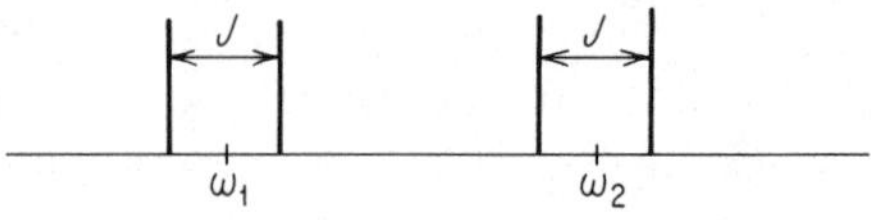

Figure 17.2 Schematic representation of the first-order N.M.R. spectrum of a two proton AX system.

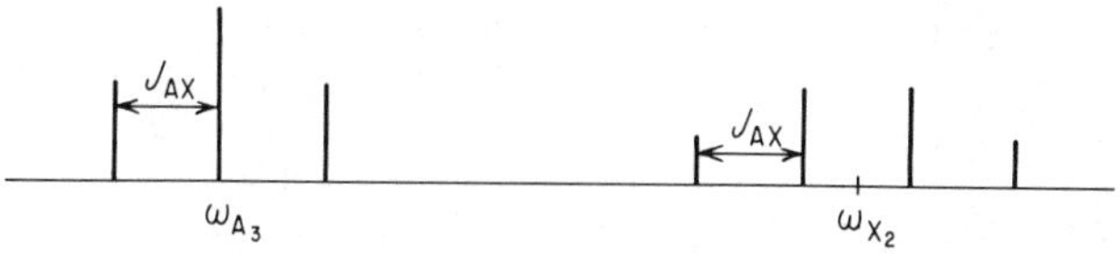

Figure 17.3 Schematic representation of the first-order N.M.R. spectrum of a five proton A_3X_2 system.

Consider what the N.M.R. spectrum of ethyl nitrate, $CH_3CH_2ONO_2$, should be expected to look like. The nitrate group is sufficiently electron-withdrawing to cause this to be an A_3X_2 system. The total signal for the methyl group (CH_3) should be one and a half times as large as that for the methylene (CH_2) group (intensity ratio of 3:2). The methyl signal is split by the two methylene protons into a triplet with an intensity ratio of 1:2:1. The methylene protons are split by the three methyl protons into a quartet with an intensity ratio of 1:3:3:1. The splittings are equal within either multiplet pattern. The methyl resonance should occur at the lower frequency, since the methylene group has the electron-withdrawing group attached to it. The spectrum should look something like that in Figure 17.3.

Intensities in the General Case. As in any spectroscopy involving direct absorption of electromagnetic radiation, the intensity of an N.M.R. signal is proportional to $[(N_1 - N_2)B\rho_\nu - N_2A]$ (see Eq. 6.75). The frequencies are sufficiently small that the A term is completely negligible (Eq. 6.78). The energy differences between states are sufficiently small that at reasonable temperatures the Boltzmann distribution gives

$$(N_1 - N_2) = N_1\left[1 - \exp\left(-\frac{\Delta E}{kT}\right)\right]$$

$$\cong N_1 \frac{\Delta E_{12}}{kT} \qquad \textbf{17.28}$$

The energy spread in resonance frequencies is of the order of a few parts per million. Thus, the populations of the levels differ only in parts per million, and the intensity for any transition is, for all practical purposes, proportional only to $B\rho_\nu$. In turn, B is proportional to the square of the transition dipole (Eq. 6.101).

The transition dipole is quite easy to calculate for a magnetic transition. The magnetic-field component of the electromagnetic radiation interacts with the nuclear magnetic moment of the system to change the m quantum number. In order to give a change in m, the radiation-field vector must be perpendicular to the z direction (see Section 8.4). Thus, if the magnetic field is the z direction, the radiation field must be in the x or y direction. The functional behavior of the x or y component of the dipole operator is that of $\hat{I}_x$ or $\hat{I}_y$. We will choose $\hat{I}_x$. But from Eqs. 17.15

$$\hat{I}_x = \tfrac{1}{2}(\hat{I}^+ + \hat{I}^-) \qquad \textbf{17.29}$$

Therefore, the transition dipole can be expressed

$$\mu_{ij} \sim \tfrac{1}{2}\langle\psi_i|I^+ + I^-|\psi_j\rangle \qquad \textbf{17.30}$$

(Normally, only the proportionality is needed, since intensities are usually reported in arbitrary units.) For a many-spin system, the $\hat{I}^+$ and $\hat{I}^-$ are the simple sums of the corresponding one-particle ladder operators. If the wave functions are linear combinations of spin-product functions, as in Eq. 17.25, the relations of Eq. 17.16 and the orthonormality of the individual spin functions can be used to obtain the transition dipole.

The N.M.R. spectrum of 2-Chloroacetonitrile. 2-Chloroacetonitrile, **1**, contains two nonequivalent hydrogens with nearly the same resonance frequencies. The

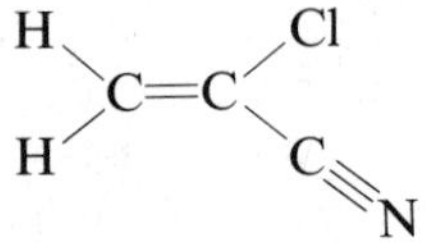

(1)

spectrum is sufficiently simple that it can easily be analyzed to obtain ω_1, ω_2, and J. There are four lines in the spectrum. On a typical 60-Mhz instrument (an instrument on which the hydrogen resonance occurs at about 6×10^7 hz) the center of the pattern lies about 372 hz to lower energy than tetramethylsilane (TMS), the usual reference signal. The observed intensities of the peaks are 1.000:1.971:1.993:1.047. (Since the intensities are not equal, this is not a first-order spectrum.) As we shall see shortly, the first and fourth of these should have the same intensity, as should the second and third. Averaging the values that should be equal gives an intensity ratio of 1.000:1.963:1.963:1.000. Relative to the center of the pattern, the peaks occur at -5.405, -2.505, 2.505, and 5.405 hz. Figure 17.4 shows the actual spectrum and a line-graph representation of the idealized spectrum.

Let us first consider the intensities. The wave functions are σ_1 and σ_4 of Eqs. 17.11 and σ^+ and σ^- of Eq. 17.25. Let us call them $\psi_1 - \psi_4$:

$$\psi_1 = \alpha(1)\alpha(2) \qquad \textbf{17.31a}$$

$$\psi_2 = c_{21}\alpha(1)\beta(2) + c_{22}\beta(1)\alpha(2) \qquad \textbf{17.31b}$$

$$\psi_3 = c_{31}\alpha(1)\beta(2) + c_{32}\beta(1)\alpha(2) \qquad \textbf{17.31c}$$

$$\psi_4 = \beta(1)\beta(2) \qquad \textbf{17.31d}$$

The allowed transitions are $\psi_2 \leftarrow \psi_1$, $\psi_3 \leftarrow \psi_1$, $\psi_4 \leftarrow \psi_2$, and $\psi_4 \leftarrow \psi_3$. The transition dipoles can be calculated by use of Eq. 17.30. For example,

$$\mu_{21} \sim \tfrac{1}{2}\langle c_{21}\alpha(1)\beta(2) + c_{22}\beta(1)\alpha(2)|\hat{I}^+(1) + \hat{I}^+(2) + \hat{I}^-(1) + \hat{I}^-(2)|\alpha(1)\alpha(2)\rangle$$
$$= \tfrac{1}{2}(c_{21} + c_{22}) \qquad \textbf{17.32a}$$

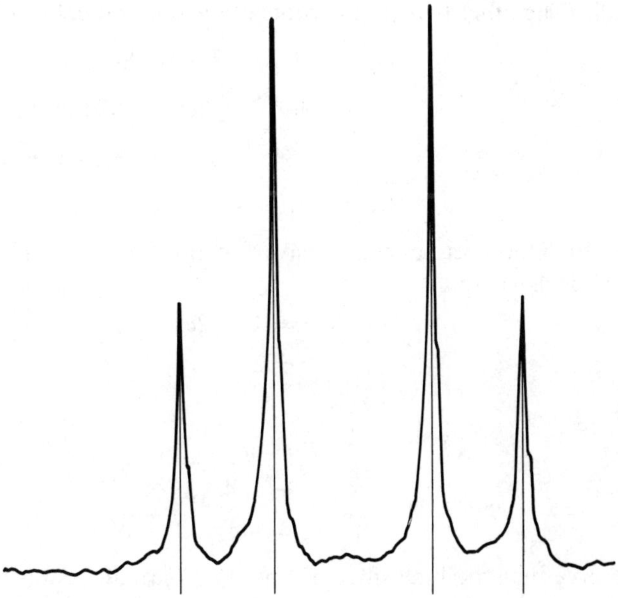

Figure 17.4 Experimental and idealized (the vertical lines) N.M.R. spectrum of 2-chloroacetonitrile.

Similarly,

$$\mu_{31} \sim \tfrac{1}{2}(c_{31} + c_{32}) \qquad \textbf{17.32b}$$

$$\mu_{42} \sim \tfrac{1}{2}(c_{21} + c_{22}) \qquad \textbf{17.32c}$$

$$\mu_{43} \sim \tfrac{1}{2}(c_{31} + c_{32}) \qquad \textbf{17.32d}$$

The intensities are proportional to the square of the transition dipole:

$$\begin{aligned} I_{21} \sim \mu_{21}^2 &\sim [\tfrac{1}{2}(c_{21} + c_{22})]^2 \\ &= \tfrac{1}{4}(c_{21}^2 + c_{22}^2 + 2c_{21}c_{22}) \\ &= \tfrac{1}{4}(1 + 2c_{21}c_{22}) \qquad \textbf{17.33a} \end{aligned}$$

$$\begin{aligned} I_{31} &\sim \tfrac{1}{4}(1 + 2c_{31}c_{32}) \\ &= \tfrac{1}{4}(1 - 2c_{21}c_{22}) \qquad \textbf{17.33b} \end{aligned}$$

$$I_{42} \sim \tfrac{1}{4}(1 + 2c_{21}c_{22}) \qquad \textbf{17.33c}$$

$$I_{43} \sim \tfrac{1}{4}(1 - 2c_{21}c_{22}) \qquad \textbf{17.33d}$$

(In Eqs. 17.33b and 17.33d, use is made of the fact that, for the coefficients arising from the 2 × 2 determinant, c_{31} equals c_{22} and c_{32} equals $-c_{21}$.)

The energies of the levels can be found as the expectation values of the Hamiltonian with respect to $\psi_1 - \psi_4$. The first and fourth of these are Eqs. 17.20a and

17.20j. (The other two are the roots of the quadratic, Eq. 17.24.) We have

$$E_1 = -\tfrac{1}{2}\omega_1 - \tfrac{1}{2}\omega_2 - \tfrac{1}{4}J \qquad \textbf{17.34a}$$

$$E_2 = \tfrac{1}{2}A\omega_1 - \tfrac{1}{2}A\omega_2 + \tfrac{1}{4}J(1 - B) \qquad \textbf{17.34b}$$

$$E_3 = -\tfrac{1}{2}A\omega_1 + \tfrac{1}{2}A\omega_2 + \tfrac{1}{4}J(1 + B) \qquad \textbf{17.34c}$$

$$E_4 = \tfrac{1}{2}\omega_1 + \tfrac{1}{2}\omega_2 - \tfrac{1}{4}J \qquad \textbf{17.34d}$$

(note that the antisymmetric wave function gives Eq. 17.34b and the symmetric, Eq. 17.34c), where

$$\begin{aligned} A &= 1 - 2c_{21}^2 \\ &= 1 - 2c_{32}^2 \end{aligned} \qquad \textbf{17.35}$$

and

$$\begin{aligned} B &= 4c_{21}c_{22} \\ &= -4c_{31}c_{32} \\ &= 4\left|c_{21}\sqrt{1 - c_{21}^2}\right| \end{aligned} \qquad \textbf{17.36}$$

[Observe that the intensities of Eqs. 17.33 are all proportional to $\tfrac{1}{2}(2 \pm B)$.] For convenience, we will assume that the E_i, the ω_i, and J are all in units of frequency (hz). The frequencies of the allowed transitions are

$$\begin{aligned} \nu_{21} &= E_2 - E_1 \\ &= \tfrac{1}{2}(1 + A)\omega_1 + \tfrac{1}{2}(1 - A)\omega_2 + \tfrac{1}{4}J(2 - B) \end{aligned} \qquad \textbf{17.37a}$$

$$\nu_{31} = \tfrac{1}{2}(1 - A)\omega_1 + \tfrac{1}{2}(1 + A)\omega_2 + \tfrac{1}{4}J(2 + B) \qquad \textbf{17.37b}$$

$$\nu_{42} = \tfrac{1}{2}(1 - A)\omega_1 + \tfrac{1}{2}(1 + A)\omega_2 - \tfrac{1}{4}J(2 - B) \qquad \textbf{17.37c}$$

$$\nu_{43} = \tfrac{1}{2}(1 + A)\omega_1 + \tfrac{1}{2}(1 - A)\omega_2 - \tfrac{1}{4}J(2 + B) \qquad \textbf{17.37d}$$

The average of the frequencies, which we will call ω_0, is

$$\omega_0 = \tfrac{1}{2}(\omega_1 + \omega_2) \qquad \textbf{17.38}$$

If we let

$$\omega_1 = \omega_0 - \delta \qquad \textbf{17.39}$$

and

$$\omega_2 = \omega_0 + \delta \qquad \textbf{17.40}$$

the transition frequencies are

$$\nu_{21} = \tfrac{1}{2}(1 + A)(\omega_0 - \delta) + \tfrac{1}{2}(1 - A)(\omega_0 + \delta) + \tfrac{1}{4}J(2 - B) \qquad \textbf{17.41a}$$

$$\nu_{31} = \tfrac{1}{2}(1 - A)(\omega_0 - \delta) + \tfrac{1}{2}(1 + A)(\omega_0 + \delta) + \tfrac{1}{4}J(2 + B) \qquad \textbf{17.41b}$$

$$\nu_{42} = \tfrac{1}{2}(1 - A)(\omega_0 - \delta) + \tfrac{1}{2}(1 + A)(\omega_0 + \delta) - \tfrac{1}{4}J(2 - B) \qquad \textbf{17.41c}$$

$$\nu_{43} = \tfrac{1}{2}(1 + A)(\omega_0 - \delta) + \tfrac{1}{2}(1 - A)(\omega_0 + \delta) - \tfrac{1}{4}J(2 + B) \qquad \textbf{17.41d}$$

Combining these, we find that

$$\nu_{31} - \nu_{42} = \nu_{21} - \nu_{43} = J \qquad \textbf{17.42}$$

This gives us J directly. Also

$$\nu_{31} - \nu_{43} = 2A\delta + \tfrac{1}{2}J(2 + B) \qquad \textbf{17.43a}$$

$$\nu_{42} - \nu_{21} = 2A\delta - \tfrac{1}{2}J(2 - B) \qquad \textbf{17.43b}$$

Either of these can give us δ, if we know A and B. These can be evaluated from the intensities. We can write, using k as the proportionality constant in Eqs. 17.33,

$$I_{21} + I_{31} = 2k$$

$$k = \frac{I_{21} + I_{31}}{2} \qquad \textbf{17.44}$$

$$= 1.482 \qquad \textbf{17.44a}$$

Also

$$k(c_{31} + c_{32})^2 = I_{31}$$

$$= 1.0$$

$$c_{31} + c_{32} = k^{-1/2}$$

$$= .8216 \qquad \textbf{17.45}$$

or

$$c_{31} = .8216 - c_{32} \qquad \textbf{17.45a}$$

but

$$c_{31}^2 + c_{32}^2 = 1.0 \qquad \textbf{17.46}$$

or

$$2c_{32}^2 - 1.6432c_{32} - .3250 = 0 \qquad \textbf{17.46a}$$

Taking the smaller root for c_{32}, we have

$$c_{32} = -.1647 \qquad \textbf{17.46b}$$

and

$$c_{31} = .9863 \qquad \textbf{17.46c}$$

from whence

$$A = .9457 \qquad \textbf{17.47a}$$

$$B = .6379 \qquad \textbf{17.47b}$$

and, using Eq. 17.43a and the observed frequency difference,

$$\delta = 3.693 \text{ Hz} \qquad \textbf{17.48}$$

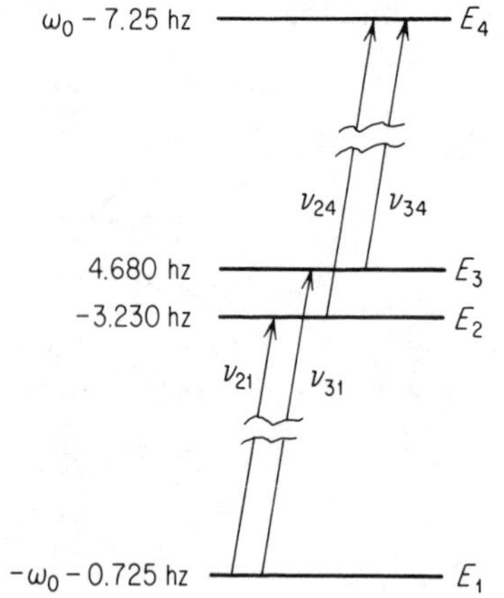

Figure 17.5 Energy levels of the proton spin states of 2-chloroacetonitrile, with the observed N.M.R. transitions.

Substituting these back into Eqs. 17.35 gives the energy levels

$$E_1 = -\omega_0 - .725 \text{ Hz} \tag{17.49a}$$

$$E_2 = -3.230 \text{ Hz} \tag{17.49b}$$

$$E_3 = 4.680 \text{ Hz} \tag{17.49c}$$

$$E_4 = \omega_0 - .725 \text{ Hz} \tag{17.49d}$$

These are shown schematically in Figure 17.5.

Systems with more than two protons cannot be treated in as direct a manner as the two-proton case. As mentioned previously, such systems must be handled by trial and error. The terms that appear in the secular determinant are adjusted until a calculated spectrum is obtained that agrees, to the desired accuracy, with that obtained experimentally.

17.5 E.P.R. SPECTRA

General. The resonance frequency of an electron is very far from that of a nucleus; consequently the E.P.R. spectrum of a system with a single unpaired electron is an extreme case of a first-order spectrum. Normally only the electron resonance is observed. For special purposes the nuclear resonance may also be observed, but the two are too widely separated for both to be observed in the same experiment. For a system with a single electron and a single proton, the energy levels are again as in Eqs. 17.26. At a given field strength, the resonance frequency for an electron is several orders of magnitude higher than that for a nucleus. Although the possible transitions again schematically resemble Figure 17.2, only the right half of the spectrum is observed.

In order to gain an understanding of what goes on in an E.P.R. experiment, let us consider in detail a system of a single unpaired electron and two protons. We will consider, initially, the magnetic Hamiltonian as written in Eq. 17.19, and later translate the various terms into the more common expressions used in E.P.R. If we use the subscript e to denote the electron and A and B to denote the two nuclei, the complete magnetic Hamiltonian is

$$\hat{H} = -\{\omega_e \hat{I}_{ze} + \omega_A \hat{I}_{zA} + \omega_B \hat{I}_{zB} + J_{eA}[\hat{I}_{ze}\hat{I}_{zA} + \tfrac{1}{2}(\hat{I}_e^+\hat{I}_A^- + \hat{I}_e^-\hat{I}_A^+)] + J_{eB}[\hat{I}_{ze}\hat{I}_{zB} + \tfrac{1}{2}(\hat{I}_e^+\hat{I}_B^- + \hat{I}_e^-\hat{I}_B^+)] + J_{AB}[\hat{I}_{zA}\hat{I}_{zB} + \tfrac{1}{2}(\hat{I}_A^+\hat{I}_B^- + \hat{I}_A^-\hat{I}_B^+)]\} \qquad \textbf{17.50}$$

There are eight possible simple spin-product functions. If we let the first entry in the product be that for the electron, the second that for proton A, and the third that for proton B, we have, omitting the indices,

$$\sigma_1 = \alpha\alpha\alpha \qquad \textbf{17.51a}$$

$$\sigma_2 = \alpha\alpha\beta \qquad \textbf{17.51b}$$

$$\sigma_3 = \alpha\beta\alpha \qquad \textbf{17.51c}$$

$$\sigma_4 = \alpha\beta\beta \qquad \textbf{17.51d}$$

$$\sigma_5 = \beta\alpha\alpha \qquad \textbf{17.51e}$$

$$\sigma_6 = \beta\alpha\beta \qquad \textbf{17.51f}$$

$$\sigma_7 = \beta\beta\alpha \qquad \textbf{17.51g}$$

$$\sigma_8 = \beta\beta\beta \qquad \textbf{17.51h}$$

The first of these has a total M value of $\frac{3}{2}$; σ_2, σ_3, and σ_5 have an M_T of $\frac{1}{2}$; σ_4, σ_6, and σ_7 have an M_T of $-\frac{1}{2}$; and σ_8 has an M_T of $-\frac{3}{2}$. If the spin states are not mixed—which, as we will see shortly, is a good approximation—the only allowed E.P.R. transitions are $\sigma_5 \leftarrow \sigma_1$, $\sigma_6 \leftarrow \sigma_2$, $\sigma_7 \leftarrow \sigma_3$, and $\sigma_8 \leftarrow \sigma_4$. There are no off-diagonal matrix elements between states of different M_T, so the 8×8 secular determinant factors into two 1×1 blocks ($M_T = \pm\frac{3}{2}$) and two 3×3 blocks ($M_T = \pm\frac{1}{2}$).

The matrix elements of the Hamiltonian are

$$H_{11} = -\tfrac{1}{2}(\omega_e + \omega_A + \omega_B) - \tfrac{1}{4}(J_{eA} + J_{eB} + J_{AB}) \cong E_1 \qquad \textbf{17.52a}$$

$$H_{22} = -\tfrac{1}{2}(\omega_e + \omega_A - \omega_B) - \tfrac{1}{4}(J_{eA} - J_{eB} - J_{AB}) \cong E_2 \qquad \textbf{17.52b}$$

$$H_{33} = -\tfrac{1}{2}(\omega_e - \omega_A + \omega_B) + \tfrac{1}{4}(J_{eA} - J_{eB} + J_{AB}) \cong E_3 \qquad \textbf{17.52c}$$

$$H_{44} = -\tfrac{1}{2}(\omega_e - \omega_A - \omega_B) + \tfrac{1}{4}(J_{eA} + J_{eB} - J_{AB}) \cong E_4 \qquad \textbf{17.52d}$$

$$H_{55} = \tfrac{1}{2}(\omega_e - \omega_A - \omega_B) + \tfrac{1}{4}(J_{eA} + J_{eB} - J_{AB}) \cong E_5 \qquad \textbf{17.52e}$$

$$H_{66} = \tfrac{1}{2}(\omega_e - \omega_A + \omega_B) + \tfrac{1}{4}(J_{eA} - J_{eB} + J_{AB}) \cong E_6 \qquad \textbf{17.52f}$$

$$H_{77} = \tfrac{1}{2}(\omega_e + \omega_A - \omega_B) - \tfrac{1}{4}(J_{eA} - J_{eB} - J_{AB}) \cong E_7 \qquad \textbf{17.52g}$$

$$H_{88} = \tfrac{1}{2}(\omega_e + \omega_A + \omega_B) - \tfrac{1}{4}(J_{eA} + J_{eB} + J_{AB}) \cong E_8 \qquad \textbf{17.52h}$$

$$H_{23} = -\tfrac{1}{2}J_{AB} \qquad \textbf{17.52i}$$

$$H_{25} = -\tfrac{1}{2}J_{eB} \qquad \textbf{17.52j}$$

$$H_{35} = -\tfrac{1}{2}J_{eA} \qquad \textbf{17.52k}$$

$$H_{46} = -\tfrac{1}{2}J_{eA} \qquad \textbf{17.52l}$$

$$H_{47} = -\tfrac{1}{2}J_{eB} \qquad \textbf{17.52m}$$

$$H_{67} = -\tfrac{1}{2}J_{AB} \qquad \textbf{17.52n}$$

The ratio of the magneton for an electron to that for a nucleus is the ratio of the proton to electron masses (see Eq. 17.2). Consequently, ω_e is approximately three orders of magnitude greater than ω_A or ω_B. The coupling constants are another several orders of magnitude smaller, when ω_e is measured at the usual field strengths. Thus, the off-diagonal matrix elements are negligible when compared to the differences between the diagonal elements, and the first-order approximation, in which the diagonal Hamiltonian matrix elements are identified with the energies, is a good approximation. The interpretation of the spectrum is, in principle, quite simple.

If the ω's and J's are expressed in frequency units, the transition frequencies are

$$\nu_{51} = \omega_e + \tfrac{1}{2}(J_{eA} + J_{eB}) \qquad \textbf{17.53a}$$

$$\nu_{62} = \omega_e + \tfrac{1}{2}(J_{eA} - J_{eB}) \qquad \textbf{17.53b}$$

$$\nu_{73} = \omega_e - \tfrac{1}{2}(J_{eA} - J_{eB}) \qquad \textbf{17.53c}$$

$$\nu_{84} = \omega_e - \tfrac{1}{2}(J_{eA} + J_{eB}) \qquad \textbf{17.53d}$$

Notice that these differ only by the coupling energies with the proton spin states. If the protons are not equivalent, four lines of equal intensity are observed. The frequency difference between the outermost lines gives $|J_{eA}| + |J_{eB}|$, while that between the inner lines gives $|J_{eA}| - |J_{eB}|$. The coupling constants can be either positive or negative. Only their magnitudes can be determined directly, and the assignment of which is A and which is B cannot be made from the E.P.R. spectrum alone. The sign of the coupling constants can be determined by N.M.R. experiments on the radical. If the nuclei are equivalent, J_{eA} equals J_{eB} and ν_{62} occurs at the same frequency as ν_{73}. The spectrum shows only three lines of intensity 1:2:1.

Since an E.P.R. transition involves a change of m for only the electron, the energies of the transitions for any system containing any number of nuclei of any spin type can be written directly as a generalization of Eqs. 17.53. The spin-product function can be written as

$$\sigma_i = \left| m_e \prod_A m_A \right\rangle \qquad \textbf{17.54}$$

and the energy of the allowed transitions as

$$\nu_{ji} = \omega_e + \sum_A m_A J_{eA} \qquad \textbf{17.55}$$

In E.P.R. experiments, the transitions and couplings are usually reported in units of magnetic-field strength, rather than frequency, as is usual for N.M.R. The coupling constants for electron-nuclear couplings are given the symbol a_A and are called *hyperfine constants.* Equation 17.55 can be rewritten in units of field strength (using Eq. 17.14, and dividing through by g):

$$\Delta E_{ji} = |H| + \sum_A m_A a_A \qquad \textbf{17.56}$$

Usually the frequency is held fixed, and the field is varied about a field at which the electron would resonate if there were no coupling, $|H_0|$:

$$\nu_{ji} = g\beta|H_0| \qquad \textbf{17.57a}$$

$$\Delta E_{ji} = |H_0| \qquad \textbf{17.57b}$$

Equating Eqs. 17.56 and 17.57b and rearranging, we obtain, as the usual expression for E.P.R. transitions,

$$|H| = |H_0| - \sum_A m_A a_A \qquad \textbf{17.58}$$

Differences of $|H_0|$ from the field expected for a single isolated electron are interpreted as a change in the effective g value for the electron in the system under study (for an isolated electron, g equals 2.0023). These deviations are due to spin-orbital interactions. For organic free radicals they are usually small ($< \pm 5$ percent), but for transition-metal ions the effective g value can be a factor of two or more different from the free-electron value.

Information about the distribution of the unpaired electron in a system can be obtained from the hyperfine coupling constants. The interaction of an electron spin and a nuclear spin can be factored into a scalar contribution and a tensor contribution. In an isotropic fluid medium the tensor contribution averages to zero. (In the solid phase an analysis of the tensor contribution can provide valuable information, but we will not consider it here.) The scalar term involves the so-called *contact interaction.* The contact interaction can be calculated. It leads to the expression, for a,

$$a = \frac{8\pi}{3} g_e \beta_e g_N \beta_N |\psi(0)|^2 \qquad \textbf{17.59}$$

where the e and N subscripts refer to the electron and nucleus, respectively, and $\psi(0)$ is the *spatial* wave function of the electron *at the nucleus.* Atomic orbitals with an l value greater than zero have nodes passing through the nucleus. Thus, at first glance, it would appear that an electron in one of these orbitals should lead to an

a value of zero. This is not the case, however, if there are occupied s orbitals in the atom. Consider, for example, the ground state of the neutral B atom, having the configuration $(1s)^2(2s)^2(2p)^1$. If the single p electron has, say, α spin, the exchange terms in the energy expression (see Section 7.10) involving this electron and the β-spin electrons in the $1s$ and $2s$ levels vanish. The result is that the $1s\alpha$ and $2s\alpha$ spin-orbitals are somewhat lower in energy than the $1s\beta$ and $2s\beta$. The energetically lower-energy orbitals are effectively more contracted than the higher-energy orbitals. Thus, the $2p\alpha$ spin-orbital induces excess α spin density in the vicinity of the nucleus, even though it itself has a node at the nucleus. Similar effects can induce spin density on atoms in a molecule adjacent to the atom bearing the spin.

In practice, the hyperfine constant is not usually calculated. It is assumed to be directly proportional to the density of the unpaired electron (the *spin density*) in the available atomic orbital closest to the nucleus with which it is interacting:

$$a = Q\rho \qquad \textbf{17.60}$$

(McConnell's relation), where ρ is the electron density in this orbital and Q is an empirically determined proportionality constant. (Sometimes, especially in ^{13}C systems, the spin densities on the atom under consideration and on adjacent atoms have to be considered.)

The E. P. R. spectrum of the butadiene anion radical. Let us consider in detail the expected E.P.R. spectrum of the butadiene anion radical. The structure is as in **2**. In Section 12.4 we determined the Hückel-level π-electron wave functions for

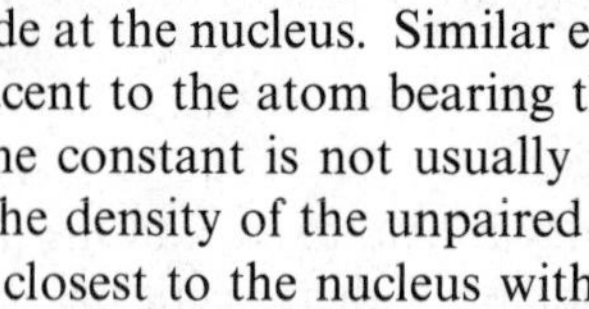

(2)

butadiene. The butadiene anion radical has a single electron in ψ_3 (Eq. 12.35c). The system has six protons, giving a total of $2^6 = 64$ proton spin states. A number of these are degenerate, however, owing to the equivalence of the four A protons and of the two B protons. The transition energies can be determined by the use of Eq. 17.58. In the absence of the equivalences there would be 64 transitions. The number actually expected is greatly reduced by the equivalences. The simplest approach to computing the spectrum is to find the substates corresponding to each equivalent set and then to combine them. For the A set we have the single $\alpha\alpha\alpha\alpha$ spin function, for an M_A value of 2; the four degenerate functions $\alpha\alpha\alpha\beta$, $\alpha\alpha\beta\alpha$, $\alpha\beta\alpha\alpha$, and $\beta\alpha\alpha\alpha$ with an M_A of 1; the six degenerate functions, $\alpha\alpha\beta\beta$ and its permutations, with an M_A of zero; the four ($\alpha\beta\beta\beta$, etc.) with an M_A of -1; and the single $\beta\beta\beta\beta$ with an M_A of -2. For the B set we have $\alpha\alpha$ ($M_B = 1$), $\alpha\beta$ and $\beta\alpha$ (doubly degenerate with $M_B = 0$), and $\beta\beta$ ($M_B = -1$). Each of the five M_A levels is combined with each of the three M_B

levels to give fifteen different proton levels. The overall degeneracy of each level is the product of the degeneracies arising from the A and the B sets. For example, the electron transition accompanying an $M_A = 1$ and $M_B = 0$ proton state should have an energy

$$\begin{aligned} |H| &= |H_0| - (3 \times \tfrac{1}{2} \times a_A - \tfrac{1}{2}a_A + \tfrac{1}{2}a_B - \tfrac{1}{2}a_B) \\ &= |H_0| - a_A \end{aligned} \qquad \textbf{17.61}$$

and a relative intensity (arising from the degeneracies of the two proton substates) of $4 \times 2 = 8$. Table 17.1 lists the energies and relative intensities for the fifteen allowed transitions.

From Eq. 12.35c we find that the calculated unpaired electron densities are .3618 for the terminal carbon p orbitals of butadiene and .1382 for the inner carbons. The empirical value for Q for a hydrogen attached to a carbon in a π system is approximately -22.5 gauss (-22.5×10^{-4} T). Thus, a_A should be about -8.14 and a_B about -3.11 gauss. The calculated resonance frequencies are listed in Table 17.1. Experimentally, fifteen lines are observed, with approximately the correct intensity ratios. However, the observed hyperfine constants are 7.62 gauss for a_A and 2.79 gauss for a_B (the absolute signs for the constants cannot be obtained from the experiment). Use of the experimental data, along with the fact that the sum of the spin

Table 17.1

RESONANCE FIELD STRENGTH AND RELATIVE INTENSITIES FOR THE E.P.R. TRANSITIONS OF THE BUTADIENE ANION RADICAL

M_A	M_B	M_T[a]	$\|H\|$[b]	Relative Intensities
2	1	3	$\|H_0\| - 2a_A - a_B \sim \|H_0\| - 19.4$	1
2	0	2	$\|H_0\| - 2a_A \sim \|H_0\| - 16.3$	2
2	−1	1	$\|H_0\| - 2a_A + a_B \sim \|H_0\| - 13.2$	1
1	1	2	$\|H_0\| - a_A - a_B \sim \|H_0\| - 11.3$	4
1	0	1	$\|H_0\| - a_A \sim \|H_0\| - 8.1$	8
1	−1	0	$\|H_0\| - a_A + a_B \sim \|H_0\| - 5.0$	4
0	1	1	$\|H_0\| - a_B \sim \|H_0\| - 3.1$	6
0	0	0	$\|H_0\| = \|H_0\|$	12
0	−1	−1	$\|H_0\| + a_B \sim \|H_0\| + 3.1$	6
−1	1	0	$\|H_0\| + a_A - a_B \sim \|H_0\| + 5.0$	4
−1	0	−1	$\|H_0\| + a_A \sim \|H_0\| + 8.1$	8
−1	−1	−2	$\|H_0\| + a_A + a_B \sim \|H_0\| + 11.3$	4
−2	1	−1	$\|H_0\| + 2a_A - a_B \sim \|H_0\| + 13.2$	1
−2	0	−2	$\|H_0\| + 2a_A \sim \|H_0\| + 16.3$	2
−2	−1	−3	$\|H_0\| + 2a_A + a_B \sim \|H_0\| + 19.4$	1

[a] The total proton M value ($M_T = M_A + M_B$).

[b] The numerical values (in gauss) arise from the Hückel value for the unpaired spin density and a Q value of -22.5 gauss, using Eq. 17.60.

densities must be unity, implies that, for the butadiene anion radical, $|Q|$ is 20.8 gauss, and that the actual charge densities are .366 and .134. The spectrum for any other system can be analyzed in a similar fashion. Techniques involving the anisotropy of the hyperfine coupling in solids, rates of electron exchange, and many other more advanced techniques are described in the books listed in the bibliography.

17.6 MAGNETIC SUSCEPTIBILITIES

Magnetic susceptibility, χ, is a measure of the susceptibility of a system to a magnetic field. It is defined as the Boltzmann average of the change of the energy of a system with a change of field

$$\chi = \left\langle \frac{\partial E}{\partial H} \right\rangle \tag{17.62}$$

On a molar basis, this becomes

$$\chi = -\frac{N \sum_i \frac{\partial E_i}{\partial H} \exp\left(-\frac{E_i}{kT}\right)}{\sum_i \exp\left(-\frac{E_i}{kT}\right)} \tag{17.62a}$$

where N is Avogadro's number, k is Boltzmann's constant, T is the absolute temperature, and the sum is over all energy states. If spin-only magnetism is being considered, the energy states are those, arising from different m values, of the particles in a magnetic field. Since the magnetic susceptibilities normally observed are due to unpaired electrons, we will discuss the phenomenon in terms of electrons only.

There can be three distinct situations. The unpaired electrons can be so far apart that there is no coupling; they can exist in clusters within which there is coupling, but with no coupling between clusters; or they can be so close together that there is significant coupling over the bulk of the material. In the first and second situations it is easy to set up the secular determinant, determine the energy levels, and then solve Eqs. 17.62 directly. In the third case there would be of the order of Avogadro's number of terms in the summations of the Hamiltonian and in the products of the spin functions. The resulting equations cannot be solved by the methods that we have presented here. They require the methods of *band theory*, as developed by solid-state scientists. The results of band theory can lead to ferromagnetism and antiferromagnetism, as well as normal diamagnetism and paramagnetism. Experimentally, ferromagnetism is the ability to retain a bulk magnetization. Theoretically, it is obtained when the state of maximum total angular momentum, for the collection of spins of the bulk material, is the ground state. Antiferromagnetism is obtained when the state of minimum total angular momentum is the ground state. It is a diamagnetic state.

BIBLIOGRAPHY

ATKINS, P. W., *Molecular Quantum Mechanics.* Clarendon Press, Oxford, 1970.

CARRINGTON A., and MCLACHLAN, A. D., *Introduction to Magnetic Resonance.* Harper & Row, Publishers, New York, 1967.

CORIO, P. L., *Structure of High-Resolution NMR Spectra.* Academic Press, Inc., New York, 1966.

EMSLEY, J. W., FEENEY, J., and SUTCLIFF, L. H., *High Resolution Nuclear Magnetic Resonance Spectroscopy.* Pergamon Press, Oxford, 1965.

FIGGIS, B. N., *Introduction to Ligand Fields.* Wiley-Interscience, New York, 1966.

KARPLUS, M., and PORTER, R. N., *Atoms and Molecules.* W. A. Benjamin, Inc., New York, 1970.

WERTZ, J. E., and BOLTON, J. R., *Electron Spin Resonance: Elementary Theory and Practical Application.* McGraw-Hill Book Company, New York, 1972.

PROBLEMS

17.1 Sketch the expected N.M.R. spectrum of the following systems, within the X approximation.

(a) AX_3. (b) A_2X_2. (c) A_3X_3. (d) A_2X_4.

***17.2** Obtain the magnetic energies from Eq. 17.21 in terms of δ (where δ is $\omega_1 - \omega_2$) for δ/J ratios of .1, .5, 1.0, 2.0, and 10.0.

***17.3** Sketch the N.M.R. spectra for the cases from the previous problem, including calculated intensities.

17.4 Calculate the molar susceptibility of a two-electron system at 1°K and at 300°K, based upon Eq. 17.21. Let $\omega_1 = \omega_2$ as defined in Eq. 17.14. Let $g = 2.000$, $\beta = 9.273 \times 10^{-21}$ erg/gauss, $\mathbf{H} = 10{,}000$ gauss, and $J = 5.000 \times 10^{-17}$ erg.

17.5 Set up the nuclear magnetic secular equation for a three-proton problem where

(a) All ω_i and J_{ij} are equal.

(b) ω_1 and ω_2 are equal, as are J_{13} and J_{23}.

(c) All ω_i and J_{ij} are unequal.

17.6 Solve for the energies for the preceding problem in the X approximation.

chapter eighteen

Chemical reactivity*

18.1 INTRODUCTION

Three fundamental questions arise when we consider chemical reactivity: (1) Will a molecule or group of molecules react at all? (2) If they react, what will the products be? (3) If they react, how rapidly will they react? The first two questions normally relate to the equilibrium between the reactants and products (the field of chemical thermodynamics) and the last to the rate at which reactants are converted to products (the field of chemical kinetics).

In principle, it is easy to relate thermodynamics to quantum chemistry. The energy of an atom or molecule, as computed by quantum chemistry, can be interpreted as the absolute internal energy, U, of that atom or molecule at absolute zero. Statistical-mechanical partition functions can be constructed from calculated energy differences (rotational, vibrational, and electronic). Any desired thermodynamic quantity can then be calculated. (In a few cases this has been done.) If the thermodynamic properties of the reactants and products are known, the position of the chemical equilibrium can be determined.

Frequently, rate data are of more interest than equilibrium data. No matter how favorable an equilibrium may be, a reaction is of no use if the rate is too slow to be observed. Also, relative rates of formation, rather than thermodynamics, can often determine the products of a reaction. In principle, rates can be related to quantum chemistry, although the theories are less well developed and the computations

* Adapted from R. L. Flurry, Jr., *Symmetry Groups: Theory and Chemical Applications*. Prentice-Hall, Inc., Englewood Cliffs, N.J., 1980. By permission.

much more complicated than for thermodynamics. The theories are less well developed because kinetics involves time-dependent phenomena, rather than steady-state phenomena, as does thermodynamics. The computations are more complicated because all reasonably complete kinetic theories require energy surfaces, which give the energy at intermediate points as the reactants proceed to products and as a function of many geometric features of the molecules and intermediates in the system. In spite of the complexities, relatively complete studies of a few simple reacting systems have been carried out (see, for example, the discussion of H_3^+ in Appendix 3).

Most attempts at applying quantum chemistry to chemical reactivity make use of many simplifications and assumptions. Reaction thermodynamics is usually based only upon the relative calculated energies of the reactants and products. The differences in these often are loosely interpreted as heats of reaction. The calculated energies may be from nonempirical or from semiempirical calculations. For closed-shell reactants and products, calculations near the Hartree-Fock limit generally give reliable results for heats of reactions. As might be expected, the results from semiempirical calculations depend upon the care with which the calculations are performed and the results interpreted. Useful results have been obtained even from simple Hückel theory, when comparisons between closely related compounds are considered.

Most quantum-chemical studies of rates of reactions have focused on some static property of the molecule or of a reaction intermediate, rather than on attempts to calculate an energy surface and solve the kinetic equations. Some specific properties that have been studied are: charge densities or electrostatic potentials (the logic is that the most favorable approach of the reactants should determine the products); energies of postulated transition states, or other intermediates (the reaction should go through the lowest-energy pathway); energies and/or charge distributions of highest occupied and/or lowest vacant molecular orbitals (these should predominate in perturbation-theory calculations); bond dissociation energies (the weakest bonds should break most rapidly, the strongest should form most rapidly); and many others. All these have had some degree of success (and some degree of failure). All have been most successful when comparing either positions of reaction within a given molecule or the relative reactivities of closely related series of molecules.

A survey of the attempts to apply quantum chemistry to chemical reactivity is beyond the scope of this text. Instead, we will concentrate on a variant of the Woodward-Hoffmann rules. These rules (first proposed in 1965) were originally based upon matching the phases of the highest-energy occupied and the lowest vacant molecular orbitals in the reacting system. (This draws on some earlier ideas of Fukui.) Basically, the reactivity argument is a kinetic one. If there is a feasible reaction pathway, the reaction is allowed. If not, it is forbidden because the energy barrier is too high to surmount. The rules provide a simple qualitative procedure for predicting the feasibility and the products of many reactants. The application of the rules is particularly simple when symmetry can be utilized. The results are remarkably accurate for such a simple treatment.

18.2 SYMMETRY CONTROL OF CHEMICAL REACTIONS

The results obtained by Woodward and Hoffmann were first given symmetry and group-theoretical justification by Longuet-Higgins and Abrahamson. The original treatment is based upon a molecular-orbital description of the reacting system. In effect, a symmetry is assumed for the transition state, and an orbital correlation diagram for the reactants and products is constructed with respect to this symmetry. If the occupied orbitals of the reactants correlate only with occupied orbitals of the products, the reaction is considered to be allowed, while if any of the occupied orbitals of the reactants correlate with unoccupied orbitals of the products and vice versa, the reaction is considered to be forbidden. Many other workers have contributed to the field. Our approach will differ from the traditional treatment in several respects.

The symmetry group of the chemical reaction is defined by the orientation of the reactants and products. The total electronic state of the reacting system is deduced within this group. The allowedness or forbiddenness of a reaction is determined by determining whether or not there is an allowed nuclear motion within this group that connects the reactants to the products. The method is independent of the theoretical description of the system. Any bonding theory that adequately describes the system can be employed. Even Lewis dot structures work for ground-state (i.e., thermal) reactions in most cases.

We will approximate the total wave function of the reacting system Ψ as a linear combination of the total wave functions of the reactants Ψ_R and the products Ψ_P:

$$\Psi = C_R\Psi_R + C_P\Psi_P \qquad \textbf{18.1}$$

where C_R and C_P are linear coefficients that vary as a function of the distance along the "reaction coordinate." The resulting energy will be "exact" for the isolated reactants and products but will lie above this "exact" energy at intermediate points along the reaction coordinate. Within the Born–Oppenheimer approximation, the energy at any point along the reaction coordinate will be the lowest root of the determinant

$$\begin{vmatrix} H_{RR} - E & H_{RP} - S_{RP}E \\ H_{PR} - S_{PR}E & H_{PP} - E \end{vmatrix} = 0 \qquad \textbf{18.2}$$

where

$$H_{RR} = \langle \Psi_R | \hat{H} | \Psi_R \rangle \qquad \textbf{18.3a}$$

$$H_{RP} = H_{PR} = \langle \Psi_R | \hat{H} | \Psi_P \rangle \qquad \textbf{18.3b}$$

$$S_{RP} = S_{PR} = \langle \Psi_R | \Psi_P \rangle \qquad \textbf{18.3c}$$

$$H_{PP} = \langle \Psi_P | \hat{H} | \Psi_P \rangle \qquad \textbf{18.3d}$$

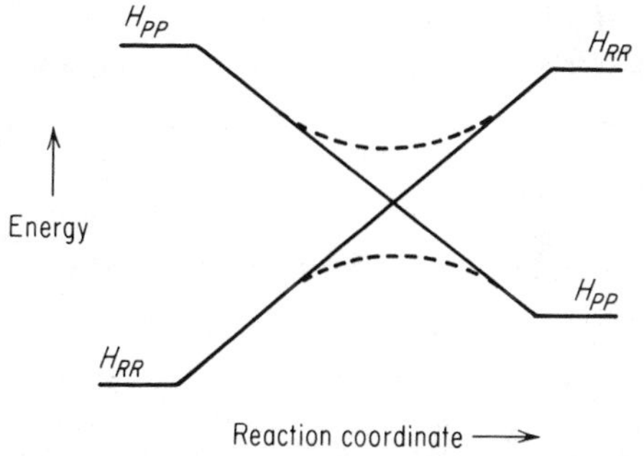

Figure 18.1 Energy as a function of the reaction coordinate. Solid lines, $H_{RP} = S_{RP} = 0$; dashed lines, H_{RP} and/or S_{RP} not equal to zero.

and the Hamiltonian is a function of the reaction coordinate. The roots of this are

$$E = (1 - S_{RP}^2)^{-1}\{\tfrac{1}{2}(H_{RR} + H_{PP}) - H_{RP}S_{RP} \pm [H_{RR}H_{PP}S_{RP}^2 - H_{RP}S_{RP}(H_{RR} + H_{PP}) + H_{RP}^2 + \tfrac{1}{4}(H_{RR} - H_{PP})^2]^{1/2}\} \quad \textbf{18.4}$$

If H_{RP} and S_{RP} are zero, the roots are H_{RR} and H_{PP}. H_{RR} will monotonically increase from the energy of the isolated reactants to some higher value as the reaction proceeds along the reaction coordinate. H_{PP} will similarly monotonically decrease from some high value to the energy of the isolated products. This is shown by the solid lines in Figure 18.1. Such a path corresponds to a forbidden reaction path. If H_{RP} and/or S_{RP} do not equal zero, there is an energy lowering at intermediate points along the reaction coordinate, and the reactants can go smoothly to products, as shown by the dashed lines in Figure 18.1. Such a path corresponds to an allowed reaction path. In order to determine whether or not a reaction is allowed, the vanishing or nonvanishing of H_{RP} or S_{RP} must be determined. These can be determined by symmetry.

It should be noted that the nonvanishing of H_{RP} and/or S_{RP} is a necessary, but not sufficient, requirement for the reaction to be allowed by a one-step process. Since no estimate of the activation energy is given by these arguments, the activation energy may be sufficiently high that the reaction may proceed by some other path, or it may not go at all.

18.3 SYMMETRY CONSIDERATIONS

We will work from the reactant side. We will consider the point symmetry of the reacting system to be the point group of the system of reactants aligned in such a way as to give a least-motion transition to the products. Furthermore, we will consider a symmetry derived from a topology that contains only the bonds important to the reaction. For example, for the butadiene–cyclobutene type of conversion, the appropriate point group is the $\mathbf{C}_{2v}$ group of *cis*-butadiene, **1**; for an S_N2 displacement

(1)

at a tetrahedral carbon atom, the appropriate point group is the $\mathbf{C}_{3v}$ group of the system with the entering and leaving groups on the C_3 axis, **2**.

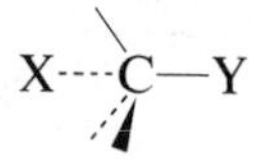

(2)

In either the molecular-orbital or the valence-bond scheme, the total electronic wave function of either the reactants or the products can be represented as a properly antisymmetrized product of one-electron spin orbitals. In the molecular-orbital scheme, these can immediately be characterized with respect to symmetry. In the valence-bond scheme, the total wave function is represented by a product of pair functions. Symmetry-adapted combinations of those pair functions may be required for the symmetry classifications. We will present our development within the molecular-orbital formalism and then point out the generalization to the valence-bond scheme and to even simpler bonding schemes.

The symmetry restrictions on S_{RP} and H_{RP} are the usual ones for overlap and Hamiltonian matrix elements involving determinantal functions. The overlap integral S_{RP} consists of a sum of permutations of terms of the form

$$\langle \phi_1^R(1)\bar{\phi}_2^R(2)\phi_3^R(3)\bar{\phi}_4^R(4)\ldots|\phi_1^P(1)\bar{\phi}_2^P(2)\phi_3^P(3)\bar{\phi}_4^P(4)\ldots\rangle \qquad \mathbf{18.5}$$

where the ϕ_i^R and ϕ_i^P are one-electron reactant and product molecular orbitals, respectively. A bar indicates a β spin function, and the absence of a bar an α spin function. As is usual with antisymmetrized product functions, permutations on only one side of the bracket are sufficient to yield all terms. Of these permutations, only those that do not interchange members of the α and β spin subsets lead to nonvanishing terms. Equation 18.5 can be rewritten in terms of a product of one-electron brackets:

$$\langle \phi_1^R(1)|\phi_1^P(1)\rangle\langle \phi_3^R(3)|\phi_3^P(3)\rangle\ldots \times \langle \bar{\phi}_2^R(2)|\bar{\phi}_2^P(2)\rangle\langle \bar{\phi}_4^R(4)|\bar{\phi}_4^P(4)\rangle\ldots \qquad \mathbf{18.6}$$

Each of the one-electron brackets has the usual symmetry restrictions associated with it; that is, $\langle \phi_i^R(i)|\phi_i^P(i)\rangle$ vanishes unless ϕ_i^R and ϕ_i^P belong to the same irreducible representation of the point group of the system. Thus, S_{RP} vanishes unless the same irreducible representations are spanned by the one-electron orbitals of the reactants and by the one-electron orbitals of the products. *The ordering of these orbitals with respect to energy is unimportant, as long as they are restricted to the occupied manifold.* If the same representations are spanned, there will be some permutation of Eq. 18.6 that is nonvanishing.

Let us now consider the off-diagonal term of the Hamiltonian, H_{RP}. We will consider this at an early stage along the reaction coordinate and assume that if the reaction starts off allowed, it will remain allowed throughout the reaction. If "concerted" is taken to apply only to a reaction that takes place during one normal vibration of the system, this must be the case, or else the motion would have to change symmetries in midcourse. This can happen only in restricted circumstances. The same conclusions result from consideration of the reverse reaction.

For small displacements from the starting geometry of the reactants, we can express the Hamiltonian as

$$\hat{H} = \hat{H}_R^0 + \hat{V} \qquad \textbf{18.7}$$

where $\hat{H}_R^0$ is the Hamiltonian for the static reactants and $\hat{V}$ represents the perturbation caused by a small displacement of the nuclei along the reaction coordinate. This displacement can be expressed in terms of the normal vibrational modes of the system. $\hat{H}_R^0$ is a totally symmetric operator within the point group of the reactants; consequently, the selection rules for $\langle \Psi_R | \hat{H}_R^0 | \Psi_P \rangle$ are the same as for S_{RP}. Any additional considerations must come from $\langle \Psi_R | \hat{V} | \Psi_P \rangle \equiv V_{RP}$. The potential-energy portion of the Hamiltonian contains the interaction of each of the electrons with each of the nuclei. Thus, $\hat{V}$ is a many-electron operator that, in the independent-particle approximation, can be expressed as a sum of one-electron operators, each term expressing the change in nuclear attraction experienced by a single electron. V_{RP} can be expressed as sums of permutations of terms of the type

$$\langle \phi_1^R(1) | \hat{V}(1) | \phi_1^P(1) \rangle \langle \phi_3^R(3) | \phi_3^P(3) \rangle \ldots \times \langle \bar{\phi}_2^R(2) | \bar{\phi}_2^P(2) \rangle \langle \bar{\phi}_4^R(4) | \bar{\phi}_4^P(4) \rangle \ldots \qquad \textbf{18.8}$$

The α and β spin sets can again be considered separately, since $\hat{V}$ is a spin-free operator. The requirement that these be nonvanishing is that there be a one-to-one correspondence between the irreducible representations of all but one of the ϕ_i^R and ϕ_i^P from each spin set, and that for the noncorrespondence the triple product $\Gamma_i^R \Gamma^V \Gamma_i^P$ must contain the totally symmetric irreducible representation of the point group (where Γ_i^R, Γ^V, and Γ_i^P are the irreducible representations corresponding to ϕ_i^R, $\hat{V}$, and ϕ_i^P, respectively). Thus, the overall selection rule for an allowed reaction is that there be at most one one-electron orbital from each spin set that differs in symmetry classification in the reactants and products. (For closed-shell systems, only one spin set need be considered, since the spatial orbitals are the same for both spin sets.) Furthermore, the product $\Gamma_i^R \Gamma_i^P$ for this mismatched orbital determines the symmetry of the allowed nuclear motion, since $\Gamma_i^R \Gamma^V \Gamma_i^P$ contains the totally symmetric irreducible representation only if Γ^V is contained in $\Gamma_i^R \Gamma_i^P$.

Molecular orbitals are sometimes approximated by linear combinations of localized two-center, one-electron bond orbitals. For neutral molecules with closed-shell ground states, the combinations of the bond orbitals that are bonding, in the two-center sense, give overall molecular orbitals that are bonding in character, while the combinations of antibonding bond orbitals give antibonding molecular orbitals. The combinations of the bonding bond orbitals are all occupied in the ground state of the molecule. Furthermore, the linear combinations of the bond orbitals have the same symmetry properties as do normal L.C.A.O. molecular orbitals constructed from the same basis set.

Consider, as an example, the *cis*-butadiene shown in **1** with the added labeling of the atomic p-π orbitals, **1a**. The valence structure drawn implies localized bonding.

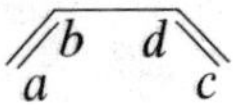

(1a)

We can construct two localized two-center bonding bond orbitals

$$\lambda_1 = \frac{1}{\sqrt{2}}(a + b) \quad \textbf{18.9a}$$

$$\lambda_2 = \frac{1}{\sqrt{2}}(c + d) \quad \textbf{18.9b}$$

and two antibonding bond orbitals

$$\lambda_3 = \frac{1}{\sqrt{2}}(a - b) \quad \textbf{18.10a}$$

$$\lambda_4 = \frac{1}{\sqrt{2}}(c - d) \quad \textbf{18.10b}$$

These can be combined to give

$$\psi_1 = \frac{1}{\sqrt{2}}(\lambda_1 + \lambda_2)$$

$$= \frac{1}{2}(a + b + d + c) \quad \textbf{18.11a}$$

$$\psi_2 = \frac{1}{\sqrt{2}}(\lambda_1 - \lambda_2)$$

$$= \frac{1}{2}(a + b - d - c) \quad \textbf{18.11b}$$

$$\psi_3 = \frac{1}{\sqrt{2}}(\lambda_3 + \lambda_4)$$

$$= \frac{1}{2}(a - b - d + c) \quad \textbf{18.11c}$$

$$\psi_4 = \frac{1}{\sqrt{2}}(\lambda_3 - \lambda_4)$$

$$= \frac{1}{2}(a - b + d - c) \quad \textbf{18.11d}$$

These have the same nodal properties (and hence the same energy ordering) and symmetries as the Hückel molecular orbitals of Eqs. 12.35. The functions of Eqs. 18.11a and 18.11b are doubly occupied, in the ground state. Qualitative, symmetry-based, reactivity arguments based on Eqs. 18.11 lead to the same conclusions as arguments based on Eqs. 12.35.

Considering structure **1a** from a different point of view, the π bonds drawn on the structure are a representation of the major contributor to the valence-bond description of the π system of butadiene. This implies that the one-electron symmetry properties of the valence-bond description for a system are the same as those of the molecular-orbital description. (This can be proved formally, but the arguments are rather involved.)

For applications using the valence-bond description of the reactants and products, the appropriate functions to use for the ϕ_i^R and ϕ_i^P are the localized two-center, one-electron bond orbitals. Even Lewis dot structures can be used if these adequately describe the system (the σ and π nature of double bonds must, however, be recognized). The symmetry-related bonds must be combined to correspond to the irreducible representations of the point group of the system. The irreducible representations thus spanned are completely equivalent to the representations spanned by the occupied molecular orbitals in a molecular-orbital description of the system. Once these symmetry-adapted functions are constructed, the selection rules are the same as already outlined. In many cases the bond formalism has a distinct advantage in that it is often easier to deduce the proper symmetry-adapted combinations of these bond orbitals than to find the symmetries of the occupied molecular orbitals.

The justification of the orbital correlation schemes is obvious from this development. If there is a one-to-one correspondence between the ϕ_i^R and the ϕ_i^P, the $\hat{V}$ must transform as the totally symmetric irreducible representation for V_{RP} to be nonvanishing. In the orbital-following schemes, the point symmetry considered is that occurring well along the reaction coordinate. The reaction coordinate, and consequently $\hat{V}$, transforms as the totally symmetric representation in this point group. Thus, for an allowed reaction path, there must be a one-to-one correspondence of the occupied orbitals. In this case, S_{RP}, V_{RP}, and $\langle \Psi_R | \hat{H}_R^0 | \Psi_P \rangle$ are all nonvanishing.

Three selection rules cause a reaction to be either forbidden or nonconcerted:

1. There are two or more orbital mismatches between reactants and products. In this case H_{RP} vanishes and the reaction is forbidden. This does not preclude the possibility of there being a multistep mechanism to go from reactants to products.
2. There is only one mismatch, but Γ^V does not correspond to a normal mode of vibration of the system. This will be a relatively rare occurrence for systems having more than a very few atoms.
3. There is only one mismatch, but Γ^V does not correspond to a motion taking the reactants to the products. The reaction may or may not be allowed, but, if allowed, the mechanism will be nonconcerted.

18.4 EXAMPLES

Electrocyclic Reactions. The first application of the Woodward–Hoffmann rules was to electrocyclic reactions. This has become the classic test for any discussion of symmetry control of reactions. The classic examples are the butadiene–cyclobutene isomerization and the hexatriene–cyclohexadiene isomerization. The topological symmetry for the reactants or the products is $\mathbf{C}_{2v}$. The orbitals to be considered are the π orbitals of the acyclic polyolefin and the π orbitals and the new σ bond for the cyclic compound. Table 18.1 shows the molecular-orbital scheme for the ground and first excited states for the first three members of the series. The symmetry labels for the orbitals are from the $\mathbf{C}_{2v}$ point group. For the butadiene–cyclobutene isomerization (A) in the ground state, there is orbital matching between the $b_2\pi_1$ orbital of butadiene and the $b_2\pi_1$ orbital of cyclobutene. There is a mismatching of the other orbitals. Thus, if the reaction occurs in a concerted fashion in the ground state (thermally), the nuclear motion must transform as $A_2 \times A_1 = A_2$. The conrotatory motion (i.e., both groups rotating in the same direction) of the two terminal CH_2 groups of *cis*-butadiene transforms as the A_2 irreducible representation within $\mathbf{C}_{2v}$. For the excited state, the two singly occupied orbitals match (providing

Table 18.1

MOLECULAR-ORBITAL SCHEMES FOR SOME ELECTROCYCLIC REACTIONS[a]

A B C

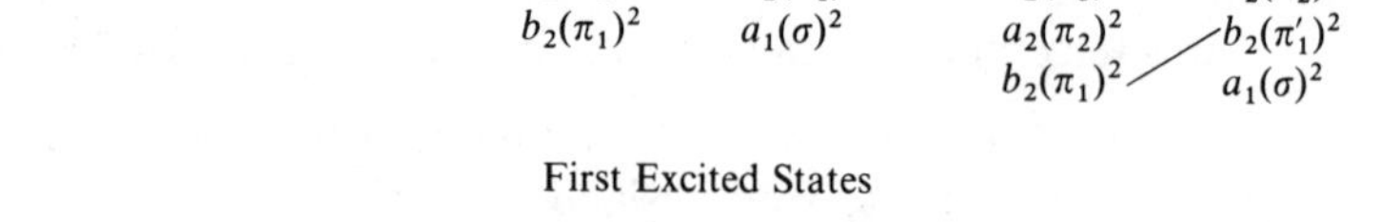

A		B		C	
Ground States					
$a_2(\pi_2)^2$	$b_2(\pi'_1)^2$	$b_2(\pi_3)^2$	$a_2(\pi'_2)^2$	$a_2(\pi_4)^2$	$b_2(\pi'_3)^2$
$b_2(\pi_1)^2$	$a_1(\sigma)^2$	$a_2(\pi_2)^2$	$b_2(\pi'_1)^2$	$b_2(\pi_3)^2$	$a_2(\pi'_2)^2$
		$b_2(\pi_1)^2$	$a_1(\sigma)^2$	$a_2(\pi_2)^2$	$b_2(\pi'_1)^2$
				$b_2(\pi_1)^2$	$a_1(\sigma)^2$
First Excited States					
$b_2(\pi_3)^1$	$a_2(\pi'_2)^1$	$a_2(\pi_4)^1$	$b_2(\pi'_3)^1$	$b_2(\pi_5)^1$	$a_2(\pi'_4)^1$
$a_2(\pi_2)^1$	$b_2(\pi'_1)^1$	$b_2(\pi_3)^1$	$a_2(\pi'_2)^1$	$a_2(\pi_4)^1$	$b_2(\pi'_3)^1$
$b_2(\pi_1)^2$	$a_1(\sigma)^2$	$a_2(\pi_2)^2$	$b_2(\pi'_1)^2$	$b_2(\pi_3)^2$	$a_2(\pi'_2)^2$
		$b_2(\pi_1)^2$	$a_1(\sigma)^2$	$a_2(\pi_2)^2$	$b_2(\pi_1)^2$
				$b_2(\pi_1)^2$	$a_1(\sigma)^2$

[a] A, Butadiene–cyclobutene; B, hexatriene–cyclohexadiene; C, octatetraene–cyclooctatriene. Symmetry labels are from the $\mathbf{C}_{2v}$ point group. Superscripts are the orbital occupancy.

the spins are matched properly), and the mismatch is between the $b_2\pi_1$ orbital of butadiene and the $a_1\sigma$ orbital of cyclobutene. The required nuclear motion for a concerted excited state (photochemical) reaction is $B_2 \times A_1 = B_2$. The disrotatory motion (opposite directions) of the two terminal CH_2 groups transforms as B_2. In reaction B we have a b_2, a_1 orbital mismatch for the thermal reaction, requiring a B_2 disrotatory motion, while the photochemical reaction has an a_2, a_1 mismatch, requiring an A_2 conrotatory motion. The motions of reaction C repeat those of reaction A. The same patterns will be followed for higher polyenes. The results are in complete agreement with the Woodward–Hoffmann rules and with experiment. It is interesting to note that, in all these reactions, the treatment requires the "promoted" electron to have opposite spins in the reactant and product if the photochemical reaction goes through an excited singlet state.

Cycloaddition Reactions. Table 18.2 shows the molecular-orbital schemes for some cycloaddition reactions. In this figure, the ethylene dimerization is characterized within the $\mathbf{C}_s$ point group, even though the true point symmetry is $\mathbf{D}_{2h}$, to emphasize the continuity within the series. The orientation of the molecules is assumed to be such that the π orbitals on atoms A, B, C, and D lie in the same plane. The localized

Table 18.2

MOLECULAR-ORBITAL SCHEMES FOR SOME CYCLOADDITION REACTIONS[a]

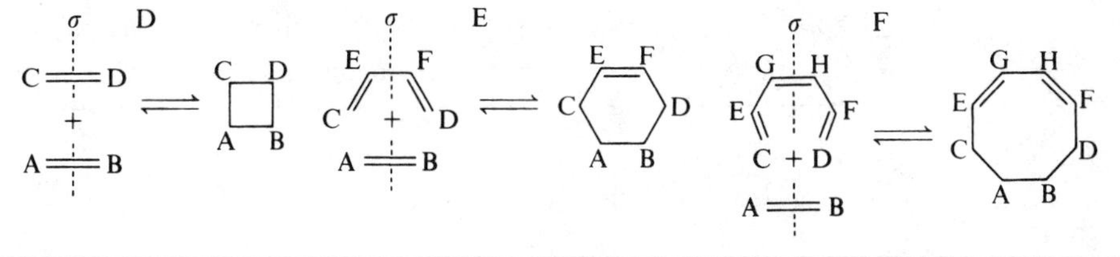

Ground States

$a'(\pi_{CD})^2$	$a''(\sigma_{AC}-\sigma_{BD})^2$	$a''(\pi_2)^2$	$a'(\pi)^2$	$a'(\pi_3)^2$	$a''(\pi_2)^2$
$a'(\pi_{AB})^2$	$a'(\sigma_{AC}+\sigma_{BD})^2$	$a'(\pi_1)^2$	$a''(\sigma_{AC}-\sigma_{BD})^2$	$a''(\pi_2)^2$	$a'(\pi_1)^2$
		$a'(\pi_{AB})^2$	$a'(\sigma_{AC}+\sigma_{BD})^2$	$a'(\pi_1)^2$	$a''(\sigma_{AC}-\sigma_{BD})^2$
				$a'(\pi_{AB})^2$	$a'(\sigma_{AC}+\sigma_{BD})^2$

Excited States

$a'(\pi^*_{CD})^1$	$a'(\sigma^*_{AC}+\sigma^*_{BD})^1$	$a'(\pi_3)^1$	$a''(\pi_2)^1$	$a''(\pi_4)^1$	$a'(\pi_3)^1$
$a'(\pi_{CD})^1$	$a''(\sigma_{AC}-\sigma_{BD})^1$	$a''(\pi_2)^1$	$a'(\pi_1)^1$	$a'(\pi_3)^1$	$a''(\pi_2)^1$
$a'(\pi_{AB})^2$	$a'(\sigma_{AC}+\sigma_{BD})^2$	$a'(\pi_1)^2$	$a''(\sigma_{AC}-\sigma_{BD})^2$	$a''(\pi_2)^2$	$a'(\pi_1)^2$
		$a'(\pi_{AB})^2$	$a'(\sigma_{AC}+\sigma_{BD})^2$	$a'(\pi_1)^2$	$a''(\sigma_{AC}-\sigma_{BD})^2$
				$a'(\pi_{AB})^2$	$a'(\sigma_{AC}+\sigma_{BD})^2$

[a] The molecules are all classified within the $\mathbf{C}_s$ point even though the true point symmetry for the ethylene dimerization is $\mathbf{D}_{2h}$.

σ orbitals in the cyclic structures are not symmetry orbitals; consequently the indicated linear combinations must be used. The orbital $(\sigma^*_{\mathrm{AC}} + \sigma^*_{\mathrm{BD}})$ of cyclobutane has the MO form $(\chi_A - \chi_C + \chi_B - \chi_D)$, where the χ's are the σ-type basis functions. From Table 18.2 it is seen that in the ground state of reaction E (the Diels–Alder reaction) and in the excited states of reactions D and F there is complete matching of orbital symmetries. These reactions are allowed by a motion that is totally symmetric within $\mathbf{C}_s$. The direct symmetric approach of the A—B ethylene to the other molecule is such a motion. In the other cases there is a mismatch between an a' orbital in one structure and an a'' orbital in the other. For the reaction to be allowed, the motion would have to transform as $A' \times A'' = A''$ (i.e., it would correspond to a motion that is antisymmetric with respect to the plane of symmetry). There is no in-plane motion of this type that can yield the indicated products; consequently the concerted in-plane reactions must be considered as forbidden. These results are in complete agreement with the Woodward–Hoffmann rules and with experiment for *suprafacial* cycloadditions. The *antarafacial* cycloadditions would require a different starting point group for the reactants. Further, the symmetry of the reactants and products would be different. However, an out-of-plane twisting motion transforms as A'' in $\mathbf{C}_s$. This would tend toward the postulated perpendicular transition state required for the *suprafacial–antarafacial* reactions to occur.

An interesting but complicated cycloreversion reaction is the isomerization of prismane to benzene, which is shown in Table 18.3. The arrows on the prismane

Table 18.3

VALENCE-BOND SCHEME FOR THE ISOMERIZATION OF PRISMANE TO BENZENE[a]

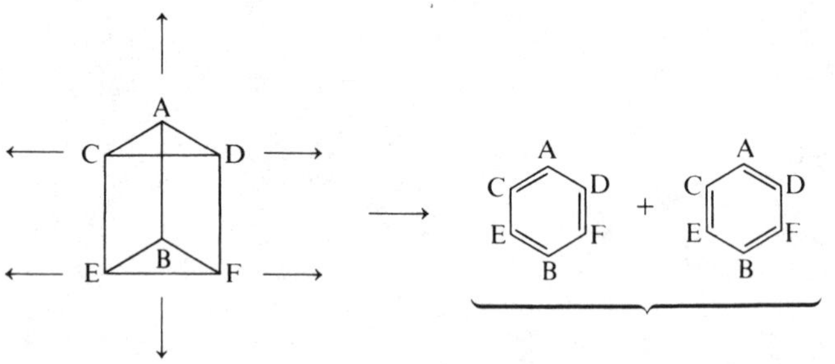

Bonds broken: $\sigma_{\mathrm{AB}}, \sigma_{\mathrm{CD}}, \sigma_{\mathrm{EF}}$ Bonds formed: $\pi_{\mathrm{AC}}, \pi_{\mathrm{BE}}, \pi_{\mathrm{DF}}$, and $\pi_{\mathrm{AD}}, \pi_{\mathrm{BF}}, \pi_{\mathrm{CE}}$

$b_2(\sigma_{\mathrm{CD}} - \sigma_{\mathrm{EF}})^2$		$b_1(\pi_{\mathrm{AC}} - \pi_{\mathrm{BE}} + \pi_{\mathrm{AD}} - \pi_{\mathrm{BF}})^2$
$a_1(\sigma_{\mathrm{CD}} + \sigma_{\mathrm{EE}})^2$	———	$a_1(\pi_{\mathrm{AC}} + \pi_{\mathrm{BE}} + \pi_{\mathrm{AD}} + \pi_{\mathrm{BF}})^2$
$a_1(\sigma_{\mathrm{AB}})^2$	———	$a_1(\pi_{\mathrm{CE}} + \pi_{\mathrm{DF}})^2$

[a] The symmetry labels are from the $\mathbf{C}_{2v}$ point group, which is a common subgroup to the point symmetries of both molecules. The C_2 axis is coincident with the C_6 axis of benzene and perpendicular to the A—B bond of prismane.

structure indicate the motion that would carry out the indicated isomerization. The valence-bond scheme is used to avoid having to determine the symmetries of the molecular orbitals of prismane. The symmetry labels are in terms of the $\mathbf{C}_{2v}$ point group, which is a common subgroup of the point symmetries of both molecules. The indicated canonical structures are symmetry-adapted linear combinations of pair functions for prismane, and of pair functions and the two Kékulé structures for benzene. Note that there is a b_2, b_1 mismatching. This means that the reaction would be thermally allowed by a $B_2 \times B_1 = A_2$ motion. The required motion transforms as A_1, however. Consequently, the thermal reaction must be considered to be forbidden or nonconcerted.

Group Transfers and Eliminations. Table 18.4 shows the MO schemes for some concerted transfers of two hydrogen atoms. The same schemes would be valid for the concerted transfer of any two σ-bonded groups. Here, the symmetrical concerted ground-state reaction is allowed for reactions G and I but not for reaction H. The excited-state reactions are allowed for reactions G and H, but not I. The reason G is both thermally and photochemically allowed is that the reactants and products are identical and have identical orientation with respect to the symmetry elements.

Table 18.5 shows the molecular-orbital schemes for two elimination reactions that give radical products. These are classified according to the $\mathbf{C}_{2v}$ point group. Here, because of the different number of occupied orbitals in the reactants and products, the α and β spin sets must be considered separately. In these, only one of a number of possible spin configurations is listed. In both ground-electronic-state reactions there is a mismatching of the α spin sets. (Interchange of the α and β spins in the product radicals would cause the mismatch to be in the β set.) The mismatch is a one-orbital a_1, b_1 mismatch. This would require that the nuclear motion accompanying the reaction be of B_1 symmetry. This corresponds to an unsymmetrical breaking of the bonds as shown in **3** and **4**. Thus, the thermal reaction is predicted

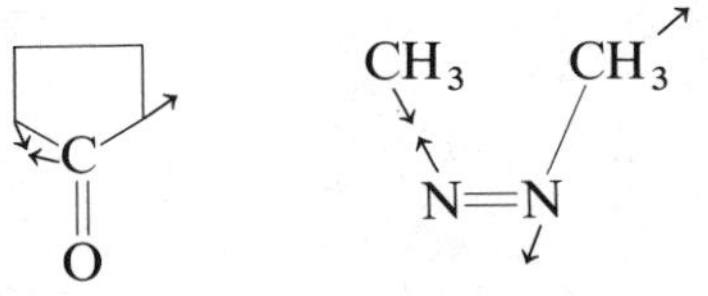

(3) (4)

not to be concerted. For the indicated spin configurations for the excited states, there is complete matching of the orbital symmetries for both spin sets for both reactions. (A number of other reasonable spin configurations are possible, of which some give complete matching and some do not.) The reaction can go in a symmetrical manner (i.e., the leaving group leaving along the C_2 axis). These results are consistent with the experimental results for the decarbonylation of cyclopentanone. The vapor-phase photolysis of cyclopentanone yields a large amount of cyclobutane, while the pyrolysis yields only acyclic compounds. It is interesting to note that, if the scheme is applied to a direct one-step conversion of cyclopentanone to cyclobutane, both the excited-state and ground-state reactions are predicted to be nonconcerted.

Table 18.4

MOLECULAR-ORBITAL SCHEMES FOR THE CONCERTED TRANSFER OF TWO H ATOMS

G | H | I

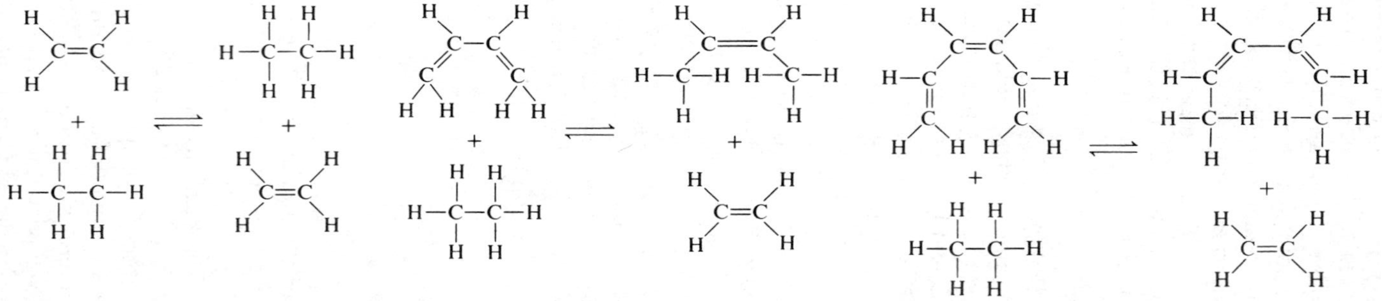

Ground States

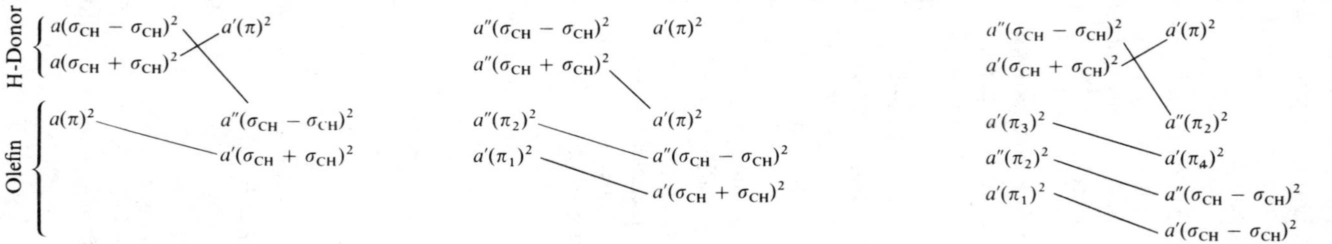

Excited States

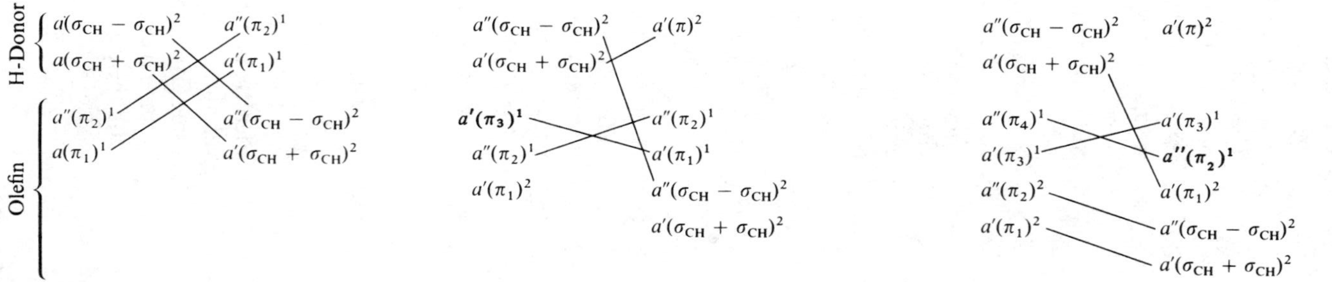

Table 18.5

MOLECULAR-ORBITAL SCHEMES FOR SOME ELIMINATION REACTIONS THAT GO TO RADICAL PRODUCTS (C_{2v} POINT GROUP)[a]

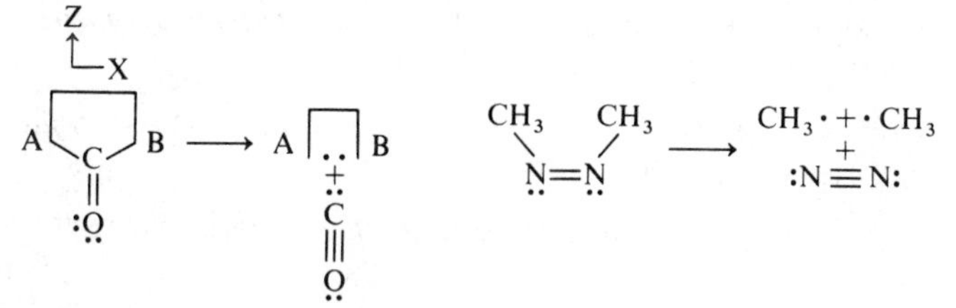

Ground States

b_1 $n_O - n_O$ $\uparrow\downarrow$	b_1 $A\cdot - B\cdot$ $\downarrow$	b_1 $n_N - n_N$ $\uparrow\downarrow$	b_1 $C\cdot - C\cdot$ $\downarrow$
a_1 $n_O + n_O$ $\uparrow\downarrow$	a_1 $A\cdot + B\cdot$ $\uparrow$	a_1 $n_N + n_N$ $\uparrow\downarrow$	a_1 $C\cdot + C\cdot$ $\uparrow$
b_2 π_{CO_y} $\uparrow\downarrow$	a_1 n_C $\uparrow\downarrow$	b_2 π_{NN_y} $\uparrow\downarrow$	b_1 $n_N - n_N$ $\uparrow\downarrow$
b_1 $\sigma_{CA} - \sigma_{CB}$ $\uparrow\downarrow$	a_1 n_O $\uparrow\downarrow$	b_1 $\sigma_{NC} - \sigma_{NC}$ $\uparrow\downarrow$	a_1 $n_N + n_N$ $\uparrow\downarrow$
a_1 $\sigma_{CA} + \sigma_{CB}$ $\uparrow\downarrow$	b_1 π_{CO_x} $\uparrow\downarrow$	a_1 $\sigma_{NC} + \sigma_{NC}$ $\uparrow\downarrow$	b_2 π_{NN_y} $\uparrow\downarrow$
	b_2 π_{CO_y} $\uparrow\downarrow$		a_1 π_{NN_z} $\uparrow\downarrow$
α: $2a_1, 2b_1, b_2$	α: $3a_1, b_1, b_2$	α: $2a_1, 2b_1, b_2$	α: $3a_1, b_1, b_2$
β: $2a_1, 2b_1, b_2$	β: $2a_1, 2b_1, b_2$	β: $2a_1, 2b_1, b_2$	β: $2a_1, 2b_1, b_2$

Excited States

b_2 $\pi^*_{CO_y}$ $\downarrow$	b_1 $A\cdot - B\cdot$ $\uparrow$	a_2 $\pi^*_{NN_y}$ $\downarrow$	b_1 $C\cdot - C\cdot$ $\uparrow$
b_1 $n_O - n_O$ $\uparrow$	a_1 $A\cdot + B\cdot$ $\downarrow$	b_1 $n_N - n_N$ $\uparrow$	a_1 $C\cdot + C\cdot$ $\downarrow$
a_1 $n_O + n_O$ $\uparrow\downarrow$	b_2 $\pi^*_{CO_y}$ $\downarrow$	a_1 $n_N + n_N$ $\uparrow\downarrow$	a_2 $\pi^*_{NN_y}$ $\downarrow$
b_2 π_{CO_y} $\uparrow\downarrow$	a_1 n_C $\uparrow$	b_2 π_{NN_y} $\uparrow\downarrow$	b_1 $n_N - n_N$ $\uparrow\downarrow$
b_1 $\sigma_{CA} - \sigma_{CB}$ $\uparrow\downarrow$	a_1 n_O $\uparrow\downarrow$	b_1 $\sigma_{NC} - \sigma_{NC}$ $\uparrow\downarrow$	a_1 $n_N + n_N$ $\uparrow$
a_1 $\sigma_{CA} + \sigma_{CB}$ $\uparrow\downarrow$	b_1 π_{CO_x} $\uparrow\downarrow$	a_1 $\sigma_{NC} + \sigma_{NC}$ $\uparrow\downarrow$	b_2 π_{NN_y} $\uparrow\downarrow$
	b_2 π_{CO_y} $\uparrow\downarrow$		a_1 π_{NN_z} $\uparrow\downarrow$
α: $2a_1, 2b_1, b_2$	α: $2a_1, 2b_1, b_2$	α: $2a_1, 2b_1, b_2$	α: $2a_1, 2b_1, b_2$
β: $2a_1, b_1, 2b_2$	β: $2a_1, b_1, 2b_2$	β: $2a_1, b_1, a_2, b_2$	β: $2a_1, b_1, a_2, b_2$

[a] The spins are indicated by arrows: $\uparrow \equiv \alpha$ spin, $\downarrow \equiv \beta$ spin.

Limitations. With the present development, the reactants and the products must have the same symmetry or a common subgroup of symmetry elements preserved in the reaction. In either case, this symmetry must be nontrivial. The bonds broken and formed must have different symmetries with respect to some symmetry element or elements of the group. In addition to the symmetry limitations, the wave functions of the reactant and product systems must each be adequately described by a single electronic configuration, or if configuration interaction is important, the important configurations must be known.

BIBLIOGRAPHY

LONGUET-HIGGINS, H. C., and ABRAHAMSON, E. W., *J. Am. Chem. Soc.*, **87**, 2045 (1965).

PEARSON, R. G., *Symmetry Rules for Chemical Reactions.* John Wiley & Sons, New York, 1976.

SIMMONS, H. E., and BURNETT, J. F., Eds., *Orbital Symmetry Papers.* American Chemical Society, Washington, D. C., 1974.

WOODWARD, R. B., and HOFFMANN, R., *The Conservation of Orbital Symmetry.* Academic Press, Inc., New York, 1970.

WOODWARD, R. B., and HOFFMANN, R., *J. Am. Chem. Soc.*, **87**, 395 (1965).

PROBLEMS

Deduce, by symmetry arguments, whether or not the following reactions are thermally or photochemically allowed.

***18.1** 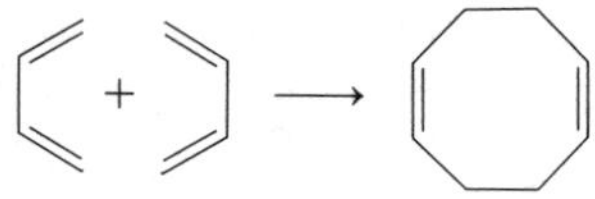

***18.2** 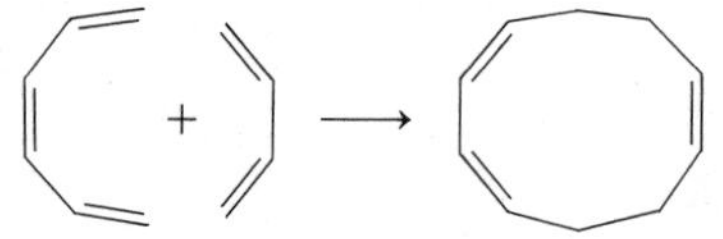

***18.3** 

***18.4** 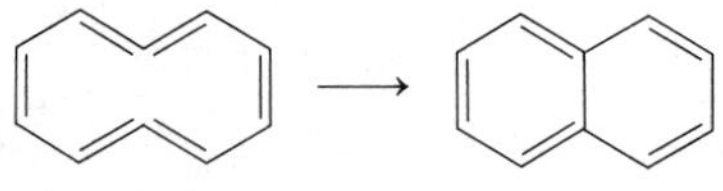

***18.5** 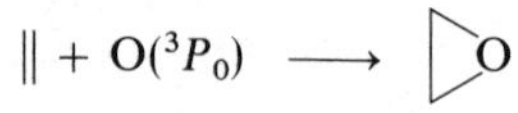

***18.6** 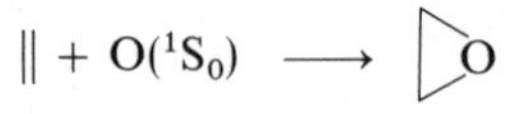

appendix one

Symbols and notation

An effort has been made in this text to use symbols and notation that are reasonably consistent with each other and with those used in the literature. This appendix lists the symbols most commonly used in the text. For alphabetizing, Greek letters are transliterated, and come after the corresponding Roman letters. Note that quantum numbers are given both as lower-case and as upper-case quantities. The lower-case letter is a single-particle quantum number, while the upper-case letter is a total quantum number.

Symbol	*Meaning*
A	A one-dimensional representation of a point group that is symmetric with respect to rotation about the principal axis.
α	A spin function having an m_s value of $+\frac{1}{2}$.
α	The one-center Hückel molecular-orbital parameter.
B	A one-dimensional representation of a point group that is antisymmetric with respect to rotation about the principal axis.
β	A spin function having an m_s value of $-\frac{1}{2}$.
β	The two-center Hückel molecular orbital parameter.
C_n	A proper rotation by $2\pi/n$.
$C(\phi)$	An arbitrary rotation.
$\chi(R)$	The character of the group operation R.
χ_μ	An atomic-centered basis function.
D^j	A representation of the three-dimensional rotation group.
δ	The variation of a quantity.
δ_{ij}	The Kronecker delta function.

∇^2	The Laplacian operator.
∂	Partial derivative.
E	The total energy of a system.
E	The identity operation of a group.
E	A two-dimensional representation of a point group.
ε	A one-electron energy level.
f	A vibrational force constant.
$\mathbf{F}$	The force-constant matrix.
$\hat{F}$	The Fock operator.
ϕ	A rotation—particularly, the angle from the positive x axis of a projection in the xy plane.
ϕ_i	A molecular orbital.
γ	The gyromagnetic ratio of a particle.
Γ	A generalized representation of a group.
h	Planck's constant.
$\hbar$	Planck's constant divided by 2π.
$\mathbf{H}$	The magnetic-field vector.
$\hat{H}$	The Hamiltonian operator.
i	The square root of -1.
i	The point of inversion.
$\mathbf{i}$	A unit vector along the Cartesian x axis.
j	The total angular momentum of a single particle.
$\mathbf{j}$	A unit vector along the Cartesian y axis.
J	The total angular momentum of a system.
$\mathbf{k}$	A unit vector along the Cartesian z axis.
K	The internal component angular-momentum quantum number for a symmetric top.
κ	The rotational-state index for an asymmetric top.
l	The orbital angular momentum of a single particle.
L	The total orbital angular momentum of a system.
λ	Wavelength.
m	The mass of a particle.
m	The z component of a single-particle orbital angular momentum.
m_j	The z component of a single-particle total angular momentum.
m_s	The z component of a single-particle spin angular momentum.
M	The z component of the total angular momentum; L, S or J can be used as subscripts.
μ	The reduced mass.
μ_{ij}	The transition dipole.
n	The atomic principal quantum number.
N	A normalizing constant.
ν	Frequency.
$\bar{\nu}$	Wave number, $1/\lambda$.
$\hat{O}$	A generalized operator.

O(3)	The three-dimensional rotation-inversion group; the group of all orthogonal unitary matrices.
ω	Angular frequency.
$\hat{P}$	A projection operator.
ψ	A generalized wave function.
q, Q	A generalized coordinate.
R	A general operation of a point group.
R(3)	The three-dimensional rotation group.
s	The single-particle spin quantum number.
S	The total spin quantum number.
S_n	An improper rotation.
σ	Reflection through a plane.
T	A three-dimensional representation of a point group.
$\hat{T}$	The kinetic-energy operator.
θ	A rotation, especially the inclination of a vector from the positive z axis.
u	An atom-centered basis function.
V	A perturbation.
$\hat{V}$	The potential-energy operator.
$*$	Complex conjugate.
$\sim$	Transpose.
$\dagger$	Conjugate transpose.
$\oplus$	Direct sum.
$\otimes$	Direct (or outer) product.
$\subset$	Contained in.
$\not\subset$	Not contained in.
$\langle\ ,\ \rangle$	Dirac brackets.
$\langle\ \ \rangle$	Expectation value.

appendix two

Vectors, matrices and tensors

Vectors, matrices, and tensors are arrayed quantities. A *vector* is a one-dimensional array—that is, a quantity that requires a single index to describe it. *Matrices* are two-dimensional arrays; they require two indexes. The general arrayed quantity is a *tensor*. The number of indexes required to specify the tensor is called its *rank*. Thus, a vector is a tensor of rank one, a matrix is a tensor of rank two, and a *scalar* is a rank-zero tensor.

Vectors are commonly used in a three-dimensional Cartesian space. Any property that requires a direction, as well as a magnitude, is a vector quantity. Vectors in Cartesian space are normally constructed relative to the Cartesian unit vectors, **i**, **j**, and **k**, directed in the x, y, and z directions. The arbitrary vector, **v**, has the form

$$\mathbf{v} = a\mathbf{i} + b\mathbf{j} + c\mathbf{k} \qquad \textbf{A2.1}$$

The **i**, **j**, and **k** are the *basis vectors* from which **v** is constructed. The concept can be extended to as many dimensions as is desired, if the proper basis set is defined.

$$\mathbf{v} = \sum_i a_i \mathbf{e}_i \qquad \textbf{A2.2}$$

where the $\mathbf{e}_i$ are the basis set. The basis itself need not be a set of vectors. It can be a set of functions or any other quantities according to which the vector is to be constructed. For example, six types of coins are in use in the United States monetary system: the one-cent piece (the penny), the five-cent piece (the nickel), the ten-cent piece (the dime), the quarter dollar, the half dollar, and the dollar. A pocket full of change can be expressed as a vector quantity, with the denominations of the coins being the basis. Let the basis have the obvious set of labels, p, n, d, q, h, and D. A typical collection of change, $\mathbf{c}$, might be

$$\mathbf{c} = 5p + 2n + 3d + 0q + 4h + 0D \qquad \textbf{A2.3}$$

If we define p as \$.01, n as \$.05, . . . , D as \$1.00, and substitute into Eq. A2.3, we obtain the dollar amount of the change: \$2.45. This combination of change has the same buying power as any other combination that gives \$2.45. Having \$2.45 worth of change is not a unique specification, but Eq. A2.3 is.

Sums of vectors, expressed in terms of the same basis, are obtained simply by adding the components of each basis from the two vectors. For example, if the change vector, $\mathbf{c}'$,

$$\mathbf{c}' = (0p + 0n + 0d + 3q + 0h + 2D) \qquad \textbf{A2.4}$$

were added to Eq. A2.3, the result would be

$$\begin{aligned} \mathbf{c}'' &= \mathbf{c} + \mathbf{c}' \\ &= (5p + 2n + 3d + 3q + 4h + 2D) \end{aligned} \qquad \textbf{A2.5}$$

This is the usual description of a vector sum. There is also another type of vector sum, the *direct sum*. The vector, $\mathbf{c}$ (Eq. A2.3), can be expressed in terms of only the $\{p, n, d, h\}$ basis, since it contains no quarter or dollar contributions. Similarly, $\mathbf{c}'$ (Eq. A2.4) can be expressed in only the $\{q, D\}$ basis. In this case the vector $\mathbf{c}''$ would be the direct sum (symbolized $\oplus$) of $\mathbf{c}$ and $\mathbf{c}'$. We would have

$$\mathbf{c} = (5p + 2n + 3d + 4h) \qquad \textbf{A2.3a}$$

$$\mathbf{c}' = (3q + 2D) \qquad \textbf{A2.4a}$$

and

$$\begin{aligned} \mathbf{c}'' &= \mathbf{c} \oplus \mathbf{c}' \\ &= (5p + 2n + 3d + 4h + 3q + 2D) \end{aligned} \qquad \textbf{A2.5a}$$

Notice that normally the bases from the second term are added on at the end. To reiterate, the usual vector sum is the sum of vectors having the same basis set, while the direct sum involves two mutually exclusive basis sets, and the result is an expanded basis set containing those from both vectors in the sum. For special purposes, sums are sometimes used in which there is a partial overlapping of the basis sets. We will not have cause to use these.

Vectors are most commonly written as arrays of the components only, with the basis being implied. They can be written as *row* arrays or as *column* arrays. If the basis is orthonormal, the two conventions are equivalent. The convention most commonly used by mathematicians is for the vector of components to be a row vector and for a basis set to be a column array.

If $\mathbf{c}$ is the row vector

$$\mathbf{c} = (5 \quad 2 \quad 3 \quad 0 \quad 4 \quad 0) \qquad \textbf{A2.6}$$

and **b** is the column vector of the basis functions

$$\mathbf{b} = \begin{bmatrix} p \\ n \\ d \\ q \\ h \\ D \end{bmatrix} \qquad \textbf{A2.7}$$

then Eq. A2.3 is a *dot product*, $\mathbf{c} \cdot \mathbf{b}$, of the vectors **c** and **b**. A dot product, p, is the product of a row vector, **a**, and a column vector, **b**, of the same dimension, constructed as

$$p = \sum_i a_i b_i \qquad \textbf{A2.8}$$

If **a** and **b** are vectors constructed from the same orthonormal basis set, the dot product produces a *scalar* quantity and consequently is often called a *scalar product*.

A second type of product of vectors is variously called the *direct product*, the *outer product*, or the *tensor product*. It is the product of a column vector and a row vector and is commonly denoted $\mathbf{a} \otimes \mathbf{b}$. The result is a *matrix*, or *second-rank tensor*. The dimensions of the two vectors need not be the same. If they are not, the matrix will not be a square matrix. The number of rows will correspond to the dimension of the column vector and the number of columns to the dimension of the row vector. The elements of the matrix are

$$M_{ij} = a_i b_j \qquad \textbf{A2.9}$$

Two indexes are needed. The outer product of a matrix with a vector yields a *third-rank tensor*, requiring three indexes. A *fourth-rank tensor* can be constructed from the outer product of a third-rank tensor and a vector, or from the outer product of two matrices, and so on. The elements of the resulting tensor should carry the appropriate number of indexes.

The *direct sum* of two matrices is similar to the direct sum of two vectors. The same symbol, $\oplus$, is used. An expansion of a basis is implied. The dimensionality of the matrix is expanded. Here, however, the direct sum implies that the new elements are independent of the others. The matrix elements in the rows of one matrix and the columns of the others, and vice versa, are all zero. For example, if we have the two matrices, **M** and **M′**,

$$\mathbf{M} = \begin{bmatrix} a & b & c \\ d & e & f \\ g & h & i \end{bmatrix} \qquad \textbf{A2.10}$$

$$\mathbf{M}' = \begin{bmatrix} p & q \\ r & s \end{bmatrix} \qquad \textbf{A2.11}$$

their direct sum, $\mathbf{M}''$, is

$$\mathbf{M}'' = \begin{bmatrix} a & b & c & 0 & 0 \\ d & e & f & 0 & 0 \\ g & h & i & 0 & 0 \\ 0 & 0 & 0 & p & q \\ 0 & 0 & 0 & r & s \end{bmatrix} \qquad \textbf{A2.12}$$

A matrix having the form of Eq. A2.12 is said to be in *block-diagonal form.* There are smaller matrices (blocks) along the diagonal, and zeros everywhere else. Any block-diagonal matrix can be expressed as the direct sum of the submatrices.

In addition to the direct sum of matrices, there is the simple *sum* of matrices having the same dimensions. This is just the matrix that results when the equivalent elements of the two matrices are added.

A *determinant* is also represented as a second-rank array. The array must be square; in this case, however, it is not the array, but rather an *expansion* of it in terms of its elements that has significance. The most direct way of expanding a determinant is in terms of its *cofactors.* The expanded determinant is

$$D = \sum_{j=1}^{n} (-1)^{(j-1)} a_{ij} A_{ij} \qquad \textbf{A2.13}$$

where the a_{ij} are the elements of one row, the A_{ij} are the cofactors of these elements, and n is the dimension of the determinant. The cofactor is the determinant remaining when the ith row and the jth column of the original determinant are deleted. The process is repeated until only 2×2 determinants are left. A 2×2 determinant, A, can be directly expanded

$$A = \begin{vmatrix} a & b \\ c & d \end{vmatrix}$$
$$= ad - bc \qquad \textbf{A2.14}$$

Any square matrix has a determinant associated with it, but it should be emphasized that they are not the same thing. When matrices and determinants are written as two-dimensional collections of numbers or elements, matrices are enclosed in *brackets* (or parentheses), while determinants are enclosed in *vertical lines.*

Inner products of two matrices, or of a vector and a matrix, are based on scalar products. The result will be a matrix or a vector, but each entry in the result is a scalar product. If the vector is on the left, it must be a row vector. If there is a matrix on the left, it is treated as a column of row vectors. The vector on the right must be a column vector. If there is a matrix on the right, it is treated as a row of column vectors. The number of columns on the left must be the same as the number of rows on the right. The result for the product, $\mathbf{M} = \mathbf{AB}$, of the $n \times p$ matrix, $\mathbf{A}$, on the left (where n is the number of rows) with the $p \times m$ matrix, $\mathbf{B}$, on the right is the $n \times m$

matrix, **M**, with elements defined as

$$M_{ij} = \sum_{k=1}^{p} A_{ik}B_{kj} \qquad \textbf{A2.15}$$

where the first index is the row label and the second is the column label. A row vector is a $1 \times p$ matrix and a column vector a $p \times 1$ matrix. Then, Eq. A2.15 reduces to the scalar product for two vectors.

Matrices are frequently used to relate two vectors. The entries in the matrix give the relation of the components of one vector to the components of the other. Two indices are required in the matrix, one from each vector. The vector

$$\mathbf{\$} = (.01 \quad .05 \quad .10 \quad .25 \quad .50 \quad 1.00) \qquad \textbf{A2.16}$$

is the vector relating the change vector of Eq. A2.6 to the dollar amount of the change. We have the scalar product

$$m = \mathbf{c}\cdot\mathbf{\$} \qquad \textbf{A2.17}$$

A matrix relating the denominations of the coins to each other can be constructed. If the change vector is a row vector, the matrix is

$$\mathbf{M} = \begin{bmatrix} 1. & .2 & .1 & .04 & .02 & .01 \\ 5. & 1.0 & .5 & .20 & .10 & .05 \\ 10. & 2.0 & 1.0 & .40 & .20 & .10 \\ 25. & 5.0 & 2.5 & 1.00 & .50 & .25 \\ 50. & 10.0 & 5.0 & 2.00 & 1.00 & .50 \\ 100. & 20.0 & 10.0 & 4.00 & 2.00 & 1.00 \end{bmatrix} \qquad \textbf{A2.18}$$

where, for example, the second row gives the worth of a nickel in pennies, in nickels, in dimes, and so forth. The product **cM** is

$$\mathbf{cM} = (5 \quad 2 \quad 3 \quad 0 \quad 4 \quad 0) \begin{bmatrix} 1. & .2 & .1 & .04 & .02 & .01 \\ 5. & 1.0 & .5 & .20 & .10 & .05 \\ 10. & 2.0 & 1.0 & .40 & .20 & .10 \\ 25. & 5.0 & 2.5 & 1.00 & .50 & .25 \\ 50. & 10.0 & 5.0 & 2.00 & 1.00 & .50 \\ 100. & 20.0 & 10.0 & 4.00 & 2.00 & 1.00 \end{bmatrix}$$

$$= (245 \quad 49 \quad 24.5 \quad 9.8 \quad 4.9 \quad 2.45) \qquad \textbf{A2.19}$$

which gives the total equivalent of the change in pennies, in nickels, in dimes, in quarters, in half dollars, and in dollars.

The *inverse*, $\mathbf{v}^{-1}$, of a vector, **v**, is defined as the vector that gives a scalar product of unity with the original vector

$$\mathbf{v}\cdot\mathbf{v}^{-1} = \mathbf{v}^{-1}\cdot\mathbf{v} = 1 \qquad \textbf{A2.20}$$

The inverse, $\mathbf{M}^{-1}$, of a matrix, $\mathbf{M}$, is defined as the matrix that gives the *unit matrix*, $\mathbf{E}$, as its inner product with the original matrix

$$\mathbf{MM}^{-1} = \mathbf{M}^{-1}\mathbf{M} = \mathbf{E} \qquad \textbf{A2.21}$$

where $\mathbf{E}$ is the diagonal matrix having $+1$ for every element along the principal diagonal. The transpose, $\tilde{\mathbf{M}}$, of a matrix, $\mathbf{M}$, is a matrix in which the rows and columns of the original are interchanged. The transpose, $\tilde{\mathbf{v}}$, of a row vector, $\mathbf{v}$, is a column vector having the same elements in the same order. The transpose of a column vector is a row vector with the same elements. The complex conjugate, signified by an asterisk, of a vector or matrix is the vector or matrix resulting when the complex conjugate of each element is taken. The conjugate transpose, signified by a dagger, †, results when the complex conjugate of the transpose is taken.

Inverses, transposes, or conjugate transposes of products of vectors or matrices are the products of the inverses, transposes or conjugate transposes, taken in reverse order. If

$$\mathbf{M} = \tilde{\mathbf{a}}\mathbf{b} \qquad \textbf{A2.22a}$$

then

$$\mathbf{M}^{-1} = \mathbf{b}^{-1}\tilde{\mathbf{a}}^{-1} \qquad \textbf{A2.22b}$$

$$\tilde{\mathbf{M}} = \tilde{\mathbf{b}}\mathbf{a} \qquad \textbf{A2.22c}$$

$$\mathbf{M}^\dagger = \mathbf{b}^\dagger(\tilde{\mathbf{a}})^\dagger = \mathbf{b}^\dagger\mathbf{a}^* \qquad \textbf{A2.22d}$$

A *unitary matrix* is a square matrix whose inverse equals its conjugate transpose. The *magnitude* of the determinant of a unitary matrix is unity. A *unimodular matrix* is a matrix whose determinant is $+1$. Unitary matrices are useful for *transforming* vectors or matrices. A *unitary transformation* preserves the magnitude of the quantity being transformed. In Cartesian space, rotations of vectors are accomplished by unitary transformations. If $\mathbf{a}$ is a row vector and $\mathbf{U}$ a unitary matrix, a transformed $\mathbf{a}$, $\mathbf{a}'$, can be written

$$\mathbf{a}' = \mathbf{aU} \qquad \textbf{A2.23}$$

If $\mathbf{b}'$ is $\mathbf{bU}$, and the matrix, $\mathbf{M}$, is the matrix $\tilde{\mathbf{b}}\mathbf{a}$, then the transformed matrix, $\mathbf{M}'$, is

$$\begin{aligned}\mathbf{M}' &= \tilde{\mathbf{b}}'\mathbf{a}' \\ &= \tilde{\mathbf{U}}\tilde{\mathbf{b}}\mathbf{aU} \\ &= \tilde{\mathbf{U}}\mathbf{MU} \\ &= \mathbf{U}^{-1}\mathbf{MU}\end{aligned} \qquad \textbf{A2.24}$$

Matrices that are *hermitian* (those that equal their conjugate transpose) can be *diagonalized* (put in diagonal form) by unitary transformations. Rotations of coordinate systems can also be accomplished by unitary transformations.

appendix three

Selected results of nonempirical quantum-chemical calculations

A3.1 INTRODUCTION

This appendix gives a few selected results of nonempirical calculations and their application to chemistry. The choice is arbitrary, but the material included was chosen to be somewhat representative of the types of work that have been done with nonempirical calculations. To a great extent, the computer programs for doing quantum-chemical calculations, at all levels of approximation, have been developed into tools, much the same as major instruments. Like instruments, they can be used wisely or misused.

Those interested in doing their own calculations should do extensive reading in the field in order to get a feeling for what is reasonable. Entry points into the literature might be the surveys that are published regularly in the *Annual Reviews of Physical Chemistry* or in the *Annual Reports of the Chemical Society of London.* Calculations are published in many diverse journals, but the majority of those of interest to chemists are probably found in the *Journal of Chemical Physics*, the *International Journal of Quantum Chemistry*, *Theoretica chimica Acta* (Berlin), or the *Journal of the American Chemical Society*.

The material in this appendix is divided into ground electronic states (the broad areas of structure and reactivity) and excited electronic states (primarily spectroscopy). Each entry gives the problem to be solved, the approach taken to solve it (mainly, the type of calculation performed), the results, any additional discussion, and the references. We have retained the units used in the original articles, rather than attempting to convert them to a common set of units.

A3.2 GROUND ELECTRONIC STATES

For most stable chemical species, the ground-state molecular structure is more easily determined by experiment than by computation. This is not the case for metastable or highly reactive species. Computations on such systems have been useful, both to obtain the structure for its own sake and to determine the course of chemical reactions. We will discuss one stable species, the hydrogen molecule, and several metastable species.

The Hydrogen Molecule. The hydrogen molecule is the prototype of all neutral molecules. Very many calculations of various types have been performed on it. As is common with molecular calculations, almost all of these employed the Born-Oppenheimer approximation (or the *adiabatic* approximation, as it is often called). Corrections are commonly applied to the adiabatic results to obtain D_0 from D_e. Most such calculations are not sufficiently accurate to tell whether this is a valid approximation or not.

Kolos and Wolniewicz performed very accurate calculations on H_2, both with and without the adiabatic approximation. They used a wave function of the type described in Section 10.5. With the adiabatic approximation, 80 terms were used in the wave-function expansion. Without it, 147 terms were used. Relativistic corrections were included. The computational accuracy of their results was felt to be greater than the experimental accuracy of the experiments.

Their calculated D_e value with the adiabatic approximation was 38,297.1 cm^{-1}. The reported experimental value was 38,292.9 $\pm$ .5 cm^{-1}. In other words, the theoretical ground-state energy, obtained from a variational wave function, came out to be lower than the experimental. Any improvements in the wave function could only make the calculated energy even lower. Without the adiabatic approximation, the calculated D_0 value was 36,114.2 cm^{-1}, while the reported experimental value was 36,113.6 $\pm$ .3 cm^{-1}.

In answer to the original question, the Born-Oppenheimer approximation does introduce a slight inaccuracy into the calculations, giving in this case an energy slightly lower than the true ground state. More surprising, however, is the fact that the calculation without the approximation gave a ground-state energy lower than the experimental value, by an amount outside the estimated error of either. This discrepancy was resolved when the experimentalists found a small error that brought the two results together, to within the accuracy of each. [W. KOLOS and L. WOLNIEWICZ, *J. Chem. Phys.*, **41**, 3663, 3674 (1964).]

HNC and HCO^+. In 1970 the first unidentified lines in the microwave spectral region that originated in interstellar space were observed. One, at 8.9189×10^{10} sec^{-1}, was observed in the galactic sources called W51, W3, Orion, and others. The other, at 9.0665×10^{10} sec^{-1}, was observed in W51 and DR21. These correspond to the emission from molecular species having rotational constants, B_e, of 4.4595 $\times$ 10^{10} sec^{-1} and 4.5333 $\times$ 10^{10} sec^{-1}, respectively. The first species, in particular, occurs in a number of sources, so it was important to the astrophysicists to identify them. Since interstellar species were known to be very small, the suggestion was made that these were HNC or HCO^+.

The rotational constants are so near to each other that rather accurate calculations are neccessary to distinguish between them. Hartree-Fock S.C.F. calculations were made using a rather large Gaussian basis set at the "double-ζ" level of accuracy. (The double-ζ level formally corresponds to a calculation in which each electron has its own effective nuclear charge, ζ, rather than each pair having its effective nuclear charge, as is implied by the double occupancy of orbitals.) Extensive configuration interaction was performed after the S.C.F. calculation was completed. The geometry was optimized. Calculations were also performed on HCN, at the same level of accuracy, so that the reliability of the calculations could be checked against a molecule for which experimental data were available.

For HCN, the CN distance was calculated to be 1.153 Å (experimental, 1.153), the HC distance 1.068 Å (1.066), and the rotation constant 4.453 $\times$ 10^{10} sec^{-1} (4.452 $\times$ 10^{10}). Using the same techniques, both HNC and HCO^+ were found to be linear. For HNC, the calculated CN distance was 1.169 Å and that for HN was .995 Å. The calculated B_e was 4.543 $\times$ 10^{10} sec^{-1}. For HCO^+, the calculated CO bond length was 1.1045 Å and that for HC was 1.095 Å. The calculated B_e was 4.466 $\times$ 10^{10} sec^{-1}. These results give strong support to the idea that the species emitting at 8.9189 $\times$ 10^{10} sec^{-1} is HCO^+ and that emitting at 9.0665 $\times$ 10^{10} sec^{-1} is HNC. [P. K. Pearson, G. L. Blackman, H. F. Schaefer, III, B. Roos, and U. Wahlgren, *Astrophys. J.*, **184**, L19 (1973). U. Wahlgren, B. Liu, P. K. Pearson, and H. F. Schaefer, III, *Nature Phys. Sci.*, **246**, 4 (1973).]

Ammonia, Hydrogen Chloride, and Ammonium Chloride. The first serious effort to obtain an accurate, nonempirical calculation of the potential surface for the reaction of polyatomic molecules was Clementi's study of the reaction of ammonia and hydrogen chloride to give ammonium chloride. The work was an S.C.F. calculation using a flexible Gaussian basis set. The most favorable course of the reaction was found to be the approach of the HCl, H first, with its C_∞ axis coincident with the C_3 axis of NH_3, up to the vertex of the ammonia pyramid. There was no activation barrier to be overcome. Clementi performed extensive analyses of the charge reorganization that accompanied the reaction. [E. Clementi, *J. Chem. Phys.*, **46**, 3851 (1967); **47**, 2323 (1967). E. Clementi and J. N. Gayles, *J. Chem. Phys.*, **47**, 3837 (1967).

H_3^+. One of the simplest possible chemical reactions is that of the hydrogen ion with the hydrogen molecule. If there is isotopic substitution, the isotopic exchange can be followed. The reaction goes through a metastable intermediate, H_3^+. The structure and stability of the intermediate are of interest in their own right; of more interest, however, is a potential-energy surface, giving the relative energies of the system as a function of the positions of the three nuclei. Such results can be used with trajectory calculations to deduce details of the reaction.

Conroy performed a calculation on H_3^+ and on the potential surface using direct numerical integration. He found the most stable structure as an equilateral triangle, $1.65a_0$ on a side, having a total energy of -1.348 Hartrees. This is probably the most accurate H_3^+ calculation available. [H. CONROY, *J. Chem. Phys.*, **51**, 3979 (1969).]

Csizmadia and co-workers also studied H_3^+, using a moderate-sized Gaussian basis and configuration interaction. They obtained the equilateral structure, with an R_e value of $1.66a_0$ and a total energy of -1.3397 Hartrees. Using the same computational techniques, their energy for infinite separation of $H_2 + H^+$ was -1.1668 Hartrees, for a dissociation energy of .1729 Hartree. This D_e value is very close to what would be obtained from Conroy's calculation, indicating that the error in the H_3^+ calculation is very near to that in the H_2 calculation. One purpose of this calculation was to use it as the starting point for for a trajectory calculation of the dynamics of the $D^+ + H_2 \rightarrow DH + H^+$ reaction. This represented the first calculation of the dynamics of a chemical reaction using a nonempirical quantum-mechanical potential-energy surface.

A number of interesting features were apparent in these calculations. Among others, reaction proceeded by way of a complex having a lifetime greater than the rotational period of the complex. Reaction occurred out to impact parameters (distances) significantly greater than the polarization limit. There was both forward and backward scattering of the molecular product. The ratio of the two depended upon the collision energy but appeared purely statistical at collision energies greater than 4.5 eV. These and other predictions can be experimentally checked by molecular-beam experiments. [I. G. CSIZMADIA, R. E. KARI, J. C. POLANYI, A. C. ROACH, and M. A. ROBB, *J. Chem. Phys.*, **52**, 6205 (1970). I. G. CSIZMADIA, J. C. POLANYI, A. C. ROACH, and W. H. WONG, *Can. J. Chem.*, **47**, 4097 (1969).

Methylcarbene. Detailed descriptions of intramolecular reactions are usually easier to obtain than those of intermolecular reactions. This is because the collision dynamics do not have to be considered. As an example, methylcarbene, **1**, is a reactive

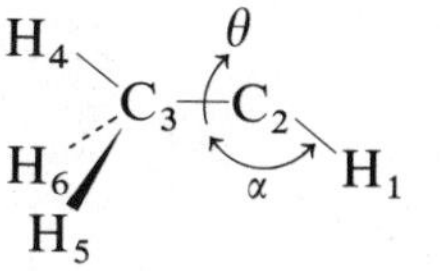

(1)

species that rearranges to give ethylene, **2**. The simplest carbene, methylene, CH_2,

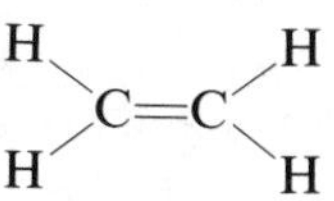

(2)

has a triplet ground state with a significant energy gap to the lowest singlet state. On this basis, it had been postulated that the rearrangement of methylcarbene started from the triplet state and initially yielded a triplet state of ethylene (which would be an excited state).

S.C.F. calculations were carried out to determine the reaction surface, using a Gaussian basis set at the double-ζ level. The molecules are sufficiently large that it is impractical to independently vary all geometrical parameters. For the ground state of **1**, only the angles α and θ were optimized. The other parameters were held fixed at values that were assumed, from experimental or other theoretical data, to be reasonable. For the reaction surface, only the transition state was considered in detail. Reasonable transition states were postulated on the basis of the singlet and triplet energies of both **1** and **2**. A trial geometry for the transition states was obtained by averaging the bond lengths of **1** and **2** and using the angles required by the transition state. These were then optimized. Overall, the reaction pathway was assumed to be that which involved the lowest-energy transition state.

The multiplicity of the ground state turned out to depend on the angle θ. When H_1 was *cis* to one of the CH bonds, the triplet was the ground state, but when it was *trans* (as shown in **1**), the singlet was the ground state. The *trans* conformation was the lowest energy for each multiplicity; consequently, the singlet state in the *trans* conformation was the lowest-energy situation, but only by about .3 kcal/mol. For the transition state, a singlet state in which H_5 has migrated halfway to C_2 was found to be the most stable. Based upon these conclusions, it should be possible to deduce the stereochemistry of the rearrangement of more highly substituted carbenes. [J. A. ALTMANN, I. G. CSIZMADIA, and K. YATES, *J. Amer. Chem. Soc.*, **96**, 4196 (1974).]

Methanol and Methoxides. Owing to the complexity of the problem of calculating potential-energy surfaces for reactions involving molecules of any size, workers frequently try to approximate the mechanisms of reactions for large systems through calculations on smaller systems. Consider, for example, the oxy-Cope rearrangement

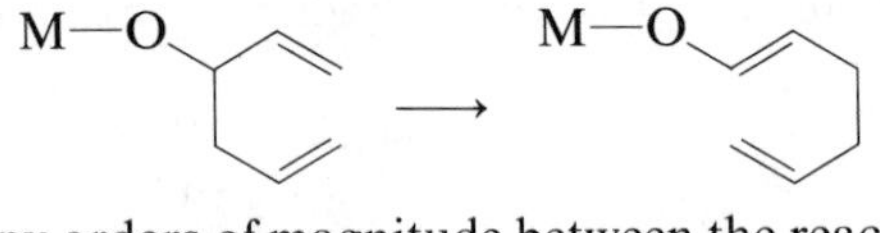

A3.1

The rate differs by many orders of magnitude between the reaction when M is hydrogen and when M is potassium.

In the reaction, a bond from the atom to which the oxygen is attached is broken. It was decided to try to rationalize the rate difference on the basis of the C—H bond-dissociation energies of methanol, CH_3—OH, and the methoxides. That for methanol was experimentally known, but it was not known for metal salts or for the free methoxide ion. A double-ζ Gaussian basis was used with a generalized valence-bond calculation and configuration interaction.

The D_e value for the C—H bond of methanol was calculated to be 90.7 kcal/mol. The experimental value was 91.8 $\pm$ 1.2. For sodium methoxide, the calculated value was 80.6 kcal/mol, for potassium methoxide it was 79.0, and for the methoxide ion it was 74.2. It was assumed that the substituent effect would follow this order for the breaking of any bond on a carbon adjacent to an oxygen. [M. L. STEIGERWALD, W. A. GODDARD, III, and D. A. EVANS, *J. Amer. Chem. Soc.*, **101**, 1994 (1979).]

A3.3 EXCITED ELECTRONIC STATES

The early discoveries in the quantum theory of the electronic structure of atoms resulted from efforts to interpret the spectra of the atoms. Many current workers in computational quantum chemistry have tended to neglect spectroscopy and excited electronic states, owing to the limitations of the variational theorem. However, the field is very important for the advances it might give in the understanding of the structure of molecules. In addition, it is useful, for studies in photochemistry, to know the energies of excited states relative to those of the ground states. It is also useful for spectroscopists to know what the molecular term symbols of the excited states are, what the orbital promotions are, and what the sources of various features in the observed spectra are.

Water. A feature is observed at 4.6 eV in the electron-impact spectrum of water that has not been explained. It has been suggested that it is the lowest triplet of water. Other workers suggested that a state found at 7.2 eV in the electron-impact spectrum was the triplet state.

S.C.F. calculations were carried out with a double-ζ Gaussian basis. This basis was augmented with a set of diffuse s and p functions, since it was believed that most of the transitions in water were primarily Rydberg transitions (transitions involving the promotion of an electron from a molecular orbital to an excited atomic orbital of one of the atoms). Extensive configuration interaction was included.

The agreement with the excitation energies for the five transitions that were experimentally well established was within $\pm .1$ eV. It was confirmed that the sixteen lowest excited states were Rydberg in character, involving promotions to oxygen $3s$ or $3p$ orbitals. The first calculated triplet occurred at 7.26 eV above the ground state, confirming the second suggestion. No excited state was found in the 4.6-eV range for the isolated water molecule. Thus, that feature must be due either to a dimer of water or to some other as yet unexplained phenomenon. [N. W. WINTER, W. A. GODDARD, III, and F. W. BOBROWICZ, *J. Chem. Phys.*, **62**, 4325 (1975).]

Ozone and Nitrogen Dioxide. Ozone, O_3, and nitrogen dioxide, NO_2, are of interest for environmental and atmospheric reasons. Both occur naturally in the atmosphere. Ozone in the upper atmosphere shields the earth from high-energy ultraviolet radiation from the sun. Nitrogen dioxide is part of the nitrogen-fixation process. It can also catalyze the decomposition of ozone. In the lower atmosphere, both are components of smog. Both molecules have problems associated with their experimental spectra. Only two excited states have been experimentally observed for ozone, owing to the low energy required for it to dissociate into the oxygen molecule and the oxygen atom (1.12 eV). On the other hand, the spectrum of nitrogen dioxide is extremely complex and irregular. It is difficult to even determine the number of different excited states in a given spectral region.

A generalized valence-bond calculation with configuration interaction was performed on the ground and fifteen excited states of ozone. Several different bases were used. The most flexible was a double-ζ Gaussian set, with added *d*-type functions on the oxygens. The least flexible was a "minimal-basis-set" Gaussian set (essentially equivalent to a single hydrogen-like function for each occupied atomic orbital in the molecule). The first three ionization potentials and the low-lying states of the ions were also determined.

The first three ionization potentials for ozone and the resulting ionized states ($\mathbf{C}_{2v}$ symmetry) are $^2A_1$, 12.91 eV; $^2B_2$, 13.03 eV; and $^2A_2$, 13.59 eV. The experimental values for the ionization energies are 12.75, 13.06, and 13.57 eV. The ordering of the $^2A_1$ and $^2A_2$ states is reversed from that expected from Koopmans' theorem. None of the vertical excitation energies are below the energy required for dissociation to O_2 and O; however, if the geometries of the excited states are optimized, three excited states are predicted to lie below the experimental dissociation energy. (The calculated dissociation energy did not agree at all well with the experimental.) These are a $^3B_2$, a $^1A_1$, and a $^3A_2$ state. [P. J. HAY, T. H. DUNNING, JR., and W. A. GODDARD, III, *J. Chem. Phys.*, **62**, 3912 (1975).]

Nitrogen dioxide was studied by a multiconfiguration self-consistent-field technique. This procedure combines an S.C.F. calculation with a simultaneous configuration-interaction calculation. The mixing coefficients for the configurations and the L.C.A.O. expansion coefficients are simultaneously determined in the variational procedure. This allows a smaller number of configurations to be used for a given degree of accuracy, particularly for properties involving energy differences in the same molecule. A double-ζ Gaussian basis was used. The geometry was optimized for the ground and several excited states.

The optimum geometry, the energy above the ground state, the vibrational frequencies for the symmetric vibrations (the symmetric stretch and the bend), and the dipole moment were determined for several states. For the ground state, the equilibrium bond length was calculated to be 1.20 Å (experimental, 1.1934); the bond angle, 134° (134.1°); the vibrational frequencies, 1351 and 758 cm^{-1} (1358 and 757); and the dipole moment is .37 D (.32). At their equilibrium geometries, the first three excited states lie 1.18 ($^2B_2$), 1.66 ($^2B_1$), and 1.84 ($^2A_2$) eV above the $^2A_1$ ground state. From this information a great many details of the spectrum can be worked out. [G. D. GILLISPIE, A. U. KHAN, A. C. WAHL, R. P. HOSTENY, and M. KRAUSS, *J. Chem. Phys.*, **63**, 3425 (1975).]

appendix four

Answers to selected problems

CHAPTER 1

1.2 Planck:

$$\rho_\nu = \frac{8\pi\nu^3 h}{c^3\left[\exp\left(\frac{h\nu}{kT}\right) - 1\right]} \quad \textbf{(a)}$$

$$= \frac{8\pi\nu^3 h \exp\left(-\frac{h\nu}{kT}\right)}{c^3\left[1 - \exp\left(-\frac{h\nu}{kT}\right)\right]} \quad \textbf{(b)}$$

$$\lim_{\nu\to 0} \text{(a)} = \frac{8\pi\nu^3 h}{c^3\left[\left(1 + \frac{h\nu}{kT} + \ldots\right) - 1\right]}$$

$$\cong \frac{8\pi\nu^2 kT}{c^3}$$

$$\lim_{\nu\to\infty} \text{(b)} = \frac{8\pi\nu^3 h}{c^3}\exp\left(-\frac{h\nu}{kT}\right)$$

1.3 (a) 1.325×10^{-34} m. (b) 6.626×10^{-32} m.
(c) 3.681×10^{-36} m. (d) 1.325×10^{-40} m.
(e) 3.722×10^{-62} m. (f) 1.039×10^{-8} m.
(g) 3.637×10^{-12} m. (h) 3.956×10^{-12} m.

1.4 $n = 1$: $v = 1.546 \times 10^6$ m sec^{-1}.
$n = 2$: $v = 7.732 \times 10^5$ m sec^{-1}.
$n = 3$: $v = 5.155 \times 10^5$ m sec^{-1}.

1.5 Li: $d = .407$, $E = 3.339$ eV $= 26\,931$ cm^{-1}.
Na: $d = 1.373$, $E = 3.166$ eV $= 25\,537$ cm^{-1}.

1.7 (a) $\psi(x) = a \exp \{(i/\hbar)[2m(E - V)]^{1/2}x\}$.
(b) If $V > E$, $\psi(x) = a \exp [-(1/\hbar)|2m(E - V)|^{1/2}x]$ (a decaying exponential).
(c) The function decays exponentially as the particle moves through the barrier. If the barrier is finite in height and thickness, there is a finite probability of the particle passing through the barrier.

1.8 (a) $\Delta\mathbf{Q}\,\Delta\mathbf{P} = -(i/2)[\mathbf{Q}, \mathbf{P}] = \hbar/2$.
(b) (1) $\Delta p = 5.27 \times 10^{-32}$ kg m sec^{-1} (2.93×10^{-34} of the total p).
(2) $\Delta p = 5.27 \times 10^{-25}$ kg m sec^{-1} (2.89×10^{-3} of the total p).

CHAPTER 2

2.1 (b) The energy is kinetic energy:

$$\frac{1}{2}mv^2 = E_n = \frac{n^2h^2}{8ml^2}$$

$$n \cong 1.572 \times 10^{32}$$

(c) $\Delta E = E_{n'} - E_n = [(n + 1)^2 - n^2]h^2/8ml^2$.
$= (2n + 1)h^2/8ml^2$.
$= 1.84 \times 10^{-33}$ J.

2.2 Under the stated assumptions, the total number of bond lengths is $(2n + 7)$, and the total number of electrons is $(2n + 6)$. Solving for the transition energy, in terms of wavelength, we find

$$\lambda = \frac{8mb^2c}{h(2n + 7)}$$

where (b is the average bond length). Plot λ against $(2n + 7)$ and extrapolate:

$$n = 11, \qquad \lambda = 5{,}300\ \text{Å}$$

$$n = 15, \qquad \lambda = 6{,}180\ \text{Å}$$

2.3 (a) $E_q = q^2\hbar^2/2mr^2$ (q is the quantum number).
$= q^2h^2/2mC^2$ (C is the circumference).
$\psi_q = (2\pi)^{-1/2} \exp(\pm iq\phi)$.
(b) Highest occupied orbital: $q = (n - 2)/4$.
Lowest vacant orbital: $q' = (n + 2)/4$.

$$C = nb$$

$$\Delta E = E_{q'} - E_q = \frac{h^2}{2mb^2n} \qquad (n \text{ is number of atoms})$$

or

$$\bar{\nu} = \frac{h}{2mcb^2n}$$

(c) Plot $\bar{\nu}$ against n and interpolate:

Anthracene: $n = 14, \quad \bar{\nu} = 2.76 \times 10^4 \text{ cm}^{-1}$

Tetracene: $n = 18, \quad \bar{\nu} = 2.58 \times 10^4 \text{ cm}^{-1}$

2.4 $E_{n_x n_y} = \dfrac{h^2}{8m}\left(\dfrac{n_x}{l_x^2} + \dfrac{n_y}{l_y^2}\right).$

For naphthalene, $l_x = 4 \times 1.4 \text{ Å} = 5.6 \text{ Å}$,
$l_y = 6 \cos 30° \times 1.4 \text{ Å} = 7.27 \text{ Å}$.

$$E_{n_x n_y} = a(n_x^2 + .593 n_y^2)$$

The first few levels are:

n_x	n_y	E (in units of the a)
1	1	1.593
1	2	3.373
2	1	4.593
1	3	6.340
2	2	6.373
2	3	9.340
2	4	13.488
3	3	14.337

There are 10 π electrons in naphthalene. The first five levels are doubly occupied. The first two transitions are

$$\Delta E_1 = E_{23} - E_{32} = 2.967a = 2.87 \times 10^4 \text{ cm}^{-1} \quad (\text{exp}, 3.2 \times 10^4)$$
$$\Delta E_2 = E_{23} - E_{13} = 3.000a = 2.90 \times 10^4 \text{ cm}^{-1} \quad (\text{exp}, 3.5 \times 10^4)$$

For anthracene, $l_x = 5.6 \text{ Å}$ and $l_y = 9.7 \text{ Å}$. There is an accidental triple degeneracy of the levels (1, 5), (2, 4), and (3, 1). The transitions are:

$$\Delta E_1 = E_{15} - E_{32} = 2.26 \times 10^4 \text{ cm}^{-1} \quad (\text{exp}, 2.6 \times 10^4)$$
$$\Delta E_2 = E_{15} - E_{14} = 2.9 \times 10^4 \text{ cm}^{-1}$$

CHAPTER 3

3.1 HCl: $I = 2.642 \times 10^{-47} \text{ kg m}^2$; $r = 1.280 \times 10^{-10}$ m.
HBr: $I = 3.303 \times 10^{-47} \text{ kg m}^2$; $r = 1.419 \times 10^{-10}$ m.
Relative populations: $(2J + 1) \exp[-BJ(J + 1)/kT]$

3.2 $^{16}O^{12}C^{32}S$: $I = 1.3800 \times 10^{-45} \text{ kg m}^2$.
$^{16}O^{12}C^{34}S$: $I = 1.4146 \times 10^{-45} \text{ kg m}^2$.
$r_{CO} = 1.16 \times 10^{-10}$ m; $r_{CS} = 1.56 \times 10^{-10}$ m.

3.3 (a) $D_g^1 \times D_u^2 = D_u^1 + D_u^2 + D_u^3$.
(b) $D_g^{3/2} \times D_g^{1/2} = D_g^1 + D_g^2$.
(c) $D_u^{3/2} \times D_u^1 \times D_g^{1/2} = D_g^0 + 2D_g^1 + 2D_g^2 + D_g^3$.
(d) $D_g^{1/2} \times D_u^{3/2} \times D_g^{5/2} = D_u^{1/2} + 2D_u^{3/2} + 2D_u^{5/2} + 2D_u^{7/2} + D_u^{9/2}$.
(e) $D_g^0 \times D_u^1 \times D_g^2 \times D_u^3 = D_g^0 + 2D_g^1 + 3D_g^2 + 3D_g^3 + 3D_g^4 + 2D_g^5 + D_g^6$.

3.5	HCN	linear (prolate)	observable microwave
	CH_2O	asymmetric	observable
	HCCH	linear	not observable
	NH_3	symmetric	observable
	H_2O	asymmetric	observable
	C_6H_6	oblate (planar)	not observable
	O_3	asymmetric	observable
	CH_3CH_3	symmetric	not observable

3.6 $Y_{0,0} = N$, $\hat{L}^2 Y_{0,0} = 0$, $\hat{L}_z Y_{0,0} = 0$.
$Y_{1,-1} = N \sin\theta e^{-i\phi}$, $\hat{L}^2 Y_{1,-1} = 2\hbar^2 Y_{1,-1}$, $\hat{L}_z Y_{1,-1} = -\hbar Y_{1,-1}$.
$Y_{2,2} = N \sin^2\theta e^{2i\phi}$, $\hat{L}^2 Y_{2,2} = 6\hbar^2 Y_{2,2}$, $\hat{L}_z Y_{2,2} = 2\hbar Y_{2,2}$.
In general:

$$\hat{L}^2 Y_{L,M} = L(L+1)\hbar^2 Y_{L,M}$$
$$\hat{L}_z Y_{L,M} = M\hbar Y_{L,M}$$

3.7 (a) $I_{a'} = 4.434 \times 10^{-47}$ kg m^2; $A' = 6.30$ cm^{-1}.
$I_b = 2.815 \times 10^{-47}$ kg m^2; $B = 9.94$ cm^{-1}.
(b)

J	K	ν'	Deg.	Rel. Pop.
0	0	0	1	1.00
1	1	16.24	4	3.70
1	0	19.88	2	1.82
2	2	45.08	10	8.04
2	1	56.00	10	7.63
2	0	59.64	3	3.75
etc.				

CHAPTER 4

4.1 The values of k are, in 10^2 N m^{-1}:
(c) 5.7613. (d) .2553 (^{7}Li). (e) .1715. (f) .0972 (^{39}K).
(g) 4.7025. (j) 9.6565. (l) 4.1153. (m) 3.1408.

4.2 (a)–(c) For isotopic substitution in the same molecule, there is essentially no change in the force constant.
(d)–(f) Going down a chemical family, the force constant gets smaller.
(g)–(i) Formally, F_2 has a single bond, O_2 a double bond, and N_2 a triple bond. The force constants increase in that order.
(j)–(m) The force constant decreases, going down the chemical family.
In general, for a like series of molecules, the force constant varies as the bonding energy between the atoms.

4.3 For H_2: (pop. $v = 0$)/(pop. $v = 1$) $= 1.67 \times 10^9$ (at 298.15°K).
For K_2: (pop. $v = 0$)/(pop. $v = 1$) $= 1.56$.

4.4 (a) $\dot{\mathbf{P}} = \mathbf{Q}^{-2}$.
(b) $\dot{\mathbf{P}} = -1$.

4.5 $E_{n_x n_y n_z} = (n_x + n_y + n_z + \frac{3}{2})h\nu_0$.

4.6 For a sphere, $x^2 + y^2 + z^2 =$ a constant. Therefore, $n_x = n_y = n_z = n$, and

$$E_n = (n + \tfrac{3}{2})h\nu_0$$

4.7 Let $\psi_{\text{total}} = \psi_{\text{vib}}\psi_{\text{rot}}$. A nonvanishing intensity requires that $\langle\psi_{\text{tot}}|\hat{\mu}|\psi'_{\text{tot}}\rangle \neq 0$. The selection rule is:

$$\Gamma_v\Gamma_r\Gamma_\mu\Gamma'_v\Gamma'_r \supset \Gamma_{\text{sym}}$$

CHAPTER 5

5.1 (a) $\langle r\rangle_{100} = a_0/Z$ (a_0 is the Bohr radius).
(b) $\langle r\rangle_{200} = (3 + \sqrt{5})a_0/Z$.
(c) $\langle r\rangle_{210} = 4a_0/Z$.
(d) $\langle r\rangle_{211} = 4a_0/Z$.
(e) $\langle r\rangle_{320} = 9a_0/Z$.
(f) $\langle r\rangle_{321} = 9a_0/Z$.
(g) $\langle r\rangle_{322} = 9a_0/Z$.

5.4 (a) $\zeta = 1.3443$.
(b) $E_{\text{tot}} = 2 \times .90355 = 1.8071$ a.u.
(c) $\langle r\rangle = .7439a_0$.

5.5 $\Delta l = \pm 1$.

5.7 $E_{1s} = -Z^2/2 = -4232 \text{ a.u.} = -1.845 \times 10^{-14}\text{ J}$
$= -\frac{1}{2}mv^2$
$v = 2.018 \times 10^8 \text{ m sec}^{-1}$.
This is two-thirds the speed of light!

CHAPTER 6

6.1 The r^3 and r^5 terms in the perturbation vanish. Only terms with even powers of r give nonvanishing first-order perturbations.

6.2 $\hat{H} = -\frac{1}{2}\nabla_1^2 - \frac{1}{2}\nabla_2^2 - Z/r_1 - Z/r_2 + 1/r_{12}$. Let $\chi(i)$ be the trial function; then

$$E = -\langle\chi(1)|\nabla_1^2|\chi(1)\rangle - 2Z\left\langle\chi(1)\left|\frac{1}{r_1}\right|\chi(1)\right\rangle + \left\langle\chi(1)\chi(2)\left|\frac{1}{r_{12}}\right|\chi(1)\chi(2)\right\rangle$$

$$\langle\chi(1)|\nabla_1^2|\chi(1)\rangle = -3\alpha$$

$$\left\langle\chi(1)\left|\frac{1}{r_1}\right|\chi(1)\right\rangle = 4\sqrt{\frac{\alpha}{2\pi}}$$

$$\left\langle\chi(1)\chi(2)\left|\frac{1}{r_{12}}\right|\chi(1)\chi(2)\right\rangle = \frac{8\alpha^3}{\pi^3} \times \left\{2\int_{r_1=0}^{\infty} 4\pi r_1^2 e^{-2\alpha r_1^2}\left[\int_{r_2=r_1}^{\infty} 4\pi r_2 e^{-2\alpha r_2^2}\,dr_2\right]dr_1\right\} = 2\sqrt{\frac{\alpha}{\pi}}$$

$$E(\alpha) = 3\alpha - 16\sqrt{\frac{\alpha}{2\pi}} + 2\sqrt{\frac{\alpha}{\pi}}$$

Minimizing w.r.t. α gives $\alpha = .76700$, and $E = -2.3010$ a.u.

6.5 (a) To first order

$$E = \frac{h^2}{8mL^2}\left\{n^2 + \frac{1}{4} + \frac{1}{n\pi}\sin\left[\left(1 - \frac{3n}{4\pi}\right)\right]\right\}$$

(c) There are no symmetry-induced restrictions on the first-order perturbation. The second-order perturbation vanishes unless n and n' differ by an even number. In general, the perturbation, which is symmetric with respect to reflection through the origin, vanishes unless ψ_n and $\psi_{n'}$ are both symmetric or both antisymmetric with respect to this reflection.

6.6 (c) The perturbation is antisymmetric on reflection through the origin. The integrals vanish unless ψ_n is symmetric and $\psi_{n'}$ is antisymmetric, or vice versa.

CHAPTER 7

7.2

C: $(1s)^2(2s)^2(2p)^2$	N: $(1s)^2(2s)^2(2p)^3$
O: $(1s)^2(2s)^2(2p)^4$	V: $[\mathrm{Ar}](4s)^2(3d)^3$
Si: $[\mathrm{Ne}](3s)^2(3p)^2$	Ni: $[\mathrm{Ar}](4s)^2(3d)^8$
Se: $[\mathrm{Ar}](4s)^2(3d)^{10}(4p)^4$	Ti: $[\mathrm{Ar}](4s)^2(3d)^2$
Cl: $[\mathrm{Ne}](3s)^2(3p)^5$	Pm: $[\mathrm{Xe}](6s)^2(4f)^5$

7.3

O^{2-}: $(1s)^2(2s)^2(2p)^6$	Zn^{2+}: $[\mathrm{Ar}](3d)^{10}$
Fe^{3+}: $[\mathrm{Ar}](3d)^5$	Ag^+: $[\mathrm{Kr}](4d)^{10}$

7.6 S(2): $\Gamma = 4[2] + 2[1^2]$

$\Gamma_R = [2] + [1^2]$

S(3): $\Gamma = [2, 1] + 2[1^3]$

$\Gamma_R = [3] + 2[2, 1] + [1^3]$

S(4): $\Gamma = [4] + 2[2^2] + 3[2, 1^2]$

$\Gamma_R = [4] + 3[3, 1] + 2[2^2] + 3[2, 1^2] + [1^4]$

Note that the regular representation contains all the representations, each multiplied by its dimension.

7.9

Li:	$(2s)^1$	$\mathbf{{}^2S_{1/2}}$
Be:	$(2s)^2$	$\mathbf{{}^1S_0}$
B:	$(2p)^1$	$\mathbf{{}^2P_{1/2}}$, ${}^2P_{3/2}$
C:	$(2p)^2$	$\mathbf{{}^3P_0}$, ${}^3P_1$, ${}^3P_2$, ${}^1S_0$, ${}^1D_2$
N:	$(2p)^3$	$\mathbf{{}^4S_{3/2}}$, ${}^2P_{1/2}$, ${}^2P_{3/2}$, ${}^2D_{3/2}$, ${}^2D_{5/2}$
O:	$(2p)^4$	${}^3P_0$, ${}^3P_1$, $\mathbf{{}^3P_2}$, ${}^1S_0$, ${}^1D_2$
F:	$(2p)^5$	${}^2P_{1/2}$, $\mathbf{{}^2P_{3/2}}$
Ne:	$(2p)^6$	$\mathbf{{}^1S_0}$

(Boldface indicates the ground state.)

7.10

Sc: ${}^2D_{3/2}$	Ti: ${}^3F_2$
V: ${}^4F_{3/2}$	Cr: ${}^7S_3$
Mn: ${}^6S_{5/2}$	Fe: ${}^5D_4$
Co: ${}^4F_{9/2}$	Ni: ${}^3F_4$
Cu: ${}^2S_{1/2}$	Zn: ${}^1S_0$

7.11 Terms: ${}^3P_{2,1,0}$; ${}^3F_{4,3,2}$; ${}^1S_0$; ${}^1D_2$; ${}^1G_4$

Ground state: ${}^3F_4$

7.12

$$\begin{aligned} E = &-\langle 1s|\nabla^2|1s\rangle - \langle 2s|\nabla^2|2s\rangle - 2\langle 1s|Z/r_1|1s\rangle \\ &- 2\langle 2s|Z/r_1|2s\rangle + \langle 1s1s|1/r_{12}|1s1s\rangle \\ &+ 4\langle 1s2s|1/r_{12}|1s2s\rangle + \langle 2s2s|1/r_{12}|2s2s\rangle \\ &- 2\langle 1s2s|1/r_{12}|2s1s\rangle \end{aligned}$$

7.13 $\hat{F}[2s(1)] = -\frac{1}{2}\nabla_1^2 - Z/r_1 + 2\langle 1s(2)|1/r_{12}|1s(2)\rangle$
$+ \langle 2s(2)|1/r_{12}|2s(2)\rangle$
$- [2s^*(1)1s(1)/2s^*(1)2s(1)]\langle 1s(2)|1/r_{12}|2s(2)\rangle$

7.18 No. 116: $[\mathrm{Rn}](7s)^2(5f)^{14}(6d)^{10}(7p)^4$ $^3P_2$
No. 119: $[118](8s)^1$ $^2S_{1/2}$
No. 120: $[118](8s)^2$ $^1S_0$

7.19
$$\psi = 1/\sqrt{6}\begin{vmatrix} 1s(1) & 1s(2) & 1s(3) \\ 1\bar{s}(1) & 1\bar{s}(2) & 1\bar{s}(3) \\ 2s(1) & 2s(2) & 2s(3) \end{vmatrix}$$
$$= 1/\sqrt{6}\,\{1s(1)1\bar{s}(2)2s(3) + 2s(1)1s(2)1\bar{s}(3) + 1\bar{s}(1)2s(2)1s(3) - 2s(1)1\bar{s}(2)1s(3) - 1\bar{s}(1)1s(2)2s(3) - 2s(1)1s(2)1\bar{s}(3)\}$$

7.20 $E(\mathrm{Li}) = -\langle 1s|\nabla^2|1s\rangle - \frac{1}{2}\langle 2s|\nabla^2|2s\rangle - 2\langle 1s|Z/r|1s\rangle$
$- \langle 2s|Z/r|2s\rangle + \langle 1s1s|1/r_{12}|1s1s\rangle$
$+ 2\langle 1s2s|1/r_{12}|1s2s\rangle - \langle 1s2s|1/r_{12}|2s1s\rangle$
$E(\mathrm{Li}^+) = -\langle 1s|\nabla^2|1s\rangle - 2\langle 1s|Z/r|1s\rangle + \langle 1s1s|1/r_{12}|1s1s\rangle$
$-\mathrm{I.P.} = E(\mathrm{Li}) - E(\mathrm{Li}^+)$
$\cong -\frac{1}{2}\langle 2s|\nabla^2|2s\rangle - \langle 2s|Z/r|2s\rangle + 2\langle 1s2s|1/r_{12}|1s2s\rangle$
$- \langle 1s2s|1/r_{12}|2s1s\rangle$
(if there is no charge reorganization in the ion). But

$$\varepsilon_{2s} = \langle 2s|\hat{F}|2s\rangle$$
$$= -\frac{1}{2}\langle 2s|\nabla^2|2s\rangle - \left\langle 2s\left|\frac{Z}{r}\right|2s\right\rangle + 2\left\langle 1s2s\left|\frac{1}{r_{12}}\right|1s2s\right\rangle - \left\langle 1s2s\left|\frac{1}{r_{12}}\right|2s1s\right\rangle$$

CHAPTER 8

8.1 (a) $^1P_1, {}^1D_2, {}^1F_3$.
(b) $^3S_1, {}^3P_1, {}^3P_2, {}^3D_1, {}^3D_2, {}^3D_3$.
(c) $^3S_1, {}^3P_0, {}^3P_1, {}^3P_2$.
(d) $^2S_{1/2}, {}^2P_{1/2}, {}^2P_{3/2}$.
(e) $^4D_{1/2}, {}^4D_{3/2}, {}^4D_{5/2}, {}^4F_{3/2}, {}^4F_{5/2}, {}^4G_{5/2}$.
(f) $^2S_{1/2}, {}^2P_{1/2}, {}^2P_{3/2}, {}^2D_{3/2}, {}^2D_{5/2}$.
(g) $^4P_{5/2}, {}^4D_{5/2}, {}^4D_{7/2}, {}^4F_{5/2}, {}^4F_{7/2}, {}^4F_{9/2}$.

8.2
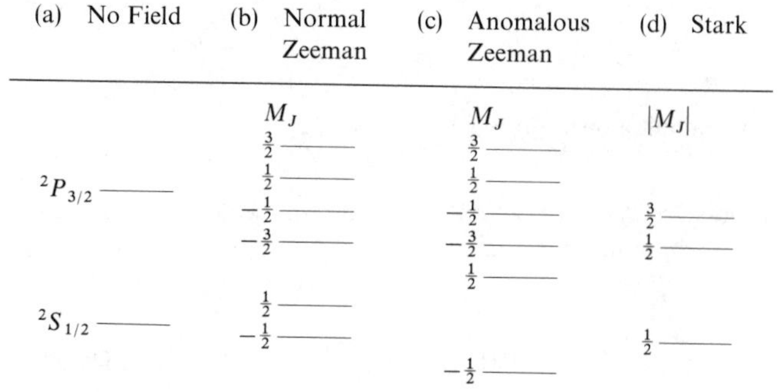

8.3 (a) One line. (b) Three lines; $\Delta M_J = -1, 0, 1$.
(c) Six lines. (d) Two lines.

8.6

Initial	Forbidden to (rules violated)
$^1S_0$	$^3P_0(\Delta S)$, $^3P_1(\Delta S)$, $^3P_2(\Delta S, \Delta J)$, $^1D_2(\Delta L, \Delta J)$
$^3P_1$	$^1D_2(\Delta S)$, $^3D_3(\Delta J)$, $^1S_0(\Delta S)$, $^1P_1(\Delta S)$
$^1D_2$	$^3P_1(\Delta S)$, $^1S_0(\Delta L, \Delta J)$
$^4P_{5/2}$	$^4D_{1/2}(\Delta J)$, $^4F_{3/2}(\Delta L)$, $^2S_{1/2}(\Delta S, \Delta J)$
$^2F_{5/2}$	$^2S_{1/2}(\Delta L, \Delta J)$, $^2P_{3/2}(\Delta L)$

8.7 Ground state $^4S_{3/2} \rightarrow {}^4P_{1/2}, {}^4P_{3/2}, {}^4P_{5/2}$ (arising from either $(2p)^2(3s)^1$ or $(2p)^2(3d)^1$ configurations.

CHAPTER 9

9.2 All excited states would lie higher in energy than ground state $H + H^+$. States that are stable with respect to dissociation to excited $H + H^+$ are theoretically predicted, but have not been observed experimentally.

9.3 There will be three bonding and three antibonding M.O.'s, one of each arising from the $|m|$ values of 0, 1, and 2. Those with $|m|$ of 1 and 2 will be doubly degenerate. They are: $3d\sigma(\sigma_g)$, $3d\sigma^*(\sigma_u)$, $3d\pi(\pi_u)$,$3d\pi^*(\pi_g)$, $3d\delta(\delta_g)$, and $3d\delta^*(\delta_u)$.

9.4 These can be deduced from the nodal properties of the functions as the nuclei go to zero separation. In general, a hydrogenic atomic orbital with a given l value has $(n - l - 1)$ radial nodes and l angular nodes, where n is the principal quantum number. If there is a uniquely defined z axis, there will be $(n - |m| - 1)$ total (radial or angular) nodes perpendicular to the z axis. For the diatomic molecules, the molecular axis is the z axis, and $|m|$ is replaced by λ. In the present case $3d$ orbitals have no radial nodes, so all the nodes are angular. There will be the same number of nodes in the united-atom limit as in the separated-atom limit. Thus: $3d\sigma \rightarrow 5g$, $3d\sigma^* \rightarrow 6h$, $3d\pi \rightarrow 4f$, $3d\pi^* \rightarrow 5g$, $3d\delta \rightarrow 3d$, and $3d\delta^* \rightarrow 4f$.

9.5 Let $\hat{h}$ be an effective Hamiltonian, and

$$h_{AA} = -\text{I.P.(Li)} = -5.39 \text{ eV}$$

$$E_{\text{Li}_2} = \frac{2}{1 + S}(h_{AA} + h_{AB})$$

$$\text{Dissociation energy} = 2E_{\text{Li}} - E_{\text{Li}_2} = \frac{2(h_{AA}S - h_{AB})}{1 + S}$$

$$\text{Transition energy} = \frac{1}{1 - S}(h_{AA} - h_{AB}) - \frac{1}{1 + S}(h_{AA} + h_{AB})$$

$$= \frac{2(h_{AA}S - h_{AB})}{1 - S^2}$$

Combining these gives

$$1.74S^2 + 1.03S - .71 = 0$$

The roots are: $S = -1.0$, and .408. Only the second root is reasonable, since $0 \leq S \leq 1$. From the dissociation energy:

$$h_{AB} = -2.92 \text{ eV}$$

$$\text{I.P.(Li}_2) = -\frac{1}{1 + S}(h_{AA} + h_{AB}) = 5.90 \text{ eV}$$

CHAPTER 10

10.1 $\psi = \det|1s_A(1)1\bar{s}_A(2)1s_B(3)| \pm \det|1s_B(1)1\bar{s}_B(2)1s_A(3)|$

10.6 The general expression for all possible configurations of all possible spins for N electrons in k orbitals is

$$\frac{(2k)!}{(2k - N)!N!}$$

[This construction counts each component of each degenerate (doublet, triplet, and so on) spin state.]

CHAPTER 11

11.1

σ_g^+ bonding	σ_u^+ antibonding
π_g antibonding	π_u bonding
δ_g bonding	δ_u antibonding
ϕ_g antibonding	ϕ_u bonding

11.2

C_2:	$(1s\sigma)^2(1s\sigma^*)^2(2s\sigma)^2(2s\sigma^*)^2(2p\sigma)^2(2p\pi)^2$	${}^3\Sigma_g^-$
P_2:	$[Ne_2](3s\sigma)^2(3s\sigma^*)^2(3p\sigma)^2(3p\pi)^4$	${}^1\Sigma_g^+$
V:	$[Ar_2](4s\sigma)^2(4s\sigma^*)^2(3d\sigma)^2(3d\pi)^4$	${}^1\Sigma_g^+$

11.3

C_2^+:	$[He_2](2s\sigma)^2(2s\sigma^*)^2(2p\sigma)^2(2p\pi)^1$	${}^2\Pi_u$
C_2^-:	$[He_2](2s\sigma)^2(2s\sigma^*)^2(2p\sigma)^2(2p\pi)^3$	${}^2\Pi_u$
N_2^+:	same as C_2^-	
N_2^-:	$[He_2](2s\sigma)^2(2s\sigma^*)^2(2p\sigma)^2(2p\pi)^4(2p\pi^*)^1$	${}^2\Pi_g$
O_2^+:	same as N_2^-	
O_2^-:	$[He_2](2s\sigma)^2(2s\sigma^*)^2(2p\sigma)^2(2p\pi)^4(2p\pi^*)^3$	${}^2\Pi_g$
F_2^{2+}:	$[He_2](2s\sigma)^2(2s\sigma^*)^2(2p\sigma)^2(2p\pi)^4(2p\pi^*)^2$	${}^3\Sigma_g^-$
F_2^+:	same as O_2^-	
F_2^-:	$[He_2](2s\sigma)^2(2s\sigma^*)^2(2p\sigma)^2(2p\pi)^4(2p\pi^*)^4(2p\sigma^*)^1$	${}^2\Sigma_u^+$

11.4 C_2^+, smaller; C_2^-, greater; N_2^+, smaller; N_2^-, smaller; O_2^+, greater; O_2^-, smaller; F_2^{2+}, greater; F_2^+, greater; F_2^-, smaller.

CHAPTER 12

12.1 $\psi_i = c_{i1}1s_1 + c_{i2}1s_2 + c_{i3}1s_3$

12.2

$$\begin{vmatrix} \alpha - \varepsilon & \beta & 0 \\ \beta & \alpha - \varepsilon & \beta \\ 0 & \beta & \alpha - \varepsilon \end{vmatrix} = 0$$

12.3

$$\begin{vmatrix} \alpha - \varepsilon & \beta & \beta \\ \beta & \alpha - \varepsilon & \beta \\ \beta & \beta & \alpha - \varepsilon \end{vmatrix} = 0$$

12.4 The cyclic form should be more stable, since there are more nonvanishing β terms.

12.6 Let the basis functions, $1s_A$, $1s_B$, $2s_A$, $2s_B$, $2p_{zA}$, $2p_{zB}$, be numbered sequentially. The **F** matrix elements for this closed-shell system have the form

$$F_{\mu\nu} = \left\langle \chi_\mu \left| -\frac{1}{2}\nabla^2 - \frac{Z}{r_A} - \frac{Z}{r_B} \right| \chi_\nu \right\rangle$$

$$+ \sum_{i=1}^{\text{occ}} \sum_{\lambda} \sum_{\sigma} c_{i\lambda} c_{i\sigma} \left[2 \left\langle \chi_\mu \chi_\lambda \left| \frac{1}{r_{12}} \right| \chi_\nu \chi_\sigma \right\rangle - \left\langle \chi_\mu \chi_\lambda \left| \frac{1}{r_{12}} \right| \chi_\sigma \chi_\nu \right\rangle \right]$$

The determinant is

$$|F_{\mu\nu} - \varepsilon S_{\mu\nu}| = 0$$

12.7 (a)
$$\begin{vmatrix} \alpha - \varepsilon & \beta & 0 \\ \beta & \alpha - \varepsilon & \beta_{CN} \\ 0 & \beta_{CN} & \alpha_N - \varepsilon \end{vmatrix} = 0$$

(c)
$$\begin{vmatrix} \alpha - \varepsilon & \beta & 0 & 0 & 0 & \beta \\ \beta & \alpha - \varepsilon & \beta & 0 & 0 & 0 \\ 0 & \beta & \alpha - \varepsilon & \beta & 0 & 0 \\ 0 & 0 & \beta & \alpha - \varepsilon & \beta & 0 \\ 0 & 0 & 0 & \beta & \alpha - \varepsilon & \beta \\ \beta & 0 & 0 & 0 & \beta & \alpha - \varepsilon \end{vmatrix} = 0$$

(e)
$$\begin{vmatrix} \alpha - \varepsilon & \beta & 0 & 0 & 0 & \beta & \beta_{CO} & 0 \\ \beta & \alpha - \varepsilon & \beta & 0 & 0 & 0 & 0 & 0 \\ 0 & \beta & \alpha - \varepsilon & \beta & 0 & 0 & 0 & 0 \\ 0 & 0 & \beta & \alpha - \varepsilon & \beta & 0 & 0 & \beta_{CO} \\ 0 & 0 & 0 & \beta & \alpha - \varepsilon & \beta & 0 & 0 \\ \beta & 0 & 0 & 0 & \beta & \alpha - \varepsilon & 0 & 0 \\ \beta_{CO} & 0 & 0 & 0 & 0 & 0 & \alpha_O - \varepsilon & 0 \\ 0 & 0 & 0 & \beta_{CO} & 0 & 0 & 0 & \alpha_O - \varepsilon \end{vmatrix} = 0$$

12.9 Using acrolein parameters: $h_O = 1$, $k_{CO} = 1$.
(c) $\varepsilon_1 = \alpha + 2\beta$, $\psi_1 = .5(p_1 + p_2 + p_3 + p_4)$, $\varepsilon_2 = \alpha + \sqrt{2}\beta$, $\psi_2 = .653(p_1 - p_4) + .271(p_2 - p_3)$, $\varepsilon_3 = \alpha$, $\psi_3 = .5(p_1 - p_2 - p_3 + p_4)$, $\varepsilon_4 = \alpha - \sqrt{2}\beta$, $\psi_4 = .271(p_1 - p_4) - .653(p_2 - p_3)$, $q_1 = q_4 = 1.354$, $q_2 = q_3 = .646$.
(d) (The carbon is atom 4.) $\varepsilon_1 = \alpha + 2.303\beta$, $\psi_1 = .230(p_1 + p_2 + p_3) + .917p_4$, $\varepsilon_2 = \varepsilon_3 = \alpha + \beta$, $\psi_2 = 1/\sqrt{6}(2p_1 - p_2 - p_3)$, $\psi_3 = 1\sqrt{2}(p_2 - p_3)$ (or any linear combination of these), $\varepsilon_4 = \alpha - 1.303\beta$, $\psi_4 = .530(p_1 + p_2 + p_3) - .398p_4$, $q_1 = q_2 = q_3 = 1.439$, $q_4 = 1.683$.

CHAPTER 13

13.1 (a) $\mathbf{C}_2$. (b) $\mathbf{D}_{2h}$. (c) $\mathbf{C}_{2v}$. (d) $\mathbf{C}_{2h}$. (e) $\mathbf{D}_{2h}$. (f) $\mathbf{D}_{3d}$. (g) $\mathbf{T}_d$. (h) $\mathbf{C}_{3v}$.
13.2, 13.3 For any integral, $\langle\psi_i|\hat{O}|\psi_j\rangle$, to be nonvanishing, the product of the irreducible representations, $\Gamma_i \times \Gamma_O \times \Gamma_j$, must contain the totally symmetric irreducible representation. The Hamiltonian operator, as well as its various parts, is totally symmetric. The σ orbitals are

symmetric with respect to the molecular plane, while the π orbitals are antisymmetric. Thus the product, $\Gamma_\sigma \times \Gamma_\pi$, is antisymmetric to at least this one operation and cannot be totally symmetric. The two-electron integrals, on the other hand, can contain two σ functions and two π functions. These can be nonvanishing if the two σ and two π functions are the same, or if one each of the σ and π functions have the same behavior with respect to all generators except the molecular plane.

CHAPTER 14

14.1 $\varepsilon^{a_2''} = \alpha + 2\beta$, $\psi^{a_2''} = 1/\sqrt{3}(p_1 + p_2 + p_3)$, $\varepsilon^{e''} = \alpha - \beta$, $\psi^{e''} = 1/\sqrt{6}(2p_1 - p_2 - p_3)$, all $q_i = 1$.

14.2 See Problem 12.9(c). ψ_1 and ψ_3 are a_u, ψ_2 and ψ_4 are b_g.

14.4 $\varepsilon_1^{a_2''} = \alpha + 2.78\beta$, $\psi_1^{a_2''} = .490(p_1 + p_2 + p_3) + .529p_4$,
$\varepsilon_2^{a_2''} = \alpha - 1.08\beta$, $\psi_2^{a_2''} = .305(p_1 + p_2 + p_3) - .849p_4$,
$\varepsilon^{e''} = \alpha - 1.70\beta$, $\psi^{e''} = 1/\sqrt{6}(2p_1 - p_2 - p_3)$,
$q_1 = 1.814$, $q_4 = .560$.
The modified α_0 value is to attempt to account for the fact that the three oxygen atoms donate a total of five electrons to the π system.

CHAPTER 15

15.1 The possible spatial representations are A_1, E, T_1, and T_2, of which T_2 is totally antisymmetric. The only allowed term is ${}^3T_2$.

15.2 $D_u^3[\mathbf{O}(3)] \rightarrow A_{2u} + T_{1u} + T_{2u}(\mathbf{O}_h)$.

15.3, 15.5 The ground-state configurations and relative energies are:

N	Configuration	Relative Energy (10^4 cm^{-1})	Ground Term
1	$(e_g)^1$	0	${}^2E_g$
2	$(e_g)^2$	0	${}^3A_{2g}$
3	$(e_g)^3$	4.5	${}^2E_g$
4	$(e_g)^4$	9.0	${}^1A_{1g}$
5	$(e_g)^4(a_{1g})^1$	35.5	${}^2A_{1g}$
6	$(e_g)^4(a_{1g})^1(b_{2g})^1$	39.8	${}^3B_{2g}$
7	$(e_g)^4(a_{1g})^2(b_{2g})^1$	44.3	${}^2B_{2g}$
8	$(e_g)^4(a_{1g})^2(b_{2g})^2$	48.8	${}^1A_{1g}$
9	$(e_g)^4(a_{1g})^2(b_{2g})^2(b_{1g})^1$	55.9	${}^2B_{1g}$

15.4

Transition		E ($\times$ 10^4 cm^{-1})
(a_{1g})	$\leftarrow (e_g)$	26.5
$(e_g)(a_{1g})$	$\leftarrow (e_g)^2$	26.5
$(e_g)^2(a_{1g})$	$\leftarrow (e_g)^3$	22.0
$(e_g)^3(a_{1g})$	$\leftarrow (e_g)^4$	22.0
$(e_g)^4(b_{2g})$	$\leftarrow (e_g)^4(a_{1g})$	4.3
$(e_g)^4(a_{1g})^2$	$\leftarrow (e_g)^4(a_{1g})(b_{2g})$ (Note that this is an orbital demotion.)	.2
$(e_g)^4(a_{1g})(b_{2g})^2$	$\leftarrow (e_g)^4(a_{1g})^2(b_{2g})$	4.3
$(e_g)^4(a_{1g})^2(b_{2g})(b_{1g})$	$\leftarrow (e_g)^4(a_{1g})^2(b_{2g})^2$	2.6
$(e_g)^4(a_{1g})^2(b_{2g})(b_{1g})^2$	$\leftarrow (e_g)^4(a_{1g})^2(b_{2g})^2(b_{1g})$	7.1

15.5 See 15.3.

CHAPTER 16

16.1 (a) $\Gamma_{\text{vib}} = 2A_1'(R) + 3E'(I, R) + A_1'' + A_2''(I) + E''(R)$
(e) $\Gamma_{\text{vib}} = A_{1g}(R) + B_{1g}(R) + B_{2g}(R) + A_{2u}(I) + B_{2u} + 2E_u(I)$
(Note that there are three infrared and three Raman allowed fundamentals here, all different. In contrast, the T_d structure has two infrared and four Raman bands, of which two are the same as the infrared.)

16.3 Use the 1 × 1 from Eqs. 16.49 and 16.50 and the antisymmetric stretching frequency to get

$$f_R = 776.0 \text{ N m}^{-1}$$

Use $\omega^2 \cong d^2 f_A g_{33}$ (from Eqs. 16.43 and 16.45, neglecting the f_{RR} and f_{RA}) to get

$$f_A = 69.98 \text{ N m}^{-1}$$

16.7 (a) A_1. (b) E. (c) A_1. (d) $A_1 + E$. (e) $A_1 + A_2 + E$.
(f) $2A_1 + 2A_2 + 4E$. (g) $A_1 + A_2 + E$.

CHAPTER 17

17.2 (E_- and E_+ are the roots of the 2 × 2 of Eq. 17.21.)

	J/δ				
	10.	2.	1.	.5	.1
$(E_1 + \omega_0)/\delta$	−2.500	−.500	−.250	−.125	−.025
E_-/δ	−2.525	−.618	−.457	−.434	−.477
E_+/δ	7.525	1.618	.957	.684	.527
$(E_4 - \omega_0)/\delta$	−2.500	.500	−.250	−.125	−.025

17.3 Relative to $(\omega_1 + \omega_2)/2 = 0$, the transition energies (relative intensities) are:

	J/δ				
	10	2.	1.	.5	.1
$E_+ \leftarrow E_1$	10.03(1)	2.12(1)	1.21(1)	.81(1)	.55(1)
$E_4 \leftarrow E_-$	.03(400)	.12(18)	.21(5.8)	.31(2.6)	.45(1.2)
$E_- \leftarrow E_1$	−.03	−.12	−.21	−.31	−.45
$E_4 \leftarrow E_+$	−10.03	−2.12	−1.21	−.81	−.55

CHAPTER 18

18.1 (C_s point group) Forbidden in the ground state, allowed in the excited.

18.2 (C_s) Ground allowed, excited forbidden.

18.3 (C_{3v}) Allowed in both. Note that the π system of benzene is orthogonal to the σ system, and to the components of the π system of acetylene that interact to form the new bonds of the σ system of benzene.

18.4 (C_{2v} along the long molecular axis) Forbidden in both. Note that in the reactant three π bonds must lie in the molecular plane.

18.5 ($\mathbf{C}_{2v}$, π orbital of ethylene contains the C_2 axis.) Assume that the oxygen has a single electron in two different p orbitals. (Can you justify this on the basis of the material in Chapter 7?) These transform as b_1 and b_2 within $\mathbf{C}_{2v}$. The π orbital of ethylene transforms as a_1. Thus the configuration of the reactants is $(a_1)^2(b_1)(b_2)$. The new bonds in the product are the two C—O bonds. The ground-state configuration is $(a_1)^2(b_1)^2$. Reaction to ground-state products is forbidden. Reaction to an $(a_1)^2(b_1)(b_2)$ excited configuration would be allowed, but there is no reasonable low-lying excited state having that configuration.

18.6 (Same orientation as above.) This oxygen configuration is best described by two electrons in a single oxygen p orbital. If this is aligned along the π orbital of the ethylene, the reactant configuration is $(a_1)^2(b_1)^2$. This is the same as the ground-state configuration of the product, so the reaction is allowed.

appendix five

Reduction of reducible representations in continuous groups

In the constructions we have used, the character under the $C(\phi)$ operation in an irreducible representation of $\mathbf{C}_{\infty v}$ or $\mathbf{D}_{\infty h}$ is

$$\chi[C(\phi)] = 1 \qquad \textbf{A5.1a}$$

for the Σ representations, and it is

$$\chi[C(\phi)] = 2 \cos \lambda\phi \qquad \textbf{A5.1b}$$

where λ is the appropriate angular-momentum value, for all other representations. Thus, the character under $C(\phi)$ for a reducible representation must be

$$\chi[C(\phi)] = a_0 + \sum_{k=1}^{\lambda_{\max}} 2a_k \cos k\phi \qquad \textbf{A5.2}$$

where a_0 is the number of Σ representations, and a_k is the number of times that the k representation occurs. These can be read directly from the character to reduce the representation. For the Σ representations, the $+$ or $-$ character can be determined by the character of the σ_v, since this has contributions from only the Σ representations. The relation is

$$a_0^+ = \tfrac{1}{2}[\chi(\sigma_v) + a_0] \qquad \textbf{A5.3}$$

where a_0^+ is the number of Σ^+ representations and a_0 is the total number of Σ representations. The g, u behavior, for $\mathbf{D}_{\infty h}$, can be determined by the sign of the various terms under $S(\phi)$ (see the character table for $\mathbf{D}_{\infty h}$ in Appendix 7).

The character under $C(\phi)$ for an irreducible representation in $\mathbf{R}(3)$ or $\mathbf{O}(3)$ is

$$\chi[C(\phi)] = 1 + \sum_{k=1}^{j} 2 \cos k\phi \qquad \textbf{A5.4}$$

If a representation in one of these groups is reducible, the character under $C(\phi)$ will be

$$\chi[C(\phi)] = \sum_{l=0}^{j_{\max}} a_l + \sum_{k=1}^{j_{\max}} \sum_{l=k}^{j_{\max}} 2a_l \cos k\phi \qquad \textbf{A5.5}$$

where a_l is the number of times that D^l appears in the reducible representation. The g, u behavior in $\mathbf{O}(3)$ can be determined by comparing the signs of the various cosine terms under $S(-\phi)$ with the corresponding terms under $C(\phi)$.

appendix six

*Correlation tables**

These tables give the representations that a given representation in a given group becomes when the symmetry is reduced to a subgroup of the given group. No redundancy is given in the tables. A given group is reduced only to the level of subgroups which themselves have already been reduced. For example, if the reduction of $\mathbf{T}_d$ to $\mathbf{C}_{2v}$ were desired, the reduction to $\mathbf{D}_{2d}$ could be found in the $\mathbf{T}_d$ table and the further reduction to $\mathbf{C}_{2v}$ would come from the $\mathbf{D}_{2d}$ table. Care should be taken when using these, or any other, correlation tables to make sure that the axes of the subgroups are properly aligned with respect to the parent group.

$\mathbf{C}_4$	$\mathbf{C}_2$
A	A
B	A
E	$2B$

$\mathbf{C}_6$	$\mathbf{C}_3$	$\mathbf{C}_2$
A	A	A
B	A	B
E_1	E	$2B$
E_2	E	$2A$

$\mathbf{S}_4$	$\mathbf{C}_2$
A	A
B	A
E	$2B$

$\mathbf{S}_6$	$\mathbf{C}_3$	$\mathbf{C}_i$
A_g	A	A_g
E_g	E	$2A_g$
A_u	A	A_u
E_u	E	$2A_u$

$\mathbf{S}_8$	$\mathbf{C}_4$
A	A
B	A
E_1	E
E_2	$2B$
E_3	E

$\mathbf{C}_{2h}$	$\mathbf{C}_2$	$\mathbf{C}_s$	$\mathbf{C}_i$
A_g	A	A'	A_g
B_g	B	A''	A_g
A_u	A	A''	A_u
B_u	B	A'	A_u

$\mathbf{C}_{3h}$	$\mathbf{C}_3$	$\mathbf{C}_s$
A'	A	A'
E'	E	$2A'$
A''	A	A''
E''	E	$2A''$

$\mathbf{C}_{4h}$	$\mathbf{C}_4$	$\mathbf{S}_4$	$\mathbf{C}_{2h}$
A_g	A	A	A_g
B_g	B	B	A_g
E_g	E	E	$2B_g$
A_u	A	B	A_u
B_u	B	A	A_u
E_u	E	E	$2B_u$

* From R. L. Flurry, Jr., *Symmetry Groups: Theory and Chemical Applications.* Prentice-Hall, Inc., Englewood Cliffs, N.J., 1980. By permission.

$\mathbf{C}_{5h}$	$\mathbf{C}_5$	$\mathbf{C}_s$
A'	A	A'
E_1'	E_1	$2A'$
E_2'	E_2	$2A'$
A''	A	A''
E_1''	E_1	$2A''$
E_2''	E_2	$2A''$

$\mathbf{C}_{6h}$	$\mathbf{C}_6$	$\mathbf{C}_{3h}$	$\mathbf{S}_6$	$\mathbf{C}_{2h}$
A_g	A	A'	A_g	A_g
B_g	B	A''	A_g	B_g
E_{1g}	E_1	E''	E_g	$2B_g$
E_{2g}	E_2	E'	E_g	$2A_g$
A_u	A	A''	A_u	A_u
B_u	B	A'	A_u	B_u
E_{1u}	E_1	E'	E_u	$2B_u$
E_{2u}	E_2	E''	E_u	$2A_u$

$\mathbf{C}_{2v}$	$\mathbf{C}_2$	(zx) $\mathbf{C}_s$	(yz) $\mathbf{C}_s$
A_1	A	A'	A'
A_2	A	A''	A''
B_1	B	A'	A''
B_2	B	A''	A'

$\mathbf{C}_{3v}$	$\mathbf{C}_3$	$\mathbf{C}_s$
A_1	A	A'
A_2	A	A''
E	E	$A' + A''$

$\mathbf{C}_{4v}$	$\mathbf{C}_4$	σ_v $\mathbf{C}_{2v}$	σ_d $\mathbf{C}_{2v}$
A_1	A	A_1	A_1
A_2	A	A_2	A_2
B_1	B	A_1	A_2
B_2	B	A_2	A_1
E	E	$B_1 + B_2$	$B_1 + B_2$

$\mathbf{C}_{5v}$	$\mathbf{C}_5$	$\mathbf{C}_s$
A_1	A	A'
A_2	A	A''
E_1	E_1	$A' + A''$
E_2	E_2	$A' + A''$

$\mathbf{C}_{6v}$	$\mathbf{C}_6$	σ_v $\mathbf{C}_{3v}$	σ_d $\mathbf{C}_{3v}$	$\sigma_v \rightarrow \sigma(zx)$ $\mathbf{C}_{2v}$
A_1	A	A_1	A_1	A_1
A_2	A	A_2	A_2	A_2
B_1	B	A_1	A_2	B_1
B_2	B	A_2	A_1	B_2
E_1	E_1	E	E	$B_1 + B_2$
E_2	E_2	E	E	$A_1 + A_2$

$\mathbf{D}_2$	$\mathbf{C}_{2(z)}$	$\mathbf{C}_{2(y)}$	$\mathbf{C}_{2(x)}$
A	A	A	A
B_1	A	B	B
B_2	B	A	B
B_3	B	B	A

$\mathbf{D}_3$	$\mathbf{C}_3$	$\mathbf{C}_2$
A_1	A	A
A_2	A	B
E	E	$A + B$

$\mathbf{D}_4$	$\mathbf{C}_4$	$\mathbf{C}_2$	C_2' $\mathbf{C}_2$	C_2'' $\mathbf{C}_2$
A_1	A	A	A	A
A_2	A	A	B	B
B_1	B	A	A	B
B_2	B	A	B	A
E	E	$2B$	$A + B$	$A + B$

$\mathbf{D}_6$	$\mathbf{C}_6$	C_2' $\mathbf{D}_3$	C_2'' $\mathbf{D}_3$	$\mathbf{D}_2$
A_1	A	A_1	A_1	A
A_2	A	A_2	A_2	B_1
B_1	B	A_1	A_2	B_2
B_2	B	A_2	A_1	B_3
E_1	E_1	E	E	$B_2 + B_3$
E_2	E_2	E	E	$A + B_1$

$\mathbf{D}_{2h}$	$\mathbf{D}_2$	$C_2(z)$ $\mathbf{C}_{2v}$	$C_2(y)$ $\mathbf{C}_{2v}$	$C_2(x)$ $\mathbf{C}_{2v}$	$C_2(z)$ $\mathbf{C}_{2h}$	$C_2(y)$ $\mathbf{C}_{2h}$	$C_2(x)$ $\mathbf{C}_{2h}$
A_g	A	A_1	A_1	A_1	A_g	A_g	A_g
B_{1g}	B_1	A_2	B_2	B_1	A_g	B_g	B_g
B_{2g}	B_2	B_1	A_2	B_2	B_g	A_g	B_g
B_{3g}	B_3	B_2	B_1	A_2	B_g	B_g	A_g
A_u	A	A_2	A_2	A_2	A_u	A_u	A_u
B_{1u}	B_1	A_1	B_1	B_2	A_u	B_u	B_u
B_{2u}	B_2	B_2	A_1	B_1	B_u	A_u	B_u
B_{3u}	B_3	B_1	B_2	A_1	B_u	B_u	A_u

$\mathbf{D}_{3h}$	$\mathbf{C}_{3h}$	$\mathbf{D}_3$	$\mathbf{C}_{3v}$	$\sigma_h \to \sigma_v(zy)$ $\mathbf{C}_{2v}$
A_1'	A'	A_1	A_1	A_1
A_2'	A'	A_2	A_2	B_2
E'	E'	E	E	$A_1 + B_2$
A_1''	A''	A_1	A_2	A_2
A_2''	A''	A_2	A_1	B_1
E''	E''	E	E	$A_2 + B_1$

$\mathbf{D}_{4h}$	$\mathbf{D}_4$	$C_2' \to C_2'$ $\mathbf{D}_{2d}$	$C_2'' \to C_2'$ $\mathbf{D}_{2d}$	$\mathbf{C}_{4v}$	$\mathbf{C}_{4h}$	C_2' $\mathbf{D}_{2h}$	C_2'' $\mathbf{D}_{2h}$
A_{1g}	A_1	A_1	A_1	A_1	A_g	A_g	A_g
A_{2g}	A_2	A_2	A_2	A_2	A_g	B_{1g}	B_{1g}
B_{1g}	B_1	B_1	B_2	B_1	B_g	A_g	B_{1g}
B_{2g}	B_2	B_2	B_1	B_2	B_g	B_{1g}	A_g
E_g	E	E	E	E	E_g	$B_{2g} + B_{3g}$	$B_{2g} + B_{3g}$
A_{1u}	A_1	B_1	B_1	A_2	A_u	A_u	A_u
A_{2u}	A_2	B_2	B_2	A_1	A_u	B_{1u}	B_{1u}
B_{1u}	B_1	A_1	A_2	B_2	B_u	A_u	B_{1u}
B_{2u}	B_2	A_2	A_1	B_1	B_u	B_{1u}	A_u
E_u	E	E	E	E	E_u	$B_{2u} + B_{3u}$	$B_{2u} + B_{3u}$

$\mathbf{D}_{5h}$	$\mathbf{D}_5$	$\mathbf{C}_{5v}$	$\mathbf{C}_{5h}$	$\mathbf{C}_5$	$\sigma_h \to \sigma(xz)$ $\mathbf{C}_{2v}$
A_1'	A_1	A_1	A'	A	A_1
A_2'	A_2	A_2	A'	A	B_1
E_1'	E_1	E_1	E_1'	E_1	$A_1 + B_1$
E_2'	E_2	E_2	E_2'	E_2	$A_1 + B_1$
A_1''	A_1	A_2	A''	A	A_2
A_2''	A_2	A_1	A''	A	B_2
E_1''	E_1	E_1	E_1''	E_1	$A_2 + B_2$
E_2''	E_2	E_2	E_2''	E_2	$A_2 + B_2$

$\mathbf{D}_{6h}$	$\mathbf{D}_6$	C_2' $\mathbf{D}_{3h}$	C_2'' $\mathbf{D}_{3h}$	$\mathbf{C}_{6v}$	$\mathbf{C}_{6h}$	C_2'' $\mathbf{D}_{3d}$	C_2' $\mathbf{D}_{3d}$	$\sigma_h \to \sigma(xy)$ $\sigma_v \to \sigma(yz)$ $\mathbf{D}_{2h}$	$\sigma_v' = \sigma_v$ $\sigma_v = \sigma_h$ $C_2 = C_2'$ $\mathbf{C}_{2v}$
A_{1g}	A_1	A_1'	A_1'	A_1	A_g	A_{1g}	A_{1g}	A_g	A_1
A_{2g}	A_2	A_2'	A_2'	A_2	A_g	A_{2g}	A_{2g}	B_{1g}	B_1
B_{1g}	B_1	A_1''	A_2''	B_2	B_g	A_{2g}	A_{1g}	B_{2g}	A_2
B_{2g}	B_2	A_2''	A_1''	B_1	B_g	A_{1g}	A_{2g}	B_{3g}	B_2
E_{1g}	E_1	E''	E''	E_1	E_{1g}	E_g	E_g	$B_{2g} + B_{3g}$	$A_2 + B_2$
E_{2g}	E_2	E'	E'	E_2	E_{2g}	E_g	E_g	$A_g + B_{1g}$	$A_1 + B_1$
A_{1u}	A_1	A_1''	A_1''	A_2	A_u	A_{1u}	A_{1u}	A_u	A_2
A_{2u}	A_2	A_2''	A_2''	A_1	A_u	A_{2u}	A_{2u}	B_{1u}	B_2
B_{1u}	B_1	A_1'	A_2'	B_1	B_u	A_{2u}	A_{1u}	B_{2u}	A_1
B_{2u}	B_2	A_2'	A_1'	B_2	B_u	A_{1u}	A_{2u}	B_{3u}	B_1
E_{1u}	E_1	E'	E'	E_1	E_{1u}	E_u	E_u	$B_{2u} + B_{3u}$	$A_1 + B_1$
E_{2u}	E_2	E''	E''	E_2	E_{2u}	E_u	E_u	$A_u + B_{1u}$	$A_2 + B_2$

$\mathbf{D}_{2d}$	$\mathbf{S}_4$	$\mathbf{D}_2$ ($C_2 \rightarrow C_2(z)$)	$\mathbf{C}_{2v}$
A_1	A	A	A_1
A_2	A	B_1	A_2
B_1	B	A	A_2
B_2	B	B_1	A_1
E	E	$B_2 + B_3$	$B_1 + B_2$

$\mathbf{D}_{3d}$	$\mathbf{D}_3$	$\mathbf{C}_{3v}$	$\mathbf{S}_6$	$\mathbf{C}_3$	$\mathbf{C}_{2h}$
A_{1g}	A_1	A_1	A_g	A	A_g
A_{2g}	A_2	A_2	A_g	A	B_g
E_g	E	E	E_g	E	$A_g + B_g$
A_{1u}	A_1	A_2	A_u	A	A_u
A_{2u}	A_2	A_1	A_u	A	B_u
E_u	E	E	E_u	E	$A_u + B_u$

$\mathbf{D}_{4d}$	$\mathbf{D}_4$	$\mathbf{C}_{4v}$	$\mathbf{S}_8$
A_1	A_1	A_1	A
A_2	A_2	A_2	A
B_1	A_1	A_2	B
B_2	A_2	A_1	B
E_1	E	E	E_1
E_2	$B_1 + B_2$	$B_1 + B_2$	E_2
E_3	E	E	E_3

$\mathbf{D}_{5d}$	$\mathbf{D}_5$	$\mathbf{C}_{5v}$
A_{1g}	A_1	A_1
A_{2g}	A_2	A_2
E_{1g}	E_1	E_1
E_{2g}	E_2	E_2
A_{1u}	A_1	A_2
A_{2u}	A_2	A_1
E_{1u}	E_1	E_1
E_{2u}	E_2	E_2

$\mathbf{D}_{6d}$	$\mathbf{D}_6$	$\mathbf{C}_{6v}$	$\mathbf{D}_{2d}$
A_1	A_1	A_1	A_1
A_2	A_2	A_2	A_2
B_1	A_1	A_2	B_1
B_2	A_2	A_1	B_2
E_1	E_1	E_1	E
E_2	E_2	E_2	$B_1 + B_2$
E_3	$B_1 + B_2$	$B_1 + B_2$	E
E_4	E_2	E_2	$A_1 + A_2$
E_5	E_1	E_1	E

$\mathbf{T}$	$\mathbf{D}_2$	$\mathbf{C}_3$
A	A	A
E	$2A$	E
T	$B_1 + B_2 + B_3$	$A + E$

$\mathbf{T}_h$	$\mathbf{T}$	$\mathbf{D}_{2h}$	$\mathbf{S}_6$
A_g	A	A_g	A_g
E_g	E	$2A_g$	E_g
T_g	T	$B_{1g} + B_{2g} + B_{3g}$	$A_g + E_g$
A_u	A	A_u	A_u
E_u	E	$2A_u$	E_u
T_u	T	$B_{1u} + B_{2u} + B_{3u}$	$A_u + E_u$

$\mathbf{T}_d$	$\mathbf{T}$	$\mathbf{D}_{2d}$	$\mathbf{C}_{3v}$	$\mathbf{S}_4$
A_1	A	A_1	A_1	A
A_2	A	B_1	A_2	B
E	E	$A_1 + B_1$	E	$A + B$
T_1	T	$A_2 + E$	$A_2 + E$	$A + E$
T_2	T	$B_2 + E$	$A_1 + E$	$B + E$

$\mathbf{O}$	$\mathbf{T}$	$\mathbf{D}_4$	$\mathbf{D}_3$
A_1	A	A_1	A_1
A_2	A	B_1	A_2
E	E	$A_1 + B_1$	E
T_1	T	$A_2 + E$	$A_2 + E$
T_2	T	$B_2 + E$	$A_1 + E$

$\mathbf{O}_h$	$\mathbf{O}$	$\mathbf{T}_d$	$\mathbf{T}_h$	$\mathbf{D}_{4h}$	$\mathbf{D}_{3d}$
A_{1g}	A_1	A_1	A_g	A_{1g}	A_{1g}
A_{2g}	A_2	A_2	A_g	B_{1g}	A_{2g}
E_g	E	E	E_g	$A_{1g} + B_{1g}$	E_g
T_{1g}	T_1	T_1	T_g	$A_{2g} + E_g$	$A_{2g} + E_g$
T_{2g}	T_2	T_2	T_g	$B_{2g} + E_g$	$A_{1g} + E_g$
A_{1u}	A_1	A_2	A_u	A_{1u}	A_{1u}
A_{2u}	A_2	A_1	A_u	B_{1u}	A_{2u}
E_u	E	E	E_u	$A_{1u} + B_{1u}$	E_u
T_{1u}	T_1	T_2	T_u	$A_{2u} + E_u$	$A_{2u} + E_u$
T_{2u}	T_2	T_1	T_u	$B_{2u} + E_u$	$A_{1u} + E_u$

appendix seven

Character tables*

These tables include the discrete axial point groups up to a rotation axis of order 6; the cubic point groups; the linear and spherical continuous rotation groups, and the symmetric permutation group up to degree 7. The point-group tables also contain the transformation properties of the Cartesian coordinates, rotations about the Cartesian coordinates, and quadratic terms in the Cartesian coordinates. The transformation properties of higher polynomials can be generated by multiplication of these representations.

The product structure for constructing the group from its most common generators is given for the point groups. This, coupled with the fact that the cyclic groups having orders that are not prime numbers have cyclic subgroups whose orders are the integer divisors of the order of the parent group, allows one to immediately find all the subgroups of a given group. For example, $\mathbf{D}_{6h}$ has the product structure $C_6 \wedge C_2' \times C_s$. Each of the groups $\mathbf{C}_6$, $\mathbf{C}_3$, $\mathbf{C}_2$, $\mathbf{C}_2'$, and $\mathbf{C}_s$, as well as their products, are subgroups of $\mathbf{D}_{6h}$.

The organization of the tables is as follows:

Group = product structure

Group	*Classes*		
Representations	Characters of representations	Cartesian coordinates and rotations	Quadratic terms

* Adapted from R. L. Flurry, Jr., *Symmetry Groups: Theory and Chemical Applications*. Prentice-Hall, Inc., Englewood Cliffs, N.J., 1980. By permission.

A7.1 THE CYCLIC GROUPS, $\mathbf{C}_n$

$\mathbf{C}_1 = C_1$

$\mathbf{C}_1$	E		
A	1	All coordinates and rotations	All

$\mathbf{C}_2 = C_2$

$\mathbf{C}_2$	E	C_2		
A	1	1	z, R_z	x^2, y^2, z^2, xy
B	1	-1	x, y, R_x, R_y	yz, xz

$\mathbf{C}_3 = C_3$

$\mathbf{C}_3$		E	C_3	C_3^2	$\varepsilon = \exp(2\pi i/3)$	
A		1	1	1	z, R_z	$x^2 + y^2, z^2$
E	{	1	ε	ε^*	(x, y)	$(x^2 - y^2, xy)$
		1	ε^*	ε	(R_x, R_y)	(xz, yz)

$\mathbf{C}_4 = C_4$

$\mathbf{C}_4$		E	C_4	C_2	C_4^3		
A		1	1	1	1	z, R_z	$x^2 + y^2, z^2$
B		1	-1	1	-1		$x^2 - y^2, xy$
E	{	1	i	-1	$-i$	(x, y)	(yz, xz)
		1	$-i$	-1	i	(R_x, R_y)	

$\mathbf{C}_5 = C_5$

$\mathbf{C}_5$		E	C_5	C_5^2	C_5^3	C_5^4	$\varepsilon = \exp(2\pi i/5)$; $\omega = \exp(\pi i/5)$	
A		1	1	1	1	1	z, R_z	$x^2 + y^2, z^2$
E_1	{	1	ε	$-\omega^*$	$-\omega$	ε^*	(x, y)	(yz, xz)
		1	ε^*	$-\omega$	$-\omega^*$	ε	(R_x, R_y)	
E_2	{	1	$-\omega^*$	ε^*	ε	$-\omega$		$(x^2 - y^2, xy)$
		1	$-\omega$	ε	ε^*	$-\omega^*$		

$C_6 = C_6$

C_6	E	C_6	C_3	C_2	C_3^2	C_6^5	$\varepsilon = \exp(2\pi i/6)$;	
A	1	1	1	1	1	1	z, R_z	$x^2 + y^2, z^2$
B	1	-1	1	-1	1	-1		
E_1 {	1	ε	$-\varepsilon^*$	-1	$-\varepsilon$	ε^*	} (x, y)	(xz, yz)
	1	ε^*	$-\varepsilon$	-1	$-\varepsilon^*$	ε	} (R_x, R_y)	
E_2 {	1	$-\varepsilon^*$	$-\varepsilon$	1	$-\varepsilon^*$	$-\varepsilon$	}	$(x^2 - y^2, xy)$
	1	$-\varepsilon$	$-\varepsilon^*$	1	$-\varepsilon$	$-\varepsilon^*$	}	

C_n (General). The elements, irreducible representations, and characters for any cyclic point group of order n can be generated as follows:

Elements: C_n^j, with j going from 0 to $n - 1$ (note that $C_n^0 = E$);

Irreducible representations:

(a) n odd: A and E_k, k going from 1 to $(n - 1)/2$.
(b) n even: A, B and E_k, k going from 1 to $(n/2 - 1)$.

Characters:

(a) A: all $+ 1$.
(b) B: alternating $+1$ and -1.
(c) E_k: $\begin{cases} \varepsilon^{kj}, \\ \varepsilon^{-kj}, \end{cases}$ j going from 0 to $(n - 1)$, $\varepsilon = \exp(2\pi i/n)$.
(Note: $\varepsilon^0 = 1$.)

A7.2 THE GROUPS $\mathbf{S}_n$

$S_2 = C_i$

C_i	E	i		
A_g	1	1	R_x, R_y, R_z	All
A_u	1	-1	x, y, z	

$S_4 = S_4$

S_4	E	S_4	C_2	S_4^3	$\omega = \exp(\pi i/4)$	
A	1	1	1	1	R_z	$x^2 + y^2, z^2$
B	1	-1	1	-1	z	$x^2 - y^2, xy$
E {	1	i	-1	$-i$	} (x, y)	(xz, yz)
	1	$-i$	-1	i	} (R_x, R_y)	

$\mathbf{S_6 = C_3 \times C_i = C_{3i}}$

$\mathbf{S_6}$		E	C_3	C_3^2	i	S_6^5	S_6	$\varepsilon = \exp(2\pi i/3)$	
A_g		1	1	1	1	1	1	R_z	$x^2 + y^2, z^2$
E_g	{	1	ε	ε^*	1	ε	ε^*	} (R_x, R_y)	$(x^2 - y^2, xy)$
		1	ε^*	ε	1	ε^*	ε		(xz, yz)
A_u		1	1	1	-1	-1	-1	z	
E_u	{	1	ε	ε^*	-1	$-\varepsilon$	$-\varepsilon^*$	} (x, y)	
		1	ε^*	ε	-1	$-\varepsilon^*$	$-\varepsilon$		

A7.3 THE GROUPS $\mathbf{C}_{nh}$

(*Note:* The $\mathbf{C}_s$ represents a horizontal plane.)

$\mathbf{C_{1h} = C_s}$

$\mathbf{C}_s$	E	σ_h		
A'	1	1	x, y, R_z	x^2, y^2, z^2, xy
A''	1	-1	z, R_x, R_y	yz, xz

$\mathbf{C_{2h} = C_2 \times C_s}$

$\mathbf{C}_{2h}$	E	C_2	i	σ_h		
A_g	1	1	1	1	R_z	x^2, y^2, z^2, xy
B_g	1	-1	1	-1	R_y, R_y	xz, yz
A_u	1	1	-1	-1	z	
B_u	1	-1	-1	1	x, y	

$\mathbf{C_{3h} = C_3 \times C_s}$

$\mathbf{C}_{3h}$		E	C_3	C_3^2	σ_h	S_3	S_3^5	$\varepsilon = \exp(2\pi i/3)$	
A'		1	1	1	1	1	1	R_z	$x^2 + y^2, z^2$
E'	{	1	ε	ε^*	1	ε	ε^*	} (x, y)	$(x^2 - y^2, xy)$
		1	ε^*	ε	1	ε^*	ε		
A''		1	1	1	-1	-1	-1	z	
E''	{	1	ε	ε^*	-1	$-\varepsilon$	$-\varepsilon^*$	} (R_x, R_y)	(xz, yz)
		1	ε^*	ε	-1	$-\varepsilon^*$	$-\varepsilon$		

$\mathbf{C}_{4h} = \mathbf{C}_4 \times \mathbf{C}_s$

$\mathbf{C}_{4h}$	E	C_4	C_2	C_4^3	i	S_4^3	σ_h	S_4		
A_g	1	1	1	1	1	1	1	1	R_z	$x^2 + y^2, z^2$
B_g	1	−1	1	−1	1	−1	1	−1		$x^2 - y^2, xy$
E_g {	1	i	−1	$-i$	1	i	−1	$-i$	} (R_x, R_y)	(yz, xz)
	1	$-i$	−1	i	1	$-i$	−1	i		
A_u	1	1	1	1	−1	−1	−1	−1	z	
B_u	1	−1	1	−1	−1	1	−1	1		
E_u {	1	i	−1	$-i$	−1	$-i$	1	i	} (x, y)	
	1	$-i$	−1	i	−1	i	1	i		

$\mathbf{C}_{5h} = \mathbf{C}_5 \times \mathbf{C}_s$

$\varepsilon = \exp(2\pi i/5)$; $\omega = \exp(\pi i/5)$

$\mathbf{C}_{5h}$	E	C_5	C_5^2	C_5^3	C_5^4	σ_h	S_5	S_5^7	S_5^3	S_5^9		
A'	1	1	1	1	1	1	1	1	1	1	R_z	$x^2 + y^2, z^2$
E_1' {	1	ε	$-\omega^*$	$-\omega$	ε^*	1	ε	$-\omega^*$	$-\omega$	ε^*	} (x, y)	
	1	ε^*	$-\omega$	$-\omega^*$	ε	1	ε^*	$-\omega$	$-\omega^*$	ε		
E_2' {	1	$-\omega^*$	ε^*	ε	$-\omega$	1	$-\omega^*$	ε^*	ε	$-\omega$	}	$(x^2 - y^2, xy)$
	1	$-\omega$	ε	ε^*	$-\omega^*$	1	$-\omega$	ε	ε^*	$-\omega^*$		
A''	1	1	1	1	1	−1	−1	−1	−1	−1	z	
E_1'' {	1	ε	$-\omega^*$	$-\omega$	ε^*	−1	$-\varepsilon$	ω^*	ω	$-\varepsilon^*$	} (R_x, R_y)	(xz, yz)
	1	ε^*	$-\omega$	$-\omega^*$	ε	−1	$-\varepsilon^*$	ω	ω^*	$-\varepsilon$		
E_2'' {	1	$-\omega^*$	ε^*	ε	$-\omega$	−1	ω^*	$-\varepsilon^*$	$-\varepsilon$	ω		
	1	$-\omega$	ε	ε^*	$-\omega^*$	−1	ω	$-\varepsilon$	$-\varepsilon^*$	ω^*		

$\mathbf{C}_{6h} = \mathbf{C}_6 \times \mathbf{C}_s$

$\mathbf{C}_{6h}$	E	C_6	C_3	C_2	C_3^2	C_6^5	i	S_3^5	S_6^5	σ_h	S_6	S_3	$\varepsilon = \exp(2\pi i/6)$	
A_g	1	1	1	1	1	1	1	1	1	1	1	1	R_z	$x^2 + y^2, z^2$
B_g	1	−1	1	−1	1	−1	1	−1	1	−1	1	−1		
E_{1g} {	1	ε	$-\varepsilon^*$	−1	$-\varepsilon$	ε^*	1	ε	$-\varepsilon^*$	−1	$-\varepsilon$	ε^*	} (R_x, R_y)	(xy, yz)
	1	ε^*	$-\varepsilon$	−1	$-\varepsilon^*$	ε	1	ε^*	$-\varepsilon$	−1	$-\varepsilon^*$	ε		
E_{2g} {	1	$-\varepsilon^*$	$-\varepsilon$	1	$-\varepsilon^*$	$-\varepsilon$	1	$-\varepsilon^*$	$-\varepsilon$	1	$-\varepsilon^*$	$-\varepsilon$	}	$(x^2 - y^2, xy)$
	1	$-\varepsilon$	$-\varepsilon^*$	1	$-\varepsilon$	$-\varepsilon^*$	1	$-\varepsilon$	$-\varepsilon^*$	1	$-\varepsilon$	$-\varepsilon^*$		
A_u	1	1	1	1	1	1	−1	−1	−1	−1	−1	−1	z	
B_u	1	−1	1	−1	1	−1	−1	1	−1	1	−1	1		
E_{1u} {	1	ε	$-\varepsilon^*$	−1	$-\varepsilon$	ε^*	−1	$-\varepsilon$	ε^*	1	ε	$-\varepsilon^*$	} (x, y)	
	1	ε^*	$-\varepsilon$	−1	$-\varepsilon^*$	ε	−1	$-\varepsilon^*$	ε	1	ε^*	$-\varepsilon$		
E_{2u} {	1	$-\varepsilon^*$	$-\varepsilon$	1	$-\varepsilon^*$	$-\varepsilon$	−1	ε^*	ε	−1	ε^*	ε		
	1	$-\varepsilon$	$-\varepsilon^*$	1	$-\varepsilon$	$-\varepsilon^*$	−1	ε	ε^*	−1	ε	ε^*		

A7.4 THE GROUPS $\mathbf{C}_{nv}$

(*Note:* The $\mathbf{C}_s$ represents a vertical plane.)

$$\mathbf{C}_{2v} = \mathbf{C}_2 \wedge \mathbf{C}_s = \mathbf{C}_2 \times \mathbf{C}_s$$

$\mathbf{C}_{2v}$	E	C_2	$\sigma_v(xz)$	$\sigma_v'(yz)$		
A_1	1	1	1	1	z	x^2, y^2, z^2
A_2	1	1	-1	-1	R_z	xy
B_1	1	-1	1	-1	x, R_y	xz
B_2	1	-1	-1	1	y, R_x	yz

$$\mathbf{C}_{3v} = \mathbf{C}_3 \wedge \mathbf{C}_s$$

$\mathbf{C}_{3v}$	E	$2C_3$	$3\sigma_v$		
A_1	1	1	1	z	$x^2 + y^2, z^2$
A_2	1	1	-1	R_z	
E	2	-1	0	$(x, y)(R_x, R_y)$	$(x^2 - y^2, xy)(xz, yz)$

$$\mathbf{C}_{4v} = \mathbf{C}_4 \wedge \mathbf{C}_s$$

$\mathbf{C}_{4v}$	E	$2C_4$	C_2	$2\sigma_v$	$2\sigma_d$		
A_1	1	1	1	1	1	z	$x^2 + y^2, z^2$
A_2	1	1	1	-1	-1	R_z	
B_1	1	-1	1	1	-1		$x^2 - y^2$
B_2	1	-1	1	-1	1		xy
E	2	0	-2	0	0	$(x, y)(R_x, R_y)$	(xz, yz)

$$\mathbf{C}_{5v} = \mathbf{C}_5 \wedge \mathbf{C}_s$$

$\mathbf{C}_{5v}$	E	$2C_5$	$2C_5^2$	$5\sigma_v$		
A_1	1	1	1	1	z	$x^2 + y^2, z^2$
A_2	1	1	1	-1	R_z	
E_1	2	2 cos 72°	2 cos 144°	0	$(x, y)(R_x, R_y)$	(xy, yz)
E_2	2	2 cos 144°	2 cos 72°	0		$(x^2 - y^2, xy)$

$$\mathbf{C}_{6v} = \mathbf{C}_6 \wedge \mathbf{C}_s$$

$\mathbf{C}_{6v}$	E	$2C_6$	$2C_3$	C_2	$3\sigma_v$	$3\sigma_d$		
A_1	1	1	1	1	1	1	z	$x^2 + y^2, z^2$
A_2	1	1	1	1	-1	-1	R_z	
B_1	1	-1	1	-1	1	-1		
B_2	1	-1	1	-1	-1	1		
E_1	2	1	-1	-2	0	0	$(x, y)(R_x, R_y)$	(xz, yz)
E_2	2	-1	-1	2	0	0		$(x^2 - y^2, xy)$

A7.5 THE GROUPS $\mathbf{D}_n$

(*Note:* the $\mathbf{C}_2$ axis is perpendicular to the $\mathbf{C}_n$.)

$$\mathbf{D}_2 = \mathbf{C}_2 \wedge \mathbf{C}_2' = \mathbf{C}_2 \times \mathbf{C}_2'$$

$\mathbf{D}_2$	E	$C_2(z)$	$C_2(y)$	$C_2(x)$		
A	1	1	1	1		x^2, y^2, z^2
B_1	1	1	-1	-1	z, R_z	xy
B_2	1	-1	1	-1	y, R_y	xz
B_3	1	-1	-1	1	x, R_x	yz

$$\mathbf{D}_3 = \mathbf{C}_3 \wedge \mathbf{C}_2$$

$\mathbf{D}_3$	E	$2C_3$	$3C_2$		
A_1	1	1	1		$x^2 + y^2, z^2$
A_2	1	1	-1	z, R_z	
E	2	-1	0	$(x, y)(R_x, R_y)$	$(x^2 - y^2, xy)(xz, yz)$

$$\mathbf{D}_4 = \mathbf{C}_4 \wedge \mathbf{C}_2$$

$\mathbf{D}_4$	E	$2C_4$	C_2	$2C_2'$	$2C_2''$		
A_1	1	1	1	1	1		$x^2 + y^2, z^2$
A_2	1	1	1	-1	-1	z, R_z	
B_1	1	-1	1	1	-1		$x^2 - y^2$
B_2	1	-1	1	-1	1		xy
E	2	0	-2	0	0	$(x, y)(R_x, R_y)$	(xz, yz)

$$\mathbf{D}_5 = \mathbf{C}_5 \wedge \mathbf{C}_2$$

$\mathbf{D}_5$	E	$2C_5$	$2C_5^2$	$5C_2$		
A_1	1	1	1	1		$x^2 + y^2, z^2$
A_2	1	1	1	-1	z, R_z	
E_1	2	2 cos 72°	2 cos 144°	0	(x, y) (R_x, R_y)	(xz, yz)
E_2	2	2 cos 144°	2 cos 72°	0		$(x^2 - y^2, xy)$

$$\mathbf{D}_6 = \mathbf{C}_6 \wedge \mathbf{C}_2$$

$\mathbf{D}_6$	E	$2C_6$	$2C_3$	C_2	$3C_2'$	$3C_2''$		
A_1	1	1	1	1	1	1		$x^2 + y^2, z^2$
A_2	1	1	1	1	-1	-1	z, R_z	
B_1	1	-1	1	-1	1	-1		
B_2	1	-1	1	-1	-1	1		
E_1	2	1	-1	-2	0	0	$(x, y)(R_x, R_y)$	(xz, yz)
E_2	2	-1	-1	2	0	0		$(x^2 - y^2, xy)$

A7.6 THE GROUPS $\mathbf{D}_{nh}$

(*Note*: The $\mathbf{C}_s$ is perpendicular to the principal axis.)

$\mathbf{D}_{2h} = \mathbf{D}_2 \times \mathbf{C}_s$

$\mathbf{D}_{2h}$	E	$C_2(z)$	$C_2(y)$	$C_2(x)$	i	$\sigma(xy)$	$\sigma(xz)$	$\sigma(yz)$		
A_g	1	1	1	1	1	1	1	1		x^2, y^2, z^2
B_{1g}	1	1	-1	-1	1	1	-1	-1	R_z	xy
B_{2g}	1	-1	1	-1	1	-1	1	-1	R_y	xz
B_{3g}	1	-1	-1	1	1	-1	-1	1	R_x	yz
A_u	1	1	1	1	-1	-1	-1	-1		
B_{1u}	1	1	-1	-1	-1	-1	1	1	z	
B_{2u}	1	-1	1	-1	-1	1	-1	1	y	
B_{3u}	1	-1	-1	1	-1	1	1	-1	x	

$\mathbf{D}_{3h} = \mathbf{D}_3 \times \mathbf{C}_s$

$\mathbf{D}_{3h}$	E	$2C_3$	$3C_2$	σ_h	$2S_3$	$3\sigma_v$		
A_1'	1	1	1	1	1	1		$x^2 + y^2, z^2$
A_2'	1	1	-1	1	1	-1	R_z	
E'	2	-1	0	2	-1	0	(x, y)	$(x^2 - y^2, xy)$
A_1''	1	1	1	-1	-1	-1		
A_2''	1	1	-1	-1	-1	1	z	
E''	2	-1	0	-2	1	0	(R_x, R_y)	

$\mathbf{D}_{4h} = \mathbf{D}_4 \times \mathbf{C}_s$

$\mathbf{D}_{4h}$	E	$2C_4$	C_2	$2C_2'$	$2C_2''$	i	$2S_4$	σ_h	$2\sigma_v$	$2\sigma_d$		
A_{1g}	1	1	1	1	1	1	1	1	1	1		$x^2 + y^2, z^2$
A_{2g}	1	1	1	-1	-1	1	1	1	-1	-1	R_z	
B_{1g}	1	-1	1	1	-1	1	-1	1	1	-1		$x^2 - y^2$
B_{2g}	1	-1	1	-1	1	1	-1	1	-1	1		xy
E_g	2	0	-2	0	0	2	0	-2	0	0	(R_x, R_y)	(xz, yz)
A_{1u}	1	1	1	1	1	-1	-1	-1	-1	-1		
A_{2u}	1	1	1	-1	-1	-1	-1	-1	1	1	z	
B_{1u}	1	-1	1	1	-1	-1	1	-1	-1	1		
B_{2u}	1	-1	1	-1	1	-1	1	-1	1	-1		
E_u	2	0	-2	0	0	-2	0	2	0	0	(x, y)	

$\mathbf{D}_{5h} = \mathbf{D}_5 \times \mathbf{C}_s$

$\mathbf{D}_{5h}$	E	$2C_5$	$2C_5^2$	$5C_2$	σ_h	$2S_5$	$2S_5^3$	$5\sigma_v$		
A_1'	1	1	1	1	1	1	1	1		$x^2 + y^2, z^2$
A_2'	1	1	1	-1	1	1	1	-1	R_z	
E_1'	2	2 cos 72	2 cos 144°	0	2	2 cos 72°	2 cos 144°	0	(x, y)	
E_2'	2	2 cos 144°	2 cos 72°	0	2	2 cos 144°	2 cos 72°	0		$(x^2 - y^2, xy)$
A_1''	1	1	1	1	-1	-1	-1	-1		
A_2''	1	1	1	-1	-1	-1	-1	1	z	
E_1''	2	2 cos 72°	2 cos 144°	0	-2	-2 cos 72°	-2 cos 144°	0	(R_x, R_y)	(xz, yz)
E_2''	2	2 cos 144°	2 cos 72°	0	-2	-2 cos 144°	-2 cos 77°	0		

$\mathbf{D}_{6h} = \mathbf{C}_6 \times \mathbf{C}_s$

$\mathbf{D}_{6h}$	E	$2C_6$	$2C_3$	C_2	$3C_2'$	$3C_2''$	i	$2S_3$	$2S_6$	σ_h	$3\sigma_d$	$3\sigma_v$		
A_{1g}	1	1	1	1	1	1	1	1	1	1	1	1		$x^2 + y^2, z$
A_{2g}	1	1	1	1	-1	-1	1	1	1	1	-1	-1	R_z	
B_{1g}	1	-1	1	-1	1	-1	1	-1	1	-1	1	-1		
B_{2g}	1	-1	1	-1	-1	1	1	-1	1	-1	-1	1		
E_{1g}	2	1	-1	-2	0	0	2	1	-1	-2	0	0	(R_x, R_y)	(xz, yz)
E_{2g}	2	-1	-1	2	0	0	2	-1	-1	2	0	0		$(x^2 - y^2, xy)$
A_{1u}	1	1	1	1	1	1	-1	-1	-1	-1	-1	-1		
A_{2u}	1	1	1	1	-1	-1	-1	-1	-1	-1	1	1	z	
B_{1u}	1	-1	1	-1	1	-1	-1	1	-1	1	-1	1		
B_{2u}	1	-1	1	-1	-1	1	-1	1	-1	1	1	-1		
E_{1u}	2	1	-1	-2	0	0	-2	-1	1	2	0	0	(x, y)	
E_{2u}	2	-1	-1	2	0	0	-2	1	1	-2	0	0		

A7.7 THE GROUPS $\mathbf{D}_{nd}$

(*Note:* The $\mathbf{C}_s$ refers to a dihedral plane of symmetry, bisecting pairs of twofold axes.)

$\mathbf{D}_{2d} = \mathbf{D}_2 \wedge \mathbf{C}_s = \mathbf{S}_4 \wedge \mathbf{C}_2$

$\mathbf{D}_{2d}$	E	$2S_4$	C_2	$2C_2'$	$2\sigma_d$		
A_1	1	1	1	1	1		$x^2 + y^2, z^2$
A_2	1	1	1	-1	-1	R_z	
B_1	1	-1	1	1	-1		$x^2 - y^2$
B_2	1	-1	1	-1	1	z	xy
E	2	0	-2	0	0	$(x, y)(R_x, R_y)$	(xz, yz)

$\mathbf{D_{3d} = D_3 \times C_i}$

$\mathbf{D_{3d}}$	E	$2C_3$	$3C_2$	i	$2S_6$	$3\sigma_d$		
A_{1g}	1	1	1	1	1	1		$x^2 + y^2, z^2$
A_{2g}	1	1	-1	1	1	-1	R_z	
E_g	2	-1	0	2	-1	0	(R_x, R_y)	$(x^2 - y^2, xy)(xz, yz)$
A_{1u}	1	1	1	-1	-1	-1		
A_{2u}	1	1	-1	-1	-1	1	z	
E_u	2	-1	0	-2	1	0	(x, y)	

$\mathbf{D_{4d} = D_4 \wedge C_s = S_8 \wedge C_2}$

$\mathbf{D_{4d}}$	E	$2S_8$	$2C_4$	$2S_8^3$	C_2	$4C_2'$	$4\sigma_d$		
A_1	1	1	1	1	1	1	1		$x^2 + y^2, z^2$
A_2	1	1	1	1	1	-1	-1	R_z	
B_1	1	-1	1	-1	1	1	-1		
B_2	1	-1	1	-1	1	-1	1	z	
E_1	2	$\sqrt{2}$	0	$-\sqrt{2}$	-2	0	0	(x, y)	
E_2	2	0	-2	0	2	0	0		$(x^2 - y^2, xy)$
E_3	2	$-\sqrt{2}$	0	$\sqrt{2}$	-2	0	0	(R_x, R_y)	(xz, yz)

$\mathbf{D_{5d} = D_5 \times C_i}$

$\mathbf{D_{5d}}$	E	$2C_5$	$2C_5^2$	$5C_2$	i	$2S_{10}^3$	$2S_{10}$	$5\sigma_d$		
A_{1g}	1	1	1	1	1	1	1	1		$x^2 + y^2, z^2$
A_{2g}	1	1	1	-1	1	1	1	-1	R_z	
E_{1g}	2	2 cos 72°	2 cos 144°	0	2	2 cos 72°	2 cos 144°	0	(R_x, R_y)	(xz, yz)
E_{2g}	2	2 cos 144°	2 cos 72°	0	2	2 cos 144°	2 cos 72°	0		$(x^2 - y^2, xy)$
A_{1u}	1	1	1	1	-1	-1	-1	-1		
A_{2u}	1	1	1	-1	-1	-1	-1	1	z	
E_{1u}	2	2 cos 72°	2 cos 144°	0	-2	-2 cos 72°	-2 cos 144°	0	(x, y)	
E_{2u}	2	2 cos 144°	2 cos 72°	0	-2	-2 cos 144°	-2 cos 72°	0		

$\mathbf{D_{6d} = D_6 \wedge C_s = S_{12} \wedge C_2}$

$\mathbf{D_{6d}}$	E	$2S_{12}$	$2C_6$	$2S_4$	$2C_3$	$2S_{12}^5$	C_2	$6C_2'$	$6\sigma_d$		
A_1	1	1	1	1	1	1	1	1	1		$x^2 + y^2, z^2$
A_2	1	1	1	1	1	1	1	-1	-1	R_z	
B_1	1	-1	1	-1	1	-1	1	1	-1		
B_2	1	-1	1	-1	1	-1	1	-1	1	z	
E_1	2	$\sqrt{3}$	1	0	-1	$-\sqrt{3}$	-2	0	0	(x, y)	
E_2	2	1	-1	-2	-1	1	2	0	0		$(x^2 - y^2, xy)$
E_3	2	0	-2	0	2	0	-2	0	0		
E_4	2	-1	-1	2	-1	-1	2	0	0		
E_5	2	$-\sqrt{3}$	1	0	-1	$\sqrt{3}$	-2	0	0	(R_x, R_y)	(xz, yz)

A7.8 THE CUBIC GROUPS

(*Note:* The $\mathbf{C}_3$ is along the xyz diagonal if the $\mathbf{C}_2$'s are along x, y, and z.)

$\mathbf{T} = \mathbf{D}_2 \wedge \mathbf{C}_3$

$\mathbf{T}$	E	$4C_3$	$4C_3^2$	$3C_2$		$\varepsilon = \exp(2\pi i/3)$
A	1	1	1	1		$x^2 + y^2 + z^2$
E	1	ε	ε^*	1		$(x^2 - y^2, 2z^2 - x^2 - y^2)$
	1	ε^*	ε	1		
T	3	0	0	-1	$(x, y, z)(R_x, R_y, R_z)$	(xy, xz, yz)

$\mathbf{T}_d = \mathbf{D}_2 \wedge \mathbf{C}_{3v}$

$\mathbf{T}_d$	E	$8C_3$	$3C_2$	$6S_4$	$6\sigma_d$		
A_1	1	1	1	1	1		$x^2 + y^2 + z^2$
A_2	1	1	1	-1	-1		
E	2	-1	2	0	0		$(x^2 - y^2, 2z^2 - x^2 - y^2)$
T_1	3	0	-1	1	-1	(R_x, R_y, R_z)	
T_2	3	0	-1	-1	1	(x, y, z)	(xy, xz, yz)

$\mathbf{T}_h = \mathbf{T} \times \mathbf{C}_i$

$\mathbf{T}_h$	E	$4C_3$	$4C_3^2$	$3C_2$	i	$4S_6^5$	$4S_6$	$3\sigma_h$		$\varepsilon = \exp(2\pi i/3)$
A_g	1	1	1	1	1	1	1	1		$(x^2 + y^2 + z^2)$
E_g	1	ε	ε^*	1	1	ε	ε^*	1		$(x^2 - y^2, 2z^2 - x^2 - z^2)$
	1	ε^*	ε	1	1	ε^*	ε	1		
T_g	3	0	0	-1	3	0	0	-1	(R_x, R_y, R_z)	(xy, xz, yz)
A_u	1	1	1	1	-1	-1	-1	-1		
E_u	1	ε	ε^*	1	-1	$-\varepsilon$	$-\varepsilon^*$	-1		
	1	ε^*	ε	1	-1	$-\varepsilon^*$	$-\varepsilon$	-1		
T_u	3	0	0	-1	-3	0	0	1	(x, y, z)	

$\mathbf{O} = \mathbf{D}_2 \wedge \mathbf{D}_3$

$\mathbf{O}$	E	$8C_3$	$3C_2$	$6C_4$	$6C_2'$		
A_1	1	1	1	1	1		$x^2 + y^2 + z^2$
A_2	1	1	1	-1	-1		
E	2	-1	2	0	0		$(x^2 - y^2, 2z^2 - x^2 - y^2)$
T_1	3	0	-1	1	-1	$(x, y, z)(R_x, R_y, R_z)$	
T_2	3	0	-1	-1	1		(xy, xz, yz)

$\mathbf{O}_h = \mathbf{O} \times \mathbf{C}_i$

$\mathbf{O}_h$	E	$8C_3$	$3C_2$	$6C_4$	$6C_2'$	i	$8S_6$	$3\sigma_h$	$6S_4$	$6\sigma_d$		
A_{1g}	1	1	1	1	1	1	1	1	1	1		$x^2 + y^2 + z^2$
A_{2g}	1	1	1	−1	−1	1	1	1	−1	−1		
E_g	2	−1	2	0	0	2	−1	2	0	0		$(x^2 - y^2, 2z^2 - x^2 - y^2)$
T_{1g}	3	0	−1	1	−1	3	0	−1	1	−1	(R_x, R_y, R_z)	
T_{2g}	3	0	−1	−1	1	3	0	−1	−1	1		(xy, xz, yz)
A_{1u}	1	1	1	1	1	−1	−1	−1	−1	−1		
A_{2u}	1	1	1	−1	−1	−1	−1	−1	1	1		
E_u	2	−1	2	0	0	−2	1	−2	0	0		
T_{1u}	3	0	−1	1	−1	−3	0	1	−1	1	(x, y, z)	
T_{2u}	3	0	−1	−1	1	−3	0	1	1	−1		

A7.9 THE ICOSAHEDRAL GROUPS

$\mathbf{I} = \mathbf{D}_3 \cdot \mathbf{D}_5$[a]

$\mathbf{I}$	E	$12C_5$	$12C_5^2$	$20C_3$	$15C_2$		
A	1	1	1	1	1		$x^2 + y^2 + z^2$
T_1	3	2 cos 36°	−2 cos 72°	0	−1	(R_x, R_y, R_z) (x, y, z)	
T_2	3	−2 cos 72°	2 cos 36°	0	−1		
G	4	−1	−1	1	0		
H	5	0	0	−1	1		$(2z^2 - x^2 - y^2,$ $x^2 - y^2,$ $xy, yz, zx)$

[a] The product here is known as the *wreath product*.

$\mathbf{I}_h = \mathbf{I} \times \mathbf{C}_i$

$\mathbf{I}_h$	E	$12C_5$	$12C_5^2$	$20C_3$	$15C_2$	i	$12S_{10}$	$12S_{10}^3$	$20S_6$	15σ		
A_g	1	1	1	1	1	1	1	1	1	1		$x^2 + y^2 + z^2$
T_{1g}	3	2 cos 36°	−2 cos 72°	0	−1	3	−2 cos 72°	2 cos 36°	0	−1	(R_x, R_y, R_z)	
T_{2g}	3	−2 cos 72°	2 cos 36°	0	−1	3	2 cos 36°	−2 cos 72°	0	−1		
G_g	4	−1	−1	1	0	4	−1	−1	1	0		
H_g	5	0	0	−1	1	5	0	0	−1	1		$(2z^2 - x^2 - y^2,$ $x^2 - y^2,$ $xy, yz, zx)$
A_u	1	1	1	1	1	−1	−1	−1	−1	−1		
T_{1u}	3	2 cos 36°	−2 cos 72°	0	−1	−3	2 cos 72°	−2 cos 36°	0	1	(x, y, z)	
T_{2u}	3	−2 cos 72°	2 cos 36°	0	−1	−3	−2 cos 36°	2 cos 72°	0	1		
G_u	4	−1	−1	1	0	−4	1	1	−1	0		
H_u	5	0	0	−1	1	−5	0	0	1	−1		

A7.10 THE TWO-DIMENSIONAL CONTINUOUS ROTATION GROUPS

In all cases there are an infinite number of classes of the form $2C(\phi)$, $2C(2\phi)$, ...

$\mathbf{R}(2) \equiv \mathbf{C}_\infty = \mathbf{C}_\infty$

$\mathbf{C}_\infty$	E	$2C(\phi)$		
Σ	1	1	z	$x^2 + y^2, z^2$
Π	2	$2\cos\phi$	$(x, y)(R_x, R_y)$	(xz, yz)
Δ	2	$2\cos 2\phi$		$(x^2 - y^2, xy)$
Φ	2	$2\cos 3\phi$		
⋮	⋮	⋮		
Γ_λ	2	$2\cos\lambda\phi$		

$\mathbf{C}_{\infty v} = \mathbf{C}_\infty \wedge \mathbf{C}_s$

$\mathbf{C}_{\infty v}$	E	$2C(\phi)$	$\infty\sigma_v$		
Σ^+	1	1	1	z	$x^2 + y^2, z^2$
Σ^-	1	1	−1	R_z	
Π	2	$2\cos\phi$	0	$(x, y)(R_x, R_y)$	(xz, yz)
Δ	2	$2\cos 2\phi$	0		$(x^2 - y^2, xy)$
Φ	2	$2\cos 3\phi$	0		
⋮	⋮	⋮	⋮		
Γ_λ	2	$2\cos\lambda\phi$	0		

$\mathbf{D}_\infty = \mathbf{C}_\infty \wedge \mathbf{C}_2$

$\mathbf{D}_\infty$	E	$2C(\phi)$	∞C_2		
Σ^+	1	1	1	z	$x^2 + y^2, z^2$
Σ^-	1	1	−1	R_z	
Π	2	$2\cos\phi$	0	$(x, y)(R_x, R_y)$	(xz, yz)
Δ	2	$2\cos 2\phi$	0		$(x^2 - y^2, xy)$
Φ	2	$2\cos 3\phi$	0		
⋮	⋮	⋮	⋮		
Γ_λ	2	$2\cos\lambda\phi$	0		

$\mathbf{D}_{\infty h} = \mathbf{D}_\infty \times \mathbf{C}_i$

$\mathbf{D}_{\infty h}$	E	$2C(\phi)$	∞C_2	i	$2S(-\phi)$	$\infty\sigma_v$		
Σ_g^+	1	1	1	1	1	1		$x^2 + y^2, z^2$
Σ_g^-	1	1	−1	1	1	−1	R_z	
Π_g	2	$2\cos\phi$	0	2	$-2\cos\phi$	0	(R_x, R_y)	(xz, yz)
Δ_g	2	$2\cos 2\phi$	0	2	$2\cos 2\phi$	0		$(x^2 - y^2, xy)$
⋮	⋮	⋮	⋮	⋮	⋮	⋮		
$\Gamma_{\lambda g}$	2	$2\cos\lambda\phi$	0	2	$(-1)^\lambda 2\cos\lambda\phi$	0		
Σ_u^+	1	1	1	−1	−1	−1	z	
Σ_u^-	1	1	−1	−1	−1	1		
Π_u	2	$2\cos\phi$	0	−2	$2\cos\phi$	0	(x, y)	
Δ_u	2	$2\cos 2\phi$	0	−2	$-2\cos\phi$	0		
⋮	⋮	⋮	⋮	⋮	⋮	⋮		
$\Gamma_{\lambda u}$	2	$2\cos\lambda\phi$	0	−2	$-(-1)^\lambda 2\cos\lambda\phi$	0		

A7.11 THE THREE-DIMENSIONAL CONTINUOUS ROTATION GROUPS

R(3) = R(3)

R(3)	E	$C(\phi, x, y, z)$		
$D^{(0)}$	1	1		$x^2 + y^2 + z^2$
$D^{(1)}$	3	$1 + 2\cos\phi$	(x, y, z) (R_x, R_y, R_z)	
$D^{(2)}$	5	$1 + 2\cos\phi + 2\cos 2\phi$		All independent combinations
$D^{(3)}$	7	$1 + 2\cos\phi + 2\cos 2\phi + 2\cos 3\phi$		
⋮	⋮	⋮		
$D^{(j)}$	$2j + 1$	$1 + \sum_{l=1}^{j} 2\cos l\phi$		

$\mathbf{O}(3) = \mathbf{R}_h(3) = \mathbf{R}(3) \times \mathbf{C}_i$

$\mathbf{O}(3)$	E	$C(\phi, x, y, z)$	i	$S(-\phi, x, y, z)$	σ		
$D_g^{(0)}$	1	1	1	1	1		$x^2 + y^2 + z^2$
$D_g^{(1)}$	3	$1 + 2\cos\phi$	3	$1 - 2\cos\phi$	-1	(R_x, R_y, R_z)	
$D_g^{(2)}$	5	$1 + 2\cos\phi + 2\cos 2\phi$	5	$1 - 2\cos\phi + 2\cos 2\phi$	1		All independent combinations
$\vdots$	$\vdots$	$\vdots$	$\vdots$	$\vdots$	$\vdots$		
$D_g^{(j)}$	$2j + 1$	$1 + \sum_{l=1}^{j} 2\cos l\phi$	$2j + 1$	$1 + \sum_{l=1}^{j} (-1)^l \cos l\phi$	$(-1)^j$		
$D_u^{(0)}$	1	1	-1	-1	-1		
$D_u^{(1)}$	3	$1 + 2\cos\phi$	-3	$-1 + 2\cos\phi$	1	(x, y, z)	
$D_u^{(2)}$	5	$1 + 2\cos\phi + 2\cos 2\phi$	-5	$-1 + 2\cos\phi - 2\cos 2\phi$	-1		
$\vdots$	$\vdots$	$\vdots$	$\vdots$	$\vdots$	$\vdots$		
$D_u^{(j)}$	$2j + 1$	$1 + \sum_{l=1}^{j} 2\cos l\phi$	$-(2j + 1)$	$-1 - \sum_{l=1}^{j} (-1)^l 2\cos l\phi$	$-(-1)^j$		
$D_g^{(1/2)}$	2	$2\cos 1/2\phi$	2	$2\sin 1/2\phi$	0		
$D_g^{(3/2)}$	4	$2\cos 1/2\phi + 2\cos 3/2\phi$	4	$2\sin 1/2\phi - 2\sin 3/2\phi$	0		
$D_g^{(5/2)}$	6	$2\cos 1/2\phi + 2\cos 3/2\phi + 2\cos 5/2\phi$	6	$2\sin 1/2\phi - 2\sin 3/2\phi + 2\sin 5/2\phi$	0		
$\vdots$	$\vdots$	$\vdots$	$\vdots$	$\vdots$	$\vdots$		
$D_g^{(j)}$	$2j + 1$	$\sum_{l=1/2}^{j} 2\cos l\phi$	$2j + 1$	$\sum_{l=1/2}^{j} (-1)^{(l-1/2)} 2\sin l\phi$	0		
$D_u^{(1/2)}$	2	$2\cos 1/2\phi$	-2	$-2\sin 1/2\phi$	0		
$D_u^{(3/2)}$	4	$2\cos 1/2\phi + 2\cos 3/2\phi$	-4	$-2\sin 1/2\phi + 2\sin 3/2\phi$	0		
$D_u^{(5/2)}$	6	$2\cos 1/2\phi + 2\cos 3/2\phi + 2\cos 5/2\phi$	-6	$-2\sin 1/2\phi + 2\sin 3/2\phi - 2\sin 5/2\phi$	0		
$\vdots$	$\vdots$	$\vdots$	$\vdots$	$\vdots$	$\vdots$		
$D_u^{(j)}$	$2j + 1$	$\sum_{l=1/2}^{j} 2\cos l\phi$	$-(2j + 1)$	$\sum_{l=1/2}^{j} (-1)^{(l+1/2)} 2\sin l\phi$	0		

A7.12 SYMMETRIC PERMUTATION GROUPS

Degree 2:

S(2)	(1^2)	(2)
[2]	1	1
$[1^2]$	1	−1

Degree 3:

S(3)	(1^3)	3(2, 1)	2(3)
[3]	1	1	1
[2, 1]	2	0	−1
$[1^3]$	1	−1	1

Degree 4:

S(4)	(1^4)	$6(2, 1^2)$	$3(2^2)$	8(3, 1)	6(4)
[4]	1	1	1	1	1
[3, 1]	3	1	−1	0	−1
$[2^2]$	2	0	2	−1	0
$[2, 1^2]$	3	−1	−1	0	1
$[1^4]$	1	−1	1	1	−1

Degree 5:

S(5)	(1^5)	$10(2, 1^3)$	$15(2^2, 1)$	$20(3, 1^2)$	20(3, 2)	30(4, 1)	24(5)
[5]	1	1	1	1	1	1	1
[4, 1]	4	2	0	1	−1	0	−1
[3, 2]	5	1	1	−1	1	−1	0
$[3, 1^2]$	6	0	−2	0	0	0	1
$[2^2, 1]$	5	−1	1	−1	−1	1	0
$[2, 1^3]$	4	−2	0	1	1	0	−1
$[1^5]$	1	−1	1	1	−1	−1	1

Degree 6:

$\mathbf{S}(6)$	(1^6)	$15(2, 1^4)$	$45(2^2, 1^2)$	$40(3, 1^3)$	$15(2^3)$	$120(3, 2, 1)$	$40(3^2)$	$90(4, 1^2)$	$90(4, 2)$	$144(5, 1)$	$120(6)$
$[6]$	1	1	1	1	1	1	1	1	1	1	1
$[5, 1]$	5	3	1	2	−1	0	−1	1	−1	0	−1
$[4, 2]$	9	3	1	0	3	0	0	−1	1	−1	0
$[4, 1^2]$	10	2	−2	1	−2	−1	1	0	0	0	1
$[3^2]$	5	1	1	−1	−3	1	2	−1	−1	0	0
$[3, 2, 1]$	16	0	0	−2	0	0	−2	0	0	1	0
$[2^3]$	5	−1	1	−1	3	−1	2	1	−1	0	0
$[3, 1^3]$	10	−2	−2	1	2	1	1	0	0	0	−1
$[2^2, 1^2]$	9	−3	1	0	−3	0	0	1	1	−1	0
$[2, 1^4]$	5	−3	1	2	1	0	−1	−1	−1	0	1
$[1^6]$	1	−1	1	1	−1	−1	1	−1	1	1	−1

Degree 7:

$\mathbf{S}(7)$	(1^7)	$21(2, 1^5)$	$105(2^2, 1^3)$	$70(3, 1^4)$	$105(2^3, 1)$	$420(3, 2, 1^2)$	$210(3, 2^2)$	$210(4, 1^3)$	$280(3^2, 1)$	$630(4, 2, 1)$	$420(4, 3)$	$504(5, 1^2)$	$504(5, 2)$	$840(6, 1)$	$720(7)$
$[7]$	1	1	1	1	1	1	1	1	1	1	1	1	1	1	1
$[6, 1]$	6	4	2	3	0	1	−1	2	0	0	−1	1	−1	0	−1
$[5, 2]$	14	6	2	2	2	0	2	0	−1	0	0	−1	1	−1	0
$[5, 1^2]$	15	5	−1	3	−3	−1	−1	1	0	−1	1	0	0	0	1
$[4, 3]$	14	4	2	−1	0	1	−1	−2	2	0	1	−1	−1	0	0
$[4, 2, 1]$	35	5	−1	−1	1	−1	−1	−1	−1	1	−1	0	0	1	0
$[3^2, 1]$	21	1	1	−3	−3	1	1	−1	0	−1	−1	1	1	0	0
$[4, 1^3]$	20	0	−4	2	0	0	2	0	2	0	0	0	0	0	−1
$[3, 2^2]$	21	−1	1	−3	3	−1	1	1	0	−1	1	1	−1	0	0
$[3, 2, 1^2]$	35	−5	−1	−1	−1	1	−1	1	−1	1	1	0	0	−1	0
$[2^3, 1]$	14	−4	2	−1	0	−1	−1	2	2	0	−1	−1	1	0	0
$[3, 1^4]$	15	−5	−1	3	3	1	−1	−1	0	−1	−1	0	0	0	1
$[2^2, 1^3]$	14	−6	2	2	−2	0	2	0	−1	0	0	−1	−1	1	0
$[2, 1^5]$	6	−4	2	3	0	−1	−1	−2	0	0	1	1	1	0	−1
$[1^7]$	1	−1	1	1	−1	−1	1	−1	1	1	−1	1	−1	−1	1

appendix eight

Glossary*

Basis functions. A set of functions from which any other function in the same function space can be constructed.

Basis vectors. A set of vectors from which any other vector in the same vector space can be constructed.

Boson. Particles whose wave functions must be symmetric under the interchange of two equivalent particles. Bosons have integer intrinsic angular momenta.

Character. The trace of a matrix representation of a group operation.

Character table. A table displaying the characters of the various operations, corresponding to the various irreducible representations, of a group.

Configuration. A particular orbital occupancy of an atom or a molecule. (Not to be confused with a state.)

Degenerate states. Independent states having the same value of the state-defining property (such as energy). The number of such states is the degeneracy.

Determinant. A quantity represented by a two-dimensional square array (conventially enclosed by vertical lines) implying a certain expansion in terms of the elements of the array.

Direct product. See *outer product.*

* Adapted from R. L. Flurry, Jr., *Symmetry Groups: Theory and Chemical Applications.* Prentice-Hall, Inc., Englewood Cliffs, N.J., 1980. By permission.

Direct sum. A sum of two vectors, matrices, or tensors that expands the dimension. In the direct sum the components have no bases in common. The new dimension is the sum of the dimensions of the components.

Eigenfunction. A function that satisfies an eigenvalue equation (see below).

Eigenvalue equation. The equation resulting when the effect of an operator, $\hat{O}$, operating on a function or vector, f, yields a constant (the eigenvalue), A, times the unchanged function or vector; i.e., $\hat{O}f = Af$.

Eigenvector. A vector that satisfies an eigenvalue equation.

Excited state. Any state of a system having an energy greater than the ground (lowest-energy) state.

Expectation value. The quantum-mechanical average value of a property in some state. The term is usually used when the property is not a state-defining property.

Fermions. Particles whose total wave function must be antisymmetric under the interchange of two equivalent particles. Fermions have half-integer intrinsic angular momenta.

Generators of a group. The simplest set of operations from which the complete group can be generated by powers and products of the operations. The set is not neccessarily unique.

Ground state. The lowest-energy (most stable) state of a system.

Group. A set of quantities having a defined law of combination (called multiplication) such that (a) the multiplication is associative, (b) the group contains an identity, (c) every element of the group has an inverse, (d) all products and powers of the elements are contained in the group.

Group element. Any member of the set forming a group. A group operation.

Hermitian matrix. A matrix that is equal to its conjugate transpose; that is, $M_{ij} = M^*_{ji}$.

Hermitian operator. An operator, say $\hat{L}$, for which $\langle u|\hat{L}|v\rangle$ equals $\langle \hat{L}u|v\rangle$, where u and v are any two quadratically integrable functions (functions for which $\langle u|u\rangle$ is finite).

Identity. The element of a group that leaves the system unchanged.

Improper axis of rotation. The axis about which an improper rotation can leave a system in a configuration indistinguishable from the original configuration. The improper rotation may be described either by a combination of a rotation and a reflection through a plane perpendicular to the rotation axis, or by a rotation and an inversion.

Irreducible representation. A member of the set of simplest possible matrix representations of a group.

Kronecker delta function. A function, δ_{ij}, that equals zero if the two indices are different, or unity if they are the same.

Matrix. A system of quantities, a_{ij}, with two indices, usually arranged in a rectangular array, with the index i labeling the rows and j the columns. A second-rank tensor.

Matrix representation of a group. A set of matrices that transform among themselves in the same manner as the operations of the group.

Mapping. An association between the members of one set of quantities and those of another.

Normalized functions. Functions (say a) for which the integral over all space $\int a^*a\, dv$ equals unity.

Normalized vectors. Vectors (say **a**) for which the scalar product $\mathbf{a}\cdot\mathbf{a}$ equals unity.

Orthogonal functions. Functions (say a and b) for which the integral $\int a^*b\, dv$ vanishes.

Orthogonal vectors. Vectors (say **a** and **b**) for which the scalar product $\mathbf{a}\cdot\mathbf{b}$ vanishes.

Orthonormal. The property of being both orthogonal and normalized.

Outer product. A product of two vectors, matrices, or tensors that expands the rank. For example, for two vectors (first-rank tensors), it is a column vector times a row vector, which yields a matrix (second-rank tensor).

Pauli-allowed state. A state of a system in which the restrictions on the interchange of the particles are properly accounted for. Usually, an antisymmetrized function of spin-$\frac{1}{2}$ particles.

Plane of symmetry. The plane through which a reflection can leave a system in a configuration indistinguishable from the original configuration.

Point group. The group describing the symmetry of a physical object.

Point of inversion. The point through which inversion can leave a system in a configuration indistinguishable from the original configuration.

Projection operator. An operator that projects out a specified component from a function, vector, or the like.

Proper axis of rotation. The axis about which a simple rotation can leave a system in a configuration indistinguishable from the original configuration.

Rank. The number of indexes required to specify an element of a tensor.

Representation of a group. A set of quantities that have the same multiplication properties as the operations of a group.

Scalar product (of vectors). A row-column product of two vectors, the result of which is a scalar. A "dot product."

Secular equation. The determinantal form of an eigenvalue equation.

State. The condition of a system arising when all arbitrariness in the properties (energy, angular momentum, and so on) defining the system is removed.

Subgroup. A subset of the set of group operations that obeys the group requirements.

Symmetric (permutation) **group** (of degree n). The set of all $n!$ permutations of n objects.

Symmetry element. A geometric entity (point, line, plane) about which a symmetry operation is performed.

Symmetry operation. An operation that carries a system into an orientation or configuration indistinguishable from the starting orientation or configuration.

Tensor. An indexed array of quantities $a_{ijk\ldots}$. The number of indexes required to define the tensor is its *rank*.

Tensor product. See *outer product*.

Term. An expression completely specifying the angular momentum state of an atom or a molecule.

Transition dipole. The expectation value of the dipole operator between two different states of a system. The direct absorption or emission of electromagnetic radiation by a system is dependent upon the transition dipole.

Vector. A system of quantities, a_i, with one index, usually arranged in a row or a column array. A first-rank tensor.

Young diagrams. Diagrams expressing the partition structure of the irreducible representations of the symmetric group as patterns of blocks.

Index

D

E

J

K

L

M

N

O

P